1st International Innovative Mining Symposium (In Memory of Prof. Vladimir Pronoza)

E3S Web of Conferences Volume 15 (2017)

Kemerovo, Russia
24 - 26 April 2017

Editors:

Maxim Tyulenev
Sergey Zhironkin
Alexey Khoreshok
Stefan Voth
Michal Cehlar
Yunliang Tan

ISBN: 978-1-5108-3894-9

Printed from e-media with permission by:

Curran Associates, Inc.
57 Morehouse Lane
Red Hook, NY 12571

Some format issues inherent in the e-media version may also appear in this print version.

Printed by Curran Associates, Inc. (2017)

For additional information, please contact EDP Sciences – Web of Conferences at the address below.

EDP Sciences – Web of Conferences
17, Avenue du Hoggar
Parc d'Activité de Courtabœuf
BP 112
F-91944 Les Ulis Cedex A
France

Phone: +33 (0) 1 69 18 75 75
Fax: +33 (0) 1 69 28 84 91

contact-edps@webofconferences.org

Additional copies of this publication are available from:

Curran Associates, Inc.
57 Morehouse Lane
Red Hook, NY 12571 USA
Phone: 845-758-0400
Fax: 845-758-2633
Email: curran@proceedings.com
Web: www.proceedings.com

TABLE OF CONTENTS

E3S Web of Conferences
Open access conference proceedings
in Environment, Energy & Earth Sciences
A flexible, fast and efficient
publication service,
for an open access recording
of your event
and a wide dissemination of
your proceedings
www.e3s-conferences.org
edp open Open Access Publishing from EDP Sciences

Aims & scope

E3S Web of Conferences is an open-access journal devoted to the publishing of conference proceedings dealing with Environment, Energy and Earth Sciences.

Coverage includes technological and scientific aspects as well as social and economic matters. Major disciplines include: soil science, hydrology, oceanography, climatology, geology, geography, energy engineering (production, distribution and storage), renewable energy, sustainable development, natural resources management, etc.

The journal aims to make conference proceedings visible, citable and widely usable by the scientific community. It offers outstanding services for the publication and dissemination of contents in order to ensure long-term archiving and maximum exposure of scientific events.

Why publish with E3S Web of Conferences?

- **Open access:** conference papers available worldwide.
- **Fast publication:** less than 3 months.
- **Total flexibility:** no page restrictions, free unlimited use of colour in the online version, any kind of scientific document welcomed (traditional articles, posters, abstracts and slideshows).
- **High visibility:** all documents indexed in main databases and easily found online.
- **Fair author rights:** CC-BY Creative Commons license.
- **Range of additional publishing options:** printed proceedings, CD-ROMs, copy-editing, etc.

Contact: Isaline Augusto
e3s@webofconferences.org

ISSN (electronic edition): 2267-1242

www.e3s-conferences.org

edp open Open Access Publishing from EDP Sciences

Organizing Committee

Aksenov Vladimir V., DSc, Prof., Head of Laboratory of Mining Geotechnical Engineering of the Institute of Coal of SB RAS, Kemerovo, Russia

Asmelash Abay, PhD, Assoc. Prof., College of Natural and Computational Sciences, Mekelle University, Ethiopia

Barysheva Galina A., DSc, Head of Economics Department, National Research Tomsk Polytechnic University, Tomsk, Russia

Cehlár Michal, PhD, Prof. Ing., Dean of Mining Faculty, Technical University of Kosice, Kosice, Slovak Republic

Demirel Nuray, PhD, Middle East Technical University, Turkey

Gasanov Magerram A., DSc in Economics, Prof., National Research Tomsk Polytechnic University, Tomsk, Russia

Gvozdkova Tatyana N., PhD, Director of Mezhdurechensk Branch of T.F. Gorbachev Kuzbass State Technical University, Mezhdurechensk, Russia

Khoreshok Alexey A., DSc, Prof., Director of Mining Institute, T.F. Gorbachev Kuzbass State Technical University, Kemerovo, Russia

Kostyuk Svetlana G., Vice-Rector, T.F. Gorbachev Kuzbass State Technical University, Kemerovo, Russia

Krechetov Andrey A., PhD, Acting Rector, T.F. Gorbachev Kuzbass State Technical University, Kemerovo, Russia

Lesin Yury V., DSc, Prof., Kuzbass State Technical University, Kemerovo, Russia

Markov Sergey O., PhD, Assoc. Prof., Surveying and Geology Department, T.F. Gorbachev Kuzbass State Technical University, Kemerovo, Russia

Misnikov Oleg S., DSc, Dean, Tver State Technical University, Tver, Russia

Myaskov Alexandr V., DSc, Dean, National University of Science and Technology MISiS, Russia

Seroni Anyona, PhD, Assoc. Prof., Jomo Kenyatta University of Agriculture and Technology, Kenya

Tyulenev Maxim A., PhD, Prof., T.F. Gorbachev Kuzbass State Technical University, Kemerovo, Russia; Yurga Technological Institute of National Research Tomsk Polytechnic University, Russia

Vöth Stefan, Dr.-Ing., Prof., TFH Georg Agricola, Bochum, Germany

Janočko Juraj, PhD, Technical University of Kosice, Kosice, Slovak Republic

Zhironkin Sergey A., DSc, Prof., National Research Tomsk Polytechnic University, Tomsk, Russia

Statement of Peer review

In submitting conference proceedings to E3S Web of Conferences, I certify to the Publisher that I adhere to the **Policy on Publishing Integrity** of the journal in order to safeguard good scientific practice in publishing.

1. All articles have been subjected to peer review administered by the proceedings editors.
2. Reviews have been conducted by expert referees, who have been requested to provide unbiased and constructive comments aimed, whenever possible, at improving the work.
3. Proceedings editors have taken all reasonable steps to ensure the quality of the materials they publish and their decision to accept or reject a paper for publication has been based only on the merits of the work and the relevance to the journal.

Title, date and place of the conference

International Innovative Mining Symposium (in memory of Prof. Vladimir Pronoza)

April 24 – 26, 2017, Kemerovo - Mezhdurechensk, Russian Federation

Proceedings editor(s):

Maxim Tyulenev, PhD, Kuzbass State Technical University

Sergey Zhironkin, DSc, National Research Tomsk Polytechnic University

Alexey Khoreshok, DSc, Kuzbass State Technical University

Stefan Voth, Dr.-Ing, TFH Georg Agricola

Michal Cehlar, PhD, Technical University of Kosice

Yunliang Tan, PhD, Shandong University of Science and Technology

Date and editor's signature

Date

17.03.2017

Preface: the role of T.F. Gorbachev State Technical University as the Flagship of Kemerovo Region Innovative Development

Andrey Krechetov[1,*]

[1]KuzSTU, Acting Rector, 650000 Kemerovo, 28 Vesennya st., Russian Federation

Abstract. Currently the Kemerovo Region is one of the most concentrated industrial agglomerations in Russia finds itself at the stage of investment modernization. This is due to the technological renovation of industries that form the core of its economy - coal, chemical, machine-building and metallurgical. The development of its human capital plays a crucial role in creating conditions for the successful innovative transformation of the regional economy. The flagship of this process is T.F. Gorbachev Kuzbass State Technical University. Alma Mater has a unique team in which creativity and experience are combined with high quality staff training for the basic branches of the regional economy and innovative activities of scientific schools, dissertation councils and the university expert community. This allows to use the existing mechanisms effectively and to create new ones for attracting external financing in all areas of activity due to the Kuzbass State Technical University scientific sector which has a significant potential for ensuring innovative development of the regional industry.

1 Introduction

The Kemerovo Region is the most densely populated part of Siberia and the entire Asian part of the Russian Federation (with a population density of 28.2 people per square kilometer) and is one of the largest industrial agglomerations of the country. The regional industrial complex provides more than a half of GRP (58.2%), it includes extracting and processing industries, mainly coal and iron ore constituting 35.1% of GRP in 2016, as well as processing industries, such as chemical and engineering, making up to 13.9% of GRP [1].

The Kemerovo region has the largest national coal cluster with an annual production volume of more than 210 million tons, which makes up 56% of coal mining in Russia (100% for especially valuable brands), while up to 75% of the coal produced pass through the first stage in coal enrichment. One of the largest regional power systems of Russia operates in the region with the installed capacity for generating electricity is 4800 MW and heat power is 7000 Gcal / hour. Its share in the energy balance of Siberia reaches 30%. In

* Corresponding author: krechetovaa@kuzstu.ru

the Kemerovo region there is the largest chemical complex of Russia, concentrating up to 30% of the national plastics production [2].

The higher education of specialists in the region is carried out by 26 state higher professional educational institutions, including 10 independent and 16 branches according to the data of January 2017. There are about 57 thousand students at the state universities of the region. Twenty nine thousand students out of total sixty one are enrolled in full-time education, which corresponds to the average in Russia [3-4]. One of the largest in a number of students is T.F. Gorbachev Kuzbass State Technical University (KuzSTU). By January 2017 there are 54 small innovative enterprises created in the universities of the region, eleven of them affiliate with KuzSTU.

2 Materials and Methods

At the present stage T.F. Gorbachev Kuzbass State Technical University is a key element of the educational, scientific and innovative system of the Kemerovo region, which trains future employees and generates technologies for leading industries of the regional economy. At the enterprises of mineral-resource and fuel-energy complex of the region more than 75% of managers are KuzSTU graduates. In engineering, construction, road transport, chemical industry of the region the share of the KuzSTU university graduates makes up to 35-50%.

The university has close integration relations with the regional business in personnel, scientific and technological support. The strongest in Russia school of minerals extraction technology has been developed on the basis of the university. The university trains employees of all main specializations for the mineral-resource complex of Kuzbass and Siberian macro-region: mining, construction, chemical and petrochemical, and machine building. According to employers poll the leaders are graduates from the Mining Institute, the Institute of Energy, Economics and Management Institute.

T.F. Gorbachev Kuzbass State Technical University is the only university in the region that carries out multi-level training of highly qualified personnel for enterprises and companies in the energy sector. Over 100 full-time bachelors and masters in six majors of KuzSTU graduates annually fully meeting the needs of the industry. Innovative development of the energy system of the region is provided by small innovative enterprises of KuzGTU.

T.F. Gorbachev Kuzbass State Technical University is the only university in the region that satisfies the need of regional engineering in high-skilled personnel. In KuzSTU more than a thousand students are trained in educational programs that provide engineering personnel for almost all production processes of machine building. The contribution of KuzSTU to the innovative development of the industry is ensured by the small innovative enterprises operating within its structure, as well as the research laboratory for quality control of machine parts and the student design bureau.

T.F. Gorbachev Kuzbass State Technical University has a unique base for training highly qualified personnel in accordance with the structure of chemical and coal-chemical industries in the region and the directions of its innovative development. More than 500 students are trained on educational programs that ensure the functioning and development of chemical industry enterprises with the subsequent almost 100% employment.

3 Results and Discussion

For many decades KuzSTU has accumulated a unique experience of mining engineers training, the annual graduate number of which exceeds 200 people. A significant contribution to the innovative development of mining industry is made by the research laboratory of mechanization tools for coal seam mining and a number of small innovative enterprises of KuzSTU. The special role of KuzSTU is in providing the personnel for mine safety.

KuzSTU actively cooperates with the academic institutes and scientific centers of the Siberian Branch of Russian Academy of Sciences, including the development of basic departments to improve the training efficiency of undergraduates and graduate students. Besides in the development of the basic departments of KuzSTU the industrial enterprises of both basic and manufacturing industries, telecommunication companies, commercial banks take part.

T.F. Gorbachev Kuzbass State Technical University has its own developed innovative infrastructure including small innovative enterprises, design centers, a center of industrial safety expertise, research laboratories, scientific and educational and resource centers, in which engineering projects are implemented, including those related to import substitution in machine building and chemical industry.

Thus, KuzSTU undoubtedly can be considered an integrator of the educational, scientific and innovative-entrepreneurial systems of the region, the core of modernization of the regional environment and advanced development through the development of practice-oriented training of competitive highly skilled personnel, consolidation of intellectual resources, consolidation of network interactions.

Further development prospects of KuzSTU are related to the formation of the flagship university complex being a provider of personnel and technologies for the positive structural shift of the regional economy and a center of commercialization and socialization of innovations, which realizes effective research, educational and innovative-entrepreneurial activity. This project could form a modern upward mobility in cooperation with the leaders of production and innovation, the scientific community, Russian and foreign partners.

The strategic tasks of the formation and development of the flagship university complex on the basis of KuzSTU are the following:

1. The formation of the regional higher school core, creating leaders of changes in the regional environment. The solution of this task will be carried out through the systematic identification, attraction and development of talents at all stages of education, the expansion of the educational portfolio in accordance with the requirements of stakeholders through the integration of project-oriented teaching and research as well as the development of the continuous education model and training of the regional elite.

2. The establishment of an innovative entrepreneurial university uniting scientific and research activity and training high-skilled personnel, successfully commercializing the intellectual property created within its walls and ensuring the modernization of the regional economy in the process of accelerating the technological breakthrough of its main clusters. The solution of this task will be carried out in the process of timely response to foresights of industrial innovative development and the formation of centers of excellence in the scientific research areas relevant to the Siberian macro-region, improving the economic efficiency of R&D, implementing the concept of the expert higher school "think tank" and the formation of a new format of university innovation ecosystem.

3. The building of a highly professional team with a global competence level and a highly effective management system that allows implementing a value approach to stimulating professional and social and cultural activity. The third task will be achieved through focusing on increasing the value of employees in the process of developing professional development programs, studying in magistracy, postgraduate and doctoral studies, introducing an incentive system for an effective contract, increasing the mobility

and prestige of scientific and pedagogical workers and implementing the electronic management system.

4. The development of modern infrastructure of talents' attraction, contributing to the concentration of potential change leaders in the Key University. This decision will be ensured by the development of a system for identifying and recruiting talented students, strengthening pre-university training, developing the university academic competition practice, developing networking with general education institutions.

5. The implementation of the concept of a modern open university which is a transparent subject of the media environment integrated into the international scientific, educational and innovative space and providing equal conditions for access to all of its systems for individuals with physical disabilities. The fifth task will be solved in the process of image breakthrough of the Key University, promotion of R&D results in the publishing environment and increase of citation index, modernization of the university campus and development of a universal barrier-free environment for inclusive continuous education and unhindered access for disabled students.

6. The emergence of a new quality of educational, research and innovative and entrepreneurial processes of the Key University in deploying its electronic platform. The solution of this task will be achieved through the development of an interactive digital image of the University, integrating processes and solutions in the field of online education and commercialization of innovation results, the attraction of talents and the development of scientific and pedagogical personnel as well as management of the University.

7. The increase of the expert role of the University in the development of local communities, urban and regional environment, in improving life safety and work. The solution of the seventh task will be reached in the process of implementing the function of the Key University as a strategic and system analyst of regional development, organization of expertise of technogenic objects safety and training of expert staff.

4 Conclusions

The transition to the target model within the framework of the University Development Program is the transformation of the educational institution into an educational and research complex that implements the prerequisites for the internationalization of Russian education, provides an attractive social environment contributing to knowledge for the regional economy in the process of systemic interaction with academic institutions, joint educational programs and scientific projects, the development of an innovative ecosystem of intellectual property commercialization.

References

1. S.A. Zhironkin, A.A. Khoreshok, M.A.Tyulenev, G.A.Barysheva, M.C. Hellmer, IOP Conf. Ser.: Mater. Sci. Eng., **142 (1)**, 012127 (2016)
2. A.A. Khoreshok, S.A. Zhironkin, M.A.Tyulenev, G.A.Barysheva, V.Yu.Blumenstein, M.C. Hellmer, S.V. Potyagailov, IOP Conf. Ser.: Mater. Sci. Eng., **142 (1)**, 012122 (2016)
3. S. Zhironkin, M. Gasanov, O. Zhironkina, EpSBS, **XIX**, 753 (2017)
4. S. Zhironkin, M. Gasanov, K. Kolotov, EpSBS, **XIX**, 761 (2017)
5. S. Zhironkin, V. Guzyr, EpSBS, **VII**, 160 (2016)

The Length Of Bearing Pressure Zone For The Flat Seams Extraction In A Linear Deformation Rock

Vjacheslav Gogolin[1,*], *Yury* Lesin[1], *Anna* Djagileva[1]

[1] T.F. Gorbachev Kuzbass State Technical University, 650000, 28 Vesennyaya St., Kemerovo, Russia

Abstract. The article describes the problem of numerical calculation of stress-strain state of the rock mass during flat seams extraction by finite element method. The distribution of bearing pressure under changing the seam thickness and depth of development is described. The length of the bearing pressure zones for different references of the pressure was defined. The independence of the bearing pressure zone length on the depth of development was proved. The analytical dependence of the bearing pressure zone length on the seam thickness was obtained. This result defines the influence of mining area on geo-mechanical condition of mine workings.

1 Introduction

The calculation of the bearing pressure zone length is one of the main tasks of geo-mechanics at coal seams mining. A lot of works are devoted to the solution of this problem, in which we underline a few different approaches to the formulation of solutions and methods [1-4]. In this paper, the method of finite element numerical simulation of the stress-strain state of the rocks was used to define the linear deformation of the rock [5-7]. This approach is implemented in solving various problems of geo-mechanics in coal seams underground mining [8, 9]. In this paper, the area of influence of mining works, depending on the depth, design and capacity of the coal seam was defined.

2 Materials and methods

The object of research is the length of the coal seam of high stress zone. To address the issue of the length of the bearing pressure zones it is necessary to analyze the stress state of the coal seam. For the calculation of the stress-strain state of the rocks near mining working the finite element method was used, and the series of numerical experiments were conducted.

* Corresponding author: gva.pm@kuzstu.ru

3 Results and discussion

The diagram of the problem of calculating the stress-strain state of the rocks in the development of mining works is shown in Fig. 1. Estimated rectangular area included the following units: 1 – the coal seam; 2 – the rock caving in goaf; 3 – the seam's roof and the soil containing rocks. The boundary conditions and loads are taken in the following form. At the upper boundary of the estimation area AB there is no load (full surface). At the lower boundary of the CD, remote from the pillar, there is no normal displacement (the border is secured). Vertical boundaries BC and AD, rather remote from the mining face do not feel the effect of the mining working and have no horizontal displacement.

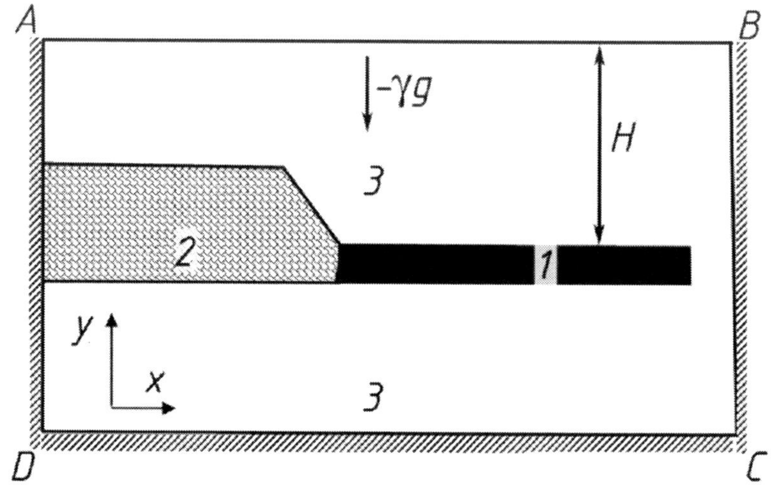

Fig. 1. The design scheme: 1 – coal seam; 2 – the goaf; 3 – host rocks

The rock of roof and soil are loaded by the volume strength $\gamma = 25000$ N/m^3 corresponding weight of host rocks; Young's module E =1010 N/m^2, Poisson ratio $\nu = 0{,}25$.

Coal seam is loaded by volumetric force $\gamma = 14000$ N/m^3, E = 109 N/m^2, $\nu = 0.25$.

Caving in goaf has a capacity of five times more than the thickness of the seam with the following characteristics: $\gamma = 20000$ N/m^3, $E = 10^9$ N/m^2, $\nu = 0.25$.

Seam thickness ranged from 1 to 10 m, the depth of development – from 200 to 1000 m. The stress-strain state of the area of estimation is described by linear deformable blocks behavior. The problem was solved by finite element method (FEM) using licensed program «ELCUT». Fig. 2 shows the example of vertical stress distribution in excess of their initial value, i.e., high rock pressure zone (lightly background) and unloading zone (dark background). In this example the seam thickness was 2 m, the depth of development – 400 m.

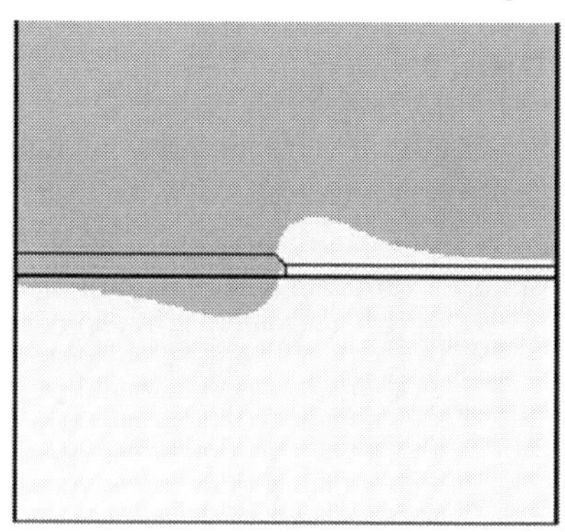

Fig. 2. The distribution of high and low vertical stresses near the working

The distribution of high and low stresses corresponds to dominating opinions about unloading zones position in rock arrays of the seam's roof and soil during coal mining [10-12]. The bearing pressure distribution in the seam for the same conditions in the linear deformed rocks is shown in Figure 3. Here the compressive stresses are taken negative.

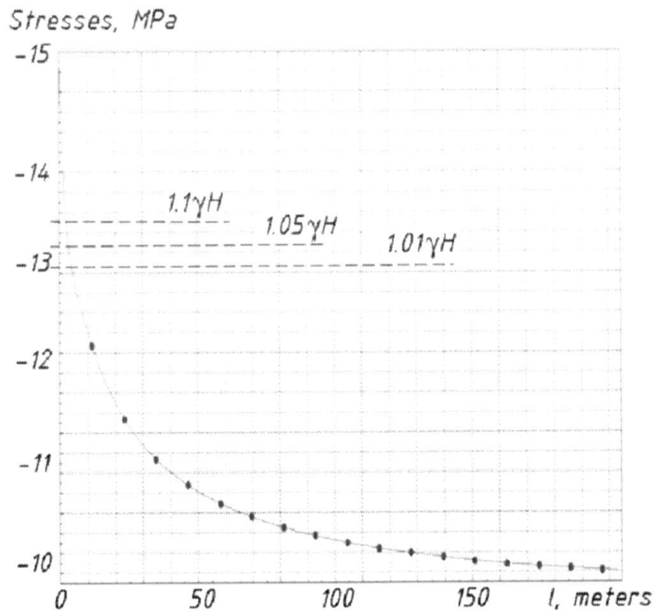

Fig. 3. Bearing pressure distribution in linear deformed rocks

The most remote point of the seam from mining stope (at which the increment of the bearing pressure relative to γH is 5%), is considered to be the boundary of the bearing pressure. Also, the bearing pressure value increments of 10% can be used as the boundary threshold value. This paper observes three possible values of the length of bearing pressure zone accordingly the increments of the bearing pressure of 10%, 5% and 1%, and designated as l_{10}, l_5, l_1. The length of the bearing pressure zones was determined with accuracy up to 1 m. The values of these lengths with seam thickness of 2 m for different depth of mining works are shown in Table 1.

Table 1. The length of bearing pressure zone

The length	The depth of mining works, m				
	200	400	600	800	1000
l_{10}, m	39	41	42	43	43
l_5, m	69	74	76	77	78
l_1, m	150	153	157	158	161

Accordingly the results of calculations of the length of the bearing pressure zones (Table 1) we concluded that during changing the depth of working of 200 to 1000 m the lengths vary by less than 10% relatively their average value. Therefore, we can assume that for linear deformed rocks the depth of mining works does not affect the length of the bearing pressure zone. And you can only specify one, the average length of each seam thickness. Thus, for seam thickness of 2 m the average bearing pressure zone lengths are 42, 75 and 155 meters for 10%, 5% and 1% increment of the bearing pressure pressures γH. This result can be explained by the fact that with increasing the depth of mining works the percentage increment of bearing pressure increases in proportion. A similar result (the independence of the bearing pressure zone length on the depth of mining works) was obtained for each seam thickness from 1 to 10 m.

According to the results of numerical experiments the regression and correlation analysis were conducted, and the dependences of bearing pressure zone length on the seam thickness *m* were obtained. From considering extend, quadratic and logarithmic dependencies the logarithmic was recognized as the most significant with the following equations:

$$l_{10} = 34.3 \cdot \ln(m) + 18.3; \quad (R^2 = 0.997) \quad (1)$$

$$l_5 = 44.1 \cdot \ln(m) + 41.8; \quad (R^2 = 0.987) \quad (2)$$

$$l_1 = 34.1 \cdot \ln(m) + 119.0; \quad (R^2 = 0.917) \quad (3)$$

The existing graphical dependence of bearing pressure zone length on the depth of mining works was obtained for rocks with the other deformation properties.

4 Conclusions

1. In the linear deformed rocks the length of the bearing pressure zone length does not depend on the depth of mining works.

2. The length of the bearing pressure zone depends logarithmically on the flat seam thickness in the limits from 1 to 10 m.

3. These dependencies form the area of geo-mechanical influence of mining works.

The authors thank JSC "SUEK-Kuzbass" for financial support of this research.

References

1. B.H.G. Brady, E.T. Brown, *Rock Mechanics for underground mining* (Springer science, 2005)
2. J. Jaeger, N.G. Cook, R. Zimmerman, *Fundamentals of rock mechanics* (Wiley Blackwell, 2007)
3. M. Pastor, C. Tamaonini, *Numerical modeling in geomechanics* (Kogan Page science, 2004)
4. J.A. Hudson, J.P. Harrison, *Engineering rock mechanics. An introduction to the principles* (Pergamon press, 2000)
5. O.C. Zienkiewicz, R.L. Taylor, J.Z. Zhu, *The Finite Element Method: Its Basis and Fundamentals (7th Edition)* (Butterworth-Heinemann, 2013)
6. O. Fusao, *Computer methods and recent advances in geomechanics* (CRC Press, 2014)
7. M. Fremond, F. Maceri, *Mechanical modeling and computational issues in civil engineering* (Springer, 2005)
8. V.A. Gogolin, Y.A. Ryzhkov, Soviet Mining Science, **13**, 17 (1977)
9. V.M. Seryakov, J. Min. Sci., **45**, 420 (2009)
10. M.V. Kurlenya, V.E. Mirenkov, V.A. Shutov, J. Min. Sci. **50**, 1001 (2014)
11. V.A. Gogolin, Y.A. Ryzhkov, Soviet Mining Science **17**, 156 (1981)
12. Y.A. Ryzhkov, Y.V. Lesin, V.A. Gogolin, N.V. Karpenko, J. Min. Sci. **32**, 188 (1996)

The influence of advancing speed of powered mining stope with single face on earth's surface displacing in Kuzbass

Alexey Renev[1*], *Sergey* Svirko[1,2], *Alexey* Bykadorov[2], *Valery* Fedorin[1]

[1]T.F. Gorbachev Kuzbass State Technical University, Department of underground facilities and mines constructing and resource deposits mining, Vesennyaya St. 28, Kemerovo city, Russian Federation
[2]LLC SIGI (Siberian Institute of Geotechnical Research), 653033, Prokopievsk, 11 Gagarina st., Russian Federation

Abstract. The researches in the area of undermined rock massifs estimate the influence of advancing speed of mining stopes on earth's surface shifting ambiguously. It reflects the lack of maturity of this trend. The aim of this paper is to establish the relationship between subsidence of individual points on the surface profile line along strike and the speed of stope's advancing, taking into account other influencing factors. Dependencies of displacement of certain points of the earth's surface undermined during mining operations can be identified by frequency instrumental observations on the profile lines. Further analysis is made by mathematical methods, including the random number theory. In the formation of the dynamic downfold of displacement the relationship between advancing speed and subsiding of individual points in the "behind the face" half-downfold cannot be traced. In the "front face" half-downfold the increase of mining stope advancing speed reduces dynamic subsidence. In the paper empirical relationships between subsidence of individual points on the surface and the position of the working face were shown.

1 Introduction

The Earth's surface is the part of a deformable earth formation, and its displacement reflects the processes that occur in undermined rock mass. Note the hypotheses that were confirmed in the current schemes and mechanisms of displacement of undermined earth formation:

- The rule of "normal", introduced by Gonot [1] and further developed by Dumont [2], is based on the seam's roof decomposition on the following components: the normal to the bedding and the normal acting along the seam, which is balanced by the reaction of the underlying rocks. Based on this postulate, the roof displacement will occur only under the influence of the normal component;

* Corresponding author: raa@kuzstu.ru

- The "arch" hypothesis, which are based on the fact that the collapse of rocks in the roof of workings is shaped like an arch. One of the first "arch" hypotheses was proposed by Ritter [3] and then developed by Fayol [4];
- "Beams" hypothesis was proposed in 1867 by Schultz [5]. Rock layers over the worked-out area represented by the solid layered rocks are treated as the beams, embedded with one end in pillar. The other ends of the beams can lean on the supports, backfilling or even hang freely. The roof, consisting of several layers of rock, is considered as a set of beams.

The final stage of the early period of displacement researches and at the same time the beginning of a new period was Briggs theory [6], where for the first time the rock mass was considered as a medium in the volumetric stressed state.

Despite a long history of domestic and foreign scientific researches in the strata displacement, there is no single scientifically sound theory, revealing the inner essence of the phenomena and derived therefrom methods for calculating the deformation of containing rocks and the earth's surface.

2 Materials and Methods

The first methods of calculation of Earth's surface subsidence at a small depth of coal seams mining were based on data obtained from the opinion made upon observations that subsidence depends on the thickness of a mined seam m and its dip angle α: $\eta_Z = m \cdot \cos\alpha$. On the mines of Essen the scale of earth's surface subsidence was estimated by the formula: $\eta_Z = n \cdot m \cdot z$ w, here the coefficients n and z were used for considered method of coal extraction (with the collapse of the roof or with backfilling of worked-out space) and the time elapsed after the ending of mining works. The coefficients considering the depth of mining works and the extent of undermined rocks disintegration were introduced later [7].

As a result of further instrumental observations the scholars began to consider the area of the mined segment in their calculations - first by introducing a "reduced" thickness of mined seam m_{red} [8], and then with use of the coefficient equal to the ratio of mined segment area to the area of the influence of mining operations on the earth's surface [9].

After World War II new directions in the science of strata displacement appeared: stochastic method, displacement calculation on standard curves, finite element method which are widely used now. Today, there are a lot of programs that implement the finite element method: NASTRAN, ANSYS, MARC and others.

Embodiments of the method of standard curves were developed in Poland, Hungary, Germany and the USSR, where it became principal and entered the current regulations on the calculation of displacements and deformations of the earth's surface.

Nevertheless the number of papers devoted to dynamic processes accompanying the active phase of undermining earth's surface displacement, is small enough in general, including the area of cognition research of geomechanical processes that indicate a lack of maturity in this direction.

The complexity of the problem of dynamic manifestations of rocks subsidence in undermining by fast moving mining faces is illustrated by the fact that even today the impact of stope advancing speed changes on the geomechanical processes occurring in the array at high speeds of undermining is estimated ambiguously.

A number of researchers, basing on the observations of the displacement of rocks and the earth's surface under the influence of mining works indicate that the increase in mining stope advancing speed in many cases reduces the deformation of the earth's surface. Others argue that the rate of stope advancing has no significant effect on the earth's surface deformation, but increases or decreases the duration and intensity of the displacement process. Taking into account that the rate of face's stope advancing over which the subsidence of Earth's surface was previously observed does not exceed 100 m per month, and varied considerably little, its unnoticed influence on the process of the earth's surface displacement can be assumed. However, if we consider the currently existing displacement speed values, the conclusion of non-essential role of the speed seems premature.

If 15-20 years ago, the stopes advancing speed rarely exceeded 150 m per month, then now they have reached 450 m per month and more. At the current load on mechanical systems, which some day reaches 20000 tons, the stopes advancing speed can be expected up to 600 m per month. In this regard, the special attention should be paid to the studies by A.S. Yagunov on displacements and deformations of the earth's surface for the high speed of face's stopes advancing (up to 15 m per day) in Kuzbass. He examined the dynamics of displacement downfold profile formation depending on stope advancing. At the same time the following terms were introduced:

- *the dynamic displacement downfold* - the state of displacement downfold in the process of its formation, depending on the position of the working stope of the face to the mounting chamber;
- *the dynamic displacement micro-downfold* - the state of displacement downfold at the withdrawal of the stope from the mounting chamber to the "flat bottom" formation;
- *the dynamic subsidence of the Earth's surface* η_d (mm, m) - vertical component of the displacement vectors of displacement the points in the dynamic downfold depending on the position of stope of the face to the mounting chamber;
- *the dynamic horizontal displacement of the Earth's surface* ξ_d (mm, m) - the horizontal component of the displacement vectors of the points in the dynamic downfold depending on the position of stope of the face to the mounting chamber;
- *the half-downfold behind the stope* L_{bs}– the half-downfold from the side of mounting chamber;
- *the half-downfold in front of the stope* L_{fs}- respectively from the side of dismounting chamber.

As a part of the development of theoretical ideas about the dynamics of displacements and deformations at flat and inclined coal seams processing by high-productive stopes in Kuzbass mines many various monitoring stations were founded. Each of the monitoring stations consisted of two profile lines of ground frames laid down in the directions of strike and across the strike of the faces. As the frames the metal rods were used (length -1.8 m and diameter - 20-22 mm) forged on the spearhead. The frequency observations were carried out with a periodicity of 1 time in 2 days. The observation results were processed using statistical methods and correlation analysis.

3 Results and Discussion

This article presents some results of studies of the influence of powered face's stope advancing speed during flat and inclined coal seams mining in Kuzbass on the dynamics of individual points of the undermined Earth's surface. The terminology introduced by A.S. Yagunov is retained. It is found that the character of displacement points on the Earth's surface in the plane $\xi 0 \eta$ the dynamic downfold profile along the strike is divided into four zones A, B, C and D (Fig. 1). Their dimensions were defined accordingly the border angles δ0, δ'0 and the angles of characteristic zones δA , δB, δC, δD, established by instrumental observations.

The regularities of points of the Earth's surface subsidence changes in the specific areas (*xs*) A, B, C and D depending on the current situation of the mining stope (*lt*) in the interval

$D_{2x} = 1.6H_{cp}$ were defined. Initial data of instrumental observations were presented in the normalized form. The transition from conventional to normalized units was done using linear normalization performed within the variables [0, 1] by formula (1):

$$x_{\mu ik} = \frac{x_{ik} - x_{\min k}}{x_{\max k} - x_{\min k}}, i = 1, 2, \ldots, n; k = 1, 2, \ldots, N \tag{1}$$

where: x_{ik}, $x_{\mu ik}$ - value of the variable in the traditional dimension and in a normalized form in the *k-th* sample; $x_{\min k}$, $x_{\max k}$ - the minimum and maximum values of a variable in the *k-th* sample; *n* - the number of data in the *k-th* sample; N – the number of samples.

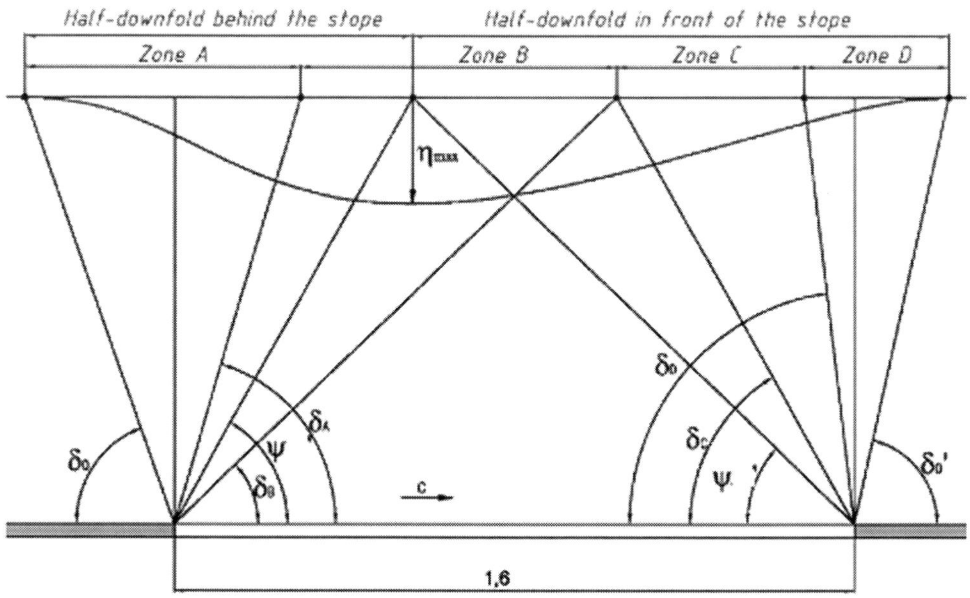

Fig. 1. Subsidence zones A, B, C and D in formed micro-downfold for mining stope position $D_{2x} = 1.6H_{cp}$ in the section across the face's strike.

Fig. 2 shows the calculation of expected subsidence and horizontal displacement of the surface point in zone B. Face with 200 m width, 1000 m length and with an average depth 200 m at advancing the stope with the speed 5 and 15 m per day is being worked. The dip angle of the coal seam is 10°, extracted thickness is 3 m.

The point being evaluated is at a distance of 270 m from the mounting chamber in zone B. Subsiding of the point begins at the position of stope at 160 meters from the mounting chamber and occurs as a sigmoid to a maximum (η_{max}) of 1450 mm for $c = 5$ m per day, and 1193 mm for c = 15 m per day. Horizontal displacement of point depending on the scale of subsidence can be treated as reciprocating. First the point moves towards dismounting chamber up to a maximum of ξ_{max} = 244 mm decreasing exponent, then increasingly shifts exponent in the direction of mounting chamber. Advancing speed of the stope does not affect analytical dependence of horizontal displacement on the point's subsiding. It depends only on the value of the final subsiding.

 DOI: 10.1051/e3sconf/20171501002

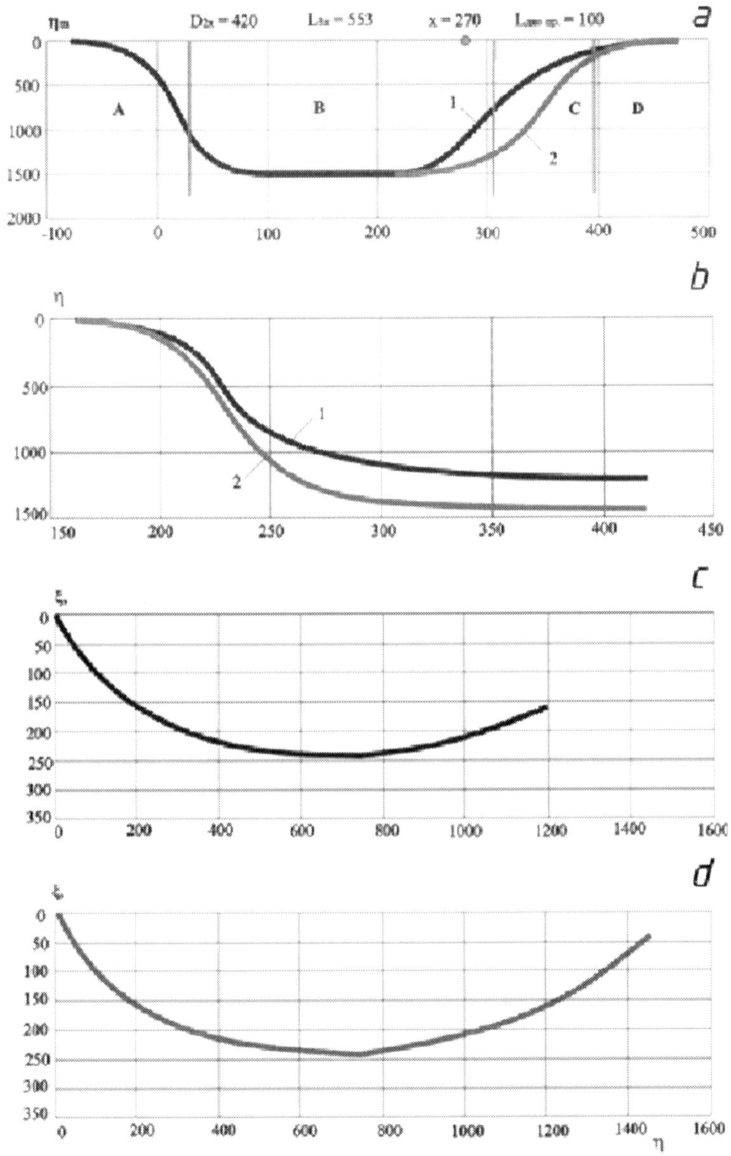

Fig. 2. The calculation for example: a) dynamic displacement downfold for mining stope withdrawal of 420 m from the mounting chamber; b) the pattern of surface points subsidence distant from mounting chamber of x = 270 m; c) horizontal displacement of the point depending on the subsidence for advancing speed of stope (c) of 5 m per day; d) horizontal displacement of the point depending on the subsidence for advancing speed of stope (c) of 15 m per day.

4 Conclusions

The growth of advancing speed of the stope (up to 15 m per day) does not have significant influence on the final value of the Earth's surface deformation, but only reduces their dynamic values in intermediate half-downfold close to mounting chamber, i.e. increase in speed changes the intensity of displacement process in "front face" half-downfold. The half-downfold by dismounting chamber flattens when advancing speed of the stope increases. That reduces the value of dynamic subsiding of the point in this area of undermined surface. Obtained analytical dependences of horizontal displacement of the surface point on the value of its subsiding in characteristic zones A, B, C and D does not change with increasing advancing speed of the stope.

References

1. Gonot, *Not sur les degradations survenues aux maisons situées le long du Quat Fragnee etc* (Liege, 1863)
2. G. Dumont, *Des affaissements du sol, produits par L'exploitation houilles* (Liege, 1871)
3. Ritter W., *Die Statik der Tunnelgewölbe* (Springer, Berlin, 1879)
4. Fayol M., Bull. de la Soc. de l'Industrie minerale, **14**, 805, 1885
5. Schulz A., *Untersuchungen über Dimensionen der Sichereitspfletlern für den Saarbrucken Steinkohlen bergbau* (Zeit. Für das Berg-Hutten und Salinenwesen im Preussische Staate, 1867)
6. Briggs H., *Mining subsidence* (London: Edvard Arnold and Co, 1929)
7. Goldreih A., *Die Theorie der Bodensenkungen in Kohlengebieten* (Berlin, J. Springer, 1913)
8. Köhne, *Unveröff. Studien, siehe Mitt. Markscheidew* (1923)
9. H. Schmitz, Mitt. Markscheidew **30**, 29 (1923)

Energy Consumption in the Process of Excavator-Automobile Complexes Distribution at Kuzbass Open Pit Mines

Ivan Panachev[1,*], *Boris* Gerike[1], *Ilya* Kuznetsov[1], *Anastasia* Shirokolobova[1]

[1]T.F. Gorbachev Kuzbass State Technical University, 650000, Kemerovo, 28 Vesennya st., Russian Federation

Abstract. Every year worldwide coal mining companies seek to maintain the tendency of the mining machine fleet renewal. Various activities to maintain the service life of already operated mining equipment are implemented. In this regard, the urgent issue is the problem of efficient distribution of available machines in different geological conditions. The problem of "excavator-automobile" complex effective distribution occurs when heavy dump trucks are used in mining. For this reason, excavation and transportation of blasted rock mass are the most labor intensive and costly processes, considering the volume of transported overburden and coal, as well as diesel fuel, electricity, fuel and lubricants costs, consumables for repair works and downtime, etc. Currently, it is recommended to take the number of loading buckets in the range of 3 to 5, according to which the dump trucks are distributed to faces.

1 Introduction

The increasingly complex operating conditions require the development of new assessment factor, which can take into account the maximum possible number of machine operating parameters (bearing capacity, bucket capacity, etc.).

The value of specific energy cost (SEC) for the transportation of 1 ton of rock mass from the open pit was accepted as the measuring energy efficiency factor of the deep pit transport systems, which was determined by the formula [1]

$$P_f^a = \frac{g}{i} \times k_{nep} \times k_{uT} \times k_d , \quad (1)$$

where P_f^a – the specific energy cost for transportation of 1 ton of rock mass to 1 m, g.s.f./t·m (grams of standard fuel/t·m); g – specific cost of diesel fuel by trucks, g/t·m; i – route inclination, ‰; k_{nep} – the processing coefficient, considering energy costs to get diesel from oil (k_{nep}=1,18÷1,20); k_d – the coefficient, considering energy costs to produce and to transport fuel (k_d=1,04÷1,10); k_{uT} – the coefficient, considering the difference of specific heat of diesel combustion and equivalent fuel (k_{um}=1,5).

* Corresponding author: pia.sm@yandex.ru

It was estimated that the specific energy cost is substantially affected by such factors as dump truck overloading, route inclination on the rise and others [2].

Transportation of blasted rock mass by heavy dump trucks can be divided into energy-consuming and energy-conservation modes according to energy indicators.

Such steps of transportation as loading and unloading, waiting for loading, shift change, lunch break are related to energy-consuming. At these stages dump trucks movement is not observed, an engine runs at idling speed, therefore, there is no significant energy consumption [5]. Only operable conditions of the engine fuel system have impact on energy consumption. If the dump truck parts are in good condition, energy cost is valid.

The main influence on energy consumption of the transportation process and metalwork operation time occurs during energy-consuming stages of operation. These stages include the machine straight technological roads traffic, upgrades and descents, turnings and traffic through the crossings. However, it should be noted that each of the given traffic parameters corresponds to its value of energy demands.

As a result of analysis of the influence of various operating parameters of the open pit automotive transport on energy consumption of transportation process, it was determined that the longitudinal inclination of open pit roads is the most informative parameter of specific energy consumption evaluation [3].

2 Results and Discussions

According to the results of investigations carried out at Kuzbass open pits, the diagram of route inclination was designed; if to reach this inclination it is possible to observe a significant change in the operating terms of dump trucks with various bearing capacities (Figure 1).

Figure 1 shows typical route inclinations for the following operating parameters: Q_{nom} – nominal value of diesel fuel consumption, at which the engine is operating at nominal power (engine life is not reduced, specified by the manufacturer); P_{fnom} - nominal value of specific energy cost, at which there is time to failure, stated by the manufacturer; σ_{nom} - nominal stresses in suspension mount of metalworking that are equal to endurance limit, at which cracks formation and growth are not yet observed; P_{fpre} - limit volume of unit costs, at which the time to failure is 20% from guaranteed by the manufacturer.

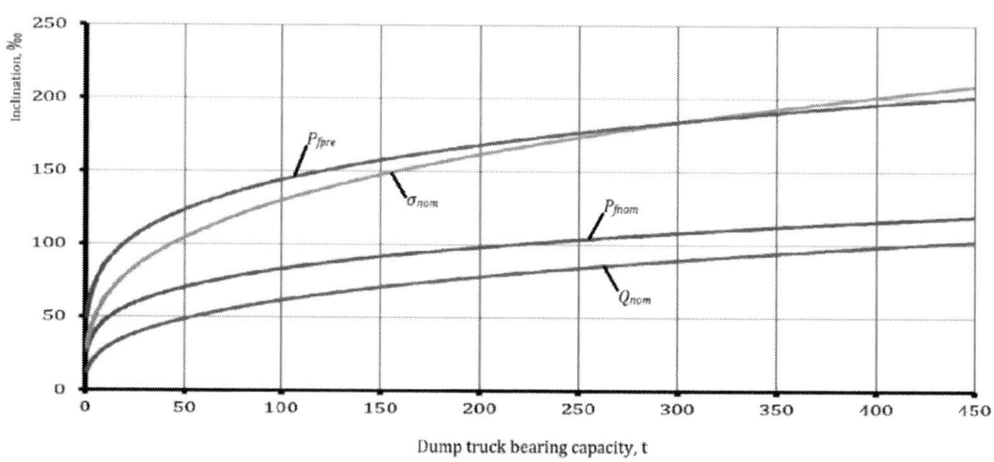

Fig. 1. The diagram of specific route inclinations

Figure 1 shows that the curves Q_{nom} and P_{fnom} are parallel in all the diagram area; and if to calculate the specific energy cost the consumption of diesel fuel is to be taken into account (see formula 1). For the dump trucks with bearing capacity from 250 to 350 t on the route inclinations from 170 to 190 ‰ the identical operating conditions influence on the values of marginal energy costs and nominal voltages in the load-bearing steel structures are observed. Moreover, the extreme point (350 t; 190 ‰) of intervals is characterized by increased stress and, consequently, reduction of structural life of heavy trucks.

In this regard, when choosing motor transport to use in developed deposits to the value of the longitudinal inclination of 190 ‰, it is necessary to be based on the change of specific energy cost, and if the value of the longitudinal inclination is more than 190 ‰ it is necessary to be based on the change of nominal voltages.

The analysis of the results allowed to get the dependence of the specific energy cost on mined rock transportation from dump truck bearing capacity. Figure 2 shows the diagram of the established relationship for heavy trucks at longitudinal inclinations from 10 ‰ to 100 ‰.

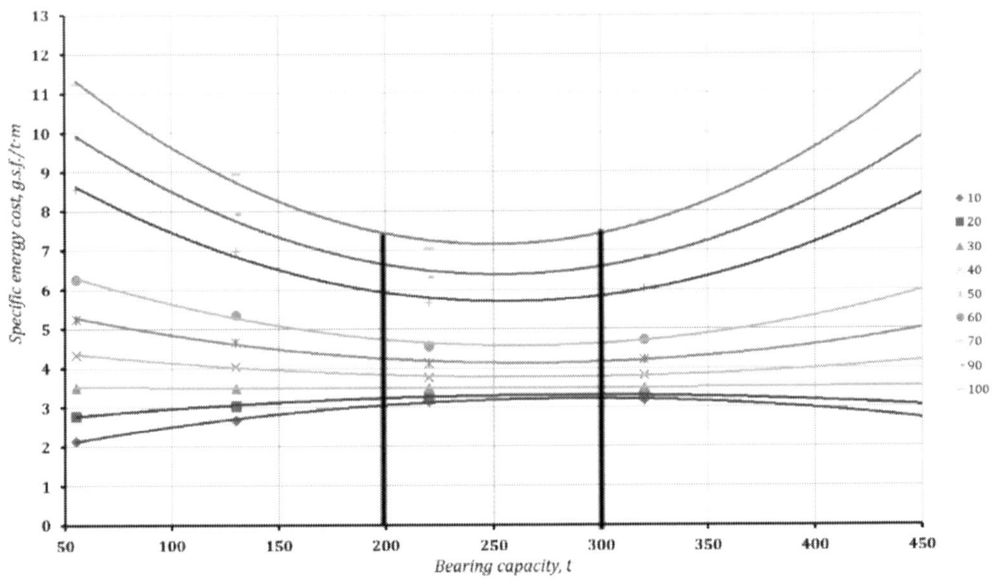

Figure 2. The dependence of the specific energy cost for transportation on dump truck bearing capacity

Figure 2 shows that the most universal indicator of energy efficiency for usage in different mining geological and technical conditions are dump trucks with bearing capacity of 200-300 t. The value of SEC is taken for the energy efficiency estimation test of excavation process [4]:

$$P_f^e = w \times k_e \times k_{loss} \times k_d, \quad (2)$$

where P_f^e – the specific energy cost for excavation of 1 m^3 of rock mass, g.s.f./t·m (grams of standard fuel/m^3); w – specific energy consumption by an excavator, kW·h /m^3; k_e – the coefficient considering energy costs to produce 1 kW·h of electricity (k_e = 310÷330 g/kW·h); k_{loss} – the coefficient considering losses of electricity under transmission and distribution ($k_{loss} \approx$ 1,09); k_∂ – the coefficient considering the cost of energy under fuel extraction and transportation (k_∂=1,04÷1,10). Specific energy consumption by an excavator is the ratio of electricity cost (E) to the machine technical performance per hour (P_{tech}):

$$w = E/P_{tech}. \quad (3)$$

The technical performance per hour is calculated by the formula:

$$P_{tech} = \frac{3600E}{T_{w.c}} \times k_{ex} \times k_f, \quad (4)$$

where E – bucket capacity, m^3; $T_{w.c.}$ – the duration of excavation working cycle, s; k_e – the excavation coefficient; k_f – the face coefficient considering the effect of auxiliary operations.

JSC "Berezovskiy" (Prokopyevsk) was chosen as an experimental base, where a large number of excavators used are the excavators with a shovel capacity of 5 m^3 (ECG-5), 10 m^3 (ECG-10) and 12 m^3 (ECG-12), the dependences of the specific energy cost on excavation from the dump truck bearing capacity (respectively) were estimated (see Fig. 3):

$$P_f^e = 0.0024q^2 - 0.2857q + 168.401, \quad (5)$$

$$P_f^e = 0.0015q^2 - 0.1252q + 154.977, \quad (6)$$

$$P_f^e = 0.0015q^2 - 0.1514q + 154.875. \quad (7)$$

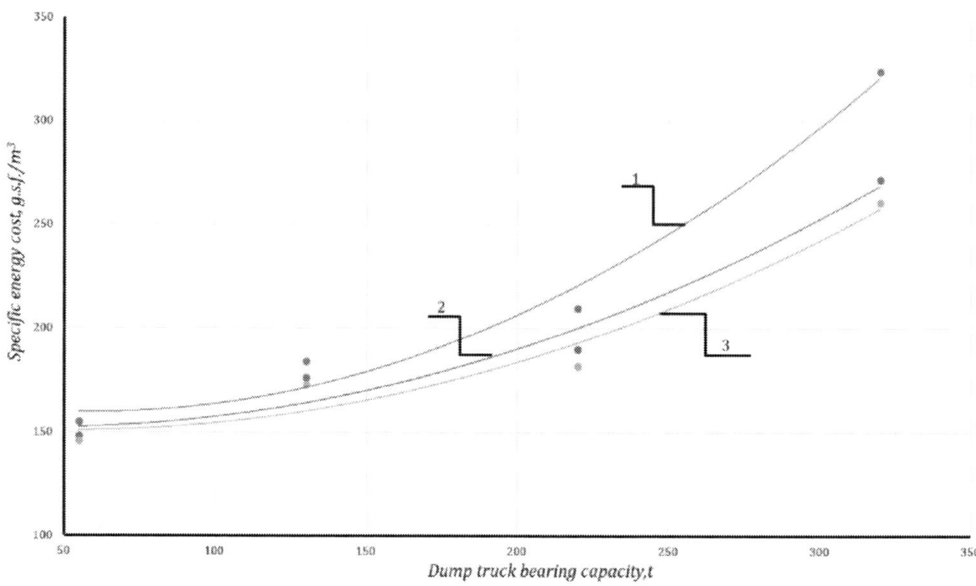

Figure 3. The dependence of specific energy cost of excavators with bucket capacity 5, 10 and 12 m^3 on excavator bearing capacity (1 – ECG-5, 2 – ECG-10, 3 – ECG-12)

Figure 3 shows that the areas of dependency in the range of 50 to 130 t are maximum close to a straight line; it indicates a slight increase in the share of energy cost due to the recommended amount of bucket loading. Further there is a substantial increase in energy consumption of the excavator with bucket capacity of 5 m^3. Consequently, the greater the deviation from the recommended amount of bucket loading is, the higher the jump in specific energy cost is. For the energy assessment of dump trucks distribution with an excavator ECG-10, a joint diagram of dependences of specific energy cost by excavator-automobile complex on dump trucks bearing capacity in operation at the maximum permissible longitudinal inclination of pit roads in 80 ‰ was built (Fig. 4).

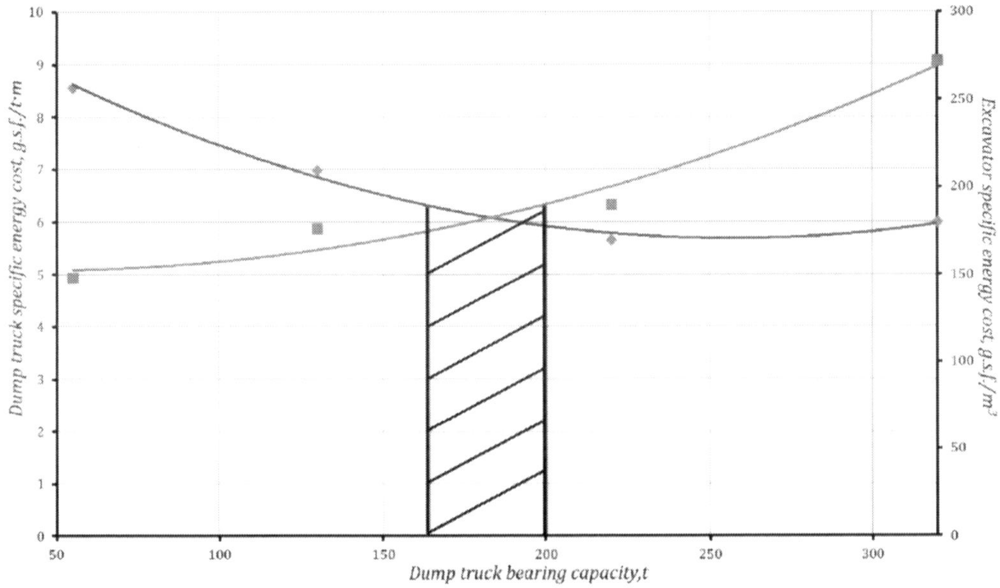

Figure 4. The dependences of specific energy cost of excavator-automobile complex on dump truck bearing capacity

Figure 4 shows that heavy dump trucks with carrying capacity of 160-200 t are the best and most effective in operation for the ECG-10 excavator according to energy parameters. In this case, the unit cost of the dump truck and the excavator are as close to their lowest values. Thus, the number of buckets loading varies from 6 to 9.

3 Conclusions

When calculating the dump truck bearing capacity corresponding to the recommended number of loading buckets (3-5), it was estimated that dump trucks with bearing capacity up to 130 t are the most effective for ECG-10 excavators. However, specific energy cost for transportation in this case takes rather high value (1.2-1.5 times higher than in energy selection of excavator-automobile complex), which has a negative impact on all the parameters of motor transport operation. According to the research results, it was estimated that energy assessment of the operational efficiency of excavator-automobile complex, based on specific energy cost, is a multi-factor characterizing fundamental parameters of the excavators and dump trucks operation in the developed deposit conditions; and it is recommended to be used when excavator-automobile complex is being chosen.

References

1. I.A. Panachev, G.V. Shirokolobov, I.V. Kuznetsov, A.G. Shirokolobova, *8th Russian-Chinese Symposium. Coal in the 21st Century: Mining, Processing and Safety*, 144 (2016)
2. I.A. Panachev, I. V. Kuznetsov, Journal of Mining Science, **51**, 267 (2015)
3. Baoying Fan, Yingju Yuan, Int. Journal of Mining Science and Tech., **26**, 653 (2016)
4. A. Soofastaei, S.M. Aminossadati, Int. J. of Min. Sci. and Tech., **26**, 745 (2016)
5. I. V. Sokolov, A. A. Smirnov, Yu. G. Antipin, J. of Mining Science, **52**, 121 (2016)
6. V.A. Solomennikov, V.I. Cheskidov, Journal of Mining Science, **51**, 1213 (2015)

7. S.V. Klishin, A.F. Revuzhenko, Journal of Mining Science, **51**, 1070 (2015)
8. Chengguo Zhang, Ismet Canbulat, Faham Tahmasebinia, Bruce Hebblewhite, International Journal of Mining Science and Technology, **27**, 43 (2017)
9. Fukun Xiao, Gang Liu, Ze Zhang, Zhiliang Shen, Fengrui Zhang, Yifei Wang, International Journal of Mining Science and Technology, **26**, 981 (2016)
10. J. Jiang, Q. Liu, J. Xu, Int. Journal of Mining Science and Technology, **26**, 1003 (2016)
11. Jesse W. Puller, Ken W. Mills, Rob G. Jeffrey, Rick J. Walker, International Journal of Mining Science and Technology, **26**, 103 (2016)
12. V.P. Efimov, Journal of Mining Science, **52**, 274 (2016)
13. S. Bornyakov, D.V. Salko, Journal of Mining Science, **52**, 338 (2016)
14. Yu. V. Lesin, S.Y. Luk'yanova, M.A. Tyulenev, J. Min. Sc., **46**, 78 (2010)
15. E.N. Sher, V.P. Efimov, Journal of Mining Science, **51**, 1108 (2015)

Slope Stability Assessment and Underground Mine Design Analysis of Achibo-Sombo Underground Conventional Coal Mine, Southwest Ethiopia

Mulata Haftu[1], *Bheemalingeswara* Konka[1], *Kifle* Woldeargay[1] and *Asmelash* Abay[1*]

[1]Department of Earth Science, College of Natural and Computational Sciences, P.O.Box.3066, Mekelle University, Mekelle, Ethiopia

Abstract. We report the slope stability assessment and underground mine design analysis results for the first of its kind conventional Achibo-Sombo underground coal mine from Ethiopia. Based on the RQD, RMR and Q classification system, the rocks in the mine are of poor to fair rock quality and are classified into category I (<10 MPa), category II (10-25 MPa), and category III (>25 MPa). Joint space and aperture are ranging from few mm to cm with E-W, NNW-SSE and N-S orientations and low to very low persistence from 50 cm to 2 m. The factor of safety calculated using CMRI, Salamon-Munro, Greenwald, Obert and Duvall, and Bieniawsk methods for the existing pillars for category I is 1.37, 4.39, 2.21, 3.58, 4.76; category II 2.25, 4.39, 7.76, 12.52, 16.72; and category III 3.21, 4.39, 13.72, 22.26, 29.58 respectively. The factor of safety calculated again for the newly proposed design using the same methods is 1.05, 3.83, 1.91, 2.99, 3.8 for category I; 1.65, 3.17, 5.48, 8.55, 10.02 for category II and 2.20, 2.63, 7.91, 8.26, 14.25 for category III respectively. The extraction percentage has increased from 24.2 % (for the existing) to 31%, 41% and 52% for I, II and III categories respectively.

1 Introduction

Ethiopia is gifted with many mineral resources. One such resource is coal. Exploration activities for coal deposits in Ethiopia have started around 1940s. Occurrence of the 4m thick coal at Yayu basin (part of the research) was reported by EMDE [1]. Review on coal reserve estimation of Yayu area was carried out by many workers [2]. In the Yayu area about 250 x 10^6 tons of coal is estimated as A and B reserve categories in 41 km^2 area [2]. About 178,576,000 tons is estimated as intrinsic economic resource for Wattete Mine field and coal rank determined to be lignite to sub- bituminous and medium to high volatile coal [2]. Though, coal mining in Achibo-Sombo (study area) is carried out by underground mining method using conventional approach, the safety of the mine and mine design, studies on structural, geotechnical and pillar design analysis are not yet done and remained a gap. The study includes, evaluation of pillar in terms size, pillar shape, strength and

* Corresponding author: asmelashby@yahoo.com

geometry, correlation of geological structures and investigate their effect on the mine, rock mass characterization and classify the rocks, study geotechnical properties of rocks, determine the safety factors and calculate the extraction ratio.

The study area is located in south western Ethiopia and bounded by geographic coordinates of $8^0$20'0''to $8^0$25'0''N and $35^0$56'0'' to $36^0$3'0''E (Figure 1).

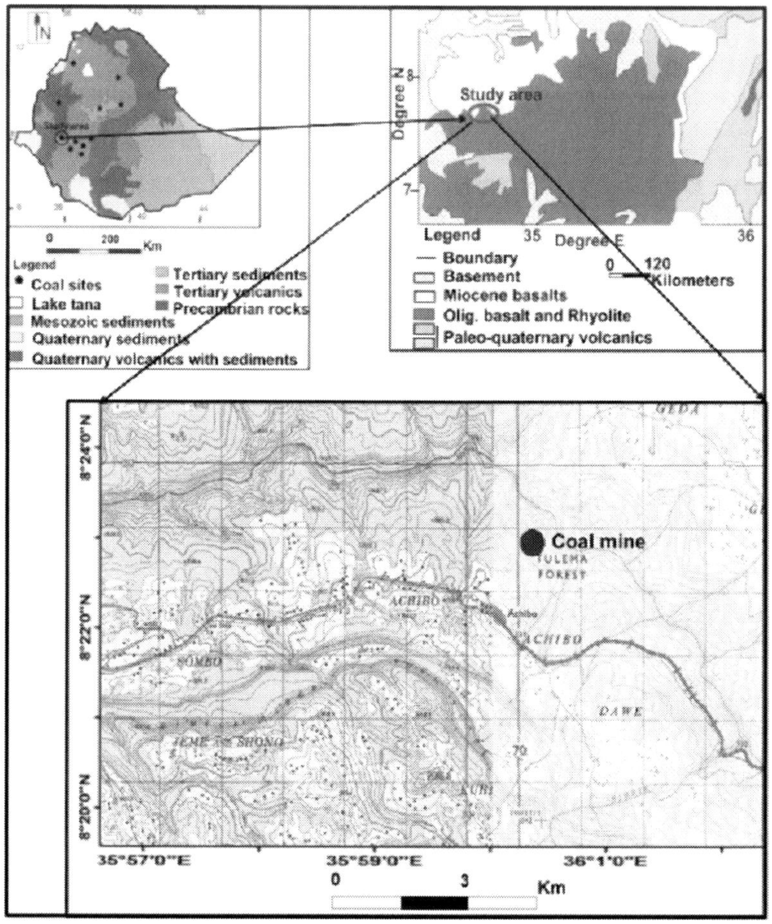

Fig. 1. Location map of the study area with reference to (A) coal map of Ethiopia (B) Regional geological map of southwestern Ethiopia (C) study area

The Achilo-Sombo mine is an underground conventional coal mine first of its kind in Ethiopia. It has three main parallel openings. At present only two are in operation, the third one is used for drainage purpose.

Published literature is very scanty for the Achilo-Sombo underground coal mine, except the unpublished reports and maps related to geology, geotechnic and mining prepared by various exploration companies and Geological Survey of Ethiopia.

2 Geology of the Study Area

The area consists of Neoproterozoic basement rocks at the base and overlain by Tertiary volcanics and coal-bearing sedimentary rocks (Figure 2). The volcanic rocks are divided into two, lower and upper volcanics based on age, stratigraphic setup and geochemical variations [1]. Upper volcanics include aphanitic basalt, porphyritic basalt, vesicular basalt, tuffs and amygdaloidal basalt. Sedimentary rocks occur between upper and lower volcanic and consists of carbonaceous shale, oil shale, sandstone, mudstone, claystone and coal.

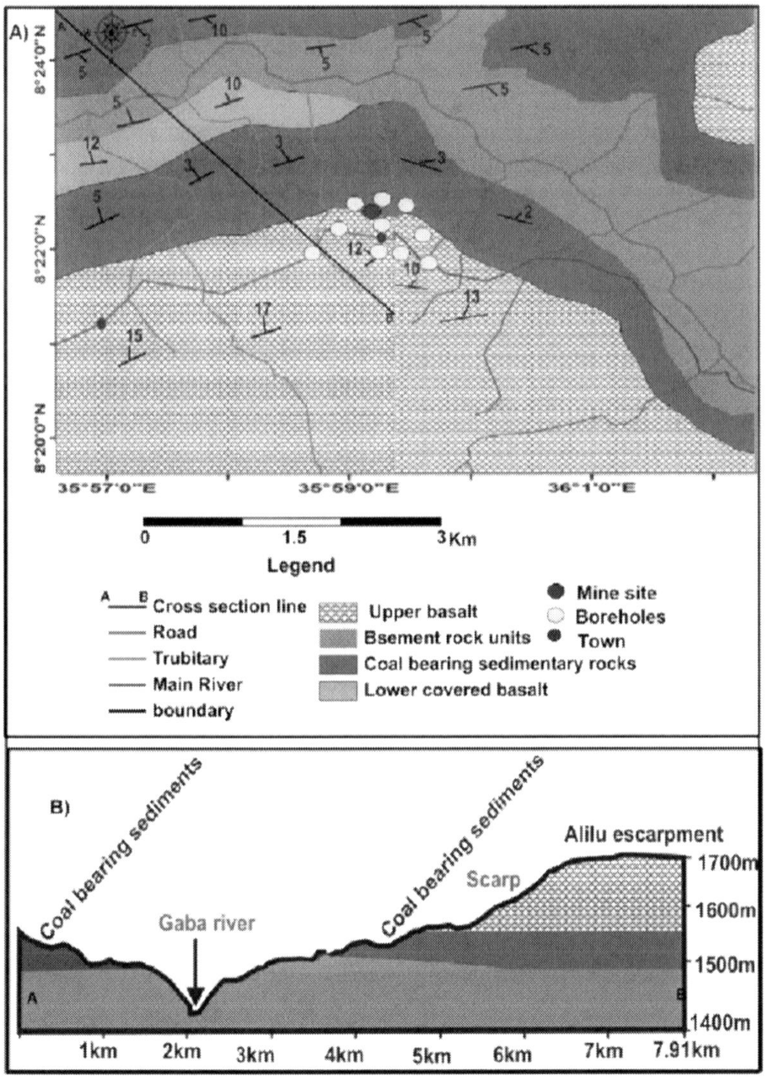

Fig. 2 (A) Geological map (B) cross section profile along the line A-B of the study area

3 Methodology

A field work was planned in February 2016 and a geological and structural map for the study area was prepared (Figure 2). All the geological units were properly studied and recorded. The borehole logs [1] were also studied to understand the subsurface and the core rock samples were collected for geotechnical analysis. The mine site was visited at different openings and studied the pillars. Rock types, number of seams, inclination of seams, pillar geometry and structures (faults, folds, structural contacts and fractures) were studied. Geological data related to coal seams location, depth of seams, seam thickness etc was collected. Mine data related to floor, roof, pillar positions, pillar dimensions and geometries etc were collected. Structural measurements were carried out and recorded them in detail. 50 rock samples from surface, core (representing all lithologies) and in the mine from pillar, roof and floor were collected to study their geotechnical properties. Out of these, 10 samples (4 from roof and 3 from floor and 3 from intermediate) were selected for compressive strength tests and conducted in Mekelle University, Ethiopia.

Rock mass was classified and characterized based on Rock Quality Designation (RQD), Rock Mass Rating (RMR) and the Rock Mass Quality System (Q-System). This is to evaluate strength of the pillars, roofs and floors as well as the support system. Based on different methods CMRI [3], Salamon-Munro [4], Greenwald [5], Obert-Duvall [6], and Bieniawski [7], factor of safety, pillar load and extraction percentage were calculated and evaluated to understand the stability of the mine. Different software packages like Rockwork, Surpack Minex 6.5, Global Mapper12, Surfer10, ARC GIS10 and Mapinfo 8.5 were employed for data processing.

4 Geological Structures

Different geological structures observed in the study area. The structures studied were normal faults (trending ENE-WSW, E-W, and N-S), strike slip faults, columnar joints, folds, weak planes and quartz veins. The structural data is presented in figure 3.

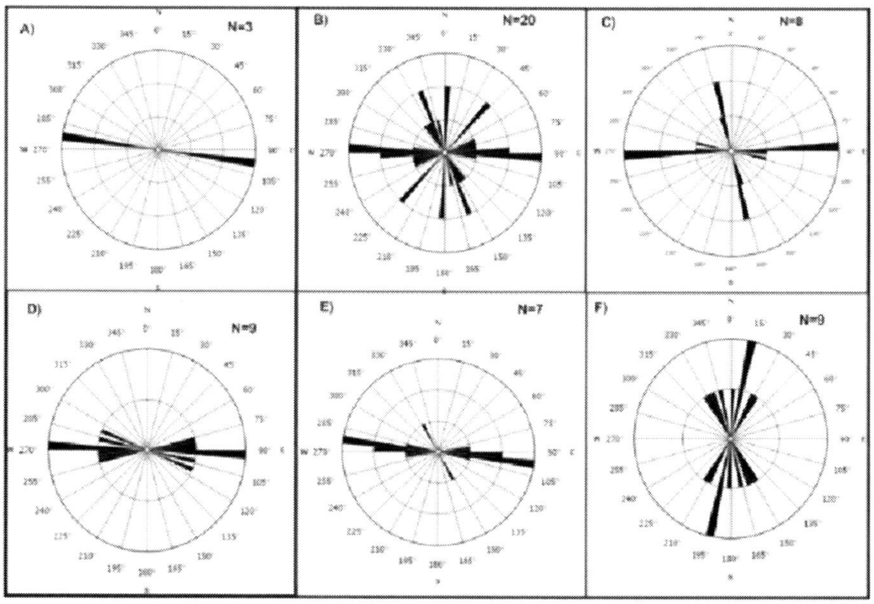

Fig. 3. Rose diagram showing (a) mining direction, (b) general joints and fractures orientation, (c) fault orientation, (d) lineament orientation (E) bed orientation, (F) quartz veins orientation.

5 Results and Discussion

The compressive strength results for the selected samples are given in table 1. Depending on the strength value the coal mine can be classified into three categories, I, II and III. Category I of <10 MPa strength, category II between 10 to 25 MPa strength and category III having >25 MPa strength.

Table 1. Compressive strength of the rock sample

S. No	Sample Code	Area (cm^2)	Area (m^2)	Weight (kg)	Load (KN)	load/area (KN/m^2)	Compressive strength(MPa)
1	FBH-5-1	13*7(91)	0.0091	0.731	155.12	17046.15	17.046

2	FBH-3-1	7.3*7(51.1)	0.00511	0.416	111.7	21859.09	21.86
3	RBH-7-4	11*5.3(58.3)	0.00583	0.411	45.01	7118.35	7.12
4	RBH-8-3	11*5.3(58.3)	0.00583	0.633	20.4	3499.14	3.5
5	AMFVE2	15*12(180)	0.018	0.818	117	6500	6.5
6	AMFMT1	10*6.3(63)	0.0063	0.493	196	31111.1	31.11
7	AMFWD1	10*4.5(45)	0.0045	0.96	169.52	59893.3	37.67111
8	AMRVE1	11*11(121)	0.012	0.833	48.54	4045	4.045
9	AMRMT1	10*7.5(75)	0.0075	0.847	39.23	5230.67	5.23067
10	AMRWD1	14*6(84)	0.0084	0.563	57.17	6805.05	6.80505

The average compressive strength values for the selected samples are given in table 2 and average values for pillars width, length and height are given in table 3.

Table 2. Average compressive strength value for three categories.

	Category I	Category II	Category III
Average Compressive strength	5.53	19.45	34.39

Table 3. Average pillar dimensions in the underground coal min

Pillar width	Pillar length	Pillar height	Average difference between pillars
12.13 m	12.9 m	1.99 m	1.8 m

Given conditions of pillar geometry to calculate pillar loading:

- L_P=12.9 m, W_P=12.13 m, W_O=1.8m, h=1.99 m where L_P is average pillar length, W_P is average pillar width, W_O is difference between pillars (width of opening), h is average pillar height (Table 3)
- Average thickness of coal seam is 1.67m.
- Depth of coal seam: 105 m

Compressive strength of the coal sample tested in laboratory: 5.53,19.45 and 34.39 MPa for category I, category II and category II respectively.

Based on the above geometries and according different workers [8][9][10] pillar loading is calculated and given below.

$\boldsymbol{\sigma_{avg}}$ = (L_P+ W_O) x (W_P+ W_O) x σv / (L_P x W_P) =stress (load) on the pillar due to over burden

σv is vertical stress = μ × H where

μ= average density of overburden rock strata (2247 Kg/m3, source from the company table of physical and mechanical Properties of rocks) and H is Depth of coal seam.

σv =105m x2247 Kg/m^3

σv=2.35 MPa

$\sigma_{avg} = (L_P + W_O) \text{ x } (W_P + W_O) \text{ x } \sigma v / (L_P \text{ x } W_P)$

σ_{avg} = (12.9m+1.8)x(13.13+1.8) x2.35 MPa/ (12.9 x12.13)

$\boldsymbol{\sigma_{avg}}$ **=3.29 MPa**= stress (load) on the pillar due to overburden.

5.1. Pillar Strength Formulas

The following researchers have calculated the pillar strength based on laboratory tests, in-situ tests, and case histories and said that the strength of the pillar is dependent on compressive strength of the pillar.

5.1.1. CMRI formula

Indian coal mine regulation [3] has developed the following formula:

$$\boldsymbol{\sigma_p = (0.27 \times \sigma c \times h^{-0.36}) + (H/160(wp/h-1))}$$

where, σ_p= Pillar strength (MPa), σc = compressive strength (MPa), h = working height or pillar height (in m), H = depth of cover (m), w = pillar width (m).

Since, there are three categories depending on compressive strength it is calculated separately as follows:

Category I

$\sigma_p = (0.27 \times 5.53\text{ MPa } x1.99m^{-0.36}m) + (105m/160(12.13m/1.99m-1))$

- $\sigma_p = 4.505$ MPa

Category II

$\sigma_p = (0.27 \times 19.45\text{ MPa } x1.99m^{-0.36}m) + (105m/160(12.13m/1.99m-1))$

- 7.43MPa

Category III

$\sigma_p = (0.27 \times 34.39\text{ MPa } x1.99m^{-0.36}m) + (105m/160(12.13m/1.99m-1))$

- 10.58MPa

5.1.2. Salamon-Munro Formula

Salamon and Munro [4] proposed the following pillar strength formula.

$$\boldsymbol{\sigma_p = 7.2 \times W_p^{0.46}/h^{0.66}}$$

Where, σp the strength of the pillar and the pillar dimensions W_P and h are width and height respectively. Since this formula does not depend on compressive strength the pillar strength value for the three categories are the same and calculated as follows.

$$\boldsymbol{\sigma_p = 7.2\ (12.13^{0.46}/1.99^{0.66})}$$

- $\sigma_p = 14.45$ MPa

5.1.3. Greenwald formula

Greenwald [5] formulated the following formula to calculate the coal pillar strength of underground coal mining.

$$\boldsymbol{\sigma_p = 0.67K\sqrt{wp}/h^{0.83}}$$

Where σ_p is the pillar strength, k is the strength of coal sample, w_p is the pillar width and h is the pillar height. The calculated pillar strength for the three categories is evaluated below:

Category I

$C_P = 0.67 \cdot 5.53\sqrt{12.13}/1.99^{0.83}$

- C_P=7.28 MPa

Category II

$C_P = 0.67 \cdot 19.45\sqrt{12.13}/1.99^{0.83}$

- 25.54MPa

Category III

$C_P = 0.67 \cdot 34.39\sqrt{12.13}/1.99^{0.83}$

- 45.33MPa

5.1.4. Obert and Duvall formula

Obert and Duvall [6] proposed the following formula:

σ_p **= C_1 [0.778 + 0.222 W / h]**

Where σ_p : pillar strength, c_1: the compressive strength of specimens, W: width of the pillar and h: height of the pillar. Because the coal mine has classified in three categories based on strength the pillar strength can be calculated as follows:

Category I

σ_p =5.53MPa [0.778+0.222 (12.13m/1.99m)]

- σ_p = 11.78 MPa

Category II

19.45 MPa [0.778+0.222 (12.13m/1.99m)]

- 41.2MPa

Category III

34.39 MPa [0.778+0.222 (12.13m/1.99m)]

- 73.25 MPa

5.1.5. Bieniawski formula

The general normalized form of the Bieniawski [7] equation is

σ_p = **σ1 (0.64+0.36(w/h))**

Where σ_p is pillar strength, w is pillar width, h is pillar height, and σ1 strength of the coal specimen.

Category I

σ_p =5.53 MPa (0.64+0.36(12.13/1.99)

- σ_p =15.67 MPa

Category II

σ_p =19.45MPa (0.64+0.36(12.13/1.99)

- 55.04 MPa

Category III

σ_p =34.39MPa (0.64+0.36(12.13/1.99)

- 97.32 MPa

6 Factor of safety

The safety factor for the existing mine design is calculated using the below mentioned formula and the values are given table 4A.

Factor of safety=**σp/ σ_{avg}**

Table 4. Factor of safety for the existing pillar width of 12.13m and b) Factor of safety for the new design (pillar width of 9m, 6m and 4m)

A)	**Method**	**Factor of safety**		
		Category		
		I	II	III
1	CMRI	1.37	2.25	3.21
2	Salamon-Munro Formula	4.39	4.39	4.39
3	Greenwald	2.21	7.76	13.72
4	Obert and Duvall	3.58	12.52	22.26
5	Bieniawski	4.76	16.72	29.58

B)	Method	I	II	III
1	CMRI	1.05	1.65	2.20
2	Salamon-Munro Formula	3.83	3.17	2.63
3	Greenwald	1.91	5.48	7.91
4	Obert and Duvall	2.99	8.55	8.26
5	Bieniawski	3.8	10.2	14.25

6.1. Rock mass classification of coal mine

This work used RQD, RMR and Q system for characterization of the underground coal mine.

6.1.1. Rock Quality Designation index (RQD)

From core log study the RQD of the rocks were calculated [8]. The results indicate that the quality of rocks in the borehole varies with lithology and depth. Accordingly, the coal-bearing sedimentary rocks together with the upper basalt rock units have very poor rock quality (<25%). The lower basalt and basement rock units show relatively fair (25-50%) to good (50-75%) quality of rocks.

6.1.2. Geotechnics Classification (RMR)

This method considers six parameters, UCS, RQD, spacing, joint condition, groundwater condition and structures orientation. While assessing the rock masses the rating value is given for each parameter. The summation of the value of all parameters of RMR is 38.

This indicates the underground coal mine is at the boundary between poor and fair rock quality.

6.1.3. Rock Tunneling Quality Index (Q)

The numerical value of the index Q is defined by= RQD/Jn*Jr/Ja*Jw/SRF where RQD is the Rock Quality Designation, Jn is the joint set number, Jr is the joint roughness number, Ja is the joint alteration number, Jw is the joint water reduction factor, SRF is the stress reduction factor.

Using these values gives Q = 0.057 which is classified under poor to extremely poor rock quality.

7 Extraction Ratio (R)

Extraction ratio is the ratio of mined volumes to total volumes of the deposit and calculated using the following formula.

$$\mathbf{R\% = (1-(W_p^2/ (W_p+W_o)^2) \times 100)}$$

Where Wp is average pillar width, W_o is the difference between pillars (width of opening). The R% calculated for the existing mine is 24.2% considering the Wp = 12.13m and Wo = 1.8m.

8 Pillar design analysis

This study proposes new mine design for the three categories, namely category I (9m), category II (6m) and category III (4m) (Figures 4, 5 and 6). The factor of safety was calculated for the new design using different approaches like CMRI [3], Salamon-Munro [4], Greenwald [5], Obert and Duvall [6] and Bieniawsk [7] and the results are shown in table 4B.

As per the new design, the extraction percentage increases from the average value of 24.2% for the existing mine to 31% for category I, 41% category II and 52% category III.

The safety factor analysis thus clearly indicates that the underground mine is safe even if the pillar width is reduced as mentioned in figures 4, 5 and 6.

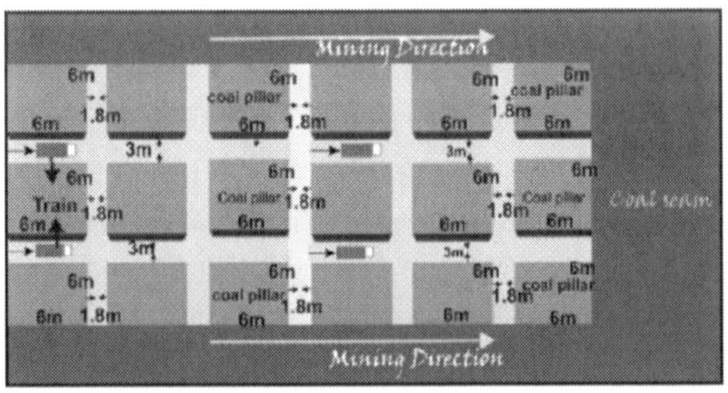

Fig. 4. New mine design with pillar width of 9m, category I.

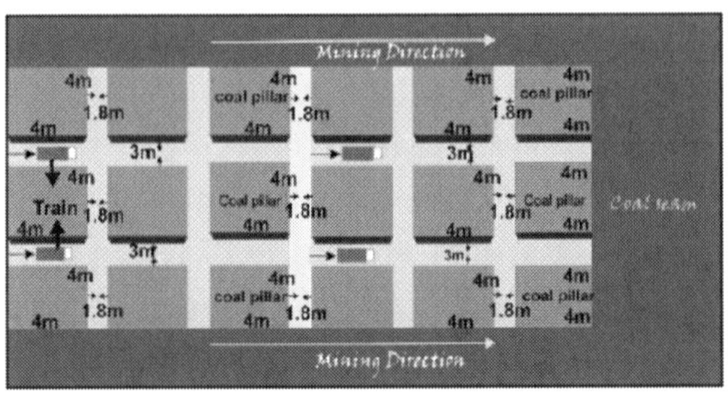

Fig. 5. New mine design with width of 6 m, category II.

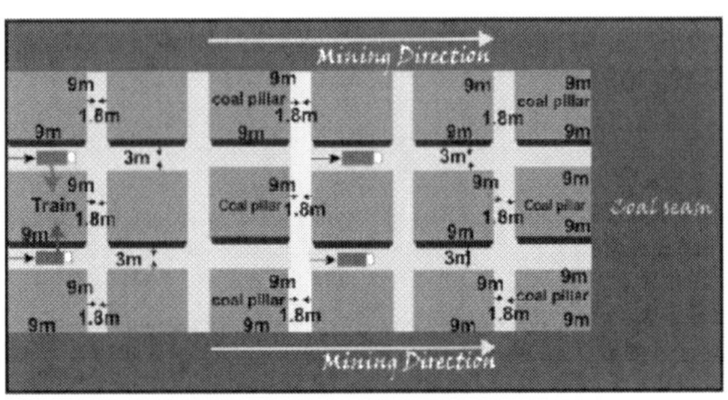

Fig. 6. New mine design with width of 4 m, category III

9 Conclusions

The geological structures in the mine site are classified into two, (1) those tangential with the mining direction; and (2) perpendicular with the underground coal mine. Weak zones are causing many problems in Achibo-Sombo coal mine. These include: (1) physical displacement of coal seams, (2) reducing the stability of roof and floor in underground workings; (3) opening of pathways for the influx of water into underground workings.

Based on rock mass RQD, RMR and Q classification, the underground coal mine rocks are characterized and classified under poor rock quality.

The factor of safety calculated for the existing pillar is 1.37, 4.39, 2.21,3.58 and 4.76 for category I, 2.25, 4.39, 7.76, 12.52 and 16.72 for category II and 3.21, 4.39, 13.72, 22.26 and 29.58 for category III by CMRI, Salamon-Munro, Greenwald, Obert and Duvall and Bieniawski respectively.

The factor of safety of the new proposed pillar design is 1.05, 3.83, 1.91, 2.99 and 3.8 for category I, 1.65, 3.17, 5.48, 8.55 and 10.02 for category II and 2.20, 2.63, 7.91, 8.26 and 14.25 for category III by the above same method.

From this investigation the extraction percentage could increase from 24.2% to 31%, 41% and 52 for the three categories respectively. This means 6.8 to 27.8% valuable coal material which could have been left as pillar could be mined. Therefore, it is recommended to use the new mine design.

References

1. Ethiopian Mine Development Enterprise (EMDE), Unpublished report (Result of Sub - Surface Exploration for Coal Deposit at Achibo-Sombo, Yayu Illubabour Area) (1998).
2. H. Gebreyohannes, Unpublished report (Geological report on coal deposit extension investigation around Wattete and Yayu blocks) 1 & 2, EIGS, Addis Ababa (2001).
3. Indian coal mine regulation (CMRI), Mine working and development sub section (1) of Act. (91) S.R.O.3419, (1957).
4. M.D.G Salamon, A.H. Munro, J. Southern African Institute of Mining and Metallurgy, 68 (1967).
5. H.P. Greenwald, H.C. Howarth, I. Hartman, Technical Paper 605 (1939), Report Investigation 3575, U.S. Bureau of Mines, (1941).
6. L. Obert, W.I. Duvall, Rock Mechanics and the Design of Structures in Rock (1967).
7. Z.T. Bieniawski. Int. J. Rock Mech. Min. Sci., 4, 325-335 (1968).
8. E. Grimstad, N. Barton, Proc. Int. Symposium Fagernes, 46 (1993).
9. D.G.F.Hedley, F. Grant, Can. Inst. Min. Metall. Bull., 65 (1972).
10. N. Krauland, P. Soder, Engineering and Mining J, 8 (1987).
11. D.U. Deere, Rock Mechanics Engineering Geology, 1, (1964)

Geomechanics of rock array for chamber system of coal deposits development on the example of finalizing by KGRP complex

Alexey Bykadorov[1,*], *Dmitry* Degtyarev[1], *Sergey* Smirnov[2], and *Oleg* Pechenegov[2]

[1]LLC SIGI (Siberian Institute of Geotechnical Research), 653033, Prokopievsk, 11 Gagarina st., Russian Federation
[2]Resourse Ltd., 654004, Novokuznetsk, 13 Schorsa st., Russian Federation

Abstract. The results of systematic instrumental observations of Earth's surface's subsidence which was undermined by SHM (Superior Highwall Miners) are given. The regularities of subsidence are revealed. The empirical relationships are established. The methodical provisions for calculating the expected displacements and deformations of the undermined earth's surface are given.

1 Introduction

This article presents the results of the first systematic instrumental observations in the section of LTD "Resource" in the coal mining on Taldinskoye coal deposit concentrated among the marginal technical and licensing boundaries. The first systematic observations were made in finalizing seam "68". Observation station consisted of five ground frames. More extensive instrumental observations were made in the finalization of seam 70.

2 Materials and Methods

The thickness of seam "70" varies from 4 m to 5.7 m, with the average 4.7 m, false roof is absent, in the immediate roof lies siltstone or sandstone with a compressive strength 44.8 MPa and 67.8 MPa respectively and unstable rock layer 1 m. In the soil of coal seam there is siltstone formation with the compressive strength 44.8 MPa, the average dip angle is 200, the depth of mining works is up to 108 m.

The coal deposit development of seam "70" by KGRP complex on the site of "Dumping Deep South 2" is carried out in 2 layers. The width of the chambers, inter-chamber and block pillars are respectively 3.5 m, 2.0 m and 5.5 m; the distance between the block pillars is 48 m (9 cameras).

Measurements on observation stations were made by coordinate measurements with the direct determination of coordinates in the field on the instrument's built-in electronic

* Correspondent author: sibigi@mail.ru

program with post-processing GPS surveying equipment in software systems "Magnet Field" [1-3] and "Credo-DAT" [4-6].

Visual and instrumental observations have revealed that the collapse of undermined rock mass occurred with some landing step during one day at almost maximum amount of subsidence. The value of the shrinkage between steps of collapse is from 2 to 6% of maximum subsidence of the array for different extent of undermining. On the Earth's surface the cauldron appears trough displacement (subsidence), a fragment of which is presented in Fig. 1. Under the subsequent subsidence the cauldron moves after the collapse, almost without changing its profile. Thus, as seen in Fig. 1, there is a part of undermining of the power lines and the road.

Fig. 1. The cauldron at the site of the first bench after the first subsidence of undermined array: 1 - the beginning of displacement between 6 and 5 benchmarks; 2 – the cauldron near the fifth frame; 3 - the cauldron's profile.

On the surface of the first bench there are concentrated deformations like cracks (Fig. 2) up to 20 cm. At the site of the second bench there is none of these deformations.

Fig. 2. Concentrated deformations of the array (cracks) at the site of the first bench in the area of the first frame.

Fig. 3 shows intact array on the site of the first bench, and Fig. 4 shows the fragment of collapsed array. Directly above the workings there are visible cracks and fractures, from which the blocks of undermined array subside. The overlying array (the side of the second bench) subsides almost without vertical and inclined cracks.

Fig 3. The intact site of the first bench.

Fig 4. The site of the first bench in the collapse area.

The example of cauldron on the Earth's surface at the site of the first bench on the profile line L1 in finalizing seam "70" is shown in Fig. 5.

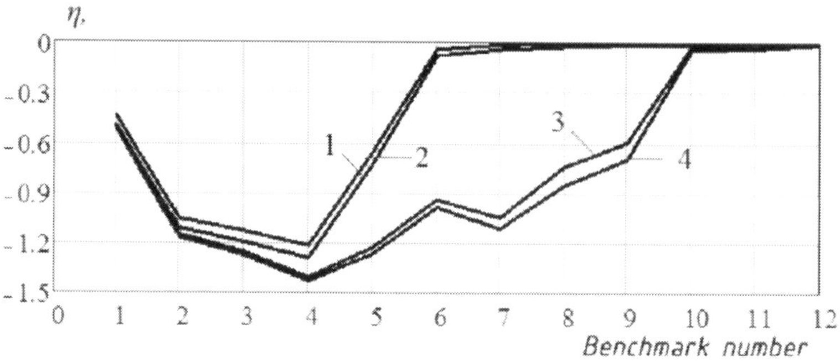

Fig. 5. Genesis of cauldron in finalizing seam "70": 1 – the cauldron in the first collapse of the array; 2 – array post-subsidence between the first and second subsidences; 3 – the cauldron of array secondary displacement; 4 - post-subsidence of the array to the next subsidence.

3 Results and Discussion

Instrumental studies showed that at different extents of array undermining, depending on the ratio of collapse along the strike of the seam to an average depth of mining, the maximum subsidence of the surface can be determined by formula (1), considering the ratio of equivalent extracting capacity K_e:

$$\eta_m = q_0 K_e m N_1 N_2 \cos\alpha; \tag{1}$$

where: m – extracting capacity of the seam, m; α – dip angle of the seam, deg.; N_1 and N_2 – the rates of array's undermining; q_0 – respective value of maximum subsidence.

The ratio of equivalent extracting capacity K_e can be defined on the results of the instrumental observations and using empirical formula (2):

$$K_e = \frac{B_k}{B_k + B_{ch}}; \tag{2}$$

where: B_k – the width of extracting chamber, m; B_{ch} – the width of inter-chamber pillar, m

The rates of array undermining N_1 and N_2 are obtained for Kuzbass conditions through approximation.

According to the results of instrumental observations the distribution function of the Earth's surface subsidence was determined (a typical curve of subsidence, Fig. 6), in the cauldron along the strike of the seam in the relative values given to the interval [0, 1].

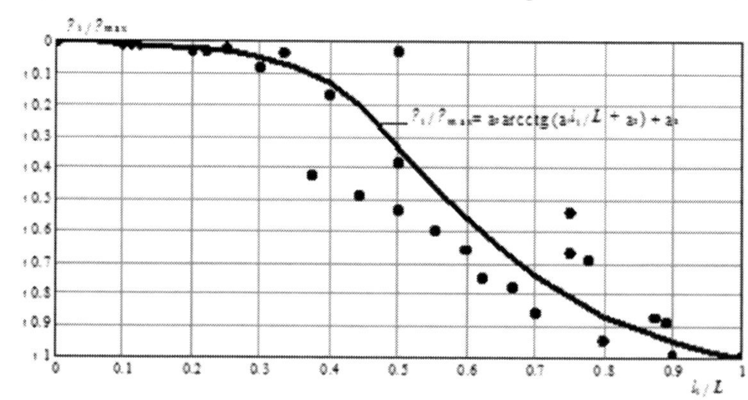

Fig. 6. The distribution function of Earth's surface subsidence in the subsidence of the main section along the strike of the formation.

Defined function of subsidence distribution was approximated by formula (3):

$$\frac{\eta_i}{\eta_m} = a_0 arcctg\left(a_1 \cdot \frac{l_i}{L_s} + a_2\right) + a_3; \qquad (3)$$

where: L_s – the cauldron of subsidence in the main section along the strike of the seam, m; η_i , l_i – current values of subsidence and cauldron's length respectively, m.

Empirical rates a_0, a_1, a_2, a_3 were defined using aggregate statistical method of the least squares.

The cauldron across the strike was calculated using formula (4):

- for incomplete undermining: (N_2< 1):

$$L_s = h_c ctg\varphi + (H_{av} - h_c)ctg\delta + \frac{D_2}{2}, \qquad (4)$$

- for complete undermining (N_2=1):

$$L_{s3} = h_c ctg\varphi + (H_{av} - h_c)ctg\delta + H_{av} ctg\psi_3; \qquad (4)$$

where: H_{av} – the thickness of cap rock, m; $\varphi = 55^0$ – the angle of displacement for dry and normal humidity cap rock; φ = 45° - for quicksand and water-bearing gravel; for Kuzbass: δ= 80°, Ψ_3= 50°.

4 Conclusion

According to methodical provisions proposed in this paper, SIGI Institute developed the algorithm of calculating of the Earth's surface expected displacements and deformations by the complex of coal seams deep processing. With the accumulation of instrumental observations the calculation algorithm will be updated.

References

1. B.G. Saksin, I.Y. Rasskazov, J. Min. Sci., **51**(2), 134 (2015)
2. G.G. Kocharyan, A.A. Ostapchuk, J. Min. Sci.,, **51**(1), 87 (2015)
3. A.A. Mylnikova, Yu.V. Yasyukevich, E.V. Kunitsyu, Results in Physics. **5**, 65 (2015)
4. C. DeMets, R.G. Gordon, D.F. Argus, Geophys. J. Int., **181**, 89 (2010)
5. R. McCaffrey, J, Geophys. Res., **110**, 07401(2005)
6. P. Bird. Geochem. Geophys. Geosys. 4(3), 33 (2003)

Justification of the Optimal Granulometric Composition of Crushed Rocks for Open-Pit Mine Road Surfacing

Viktor Shalamanov[1,*], *Vladimir* Pershin[1], *Serguey* Shabaev[1], *Dmitry* Boiko[1]

[1]T.F. Gorbachev Kuzbass State Technical University, Mining Institute, 650000, Kemerovo, Russia

Abstract. The specific weight of the open coal mining method in Kuzbass is constantly growing due to higher hazard and low profitability of the underground mining method. It is possible to reduce the cost of open-pit coal mining using overburden rock in constructing temporary pit roads. The purpose of the work is to justify the possibility of using overburden rock for the construction of temporary technological roads. To this end, we set the tasks to determine the rational granulometric composition of crushed rock, dependences of its unit weight, static elastic modulus, and ultimate shear resistance. This study justifies the most efficient granulometric composition of crushed rock based on theoretical studies of the optimal structure of granular materials, taking into account the probability theory of particle-particle packing, laboratory studies of the dependence of the unit weight and the static elastic modulus on the content of particles of different sizes, and also the calculation of ultimate shear resistance of crushed rocks. As a result of the conducted studies, it has been established that the optimal granulometric composition of crushed rock, which at the same time ensures high values of unit weight, static elastic modulus and ultimate shear resistance, is achieved when the content of particles larger than half of the maximum is 40 to 55% and increases integrally with decreasing of the average particle diameter.

1 Introduction

Road surface is one of the most important elements of the pit road dressing, the average speed of pit transport and, consequently, its efficiency largely depend on the condition of roads. Taking into account that transportation costs make up 40 to 70% of total costs, improving the quality of road topping is one of the main ways to reduce the cost of mineral mining by the open-cast method.

The experience of coal mines operation in Kuzbass has shown that construction of asphalt-concrete and cement-concrete pavements is not justified economically when it comes to roads in a dynamically developing quarry space, where temporary roads make up to 80% of their total length. The most suitable for construction of road dresses of both permanent and temporary technological roads is crushed rock.

[*] Corresponding author: shalamanov49@mail.ru

The analysis of previous studies [1-11] shows that the crushed rock of the optimal granulometric composition should consist of various size particles taken in a certain quantitative ratio, which provides increased internal friction and adhesion between the particles, high unit weight and an insignificant intensity of the grain grinding process [12 - 15]. These conditions can be achieved with the largest possible number of coarse particles and minimal content of fine and silty particles necessary to create high unit weight and to ensure the cohesion of large particles.

2 Materials and methods

These conclusions which take into account the main theoretical provisions of A.V. Biryukov, B.S. Radovsky, and I.I. Kandaurov, were laid in the basis of the theoretical model of optimal packing of different-sized particles, the following assumptions and conjectures were introduced in it:

The entire crushed rock consists of N-components, and a certain fraction characterized by an average diameter is taken as a component. It is inconvenient to use the absolute values of the average diameters of the particles, so in the theoretical model, the average particle diameters of a certain fraction were replaced by relative dimensions, such as Dmax, Dmax / 2, Dmax / 4, and so on. The particles of the largest component are evenly distributed throughout the volume, which allows us to consider not the entire set of particles (a macro object), but only one of them (a discrete object). All particles have the shape of a sphere. If we take the shape of the particles in the form of a cube, then theoretically the minimum voidness of a layer arranged from such a fragmented rock will tend to porosity of the rock itself, since individual particles can closely touch each other on all sides. It is difficult to consider a form in the form of a tetrahedron using random packing of particles, since grains can occupy one position, but they always have a different orientation in space. But the shape of particles in the form of a sphere satisfies both conditions: between particles there is always space (voids) that cannot be completely filled; regardless of the location of an individual particle, its orientation in space doesn't affect the package. To test the adequacy of the theoretical model, a laboratory study was made of the dependences of the unit weight and the static modulus of elasticity of crushed rocks of different granulometric composition. In addition, the ultimate shear resistance of that crushed rock was calculated.

All laboratory studies were performed on samples of crushed rock of different granulometric composition presented in Table 1.

Table 1. Selected grain size distributions of crushed rock

Blend #	Full residue, %, on sieves, mm:					
	10	5	2.5	0.63	0.16	0.05
# 1	9.70	24.09	33.08	48.91	56.64	71.83
# 2	12.93	28.91	46.90	55.70	66.86	77.47
# 3	33.95	48.33	61.82	61.82	81.14	85.48
# 4	48.50	58.09	62.58	66.54	69.63	80.92
# 5	29.10	43.48	61.47	69.39	77.11	83.63
# 6	19.40	38.58	52.07	67.90	79.49	83.83
# 7	38.80	57.98	66.97	66.97	86.29	88.46
# 8	58.19	72.58	81.57	89.49	93.35	93.35
# 9	48.50	58.09	67.08	75.00	78.86	85.37
# 10	58.19	67.79	76.78	80.74	83.06	88.26

3 Results

The proposed theoretical model was implemented by means of computer simulation. The simulation results are shown in Table 2.

Table 2. Optimal grain size distribution of crushed rock based on theoretical studies

Component dia., mm	Dmax	Dmax/2	Dmax/4	Dmax/8	Dmax/16	Dmax/32	Dmax/64	Dmax/128	Dmax/256	Dmax/512	Dmax/1024	Dmax/2048	Dmax/4096	Dmax/8192
The total content of the components (total residue),% by volume	52.4	58.95	64.68	71.12	74.97	77.71	80.66	83.43	85.76	87.87	89.61	91.03	92.31	93.07

The obtained results made it possible to determine the theoretical dependence characterizing the optimal granulometric composition of the crushed rocks:

$$V_D = 52.4 + \int_D^{0.5D\max} \left(\frac{\frac{30}{(D\max + 25)^{0.5}} + 1.84}{D^{0.8}} \right) dD, \quad (1)$$

where V_D – total content by volume of the i–th component and all the previous larger components, %; D – average diameter of particles of the i–th component, mm; D_{max} – average diameter of particles of the largest component, mm.

The results of the laboratory tests are shown in Figures 1 and 2 with reference to crushed rock rocks of coarse-grained sandstone with the maximum grainsize D_{max} = 20 mm.

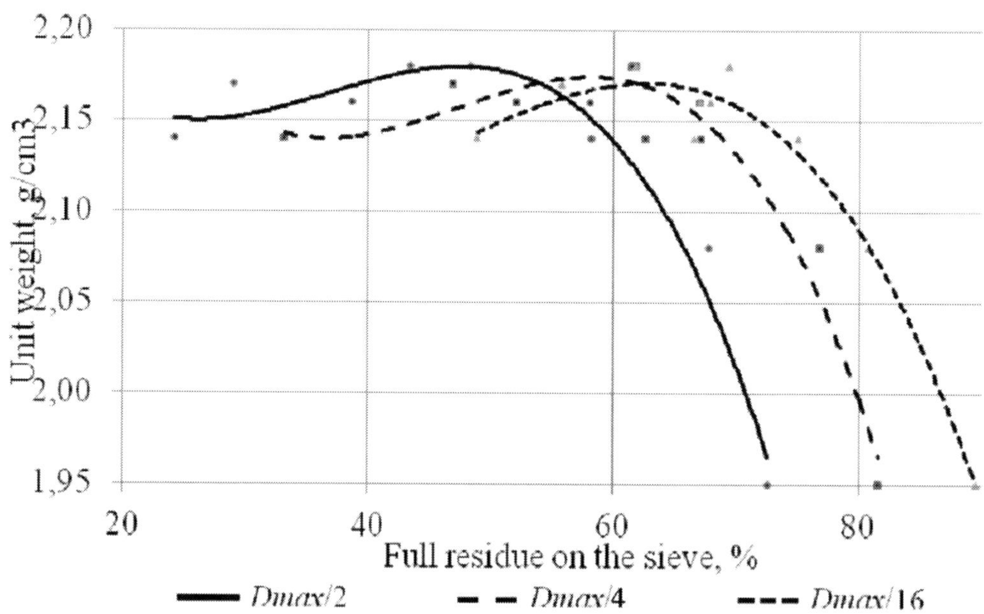

Fig.1. Dependences of unit weight on full residue of the sieves of a given size

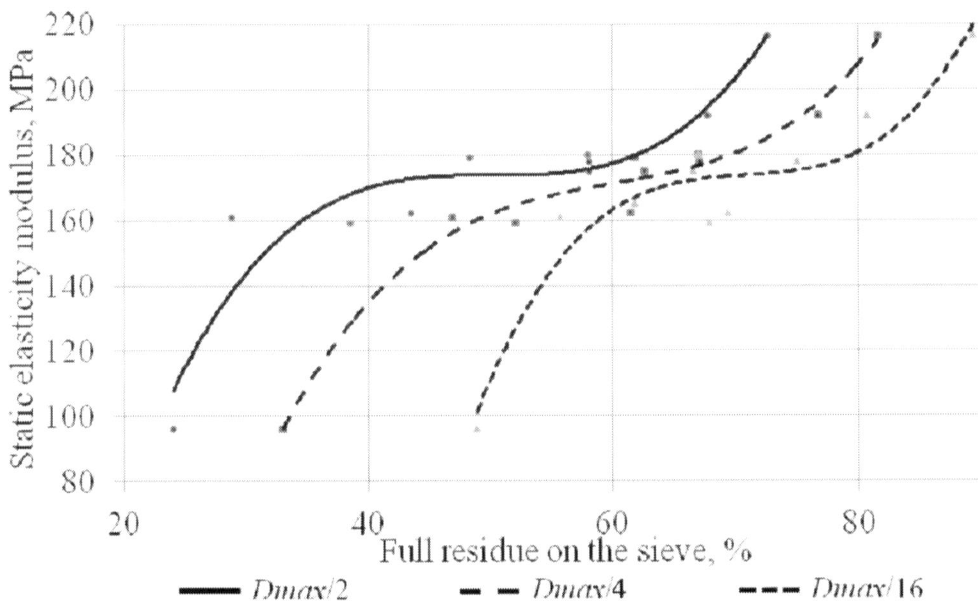

Fig. 2. Dependences of elasticity modulus on full residue on the sieves of a given size

The response functions of the unit weight (ρ) and the static modulus of elasticity (E) of the crushed rocks from the total residue on the sieve of a given size have the form (the values of the coefficients of the regression equations are given in Table 3):

$$\rho(V_D) = 0{,}79 \cdot \rho_0 \cdot \left[\cos\left(\frac{V_D^2 - k_1}{k_2}\right)\right]^2, \quad (2)$$

$$E(V_D) = \int_0^{V_D} k_3 \cdot V_D^{k_4} \cdot \left[1 - \cos(k_5 \cdot V_D)\right]^{k_6} dV_D + 40, \quad (3)$$

where V_D – content of particles over D mm in size in the crushed rock, %; ρ_0 – density of the initial rock mass, g/cm³; $k_1, k_2, k_3, k_4, k_5, k_6$ – coefficients of regression equations.

Table 3. Coefficients of regression equations

Average diameter	Regression equation coefficient value					
	k_1	k_2	k_3	k_4	k_5	k_6
Dmax/2	800	9000	13.5	–0.537	0.091	0.721
Dmax/4	1800	10000	0.838	0.223	0.075	0.822
Dmax/8	2800	13000	0.032	1.06	0.070	0.700
Dmax/16	3200	14000	0.0028	1.60	0.060	0.600

The DalNII has developed a technique for estimating the strength characteristics (ultimate shear resistance) for coarse clastic soils, which in fact differ little from fractured rocks. It is based on the dependence of the angle of internal friction and the specific cohesion of crushed rocks on the content of particles (debris) having a size of more than 2 mm. Further, according to the Coulomb law, the ultimate shear resistance is calculated. Graphical representation of the results of the performed calculation is shown in Figure 3.

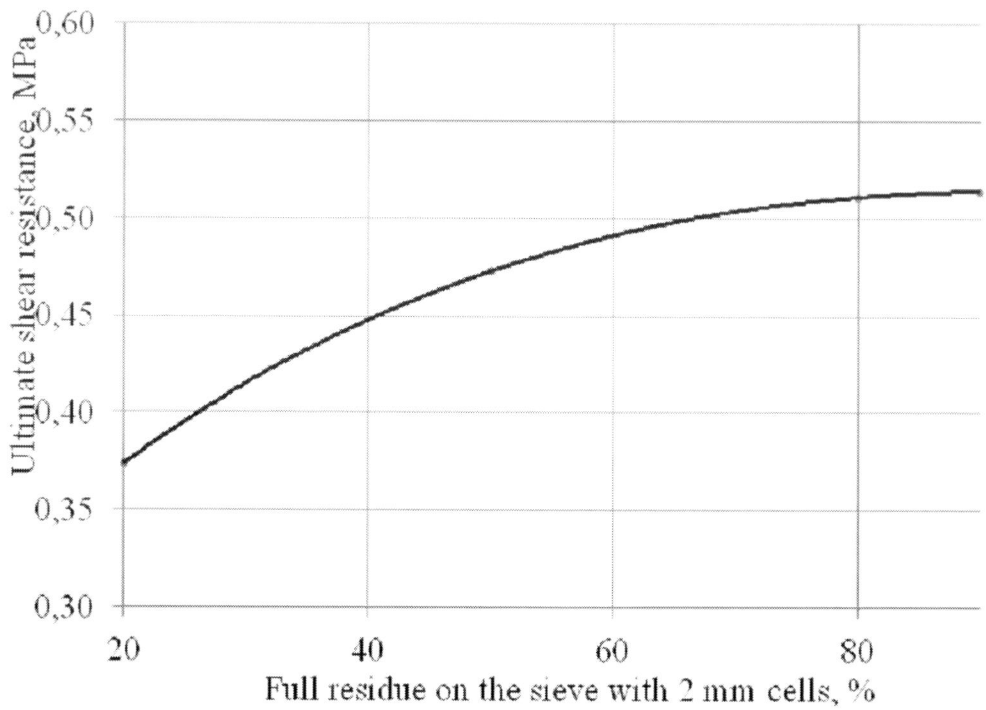

Fig. 3. Dependence of the ultimate shear resistance of crushed rocks on the content of particles over 2 mm in their composition

The obtained dependences of the volumetric mass, the static elastic modulus and the ultimate shear limiting resistance of the crushed rocks on their granulometric composition made it possible to optimize it.

Comparison of empirical data with the results of theoretical studies shows that they almost completely coincide. The difference is that according to the results of the experimental data, the optimum particle size distribution represents a region, and in theoretical studies, it is the only value in this optimal region. Consequently, the dependence characterizing the optimal granulometric composition of crushed rocks can be written in the form:

$$V_D = (40...55) + \int_{D}^{0.5D\max} \left(\frac{\frac{30}{(D\max + 25)^{0.5}} + 1.84}{D^{0.8}} \right) dD . \quad (4)$$

4 Conclusions

The optimal granulometric composition of the crushed rock, which at the same time ensures high values of unit weight, static elastic modulus and ultimate shear resistance, is achieved when the content of particles larger than half of the maximum is 40 to 55% and integrally increases with a decrease in the average particle diameter.

References

1. Czermak V, Huckenholz HG, Rybach L, et al., *Physical properties of rocks* (1982)
2. Monicard RP., *Properties of reservoir rocks: core analysis* (1980)
3. Bucher, Kurt, Grapes, Rodney, *Petrogenesis of metamorphic rocks* (2011)

4. Farmer IW (Ed.), *Engineering behaviour of rocks* (1983)
5. LianYing Zhang, XianBiao Mao, AiHong Lu., Science in China Series E: Technological Sciences, **52**, 641 (2009)
6. A. Kılıc, A. Teymen., Bulletin of Engineering Geology and the Environment, **67**, 237 (2008)
7. P. Hartlieb , M. Toifl, F. Kuchar et al., Minerals Engineering. 91, 34 (2016)
8. Chandong Chang , Mark D. Zoback , Abbas Khaksar, Journal of Petroleum Science and Engineering, **51:3–4**, 223 (2006)
9. Gioacchino Francesco Andriani , Mario Parise, Journal of Rock Mechanics and Geotechnical Engineering, **19** (2017)
10. Mohammadreza Hemmati Nourani, Mohsen Taheri Moghadder, Mohsen Safari, Journal of Rock Mechanics and Geotechnical Engineering, **8** (2016)
11. Jonny Rutqvist , Antonio P. Rinaldi, Frederic Cappa et al., Journal of Rock Mechanics and Geotechnical Engineering, **8:6**, 789 (2016)
12. Zong-Xian Zhang, Journal of Rock Mechanics and Geotechnical Engineering, **8:6**, 886 (2016)
13. Yaodong Jiang , Yixin Zhao, Hongwei Wang et al, Journal of Rock Mechanics and Geotechnical Engineering, **24** (2016)
14. L. Dormieux, E. Lemarchand , D. Kondo et al., Journal of Rock Mechanics and Geotechnical Engineering, **20** (2017)
15. Askeri Karakus, Mahmut Akatay, Nondestructive Testing and Evaluation, **28:4**, 342 (2013)

Optimization of transportless technological schemes for coal seams quarrying

Tatiana Gvozdkova[1,*], *Eugene* Plotnikov[1], and *Elena* Usova[1]

[1]Kuzbass State Technical University, Mezhdurechensk branch, Mezhdurechensk, 652881, Stroitelei Avenue, 36, Russia

Abstract. The main way to increase the competitiveness of Kuznetsk coals produced by the open method is to reduce overburden works. On the open pits of the southern Kuzbass the flat coal seam strata are developed using transportless technology during overburden works. Open-pit fields are developed in combination: the lower part to the height of 32-44 m is worked out by low-cost transportless technology (producing cost of 1 m^3 of overburden is 22-24 rubles, the upper part - by high-cost transport technology (producing cost of 1 m^3 of overburden is 35-44 rubles) using shovels EKG-10, EKG-12.5, EKG-20 excavators and quarry dump trucks with the carrying capacity of 80-180 tons. The volume increase of the rock that is transferred to the internal dump will improve the economic indicators of coal mining. In addition, due to the flat bedding, the height of the rock mass, which is processed by transport technology, is constantly increasing. At the present time, this has led to the speed decrease in the movement of the working front of the transport zone and, as a consequence, to slowing down the development of transportless. The redistribution of overburden volumes between zones in the direction of increasing the volume by transportless technology will increase the speed of movement of the transport zone (while maintaining the same complex of mining and transport equipment).

1 Introduction

In Tomusinsk and Mrassk geological and economic areas of Southern Kuzbass coal deposits are characterized by strata seams bedding in the rocks. The deposits are developed by modern large pits: "Krasnogorsky" (geological sites "Sibirginsky 8", "Sibirginsky 7", "Kiyzaksky 8", "Kiyzaksky 9"); "Mezhdurechensk" (sections "Kiyzaksky 5-7", "Sibirginsky 4-6"); "Sibirginsky" (section "Sibirginsky 1-3"), and others [1].

2 Material and Method

The overburden rocks are mudstone and siltstone with Protodyaconov hardness coefficient 5-7, as well as large-sandstones with hardness coefficient f = 8-10 [2].

* Corresponding author: kuzstu@kuzstu.su

The mining of open pit field in Tomusinsk sections is made by using two technologies: transportless - to extract the lower part combining two, less often one seam; transport - to extract the upper part consisting of two or three seams (Fig. 1).

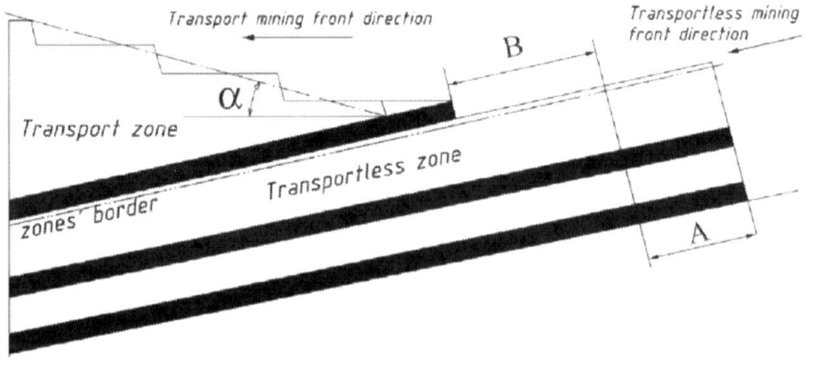

Fig. 1. The borders of the transport and the transportless zones of coal quarrying in Tomusinsk sections.

The height of cut bench on the third (top) strata seam in piling four-tier dumps depends on the piling scheme, the total height of the lower partings, bedding angle and does not depend on the width of the stope, the relation of which to the discharge radius is constant; at that cut bench height for draglines with beam length of 90-120 m is 24,7-26.9 m for dumping scheme without preparing capacity and 20.7-22.8 m for dumping with preparing capacity [3].

3 Results and Discussion

Structures of excavation schemes for the development of three flat seams strata are divided into two types, which differ in the number of stages (elements) and the work of mining equipment. The attribution criterion to the type is the ratio of the rock volume of the upper and middle ledges (taking into account rock loosing coefficient) and the capacity of the first and the second tiers [4].

Using dumping schemes with preparing capacity during the work of two excavators allows combining work of overburden and dump draglines in time [5-6]. It reduces the time of stope mining and provides the increase in the annual rate of work front movement by 12-15% and reduces the unit cost of overburden on 7-10% for all types of bedding seams in strata.

At the section "Main Field" flat seams III, IV-V and VI by mixed down-hole-continuous excavation system are developed. The power of partings of seams IV-V and VI varies from 0.2 to 14 m, with an average of 7-5 m; the power of partings of seams IV-V and III varies from 28.3 m to 49.5 m, with an average of 42-45 m. The bedding angle of the seams is from 6 to 16°, on average 10°.

The rock above seam III is worked by the transport technology. The bench on seams parting IV-V and III is worked by transportless technology by two draglines ESH 15.90 B. The parting of seams VI and IV-V, due to small average power is worked by transport technology in the process of mining operations on the seams. Transportation of rock and coal is made by trucks.

The task of choosing the technological scheme is solved on the basis of graph-analytical modeling of the development of transportless overburden bench with height H = 45 m along partings of seams III and IV-V. The model includes: calculation of parameters and outlining excavation schemes for values of the stope width A = 40 m; 45 m; 50 m; 55 m and 60 m; outlining planograms of organization of the drilling, stripping and mining equipment; the formula for the calculation of the basic technical indicators. Excavation schemes for A = 40 m; 50 m; 60 m, as an example, are shown in Fig. 2 a, b, c.

The upper ground of the bench "levels" as seams bedding reaches 10°, and for the installation of heavy drilling machines SBSH-250 MN for drilling deep holes (35-45 m) there is a need for horizontal space. Blasted rock of the "cleet" is cleared from the bench by the shovel, bulldozer and placed in the internal dump. With enlarging the width of stope the volume of rock "cleet" increases and, consequently, the volume of bench rock excavated by the dragline decreases. This fact has an impact on the development of technical indicators of the transportless bench development.

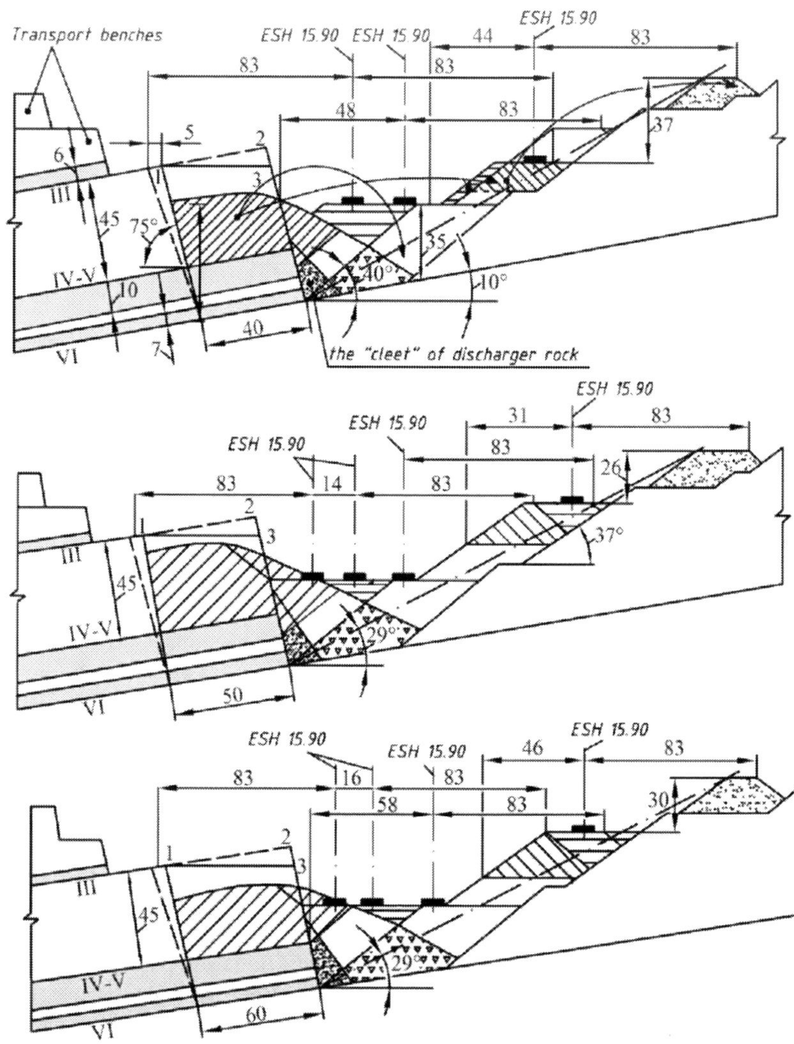

Fig. 2. The excavation schemes for the development of overburden bench with height 45 m along parting of seams IV-V and III: a, b, c – stope width 40, 50 and 60 meters, respectively.

Internal dump parameters are the following: the height of the first tier is equal to 30 m (according to the recommendations of the Siberian branch of All-Russian Research Institute of Mining Geomechanics and Survey in Prokopyevsk) with the general angle of its cut – 40° (Fig 2-b), the general slope angle of the dump – 29 ° (Fig 2-b), loosening rock coefficient in the dump – 1,4. The natural slope angle of the rock of the second layer and overlying is – 37 ° (Fig. 2-b).

When building excavation schemes the presence of the road at the base of the dump which is poured in the process of piling is taken into account [5, 6].

When dumping four-tier dumps to the listed volumes of re-excavation add the following: two rock volumes transferred to the fourth tier; the track of dragline travel when transferring rock from the third to the fourth tier.

Fig. 3 shows a graph of the dependence of explosive discharge (K_{ed}) from stope width (A).

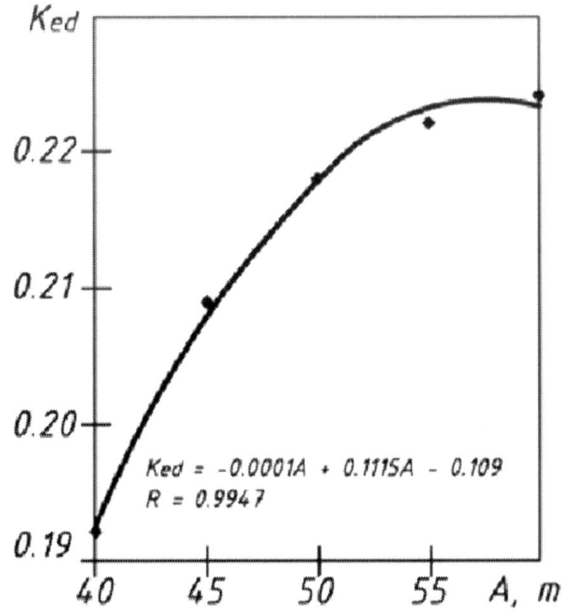

Fig. 3. The excavation schemes for the development of overburden bench with height 45 m along parting of seams IV-V and III: a - stope width 40 m; b - the same 50 m; c - the same.

Coefficient of explosive discharge increases in a parabolic dependence. When stope width is 60 m the discharge coefficient is greater than 16% relative to its value when the stope width is 40 m that improves the development of the overburden ledge. K_{ed} increasing with some slowdown is going to a value approximately equal to the width of the stope 55 m, and then stabilizes. This is due, as shown by analysis, to a decrease in the spall zone on the slope of the blasted bench because of leveling its upper platform.

To calculate the general re-excavation coefficient (K_{reg}) it was considered the work of equipment, allowing defining additional volumes of re-excavation connected with creating routes for up and down dragline movement to the working site of the second tier for dumping the third tier. K_{reg} ratio is calculated by the formula (1)

$$K_{reg} = K_{re} + \frac{n_{n(c)} \cdot V_{n.c}}{A \cdot H \cdot K_l \cdot (1 - K_{ed})}, \tag{1}$$

where: K_{re} - re-excavation coefficient; $n_{n(c)}$ – the number of up (down) tracks of dragline to dumping of the third tier: two tracks (up, down) on the left and right flanks of the stope; $V_{n.c}$ - the volume of re-excavated rock when creating or eliminating one track (220 000 m^3); A - the stope width, m; H – the height of the tier; K_l - rock loosing coefficient.

The general coefficient of re-excavation (K_{reg}) also varies according to the parabolic dependence and has a minimum in the range of 0.7 with the stope width 50-52 m. Thus, due to the amount of the total re-excavation coefficient, the stope width at the section "Sibirgin field" should be increased to 50- 52 m.

4 Conclusion

The structures of excavation schemes for the development of three flat seams strata are divided into two types, differing by the number of stages (elements) and organization of the mining equipment operation. The attribute criterion is the ratio of rock volume of the upper and middle ledges (taking into account the rock loosing coefficient) and the capacity of the first and the second tiers.

Using of dumping schemes with the preparation of capacity during the work of two excavators makes it possible to combine the operation of overburden and dump draglines in time. It reduces the time of stope mining and provides the increase in the annual rate of work front movement by 12-15% and reduces the unit cost of overburden on 7-10% for all types of bedding seams in strata.

References

1. B.N. Lokhanov, Y.A. Zakharov, M.M. Bereznyak, A.V. Kalinin, Sov. Min., **3(5)**, 523 (1967)
2. A.S. Tanaino, V.I. Cheskidov, J. Min. Sci., **35(3)**, 304 (1999)
3. M.M. Bereznyak, A.V. Kalinin and V.G. Pronoza, Soviet Mining **6(6)**, 638 (1970)
4. I.V. Nazarov, J. Min. Sci., **48(1)**, 55 (2011)
5. N. Demirel, J. Min. Sci., **47(4)**, 441 (2011)
6. I. Vukotic, V. Kecojevic, W. Zhang, Q. Cai, Int. J. Min. Sci. Tech., **23(6)**, 901 (2013)

The features of three- and four-tier internal dumps capacity calculation with the additional capacity preparation in the dump tiers

Tatiana Gvozdkova[1,*], *Eugene* Kuznetsov[1], *Anna* Rudakova[1], *Sergei* Markov[2]

[1]T.F. Gorbachev Kuzbass State Technical University, Mezhdurechensk branch, Mezhdurechensk, 652881, Stroitelei Avenue, 36, Russia
[2]T.F. Gorbachev Kuzbass State Technical University, Kemerovo, 650000, 28 Vesennyaya st., Russia

Abstract. The paper deals with the task of increasing the volume of rock placed in the internal dump. For that it is necessary first to analyze the existing technology of internal dumping, and then to determine the height of rock tiers being cut in the transport area. Analyzing the mining on the southern pits of Kuzbass it was found that now mostly three-tier dumps are piled. But for placing the extra volume of rock in the internal dump it will be required to pile dumps with more than three tiers.

1 Introduction

Transportless technology is now widely used on Kuzbass pits and in other coal producing regions. Its main advantage is the reduction of land capacity due to the possibility of placing internal dump in the worked out pit area, as well as lower cost of one cubic meter of overburden in comparison with truck-and-shovel technology [1]. As a result of low-dipping strata (6-14°) three-tier dumps are usually piled. In case of coincidence of two mutually influencing factors – the dip angle and the dragline beam length - there is a need for dumping piles with a large number of tiers [2]. There is an experience of the four-tier dumps, and it showed a certain complication of draglines' work on the dump and an increase in total coefficient of re-excavation by creating additional volumes of the secondary rehandling.

2 Material and Method

When piling five-tier dumps it is possible to expect a substantial increase in total coefficient of re-excavation and complication of draglines' work on the dump [3].

The height of cutting seam is the difference between the capacity of four-tier and three-tier dumps. Dumps' capacity is represented as a height of the rock seam in an array equivalent to this capacity [4].

The height of cutting seam H_{cut} is, m:

* Corresponding author: kuzstu@kuzstu.su

$$H_{cut} = \frac{E_{o.4} - E_{o.3}}{A \cdot K_p}, \tag{1}$$

where: $E_{0.4}$ – the four-tier dump capacity at 1 meter of mining front, m^3; $E_{0.3}$ – three-tier dump capacity, m^3; A – the stope width on the face side, m; K_p – the coefficient of rock loosening in the dump.

For more detailed analysis of the internal multi-tier dumps' capacity their classification on the following criteria was made: the number of tiers (starting with two) and piling scheme of dump tiers (with the preparation of additional capacity for a particular tier or without preparation).

For the four-tier dump it is necessary to take into account the possibility of preparing the additional capacity in the second, third or in each tier.

In the first tier there is no need on preparing capacity, since the rock at the bottom of the first tier is thickened due to its stop by an explosive move. This thickening of the prism zone contributes to the dump stability.

3 Results and Discussion

Internal four-tier dumps, being piled in the development of three flat seams strata can be classified by four types:

Type 1 – the capacity of dump tiers is formed during working the regular stope at the face side.

Type 2 – the capacity of dump tiers is formed during working the regular stope at the face side with the preparation of additional capacity on the second tier (Fig. 1).

Type 3 – the capacity of dump tiers is formed during working the regular stope at the face side with the preparation of additional capacity on the third tier.

Type 4 – the capacity of dump tiers is formed during working the regular stope at the face side with the preparation of additional capacity on the second and third tiers.

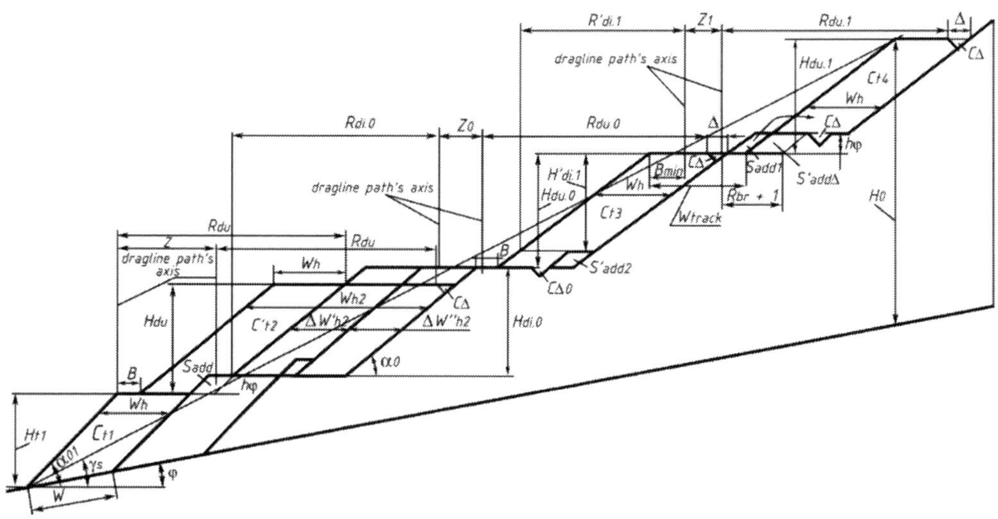

Fig. 1. The scheme of work on the dump side with the preparation of additional capacity on the second dump tier.

The schemes of dumping, being different in the preparation of additional capacity for any dump tier, have smaller capacity than schemes without capacity preparation. This provision is explained by the following.

The width of drilling and blasting stopes, the minimum value of which is determined by the condition of the turn of the coal carriers in mining faces (35-40 m), does not allow fully using the radius of loading of the currently used draglines when piling up the dumping layers [5]. Thus, Fig. 2 shows the scheme of tier filling for increasing the stacking capacity of which the additional capacity is prepared.

When preparing the capacity from a permanent dump array a cutting layer of width l_{cut} is taken out.

The unloading radius of used draglines (with beam length of 85-90 m) makes it possible to pile the tier with the width of its upper platform along the line 6-7. When the dump is being formed at the angle of the natural slope grade α_o ($\alpha_o = 37^0$), the prepared capacity is not completely filled (Figures 8-9-3-4 in Fig. 2). When using, for example, long-beam draglines, the rock unloading could be made in 7', which essentially increases the volume of rock placed in the prepared capacity (Figures 8'-7'-3-4 in Fig. 2). With the existing parameters of used draglines (ESH 15.90, ESH 20.90, that is, the beam length is 90 m), it is possible to fill the capacity ("container") (Figures 7-7-9 in Fig. 2) only in the following way: to raise the dumping dragline to the working platform of the tier (line 6-7 in Fig. 2) and to carry out an additional working stroke of the excavator. Cutting by this stroke the tier slope, it is possible to pile "the container" [6].

This method is used in the development of dumping schemes of types 2 and 4 for the piling of the second tier, since the main volumes of the rock entering the dump are accounted for it.

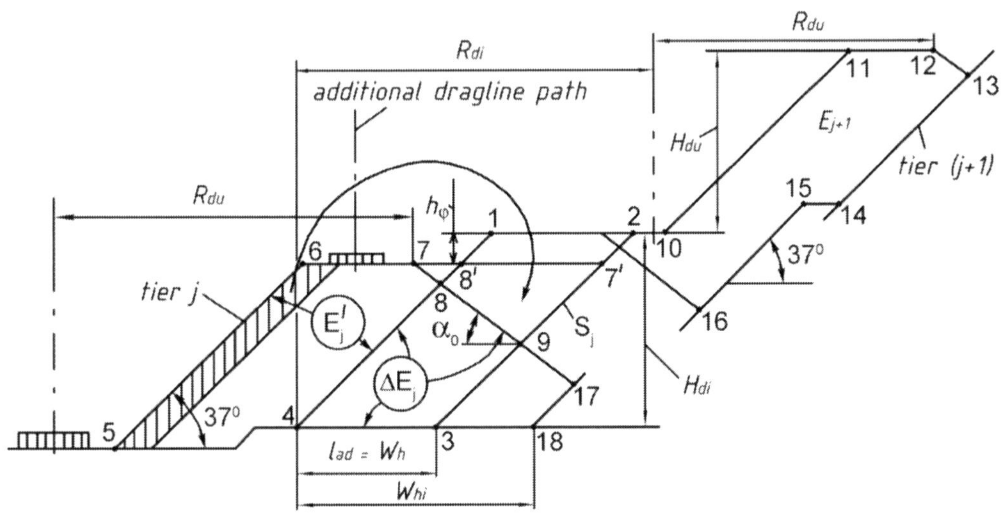

Fig. 2. The scheme of filling the tier with the preparation of additional capacity for it.

Consequently, in dumping schemes without capacity preparation overlying tier capacity ($j + 1$) is used entirely for the rock from the face side. In the scheme with capacity preparation for the incoming rock only a part of the prepared capacity is used $\Delta E_j < E\ (j+1)$.

In connection with this, the volume of the rock placed in the dumps with prepared capacity is smaller than in dumps without capacity preparation.

In case of equal withdrawal volume in the capacity preparation S_j (aeas 2-10-11-12-13-14-15-16-7' in Fig. 2), i.e. $S_j = E\ (j + 1)$ and the cutting width $l_{cut} = A_t$ the maximum increase in the volume of the rock, located in the prepared capacity $\Delta Ej = max$, is achieved.

To manage the volume ΔEj is possible only towards its reduction, for which l_{cut} or digging depth H_v should be reduced (Fig. 2). Controlling the volume ΔEj towards its reduction is used in this study in the development of the structures and parameters of excavation schemes.

The volume value ΔEj is affected by the dip angle, with increasing of which parameter A decreases and parameter h_φ increases (Fig. 2). Their combined effect reduces the capacity of the overlying tier $E(j+1)$ and, consequently, the value S_j and ΔEj.

For quantitative estimation of the effectiveness of additional capacity preparation it is offered to estimate $K_{i.\Delta E}$ – utilization coefficient of prepared capacity, equal to the ratio of the rock volume, located in the capacity, to the total volume of worked out rock while creating the capacity.

As the graphics of piling scheme, built for draglines with various beam length, are geometrically similar figures (parameter "Δ", combining unloading radius and oval dump contour depend only on the width of the stope), the value of the coefficient $K_{i.\Delta E}$ has a similar value for all models of draglines.

As an example, Tab. 1 shows the values $K_{i.\Delta E}$ for draglines ESH 20.90, ESH 20.100, ESH 30.110, ESH 25.120 calculated for the piling scheme of type 2 with a dip angle of 10^0.

Table 1. $K_{i.\Delta E}$ values for various models of draglines

Stope width (A), m	$K_{i.\Delta E}$ value for draglines			
	ESH 20.90	ESH 20.100	ESH 30.110	ESH 25.120
40	0.826 (±0%)	0.831 (+0,5%)	0.817 (–1,09%)	0.838 (+1,45%)
45	0.808 (±0%)	0.821 (+1,6%)	0.804 (–0,4%)	0.798 (–1,02%)
50	0.786 (±0%)	0.801 (+1,5%)	0.797 (+1%)	0.802 (+1,3%)

As it can be seen from Tab. 1 when calculating $K_{i.\Delta E}$ for draglines of different models, the deviation of the coefficients $K_{i.\Delta E}$ does not exceed ±1.5%, which provides sufficient accuracy and reliability of calculations by the average value of this coefficient.
The average value of $K_{i.\Delta E}$ and the calculation formulas for the four-tier dumps are given in Tab. 2, and the graphs of its dependence on angle φ and stope width A are in Fig. 3-a, b.

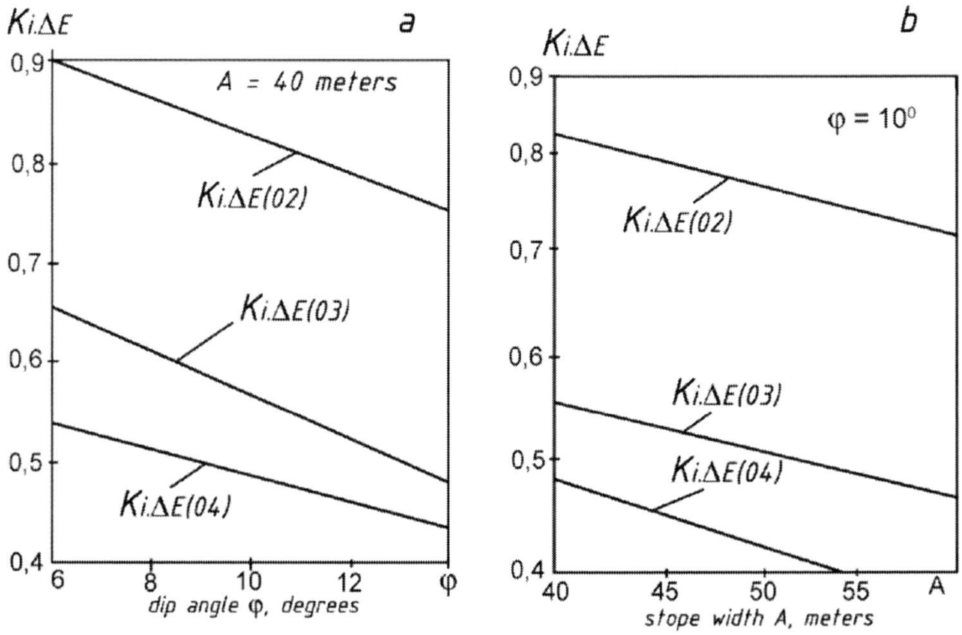

Fig. 3. The dependence of using coefficient of the prepared capacity on the basic factors: a – bedding angle of the seam; b – stope width.

4 Conclusions

As it is seen from the graphs, $K_{i.\Delta E}$ coefficient decreases both with increasing the bedding angle of seams and increasing the stope width.

The factor causing the decrease of $K_{i.\Delta E}$ in increasing the bedding angle of seams is the increase of parameter h_φ and therefore reduction of digging possibility of a dragline (H_v – h_φ). As a result, the amount of additional prepared capacity reduces.

The factor causing the decrease of $K_{i.\Delta E}$ in increasing the width of the stope is a disproportionate decrease in the volume of rock coming into the prepared capacity, relatively to the excavated volume in the prepared capacity.

References

1. I. Vukotic, V. Kecojevic, W. Zhang, Q. Cai, Int. J. Min. Sci. Tech., **23(6)**, 901 (2013)
2. M.M. Bereznyak, A.V. Kalinin and V.G. Pronoza, Soviet Mining **6(6)**, 638 (1970)
3. B. Lokhanov, Y.A. Zakharov, M.M. Bereznyak, A.V. Kalinin, Sov. Min., **3(5)**, 523 (1967)
4. A.S. Tanaino, V.I. Cheskidov, J. Min. Sci., **35(3)**, 304 (1999)
5. N. Demirel, J. Min. Sci., **47(4)**, 441 (2011)
6. I.V. Nazarov, J. Min. Sci., **48(1)**, 55 (2011)

The Relationship Between the Manifestations of Rock Pressure and the Relative Deformation of Surrounding Rocks

Svetlana Kostyuk[1], *Nikolay* Bedarev[2], *Oleg* Lyubimov[3*], and *Arthur* Shaikhislamov[4]

[1]T.F. Gorbachev Kuzbass State Technical University, the Administration, 650000, Kemerovo, Vesennya str., 28, Russian Federation
[2]T.F. Gorbachev Kuzbass State Technical University, Prokopyevsk branch, Department of technology and complex mechanization of mining, 653033, Prokopyevsk, 19a Nogradskaya st., Russian Federation
[3] T.F. Gorbachev Kuzbass State Technical University, Department of information and automated manufacturing systems, 650000 Kemerovo, Vesennya str., 28, Russian Federation
[4]West Siberian Industrial Engineering, the Directorate, 654000 Novokuznetsk, Russian Federation

Abstract. On the basis of analysis of the manifestations of the rock pressure in domestic and abroad practice the authors have the aim to research above named manifestations in conditions of new coal deposits of Kuzbass (Erunakovskoe, Kazankovskoe etc.) and to form the normative document for mining in these conditions. Presents the results of the displacement contour of the benchmarks on the conveyor and ventilation drifts during the testing of the 34 layer on the "Tagaryshskaya" mine. The formed system of measuring points and applicable instruments is described. The relative deformation of the coal mass on the conveyor drift at the approach of the coal face is given. Thus, the relationship of the displacements and relative deformations with the manifestation of rock pressure is attempted by the authors.

1 Introduction

The development of new coal deposits of Kuzbass (Erunakovskoe, Kazankovskoe etc.), which are characterized by low strength and deformation properties of the surrounding rocks, increased water content etc. [1], makes it necessary to clarify of the geomechanical parameters recommended by the normative document [2], since it is not taken into account of geological and mining conditions of these new deposits. On this basis, the authors set an aim for further research in relation to the above-named new deposits to justify the rational parameters of coal mining technology in the suite and in the presence of undermining and overmining layers. Consequently, the development of similar normative document for these fields should be carried out more like mine and laboratory tests in compliance with similarity conditions in simulated mining suite's layers on physical models.

* Corresponding author: oleglyub@gmail.com

Researches of the manifestations of the rock pressure in domestic and abroad practice are usually performed by means of the complex method, comprising an analytical stage, the mining operations modeling and field experiments.

As an example of international experience, the authors note [3], which used mathematical modeling system linking the drift movement after the coal face. And in [4, 5] the modeling of the excavation site and study of the process of forming the reference pressure zone ahead of the face of the complex are described. We have the foreign authors [6, 7, 8] for the local full-scale studies, it was noted that the length of the reference pressure zones do not exceed 15-20% of the depth of the development, and the values of the reference pressure may exceed the hydrostatic in 2-4 times.

From domestic work we should stay at work [9], which attempts to establish of the influence of coal face movement on the character of manifestations of rock pressure. But to justify such dependencies more research with the usage of physical models should be carried out.

In [10] the results of the joint influence of the high rock pressure zones from pillars (edge portions) of overmining layer and the basic pressure zone formed when the overmined layer is developing. It was noted that the greatest danger in the level of stress in the high rock pressure zones are indestructible interface pillars. On the basis of the research the location of developments outside the high rock pressure zones is recommended for specific conditions, however, such work is not enough for the development of normative documents.

Noteworthy is the work [11], which studied in detail the temperature field around the array of preparatory development and near the coal face. And recommendations on the usage of temperature changes to justify manifestations of rock pressure are issued, but this experience is hardly suitable when overmining or undermining layers in the suite are present.

Also the possibility of linking the manifestations of rock pressure with changes in relative deformations in the rock mass in the case of mining suites deserves the special attention. Under these conditions, when production face alignment is passing near of the undermining (overmining) measuring station, the recession of the relative deformations is indicating of the presence of tensile strain (unloading). Thus, the margin deformation of the overmined and undermined surrounding rocks is justified [12].

Based on the above analysis, the aim to research of manifestations of the rock pressure (on "Tagaryshskaya" mine for example) was set with a task of obtaining new data for comparison with the early known, as well as research data on physical models, to adjust and update the normative document.

Besides, in our studies, in the absence of development workings in the adjacent layers of suite, the possibility of studying the influence of the rock pressure manifestations on undermined and overmined developments was exclude, therefore, there is a need for additional research on physical models.

2 Materials and Methods

On the "Tagaryshskaya" mine researches were conducted in the period of mining of 34-3 layer (coal face 3) in the presence of the undermined layer 35 (interlayer thickness is 80 m) and overmined layer 33 (interlayer thickness is 90 m). The measurements were carried out on the path of displacement of the conveyor and ventilation drifts. In developing the layer 34 (m = 2.8 m, a = 5-8 grad, H = 320-340 m) in the coal face (Lv = 250m, Lpr = 2500m) by means of complex MKD. Coal mining was carried by KSV-460 miner. The immediate layer roof is the fine-grained siltstone (m = 9.0 m, f = 3-4), the main layer roof is also the fine-grained siltstone (m = 20.0m, f = 3-4). The average speed of the coal face movement was 2.5-3.5 m / day.

Conveyor and ventilation drifts in the roof are fixed by anchors with length 2.2 m and amount 6 anchors on the roof bar (l = 4.4 m), including 2 key anchors (with the presence of compliant washers) at the side of the ventilation drift and 1 key anchor at the side of the conveyor drift. The drift boards are fixed by SK anchors with length 1.8 m, one anchor on the sides with a distance along strike 1.0 m in alignment with the roof bars.

The displacement on the contours of the conveyor and ventilation drifts in the direction of "roof - soil" was carried out on the measuring point (MP1, MP2) 2 on each drift, and in the direction of the "side-by-side" it was carried out at 2 measuring points (MP3) [13]. Measuring of the displacement of deep and contoured frames carried out by the VNIMI method [14].

It should be noted that at the same time the observations on the caving pressure formation on the layer soil and on formation (bundles) of the layer roof up to the collapse (initial and subsequent), as well as monitored pressure change's character in the timber stands of the complex was executed.

These results are presented in other studies and considered in this case, since we consider only the relationship with the displacement on the contour of the conveyor and ventilation drifts.

Instrumental measurements of the deformation of rocks, surrounding of the development, carried out by means of the MIS-11 display rack with attachment dial (measurement accuracy ± 0.01 mm) or with a VNIMI micrometer stand.

The VNIMI micrometer stand is a rigid tubular unit, one end of which is pointed, on the other end the micrometer with division 0.01 mm is mounted. Rigid stand base is 800 mm and pull-out micrometer part is 25 mm.

The deformation measurements were carried out on a special measuring point (SMP) at a distance of about 500 m from the mining chamber, about 1.1 - 1.2 meters from the MP3 [13].

To equip SMP the pressed coal production in the wall was robed on depth 0.8 - 1.1 m. In the formed recess two boreholes were drilled up to 1.1 m. The distance between the holes was determined by base of micrometer rack or IMS-II rack and was 870 mm. The boreholes are concreted with metal benchmarks (d = 16 – 18 mm, l = 1.0 - 1.2 m) by means of BRC cement (fast setting, expanding).

The benchmarks had a recess with a diameter of 3 - 4 mm to fix the rack position and placed into the niche so that during the observations deformations normal to bedding were measured [12].

3 Results and Discussion

Displacement of the contour benchmarks on the conveyor drift began to manifest itself at the approach of coalface to the MP1 at a distance of 27 m, reached 33 mm and 29 mm respectively on the ventilation and conveyor drifts, and when coal face was approaching at 29 m to the MP2 displacement, reached 47 and 50 mm respectively on the conveyor and ventilation drifts.

On the MP3 displacements accounted for 11 and 15 mm respectively on the conveyor and ventilation drifts [13].

Table 1 shows the relative deformation $\varepsilon{\cdot}10^{-4}$ of the coal mass, depending on the distance from the SMP to coal face, as on fig. 1 the nature of their manifestations is shown. Fig. 1 shows that the relative deformation is not exceeded the 25 value, that in comparison with extremely high performance for coal samples lower than in [13-15], since there is such an order of magnitude lower, therefore the relative deformations of the coal mass on the conveyor drift were in the range of elastic deformation, so there was not be destruction.

Table 1. The value of the relative deformation $\varepsilon \cdot 10^{-4}$ of the coal mass, depending on the distance from the special measuring point (SMP) to coalface L, m

The distance from the SMP to coal face L, m	-5	0	5	10	15	20	25	30	35
$\varepsilon \cdot 10^{-4}$	25	25	24	21	17	12	8	4	2

The transition of coal face over the SMP marked the relative deformations decline that indicates the presence of tensile strain (unloading). The considered nature of the deformations of the coal array is given as the ability to search control options and valid deformations of undermined and overmined rock mass in mining of suites of layers.

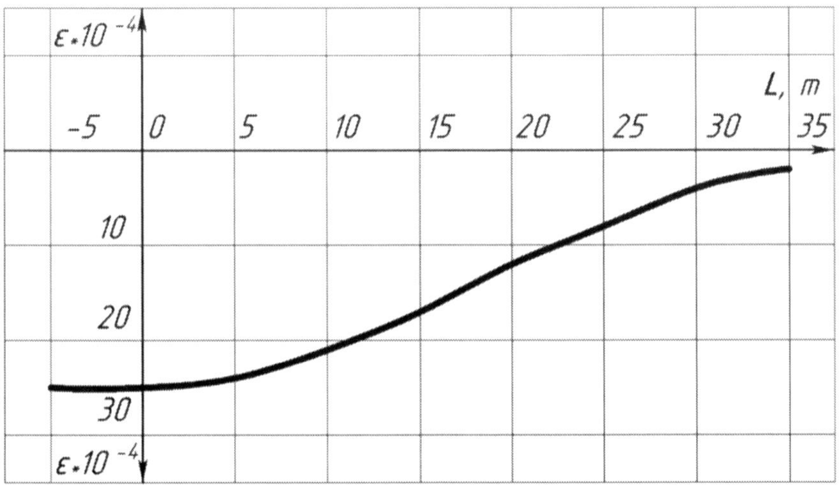

Fig. 1. The formation of the relative deformation $\varepsilon \cdot 10^{-4}$ of the coal mass, depending on the distance from the SMP to coal face L, m: $\varepsilon = \Delta l / l$, where Δl is the rock deformation, mm; l is the distance between benchmarks, mm.

It should be noted that the relative deformation compared with the displacement on the contour of the conveyor and ventilation drifts began to emerge not at 27-29 m from coal face, and at 36 m. It can be explained by the fact that the displacements on the MP1 were fixed at a distance of 240 m from the mining chamber and on the MP2 at a distance of 500 m from the mining chamber, so this fact can be explained that the movement came to the main roof of the layer.

4 Conclusion

The considered nature of the deformation of the coal array in combination with other methods is possible when selecting control options of valid deformations of undermined and overmined rock mass in mining suites beds is present, as concreted benchmarks for fixing of the strain can be stored in different places long period of time and can come in suitability for fixing the manifestations of the rock pressure and to prevent of destruction of the surrounding rocks.

In developing the normative document for the new Kuzbass deposits it needs to spend more like mine, and laboratory tests in compliance with similarity conditions in simulated mining suite's layers on physical models, so as the separate researches in the mining conditions (complex, lengthy and costly in time) is not enough to study objective lows of rock pressure manifestations at a wide variety of geological and mining factors.

References

1. L. Barron, *Proc. of 9th Intern. Conf. On Ground Control in Mining*, 142 (1990)
2. Z. Bieniawski, *Rock mechanics design in mining and tunneling* (Rotterdam, 1984)
3. K. Stefan, Glukauf, **1**, 50 (2004)
4. V. Sotskov, I. Saleev, Min. of miner. depos., **4,** 165 (2013)
5. V. Sotskov, O. Gusev, Min. of miner. depos., **6,** 197 (2014)
6. K. Erer, A. Heidareh-Zadeh, Min. Sc. and Tech., **3**, 191 (1985)
7. P. Lane, Rept. Invest Bur Mines US Dep. Intez. **89**, 29 (1985)
8. *Prediction of shears zones associated with outbursts in coal mines* (Lama R.D.S., 1987)
9. R. Scigala, Gosp. Sur. Min., **29**, 191 (2013)
10. C. Mark, G. Molinda, T. Barton, *Proc. of 21st Intern. Conf. On Ground Control in Mining*, 294 (2002)
11. Y. Zhang, Zh. Wan, B. Gu, Ch. Zhou, Therm. Sc., **20**, 2149 (2016)
12. A. Sidorenko, I. Gerasimova, Res. J. of Pharm., Biol. and Chem. Sc., **7,** 1844 (2016)
13. L. Siska, L. Dlouhy, Int. Rock Mech. Mtn. Scl., **4**, 431 (1967)
14. D. Oyler, C. Mark, W. Gale, J. Chen, *Performance of Roof Support Under High Stress in a U.S. Coal Mine* (Littleton, 2004)
15. Q. Chen, K. Zhou, T. Long, F. Gao, Journ. of Inst. of Sc. and Tech. Inf. of China, **5**, 200 (2008)

Computer Simulation of Electroosmotic Soils Treatment

Maxim Gucal[1*], *Andrey* Pokatilov[1]

[1] T.F. Gorbachev Kuzbass State Technical University, Mining Institute, 650000, Kemerovo, Russia

Abstract. The technique of experimental and analytical assessment of electrical and hydrodynamic parameters of clay soils as a result of physical and computer modelled processes in electroosmotic treatment zone is presented. The algorithm is given for calculation of the coefficients of hydraulic permeability, electroosmotic permeability and electroosmotic activity comprising the determination of geometric boundaries of the zones subjected to electric treatment, the initial filtration rate, and the calculation of the electric field and pressure. Spatial and temporal projections of dependent parameters at different stages of electric treatment are obtained.

1 Introduction

Electroosmotic treatment is one of the most effective ways to change the properties of water-saturated low permeability clay soils. Associated electroosmotic treatment processes are such as electrolysis, various exchange reaction, processes of formation and accumulation of new chemical compounds [1-4]. The classical equation of electroosmotic filtering is as follows [5]:

$$\overline{V} = k_e \overline{grad\varphi} + k_f \overline{gradh} = k_f \overline{grad}\left(k_a\varphi + h\right) \tag{1}$$

where $\overline{V}$ - velocity of particles (ions); φ – electric potential; h – water head; , k_f, k_e, k_a – coefficients of hydraulic permeability, electroosmotic permeability and electroosmotic activity, respectively.

Taking into account the non-stationary physical processes, heterogeneity and anisotropy of the properties of the soil in the area of electric treatment, the analytic solution of equation (1) with a fairly acceptable accuracy is not possible. The parameters characterizing the process of electroosmotic treatment of soil include a set of parameters $\overline{V}$, φ, h k_f, k_e, k_a from the equation (1). According to the standards, for the experimental determination of the filtration characteristics it is necessary to have special laboratory equipment described in a number of papers [6-13]. In particular, with the equipment described in these works it is possible using soil samples to determine in the laboratory the dependences of the parameters k_f, k_e, k_a on the porosity, the soil moisture, an electric field intensity, the processing

* Corresponding author: gmv.tgm@kuzstu.ru

time and the consumption of electrical current, and relevant migrations of liquids filling the pores in the inter-electrode space.

Often with the purpose of detailing the data, researchers resort to mathematical modeling of various stages of physical experiments [9, 14, 15]. In particular, the computer modeling is one of the most effective methods for the study of complex systems. To explore computer models is easier and more convenient than to run live experiments, as the latter can be very labor- and resource-intensive and may give unpredictable results. Computer models help us to identify the main factors that determine the properties of the object under study or process, to forecast the impact of the results from the taken decisions, see the problem in its development.

For a long time, the researchers of the Kuzbass State Technical University have been conducting investigations related to geophysical monitoring of processes in the areas of deliberate impact on the physical and mechanical properties of rocks [9, 14], wherein the specific electrical resistance (SER) of soil is the main informing parameter. They used both the integral measurement circuits (electrical sounding and profiling, georadar) and the differentiated ones to determine the true SER (downhole electric and induction logging, microsensors). In the interpretation of geophysical data, the main difficulty lies in the unsteadiness of the processes under review: during processing of the solid mass its electro- and hydrodynamic parameters are continuously changing. To enhance the monitored information content, it is advisable to use an approach similar to the method of successive change of stationary states: the physical processes in the processing zone to be considered as stationary in the intervals between the geophysical measurements, and changes in the massif properties to be evaluated by the data of discrete measurements.

In this paper, we consider the indirect assessment of the above-mentioned electric and hydrodynamic parameters using the results of physical and computer modeling of the area of electroosmotic treatment of the clay soil.

The block-diagram of experimental computer model of electroosmotic filtration is shown in Fig. 1. The physical basis of the proposed method is the equation of electroosmotic fluid displacement velocity in the filtration channel, subordinated to the theory of Helmholtz-Smoluchowski [11, 12, 14, 16, 17]:

$$\overline{V} = -\frac{\varepsilon\varsigma}{\mu} grad\phi \qquad (2)$$

where ε - absolute dielectric constant, F/m; μ - the coefficient of dynamic viscosity of the solution, Pa·s; ς - zeta potential, V.

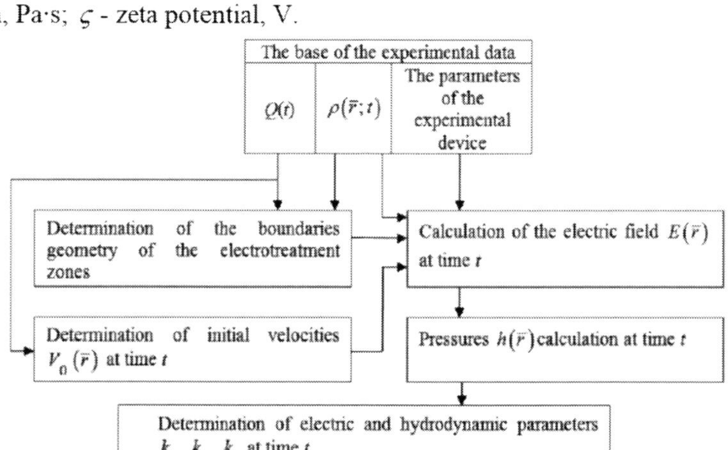

Fig. 1. Block-diagram of experimental computer model: Q - flow; ρ - SER; $\overline{r}$ - coordinate; I-V - calculation stages

When filtering in a porous medium, the module of effective velocity V_{ef} (flow) is produced on the basis of dimensional analysis:

$$V_{ef} = C\sigma_0 \frac{m^2 R}{\mu} E = \frac{m\varsigma\varepsilon}{\mu} E = k_e E \tag{3}$$

where C - constant; σ_0 - specific volume density of diffuse layer ions charge, C/m^3; t - soil porosity; R - hydraulic pore radius, m; E - electric field intensity, V/m.

The calculation is divided into five stages (I-V).

At the first stage, based on analysis of changes in the relative values of SER $\rho(\overline{r})$, the sizes of characteristic zones formed during soil treatment are determined.

The second stage is responsible for the calculation of the initial electroosmosis velocity taken as a reference. The liquid output Q from the electrode-injector recorded in the course of the experiment serves as the data for calculation.

At the third stage, the calculation of the velocity of the electroosmosis process change is made, taking the difference of the areas under the graphs SER $\rho(\overline{r})$ as the characteristic of the process change. The calculation is made using programs of PDE MATLAB library section that implement the finite element method. The task is reduced to the solution of the Laplace equation with the parameters of the massif corresponding to the real conditions:

$$\Delta\varphi(x,z) = 0 \tag{4}$$

where x, z - horizontal and vertical coordinates.

The conditions at infinity and at the interface are taken as traditional:

$$\left.\varphi_1\right|_{x^2+z^2\to\infty} \to 0; \quad \left.\varphi_i\right|_{x=x_i} = \left.\varphi_{i+1}\right|_{x=x_i}; \quad \left.\frac{\partial\varphi}{\partial z}\right|_{z=z_0} = 0; \quad \left.\frac{\partial\varphi_i}{\partial x}\right|_{x=x_i} = \left.\frac{\partial\varphi_{i+1}}{\partial x}\right|_{x=x_i}. \tag{5}$$

Calculation of pressures (phase IV) is based on the electric- hydrodynamic analogy, previously established for the progressive and flat-radial flow of filtered fluid. The dependence of the potential function of the electric field on the pressure function can be expressed as follows:

$$\varphi(\overline{r}) = K_2\left[K_3 - h(\overline{r})\right] \tag{6}$$

where $\varphi(\overline{r})$ - the function determined by integrating function $E(\overline{r})$; K_2 - the constant depending on the electrochemical, dielectric, conductive properties and porosity of the environment; K_3 – the constant depending on the limit value of pressure. Determination of a group of electric- and hydrodynamic parameters in the settlement times (Phase V) is made by solving the equations (1) - (6) with the substitution of intermediate results.

2 Results and discussion

Fig. 2 and 3 show initial data for calculations of stages I and II as experimental plots obtained on a 3-D physical model. Fig. 2 shows plots of the relative SER $\rho(x, I\times t)$, where x - coordinate on the axis of the apparatus, I - current, t – time, taken as the initial basis of experimental data, and the boundaries of the main zones of electroosmosis (drainage and accumulation of moisture). Fig. 3 shows the data of the fluid output Q from electrode-injector, which served as the baseline for the calculation of the reference filtration velocity.

Fig. 4 shows the results of calculation of the relative electroosmosis velocity $V/V_0(t,x)$.

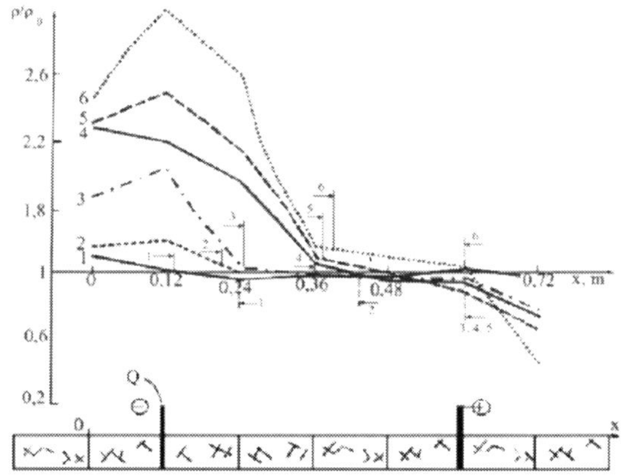

Fig. 2. Dependence of the relative SER ρ/ρ_0 on $I{\cdot}t$ and coordinates x of the sensor installation in the electroosmotic treatment area: 1 - $I{\cdot}t$ = 5.6 A·h; 2 - 8; 3 - 18; 4 - 22; 5 - 30; 6 – 54
⟶ the border of the drying zone; ⟵ - the border of moisture accumulation area

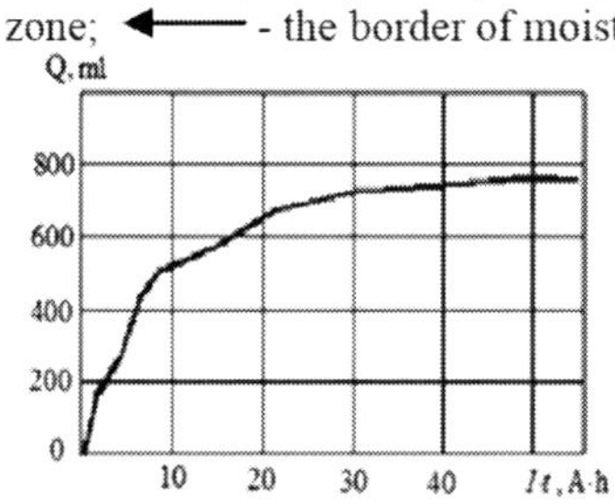

Fig. 3. The output of soil liquid Q

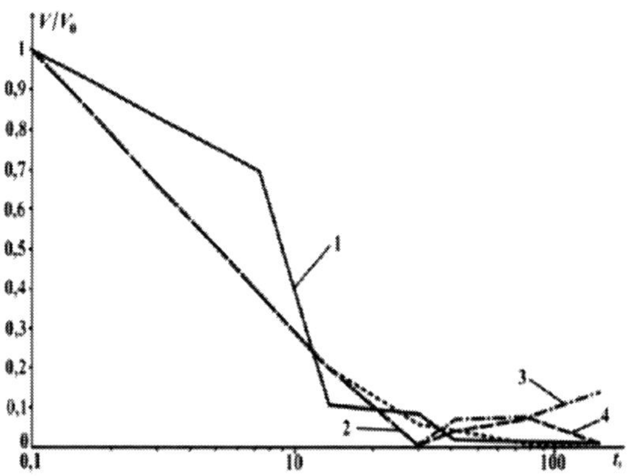

Fig. 4. Changes in the relative velocity of electroosmosis: 1 - the velocity calculated by the output of the liquid; 2, 3 and 4 - the calculated velocity in the drying zone, the intermediate zone and the zone of the accumulation of moisture, respectively.

Calculation of the electric field of the plant at the third stage is carried out using programs of section PDE of MATLAB libraries that implement finite element method, wherein the task in the heterogeneity model is implemented based on the following assumptions:

- the original massif had specific inductive capacitance (ε = 27) corresponding to the value taken for the wet clay;
- in the course of treatment, the above indicated characteristic zones are formed with value ε corresponding to wet, moisturized and dry (ε = 27, 18 and 7) clay;

- the dimensions of the computer model and the processing mode complied with the physical model, an example of which is given in the work [9], and the dimensions of the three zones varied in accordance with the data obtained at the first stage.

Fig. 5 shows the plots of changes in the electric field intensity E along the main axis x of the unit in the transverse horizontal plane located at the middle section of the immersed electrode-injectors.

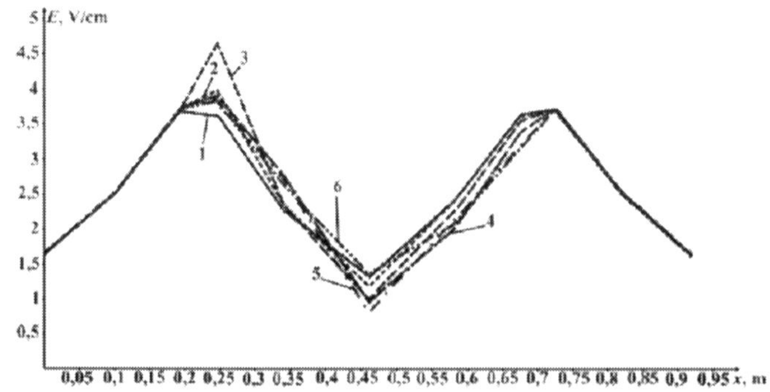

Fig. 5. Changing of the electric field intensity in the interelectrode space with time at depth z = 0.125 m: 1 - 5.6 A·h; 2 - 8 A·h 3 - 18 A·h; 4 - 22 A·h; 5 - 30 A·h; 6 - 54 A·h

Having taken, in accordance with the experimental data, the initial value k_{e0}=1.3·10^{-4} cm^2/V·s, and taking into account the average values of the relative filtration velocity by zones, we calculated the dependencies of electroosmotic and hydrodynamic parameters k_f, k_e, k_a for each zone with time t (Fig. 6) and depending on the coordinate x (Fig. 7).

a b

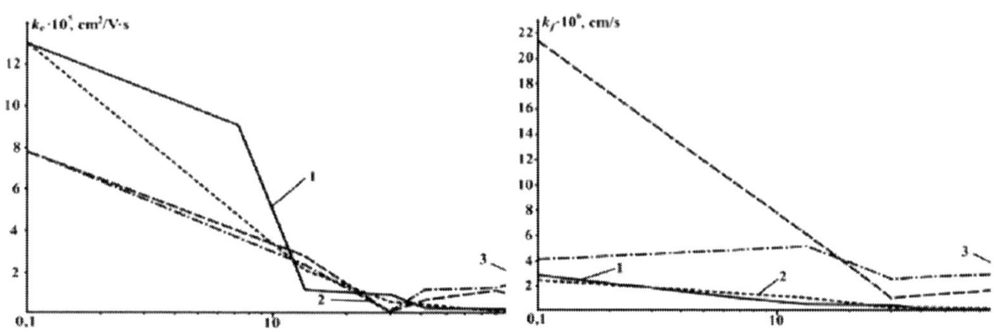

Fig. 6. Coefficients of electroosmotic permeability k_e (a), hydraulic permeability k_f (b) the versus time: 1 - calculation results for the output of liquid; 2 - drainage area; 3 - the transition zone; 4 - moisture accumulation zone

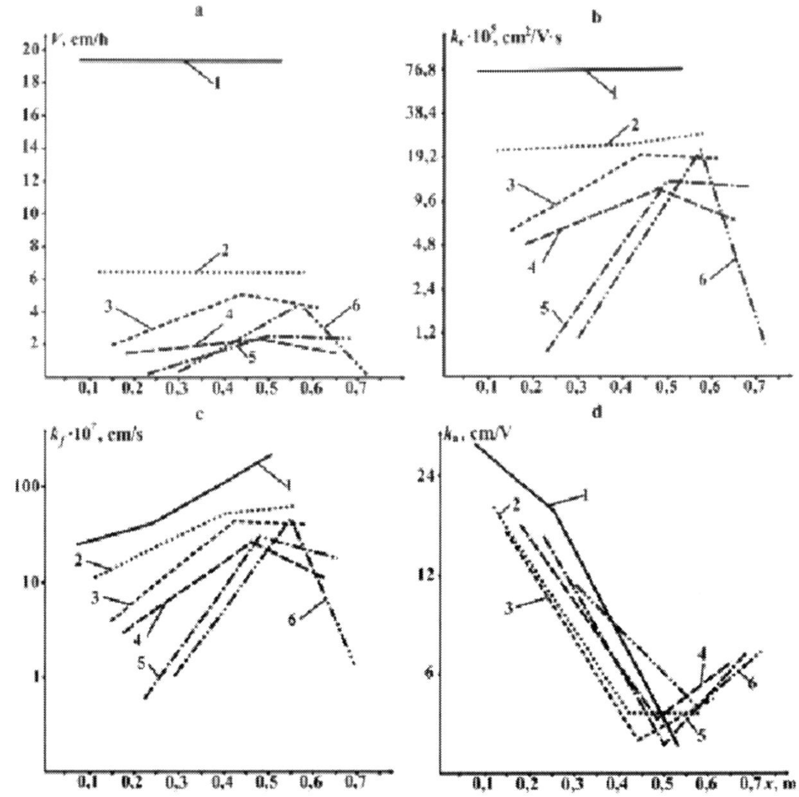

Fig. 7. Velocity of electroosmosis (a) and coefficients of electroosmotic permeability k_e (b), hydraulic permeability k_f (c) and the electroosmotic activity k_a (d) versus the coordinate x along the length of the model: 1 - 5.6 A·h; 2 - 8 A·h 3 - 18 A·h; 4 - 22 A·h; 5 - 30 A·h; 6 - 54 A·h

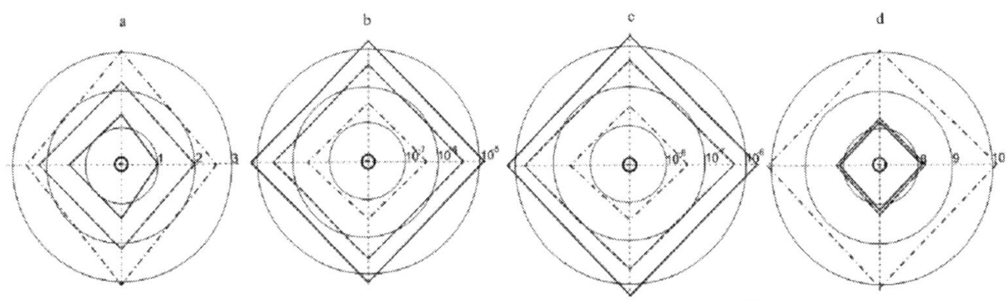

Fig. 8. The diagrams of change in the relative resistivity ρ/ρ_0 (a), k_e, cm²/V·s (b), k_f, cm/s (c), k_a, cm/V (d): ——— - I·t = 8 A·h; - - - - - - I·t = 22 A·h; — · — I·t = 54 A·h

3 Conclusion

The analysis of the results led to the following conclusions.

1. The developed method allows us to simulate unsteady hydrodynamic processes in the zone of soils subjected to electroosmotic treatment in the one-dimensional pattern, in the plane, and in space.

2. The differences in the ranges of values of electroosmotic permeability coefficient k_e for the zones of humidification and dehumidification are negligible, but, in view of the results found by the analysis of the dynamic changes in the zones, these values are of the op-

posite nature: in the course of drying the ke value decreases with time, and during moisture accumulation - it increases.

3. The estimated values of the electroosmotic permeability coefficient k_e with the decrease in the average value of electric field intensity in the range from 3.8 to 0.8 V/cm also decrease in the range from $1.3 \cdot 10^{-4}$ cm^2/V·s to $1.11 \cdot 10^{-6}$ cm^2/V·s.

4. The calculated values of hydraulic permeability coefficient k_f in the range of 0.005-0.02 m/d correspond to the range of values from 0.005 to 0.05 m/d, in which using of the electroosmotic treatment of soils is more efficient than conventional methods of dewatering.

5. In the area of active electroosmosis in the process of electroosmotic filtration the depression is created with the movement of fluid to the electrode being relatively uniform in all the directions, as evidenced by symmetrical changes in the electrical and hydrodynamic parameters obtained by an experimental and analytical way (Figure 8, a), which is consistent with the nature of the dependencies presented in the study [18].

References

1. Khaled, The Application of Electroosmosis in Clay Improvement, 55 (2016)
2. Asadi, B. B. K. Huat, H. Nahazanan, H. A. Keykhah, Int. J. Electrochem. Sci., Theory of Electroosmosis in Soil, **8** (2013)
3. J.-L. Chen, L. Murdoch, ASCE, **5** (1999)
4. S. A. A. Tajudin, *Electrokinetic stabilisation of soft clay* (2012)
5. M. F. Buehler, J. E. Surma, J. W. Virden, *Thirty-Third Hanford Symposium on Health,* 36 (1994)
6. V. dos Santos, M. O. Medeiros, A. S. D. dos Anjos, C. A. Martínez-Huitle, D. R. da Silva, Chemical Engineering Transactions, **41**, 6 (2014)
7. R.Z. Velten, D.C. Lima, M.P.F. Fontes, C.A.B. Carvalho, Soils and Rocks, **32**, 11 (2012)
8. D. V. Morris, S. F. Hillis, J. A. Caldwell, Can. Geotech. J., **22**, 8 (1985)
9. S. M. Prostov, E. A. Shabanov, *The 8th Russian-Chinese Symposium. Coal in the 21st Century: Mining, Processing and Safety,* 9 (2016)
10. Shao-Chi Chiena, Pio-Go Hsiehb, Chang-Yu Ou, ACEM **12**, 11 (2012)
11. Y. Guo, *Electrokinetic dewatering of oil sands tailings*, 112 (2012)
12. F. Eriksson, L. Gemvik, *Electro-Osmotic Treatment of Soil,* 116 (2014)
13. S. Hansbo, *Sixth International Conference on Case Histories in Geotechnical Engineering,* 15 (2008)
14. S. M. Prostov, M. V. Gucal, E. A. Shabanov, *Taishan Academic Forum*, 8 (2014)
15. J. Yuan, M.A. Hicks, J. Dijkstra, *ISSMGE - TC 211*, 8 (2012)
16. *Kit Design and modelling of electroosmotic dewatering,* 2014 (2006)
17. S. R. Maduar, A. V. Belyaev, V. Lobaskin, O. I. Vinogradova, PRL, **114**, 5 (2015)
18. Zhi-ming Liu, Jian-gui Yang, Qi-meng Li, EJGE, **21**, 13 (2016)

Innovative numerical modelling of technogenic rock arrays structure

Sergey Markov[1], *Maxim* Tyulenev[2*], *Oleg* Litvin[2] and *Ekaterina* Tyuleneva[1]

[1] Department of Geology and Survey, T.F. Gorbachev Kuzbass State Technical University, 28, Vesennyaya, 650000, Kemerovo, Russia
[2] Department of Open Pit Mining, T.F. Gorbachev Kuzbass State Technical University, 28, Vesennyaya, 650000, Kemerovo, Russia

Abstract. Numerical simulation of bulk technogenic rock arrays, unlike the full-scale, has a number of advantages. They are: the large amount of low cost measurements; operative selection of parameters necessary for the environment with the given properties; calculation of stress-strain state and assessment of the array sustainability; calculation of rock arrays filtration properties that are used as filters for mining enterprises sewage, etc. The article describes the method of numerical simulation of structure of bulk disjointed rock arrays, describes the basic initial and boundary conditions of the model; represents the numerical model visualization of the array, created with the use of peripheral dumping (as an array with more textured and visual segregation of granulometric composition adjustment). The basic applications of numerical modeling of technogenic rock arrays in the mining processes are described. Comparative analysis of the real and simulated array structure shows the adequacy of the numerical model and the developed for its creation algorithm, which in turn allows judging the quality of the study. Also, the adequacy of building models is supported by a high convergence of the experimental and calculated figures depending on the filtration rate values of the particle size distribution of technogenic rock arrays.

1 Introduction

The relevance of research lies in the fact that mathematical models of technogenic rock masses can predict and evaluate the operational structural and mechanical properties, which are very difficult or impossible to study instrumentally [1, 2]. Simulation of man-made structures rock masses is convenient to carry out by the method of discrete elements, where each particle in the array structure is described separately. The study of the internal geometry (structure and texture) of man-made rock masses on mathematical models [3, 4] is largely a solution for determining their physical and strength properties that may be of practical use to calculate the filtration characteristics of the dams, the stability of dumps and bulk structures.

* Corresponding author: tma.geolog@kuzstu.ru

2 Material and Method

In general, when an artificial rock array has initially a stable, static state, its numerical simulation means stable packaging of each of its particles in a given half-plane or hypothetic "container". This "container" dimensions correspond to the dimensions of simulated array, and the choice of the initial position of the packed particles reminds the dumping process. The particle shape in the plane case is taken as a circle because it reduces the total number of coordinates describing the place of particles in the space (the center coordinates x, z and radius r). Physical parameters (particle's density, friction rate in the contacts between the particles) are given separately on the basis of the field studies.

The research methods include photomapping, statistical analysis, field observations (used to set the initial conditions), the method of discrete elements (implemented in the algorithm of particles' packing).

3 Results and Discussion

Simulation algorithm consists of the following steps.

Step 1. Choosing the particle size.

The choice of particle size (r_0 radius) is based on granulometric distribution of simulated rock array. For overburden rock arrays of Kuzbass open pits the granulometric distribution for upper part of the array is determined by sieve method (particle size is up to 0.1 m), for lower and middle parts – by oblique photomapping method [5-7] using "Split Desktop" software available on the official website of the "Split Engineering" company.

The granulometric distribution study revealed that the total content of fractions (based on the number of particles) in the artificial rock arrays is in the range from $2.99 \cdot 10^{-4}$% for the pieces of 0.7-1 m, and 91.42% for the size of the pieces less than 0.003 m. The values for the fractions of 0.5-0.7 m ($4.02 \cdot 10^{-4}$%)); 0.25-0.5 m ($1.32 \cdot 10^{-3}$%); 0.1-0.25 m ($6.39 \cdot 10^{-3}$%); 0.025-0,1 m ($8.78 \cdot 10^{-2}$%); 0.01-0,025 m (0.37%); 0.007-0.01 m (0.56%); 0.005-0.007 m (1.46%); 0.003-0.005 m (6.10%) were calculated for the highest quality modeling.

Step 2: Determining the initial coordinates of the particle.

In general, when modeling rock mass, limited only by the bottom, initial coordinates of the particle are x_0 and z_0. Here x_0 is within the width B (if only the cross-section of the array is being simulated) or the length L (if only longitudinal section of the array is being simulated) of the upper platform of piling rock array in the simulation of spread technology, i.e., it varies from $z_0\, x\, ctg\, \beta$ to $z_0 x\, ctg\, \beta + B$ up to ($z_0 \cdot ctg\, \beta$ + L), respectively (β - natural angle of slope, degrees).

Coordinate z_0 shall be equal to the maximum of height of the bench. When z_0 is selected, the condition of non-intersection of packed particles with each already packed *i-th* particle must be fulfilled:

$$(x_i - x_0)^2 + (z_i - z_0)^2 \geq (r_i + r_0)^2, \quad (1)$$

where x_i, z_i, r_i - generalized coordinates of previously packed particles m.

If this condition is not fulfilled a number of times equal to the number of particles with z coordinate close to the height of piling array, the dumping process should be considered as finished for spread technology. For peripheral dumping simulation or for the dump piling using transportless technology, the value of initial coordinate x_0 at the beginning of the simulation is calculated as $z_0\, x\, ctg\, \beta$ and when the simulated array height reaches the full height of a dump, x_0 is shifting by the value of Δx (which is equal to r_0 for bulldozer dumping, or to the width of n-s dumping stope for excavator dumping) up to the maximum length (width) of the upper dump's platform (Fig. 1).

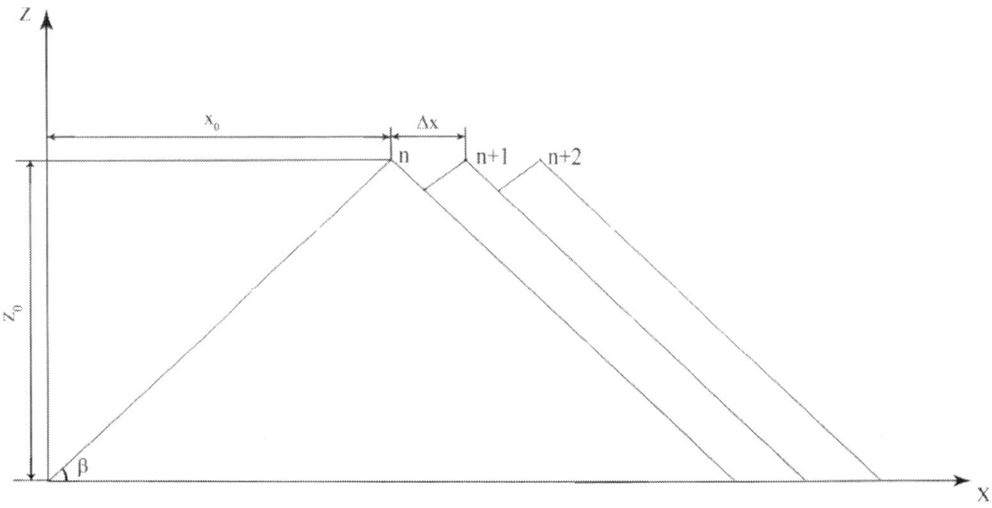

Fig. 1. The selection of initial parameters of the simulation

Step 3: Determining the conditions of contact of the particles

Simulation of particle packing process does not involve considering the dynamics of their collisions. After determining the initial coordinates of the particle (Fig. 1, point *n, n + 1, n + 2 ...*) at its moving straight down the minimum-valued coordinate *z* can be found. If the particle does not meet other particles, the coordinate z_0 value becomes equal to the particle radius r_0.

In order for the particle being packed could "meet" the previously packed *j*-th particle, the following three conditions must be observed (Fig. 2):

- the sum of the particles' radiuses must be greater than the horizontal projection of the distance between them:

$$(r_0 + r_j) > |x_0 - x_j|; \tag{2}$$

- the distance between them d_j should be minimal (it can be calculated by the search of previously packed particles):

$$d_j = \sqrt{(x_0 - x_j)^2 + (z_0 - z_j)^2} - (r_0 + r_j) \rightarrow \min; \tag{3}$$

- the coordinate z_j for previously packed particle must be less than the coordinate-ordinates z_0 of particle being packed:

$$z_j < z_0. \tag{4}$$

To find the minimum coordinate z_0, corresponding to the touch of particle being packed the previously packed particle, i.e. under the conditions (2) - (4), we consider the triangle ABC (Figure 2.). At an angle:

$$\varphi = \arccos\left(\frac{|x_0 - x_j|}{r_0 + r_j}\right) \tag{5}$$

coordinate z_0 value will be:

$$z_0 = z_j + (r_j + r_0)\cdot\sin\varphi. \tag{6}$$

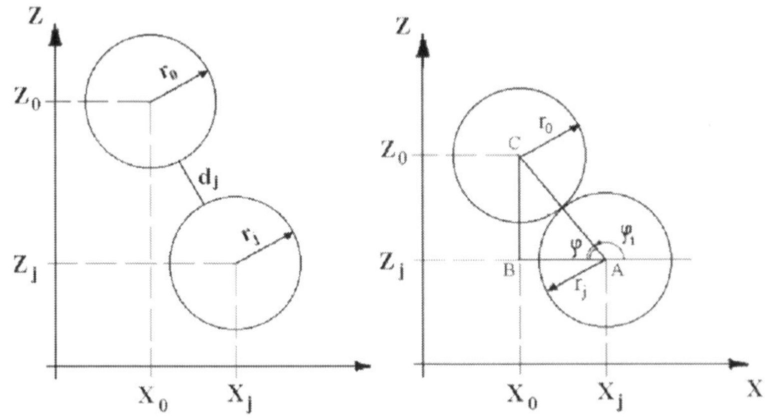

Fig. 2. The scheme of finding a sustainable position of the particle

In the particular case when $x_j = x_0$ coordinate z_0 is set to:

$$z_0 = z_j + (r_j + r_0). \tag{7}$$

Step 4: Determining the stable position of the particle.

Having calculated z_0 under the condition of contact between the particles (6) or (7), we begin to "roll" particle being packed around the previously packed one, reducing the angle φ to $d\varphi$ until the angle $\varphi = 0$. It was empirically found that for the array with particles size of the same order, optimum value $d\varphi = 0{,}1 \cdot r_{min}$ (r_{min} - the minimum value of the radius of packed particles), and $d\varphi = 0{,}0001 \cdot r_{min}$ for arrays with higher inhomogenuity of granulometric distribution.

The condition of contact of the particles being packed with i-th particle (i.e. for $i \neq j$) can be written as:

$$d_i = \sqrt{(x_0 - x_i)^2 + (z_0 - z_i)^2} - (r_0 + r_i) \leq 0, \tag{8}$$

where: d_i – the distance between the particle being packed and i-the packed particle.

This condition is checked every time the angle φ is getting changed respectively to all the particles lying near the particle being changed.

"The rolling" is stopped as soon as the condition (8) becomes equal to 0 (which means that the rolling particle contacts with the two previously packed particles), or becomes lower than 0, which means the intersection of the second from the previously packed particles.

The condition of the solid packing in this case is solving simultaneous equations:

$$\begin{cases}(x_i - x_0)^2 + (z_i - z_0)^2 - (r_i + r_0)^2 = 0 \\ (x_j - x_0)^2 + (z_j - z_0)^2 - (r_j + r_0)^2 = 0\end{cases}, \tag{9}$$

where: x_i, z_i, r_i, x_j, z_j, r_j – coordinates and radius of packed particle that touches the one being packed, $i \neq j$; x_0, z_0, r_0 – the radius and the coordinates of the particles being packed.

The numerical solution of this system of equations simply implemented by Newton-Ruffson iterative method, with a given relative error $\varepsilon = 10^{-7}$ and the number of equations $N = 2$, the maximum number of iterations $M = 1000$, and the vector of initial approximations X_{i0} (with components $x_{10}, x_{20}, \ldots, x_{N0}$). In our case, $x_{10} = x_i + x_j$, $x_{20} = z_i + z_j$.

Decomposition of nonlinear equations $F_i(X_i)$ in a Taylor series formed Jacobian matrix $[\partial F_i / \partial X_i]$ necessary to calculate the increments of $F_i(X_i)$ at the small changes of the variables. The expanded form of the Jacobian matrix can be written as:

$$\begin{bmatrix} \partial F_1/\partial x_1 & \partial F_1/\partial x_2 & \dots & \partial F_1/\partial x_N \\ \partial F_2/\partial x_1 & \partial F_2/\partial x_2 & \dots & \partial F_2/\partial x_N \\ \dots & & & \\ \partial F_N/\partial x_1 & \partial F_N/\partial x_2 & \dots & \partial F_N/\partial x_N \end{bmatrix} \tag{10}$$

Analytical differentiation $F_i(X_i)$ can be avoided by replacing the partial derivatives in the matrix of their approximate values:

$$\frac{\partial F_i}{\partial X_i} \approx \frac{F_i(X_i + H_i) - F_i(X_i)}{H_i}, \tag{11}$$

where: $H_i = \varepsilon |X_i|$ – the small increment X_i.

The solving $\Delta x_1, \Delta x_2, \dots, \Delta x_N$, i.e. ΔX_i can be found from the linear equation system:

$$\begin{bmatrix} \partial F_1/\partial x_1 & \partial F_1/\partial x_2 & \dots & \partial F_1/\partial x_N \\ \partial F_2/\partial x_1 & \partial F_2/\partial x_2 & \dots & \partial F_2/\partial x_N \\ \dots & & & \\ \partial F_N/\partial x_1 & \partial F_N/\partial x_2 & \dots & \partial F_N/\partial x_N \end{bmatrix} \cdot \begin{bmatrix} \Delta x_1 \\ \Delta x_2 \\ \dots \\ \Delta x_N \end{bmatrix} = \begin{bmatrix} -F_1 \\ -F_2 \\ \dots \\ -F_N \end{bmatrix} \tag{12}$$

Then the corrected values can be calculated:

$$\begin{aligned} X_{1(n+1)} &= x_{1(n)} + \Delta x_1, \\ X_{2(n+1)} &= x_{2(n)} + \Delta x_2, \\ &\dots \\ X_{N(n+1)} &= x_{N(n)} + \Delta x_N. \end{aligned} \tag{13}$$

For every ΔX_i one of the following conditions must be checked:

$$\begin{aligned} &|\Delta X_i| > \varepsilon, \\ &\left|\frac{\Delta X_i}{\Delta X_j}\right| > \varepsilon. \end{aligned} \tag{14}$$

With its completion another iteration must be performed, otherwise the vector $X_{i(n+1)}$ is considered to be the solution.

After accurate calculation of the coordinates the stability of the particles to be packed must be checked: if its coordinate x_0 is within the interval $[x_i - x_j]$ then the position of the particle is stable (only in the absence of collision dynamics and the impact of inertia).

Under the conditions:

$$\begin{cases} |x_0 - x_i| < |x_j - x_i| \\ |x_0 - x_j| < |x_j - x_i| \end{cases} \quad (15)$$

it is believed that the particle be packed has a steady position, otherwise it continues to "roll down" over the i-th particle.

When the model of technogenic rock array corresponds to its real size the simulation of its structure is considered to be completed. After modeling the structure it is possible to study density, strength and filtration properties of technogenic rock array. Fig. 3 shows a fragment of the visualized model technogenic rock array - the waste dump of 30 m height, pilled by bulldozer peripheral technology.

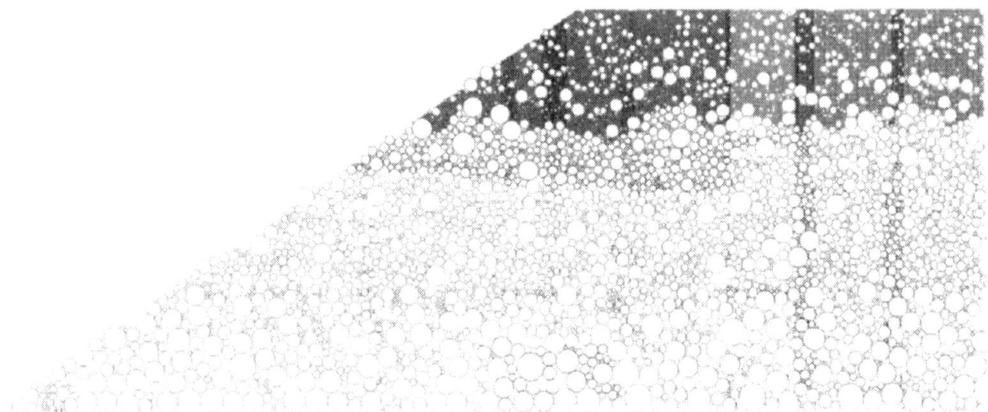

Fig. 3. The fragment of the dump array

4 Conclusion

The developed algorithm of technogenic rock array simulation has a sufficiently high reliability because reconstituted structure almost completely matches the real dumps.

The numerical modeling of granular media and technogenic rock masses which are their particular case can be used in the study of their physical and strength properties. This is due to the fact that the modeling algorithms allow flexibly change the input parameters of the simulated array. It allows selecting the parameters of dumping technology, simulating the arrays with any defined properties.

References

1. M. A. Tyulenev, T. N. Gvozdkova, S. A. Zhironkin, E.A. Garina, Geotech. Geol. Eng., **35**, 203 (2017)
2. M.Y. Blaschuk, A.A. Dronov, S.S. Ganovichev, IOP Conf. Ser. Mater. Sci. Eng., **142**, 012128 (2016)
3. A. V. Koperchuk, A. V. Murin, V. V. Filonov, IOP Conf. Ser. Mater. Sci. Eng., **127**, 012040 (2016)
4. M.Y. Blashchuk, A.A. Kazantsev, R.V. Chernukhin, App. Mech. Mat, **682**, 418 (2014)
5. J.M. Kemeny, A. Devgan et al., Journal of Geotechnical Engineering, **119**, 1144 (1993)
6. J.M. Kemeny, Mining Engineering, **46**, 1281 (1994)
7. J.C. Russ, *The Image Processing Handbook* (CRC Press, 1995)

Definition of the form of coal spontaneous combustion source as the inverse problem of geoelectrics

Dmitry Sirota[1], *Vadim* Ivanov[1*], *Veniamin* Khyamyalyaynen[1]

[1] T.F. Gorbachev Kuzbass State Technical University, 650000, 28 Vesennyaya St., Kemerovo, Russia

Abstract. The paper reviews the method of determining the shape and size of the coal self-heating source on coal pit benches and in coal piles during mining of coal by the open method. The method is based on the regularity found in the 1970s of the previous century and related to the distribution of potential of the natural electrical field arising from the temperature in the vicinity of the center of self-heating. The problem is reduced to the solution of inverse ill-posed problem of mathematical physics. The study presents the developed algorithm of its solution and the results of numerical simulation.

1 Introduction

Spontaneous combustion of coal is one of the main causes of fires in open and underground development of coal seams. Geological and mining factors play a significant role in the process of spontaneous combustion of coal. In case of open mining the tectonic disturbances of coal benches and enclosing strata can be attributed to the first factors when heavily fragmented and cracked coal can lead to increase in the porosity and surface area of coal exposed to oxidation, as a result even at normal temperatures of the ambient air the heat is generated. The order of mining operations in relation to the strike of seams, method of conducting drilling and blasting operations, parameters of the system of mining, the height and time of bench standing can be attributed to the second group of factors. The total influence of all factors shows itself in formation of the loose accumulation of coal, in some cases exceeding the critical mass, which ensures accumulation of heat due to oxidation of coal than its consumption by convection, conduction, thermal radiation and other causes.

For control and diagnostics of the processes of self-heating and spontaneous combustion of coal it is required to promptly assess the power of the heat source, its size and shape. These parameters allow the proper positioning of the wells for injection of antipyrogens in the rock mass and their necessary volume in order to retard coal oxidation process. The theoretical basis of the process of coal self-heating control and definition of the form of the heat emitting source is detected in the 1970s steady relationship between the emergence of pockets of heat and the formation on the surface of coal accumulations of changes in the potential appearing in the process of self-heating of quasi-stationary electric field. There-

* Corresponding author: vvi@kuzstu.ru

fore, the problem of determining the shape of the source by measuring the field potential on the surface of the accumulation is very important.

From the point of view of applied mathematics, the problem of determining the shape and dimension of the sources of fields having different nature (electrical, gravitational, thermal, magnetic or other) is the inverse and ill-posed problem of mathematical physics [1, 2]. Most of the existing theoretical methods and numerical algorithms are devoted mainly to the solution of linear inverse problems, where the unknown function is a co-factor of some well-known expressions characterizing the environment [3-6]. Much less papers are devoted to development of algorithms for solving non-linear inverse problems [7-11].

2 Results and discussions

2.1 Formulation and solution of the direct problem

Let us assume that the natural electric field (NEF) is generated by a horizontal cylindrical source of current with arbitrary but constant cross section. We assume as well that the length of the source and depth in the lower layer of two-layer homogeneous isotropic space with plane-parallel boundaries are known (Fig. 1).

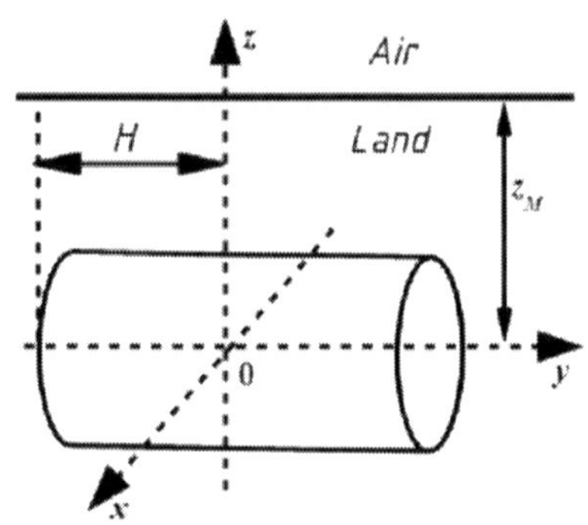

Fig. 1. The scheme of the containing space and cylindrical field source

In this case, the potential of the point source will be determined by the known formula

$$u_M = \frac{C}{\left[x_M^2 + y_M^2 + z_M^2\right]^{1/2}}, \quad (1)$$

where C – coefficient characterizing the force of the electric current source; M - the point of measurement of the field potential on the earth's surface.

To obtain the computing formula for the value of the cylinder potential, we integrate the expression (1) by the volume of cylinder V_P:

$$U_M = C \int_{V_P} \frac{dV_P}{\left[(x_P - x_M)^2 + (y_P - y_M)^2 + (z_P - z_M)^2\right]^{1/2}}, \quad (2)$$

where P – the point located in the cylinder volume.

We pass in the integral (2) to non-dimensional variables by the general formula $w \cdot z_M^{-1} = \overline{w}$, whereby w we understand coordinates of the point of the field M and the integration variables of cylinder P. Then, (2) will take the form

$$U_M = C \cdot z_M^2 \int_{\bar{V}_P} \frac{d\bar{V}_P}{\left[(\bar{x}_P - \bar{x}_M)^2 + (\bar{y}_P - \bar{y}_M)^2 + (\bar{z}_P - 1)^2\right]^{1/2}} = C \cdot z_M^2 \cdot \bar{U}_M \tag{3}$$

Then, to simplify the form of the formulas we will not draw the horizontal line at dimensionless values. Let's integrate (3) over the interval $y_P \in [-H; H]$:

$$U_M = \iint_{S_P} \left[\int_{-H}^{H} \frac{dy_P}{\left[(y_P - y_M)^2 + D\right]^{1/2}} \right] dS_P = \iint_{S_P} F(M,P)\, dS_P, \tag{4}$$

where $F(M,P) = \ln(Q_1) - \ln(Q_2)$, $Q_1 = H - y_M + \sqrt{(H - y_M)^2 + D}$, $Q_2 = -H - y_M + \sqrt{(H + y_M)^2 + D}$, $D = (x_P - x_M)^2 + (z_P - 1)^2$.

To calculate the double integral, we define the boundary of the region as a function $\rho(\varphi)$ in the polar coordinate system. The transition to the latter is performed in the standard way: $x_P = r\cos\varphi$, $y_P = r\sin\varphi$, where $\varphi \in [0; 2\pi]$ and $r \in [0; \rho(\varphi)]$ where $\rho(\varphi)$ is the periodic function.

Then (4) will assume the form $U_M = \iint_{S_P} F(M,P) dS_P = \int_0^{2\pi} \left[\int_0^{\rho(\varphi)} F(M,\varphi,r) \cdot r dr \right] d\varphi$.

Thus, the problem of determining the form of the source of the field from measurements of the potential on the day surface has been reduced to the Fredholm-Uryson integral equation of the first kind (hereinafter referred to as EI) with respect to the unknown function $\rho(\varphi)$:

$$\int_0^{2\pi} R(M, \rho(\varphi)) d\varphi = U^*, \tag{5}$$

where $R = \int_0^{\rho(\varphi)} F(M,\varphi,r) \cdot r dr$ – nonlinear kernel; U^* – values of potential measured on the day surface (NEF).

The formula (5) allows us to compute the potential of NEF of the cylinder in any point on the surface of earth. Further on, to solve the inverse problem of restoration of the cylinder form, we will use only the central axis of symmetry $y_M = 0$.

2.2 Solution of the inverse problem

For the definition of function $\rho(\varphi)$ we consider the problem of finding the minimum of regularized functional of A.N. Tikhonov [1, 2, 14].

$$\Phi(\rho) = W(x_M, \rho) + \alpha \cdot \Omega(\rho) = \int_c^d \left[A(x_M, \rho) - U^*(x_M) \right]^2 dx_M + \alpha \cdot \Omega(\rho) \to \min, \tag{6}$$

where $A(x_M, \rho) = \int_0^{2\pi} R(x_M, \rho) d\varphi$ – non-linear integral direct modeling operator (4); $\Omega(\rho) = \int_0^{2\pi} \left[\rho^2 + (\rho'_\varphi)^2\right] d\varphi$ – stabilizing second-order functional; $U^*(x_M)$ – the right part of the IE, which in practice is set experimentally, and for test problems is determined by solving a direct problem using formulas (4) with the addition of a random correction simulating inaccuracies in full-scale measurements; α is the regularization parameter.

We shall seek the minimum of the functional (6) by the method of conjugate gradients [13], whose general iterative scheme has the form

$$\rho^{(q+1)} = \rho^{(q)} + k^{(q)} \cdot I^{(q)}, \tag{7}$$

where $I^{(q)} = -\Phi'_{\rho} + \gamma^{(q)} \cdot I^{(q-1)}$, q – iteration index; k – minimization step; I – linear combination of the derivatives (gradients) in the previous steps; coefficient $\gamma^{(q)}$ is determined by one of the formulas [13], Φ'_{ρ} – the Frechet derivative of the functional (6).

Let's consider some aspects of numerical realization of formulas (5 – 7).

The integrand of the kernel R is a smooth monotonically increasing function on the interval $[0; \rho(\varphi)]$, where $\varphi \in [0; 2\pi]$ and $\rho(\varphi)$ – fixed consistent values. To calculate the integral in (5), we use the 5-point Gauss formula.

To calculate the integrals and the derivative in (6), we define a grid with respect to the variables, x_M, φ. The step of the grid with respect to the variable φ will be equal to $h_1 = \frac{2\pi}{n}$, the step by the variable x_M – $h_2 = \frac{d-c}{m}$. All integrals in (6) will be calculated by the Simpson formula. In the work [16] in case of the linear IE, the derivative ρ'_φ is approximated by the right difference formula. Numerical experiments have shown that with a relatively small (about 100) number of nodes $\varphi[j]$, the accuracy of such an approximation is low ($\approx 0{,}06$) and in case of the non-linear IE, the resulting errors have a significant effect on the computational process. Therefore, we approximate the derivative ρ'_φ by a central difference relation of the second order of accuracy $\rho'_\varphi(\varphi) = \frac{\rho(\varphi + h_1) - \rho(\varphi - h_1)}{2h_1}$. Due to the periodicity of the function $\rho(\varphi)$ this approximation can also be used for boundary points. In fact, let the function be defined on an interval $[0; \Lambda]$, where Λ is the period of the function $\rho(\varphi)$: $\rho(\varphi + \Lambda) = \rho(\varphi)$; then this equality will be true $\rho'_\varphi(0) = \rho'_\varphi(\Lambda)$, where $\rho'_\varphi(0) = \frac{\rho(h_1) - \rho(-h_1)}{2h_1} = \frac{\rho(h_1) - \rho(\Lambda - h_1)}{2h_1}$. Thus, the derivative on the borders of the grid will be equal to $\rho'_\varphi(0) = \rho'_\varphi(2\pi) = \frac{\rho[2] - \rho[K-1]}{2h_1}$, where the number of the array of values is indicated in square brackets $\rho[j]$; K- is the number of elements in the grid with respect to the variable φ. Note that if the region has a symmetry about the Ox axis, then this derivative will be zero.

Taking all the above into account, we obtain a discrete analogue of the expression (6) of the form

$$\Phi(\rho) = \sum_{i=1}^{N} \sigma[i] \cdot \left[\sum_{j=1}^{K} \mu[j] R(x_M[i], \rho[j]) - U^*(x_M[i])\right]^2 + \alpha \sum_{j=1}^{K} \mu[j] \cdot \left[\rho^2[j] + T^2[j]\right] \to \min, \tag{8}$$

where $T = \rho'_{\varphi}$ – the difference scheme indicated above; σ, μ - are the coefficients of the Simpson formula.

To implement the scheme (7), we calculate the derivative of (8) with respect to the variable $\rho[t],\ t \in [1; K]$:

$$\Phi'_{\rho[t]} = 2\mu[t] \times \left(\sum_{i=1}^{N} \left(\begin{array}{l} \sigma[i] \cdot R'_{\rho[t]}\left(x_M[i], \rho[t]\right) \times \\ \times \left[\begin{array}{l} \sum_{j=1}^{K} \mu[j] \cdot R\left(x_M[i], \rho[j]\right) - \\ -U^*\left(x_M[i]\right) \end{array} \right] \end{array} \right) + \alpha\left[\rho[t] + TT[t]\right] \right),$$

where the derivative $R'_{\rho[t]} = F(x_M[i], t, \rho[t])$, and expressions $TT[t]$ are implemented in the following way: $TT[t] = \dfrac{2\rho[t] - \rho[t-2] - \rho[t+2]}{4h_1^2}$ for $t \in [3, \ldots, K-2]$; for the rest four values the following expressions are used: $TT[1] = \dfrac{\rho[1] - \rho[3]}{4h_1^2}$; $TT[2] = \dfrac{2\rho[2] - \rho[K-1] - \rho[4]}{4h_1^2}$; $TT[K-1] = \dfrac{2\rho[K-1] - \rho[K-3] - \rho[2]}{4h_1^2}$; $TT[K] = \dfrac{\rho[K] - \rho[K-2]}{4h_1^2}$.

To find the step of minimizing k, we will use the "golden section" method for the objective function $\Phi(\rho + k \cdot I)$. Numerical experiments have shown that this function is unimodal on the interval $k \geq 0$. Since it is impossible in advance to determine the interval at which the minimum point is contained, we adhere to the following algorithm.

2.3 Algorithm

Step 1. We set the starting interval for the change of *k*: $a = 0; b = \varepsilon << 1$ and the error of the "golden section" method $\varepsilon_1 << \varepsilon$.

Step 2. Using the "golden section" method, we define the minimum point of the objective function – $k_{\min}$. We calculate the auxiliary value $\Delta = \dfrac{b}{k_{\min}} - 1 > 0$.

Step 3. We verify the truth of the inequality $\Delta < 0{,}1$. If "yes", then increase the right boundary of the minimum search interval $b = b + 1$ and go to step 2. If "no", then the value of $k_{\min}$ is the required minimum of the objective function.

The regularization parameter α will be determined, following the studies [13-15], by the following iterative way: $\alpha^{(q)} = f\left(\alpha^{(q-1)}, q\right)$, while in the first step of the conjugate gradient method we set $\alpha^{(0)} = 0$, and on the second step – $\alpha^{(1)} = \dfrac{W\left(x_M; \rho^{(1)}\right)}{\Omega\left(\rho^{(1)}\right)}$. In the works [9, 10], a saw-tooth effect of "blurring" an approximate solution is observed, depending on the choice of the regularization parameter α. In case of a linear IE, as these studies show, it is possible to select such a constant value of the parameter $\alpha^{(q)} = const$ that the indicated 'saw-tooth' formation will not be noticeable. In case of a nonlinear IE, such a value of the α parameter could not be chosen, so to neutralize the "blurring" after each step of the conjugate gradient method, the values were smoothed using the arithmetic mean of two neighboring values $\rho[j]$.

3 Implementation of a numerical

Considering the above solution method, it can be noted that it depends on the choice of the starting values and the patterns of variation of the following parameters $\gamma^{(q)}$, $\alpha^{(q)}$, ε, ε_1, $\rho^{(0)}$. Numerical experiments have shown that the choice of concrete realization of each of them has a significant effect on the efficiency of solving the inverse problem.

Here is an example of the implementation of a numerical algorithm.

Let us consider a test region in the form of a cylinder 100.0 m long and a cross-section in the form of an ellipse, the dimensionless equation of which in the polar coordinate system has the form

$$\rho(\varphi) = \frac{ab}{z_M \cdot \sqrt{a^2 \sin^2 \varphi + b^2 \cos^2 \varphi}}, \tag{9}$$

where we assume $z_M = 150.0$; $a = 130.0$; $b = 70.0$ m.

To simulate errors in full-scale measurements, we will add a random correction to the potential values calculated from formula (2) as follows: $U^* = U^{ex} + 0{,}5 \cdot (rand - 0{,}5)$, where *rand* are uniformly distributed numbers on the interval [0; 1], U^{ex} – the exact calculated values of the potential according to the formula (4).

The choice of the selected parameters will be the following: $\rho^{(0)} = 20{,}0 / z_M$; $\varepsilon = 10^{-2}$; $\varepsilon_1 = 10^{-6}$; $\gamma^{(q)} = \frac{\left(\Delta g - 2 \cdot \Delta\rho; G^{(q)}\right)}{\left(\Delta g; I^{(q-1)}\right)}$, $\Delta g = G^{(q)} - G^{(q-1)}$, $G = \Phi'_\rho$, $\Delta\rho = \rho^{(q)} - \rho^{(q-1)}$; $\alpha^{(q)} = \alpha^{(q-1)} \cdot 0{,}85^q$.

We will show in Figures 2 and 3 the graphic result of restoration of the cylinder cross-section form after 200 iterations.

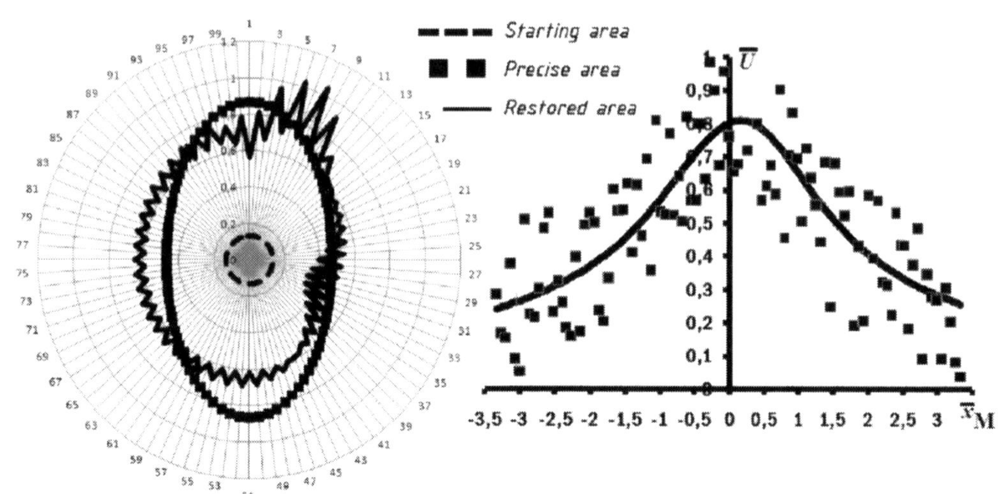

Fig. 2. Restoration of the area and value of the potential after 200 iterations and smoothing only after the first iteration of the conjugate gradient method

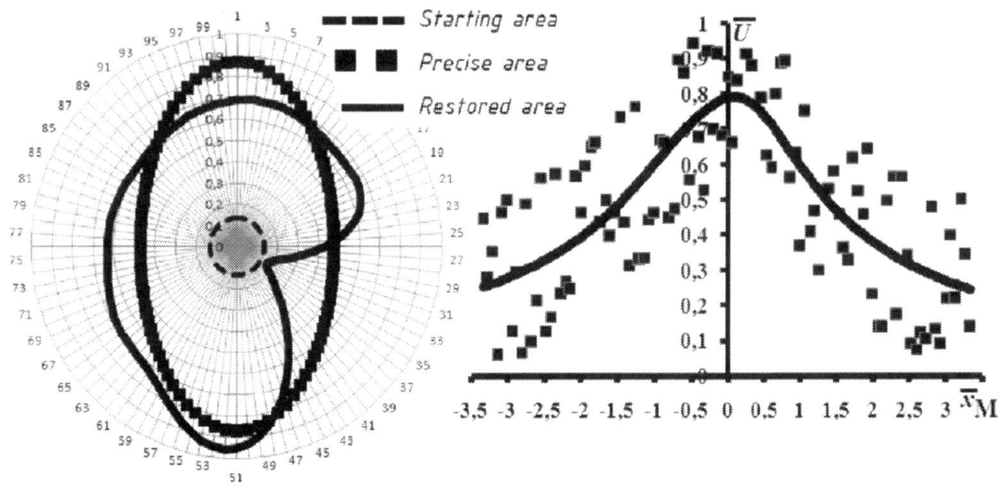

Fig. 3. Restoration of the area and value of the potential after 200 iterations and smoothing after each iteration of the conjugate gradient method

4 Conclusion

The numerical method of interpreting geophysical data of measurements of the potential of a quasi-stationary electric field is developed in the article in order to determine the form of the heat release source. Let's give practical recommendations on the use of the developed methodology.

A coordinate grid is constructed on the surface with a step of about 25-50 meters. In each node of this grid, the field potential is measured. The contour map is constructed, along which the elongation line of the proposed source is determined. More accurate measurements in the amount of at least 100 are taken along this line. These data will be used as initial data. If detailed measurements are impossible, then an approximate interpolation polynomial is constructed, along which the required values of the potential are determined. Further, with the help of the developed program, the form of the section of the cylindrical heat source is defined.

References

1. A.N. Tikhonov, V.B. Glasko, J. Comp. Math. Math. Phys., **4**, 564, (1964)
2. A.N. Tikhonov, V.B. Glasko, J. Comp. Math. Math. Phys., **5**, 463, (1965)
3. J.P. Butler, J.A. Reeds, S.E. Dawson, J. Numer. Anal., **18**, 546 (1981)
4. A. Tarantola, *The theory of inverse problem and methods for estimation of model parameters* (SIAM, 2005)
5. V.A. Morozov, *Regularization methods fo ill-posed problems* (CRC Press, 1993)
6. F.D.M. Neto, A.J.S. Neto, *An introduction to inverse problems with application* (Springer, 2013)
7. H.W. Engl, M. Hanke, A. Neubauer, *Regularization of inverse problems* (KAP, 1996)
8. C.R. Vogel, *Computational methods for inverse problems* (SIAM, 2002)
9. A.B. Bakushinsky, M.Yu. Kokurin, A. Smirnova, *Iterative methods for ill-posed problems* (Walter de Gruyter GmbH & Co. KG. Berlin/New York, 2011)

10. Y. Wang, A.G. Yagola, C. Yang, *Optimization and regularization for computational inverse problems and applications* (Springer, 2011)
11. R.C. Aster, B. Borchers, C.H. Thurber, *Parameter estimation and inverse problems* (Elsevier Academic Press, 2005)
12. G. Chavent, *Nonlinear least squares for inverse problems* (Springer, 2009)
13. W.W. Hager, Pacific J. Optim. **2,** 35, (2006)
14. M. S. Zhdanov, Geophysics, **76**, F77 (2011)
15. R. Kress, Inverse problems, **21**, 1207 (2005)

Numerical Simulation of Primary Roof Collapse in Production Workings

Inna Ermakova[1*], *Viktor* Klimov[2]

[1]T. F. Gorbachev Kuzbass State Technical University, 650000, Vesennaya st., 28, Kemerovo, Russia
[2]JSC "SUEK-Kuzbass", 652507, Russia, Leninsk-Kuznetskiy, Vasilieva street, 1.

Abstract. The study presents the formulation of the problem that is a numerical calculation of stress-strain state of the rock mass in the vicinity of the production working by the finite element method. The estimated area had a rectangular shape and included the coal seam, the roof and soil strata, and the production working. All the blocks of the estimated area being in the state of plain strain were taken as linearly deformable. The area of roof rock destruction was established by Mohr criterion. The analytical dependence of the height of primary roof strata collapse on the roof strata ultimate compression strength, the depth and the length of the mined area has been obtained. This result allows us to estimate the initial step of roof collapse.

1 Introduction

The estimation of the initial step of roof caving is one of the main tasks of rock mechanics during coal-face operations. Many studies were devoted to this issue using various approaches to the task formulation and methods of solution [1-7]. This paper uses the method of finite element numerical simulation of the stress-strain state of the rock strata in the vicinity of a coal face [8]. This approach is implemented in solving various problems of rock mechanics in underground mining of coal seams [9-15]. In this study, an assessment of the strength condition of the roof rocks strata is made, which made it possible to calculate the height of the primary roof collapse area.

2 Materials and methods

The object of the study is roof strata of production faces with different strength characteristics.

To address the issue of the primary roof strata caving, it is required to conduct analysis of their stress-strain and strength state and to set the dimensions of the disturbed areas. The finite element method was used to calculate the stress-strain state of the roof rock. The area of roof rock collapse was established by Mohr criterion.

* Corresponding author: eia.pm@kuzstu.ru

3 Results and discussions

We considered the stress-strain state of the rocks in the first period of coal-face operations before the collapse of roof rocks. The estimated rectangular area included the following blocks: coal seam 1; goaf 2; enclosing roof strata and seam floor strata 3 (Fig. 1). The boundary conditions and loads were taken in the following form. At the upper boundary of the computation area AB there is no load (daylight area). At the lower boundary CD remote from the pillar, there are no normal displacements (the border is fixed). The vertical boundaries BC and AD, quite remote from the mined-out area, are not influenced by the coal face and have no horizontal movements.

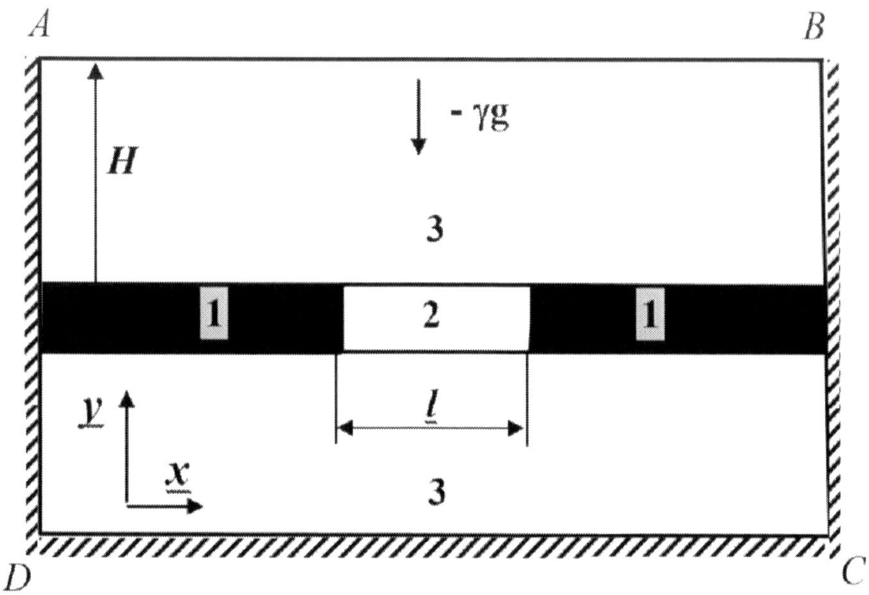

Fig. 1. The estimated area: 1 – mined-out area; 2 – coal seam; 3 – enclosing rock strata

The coal seam is loaded with body force – $\gamma = 14000$ N/m^3, E = 10^9 N/m^2 , ν = 0,25.

The roof rock strata is loaded with the body force – $\gamma = 25000$ N/m^3 which corresponds to the weight of the enclosing strata; Young modulus E = 10^{10} N/m^2, Poisson ratio ν = 0,25.

The floor strata is also loaded with the body force – $\gamma = 25000$ N/m^3 corresponding to the weight of the enclosing strata: Young modulus E = $4{\cdot}10^9$ N /m^2, Poisson ratio ν = 0,25.

The limits of the enclosing strata compressive strength ranged from 20 to 60 MPa, and ratio of rock tensile strength limits to compressive strength was 1:10. The length of the production face ranged from 20 to 50 m, the depth of mining varied from 200 to 500 m, and the seam thickness -from 2 to 5 m. Total 80 numerical experiments were carried out. The stress-strain state of the computational area was described by linear deformable blocks behavior. All the blocks were loaded by their own weight. The problem is solved by finite element method (FEM) using the licensed program ELCUT.

Fig. 2 shows the distribution of vertical and horizontal stresses exceeding their initial value, i.e. high rock pressure zone (light background) and pressure release zone (dark background).

The distribution of increased and decreased stresses agrees in quality with the available views on the situation of the pressure release zones in the blocks of roof and floor rock strata during coal-face operations.

a) b)

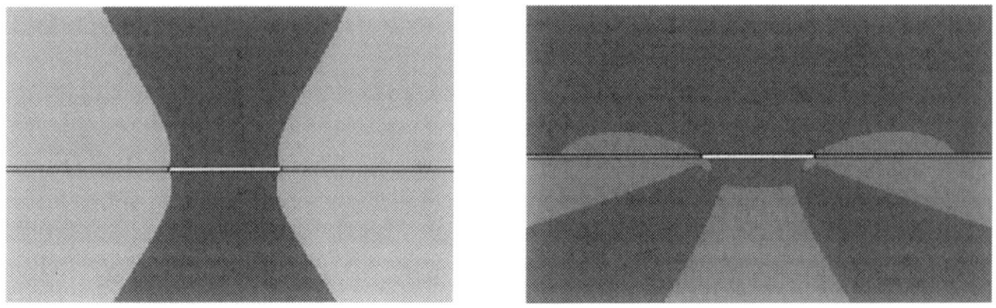

Figure 2. Distribution of increased and decreased vertical (a) and horizontal (b) stresses in the vicinity of the mine working

To assess the strength (cavability) of roof rock, the strength criterion of Mohr-Coulomb was used that was assigned in the form of principal stresses and ultimate strength as follows:

$$\sigma_1 - \frac{\sigma_t}{\sigma_s} \cdot \sigma_2 < \sigma_t, \tag{1}$$

where σ_1 and σ_2 – the highest and the lowest principal stress; σ_s, σ_t – rock strength and tensile limits.

The strength parameters are expressed via the angle of internal friction ϕ and engagement k as follows

$$\sigma_s = 2k\cos\phi/(1-\sin\phi); \ \sigma_t = 2k\cos\phi/(1+\sin\phi), \tag{2}$$

where k – engagement; ϕ – angle of internal friction.

Fig. 3 shows the example of calculation of the roof caving zones above the mined-out areas having seam thickness 3 m, depth of mining 500 m, length of production face 50 m. At limit rock compression strength σ_s = 60 MPa roof collapses in the area where $\sigma_{MO} > 6$ MPa (dark contour). In case σ_s = 20 MPa, roof collapses in the area where $\sigma_{MO} > 2$ MPa (the lightest contour).

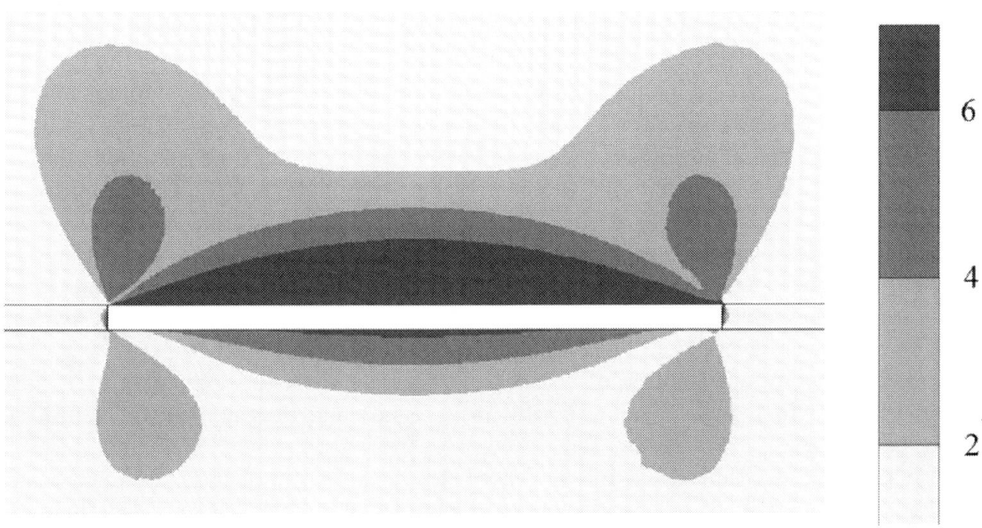

Figure 3. Areas of roof strata collapse in the seam 3 m thick at the mining depth 500 m and production face length 50 m

The area of roof strata collapse was characterized by height (thickness) of this area at the level of the middle of the mine working. The performed numerical experiments have established the dependence of the change in the height of the collapsed area on the parameters of the problem. In accordance with the theory of similarity the sought dependence was taken in the following form:

$$\frac{h}{l} = F\left(\frac{m}{l}; \frac{\sigma_t}{\gamma H}\right), \tag{3}$$

where h – thickness of collapsed rock; l – length of the production face; m – seam thickness; H – depth of mining operations; σ_t, γ – limit tensile strength and the bulk weight of roof strata respectively.

The statistical processing of the results of numerical experiments revealed the following. The dependence of the height of collapsed rock on the seam thickness is not significant within the considered seam thickness 2 to 5 m. This circumstance is due to the fact that the reviewed formulation of the problem does not take into account the scale factor, i.e. the dependence of coal deformation and strength characteristics on the coal seam thickness. The rest of the problem parameters are significant and determine the dependence of the relative height of the collapse on the considered parameters as shown in Fig. 4.

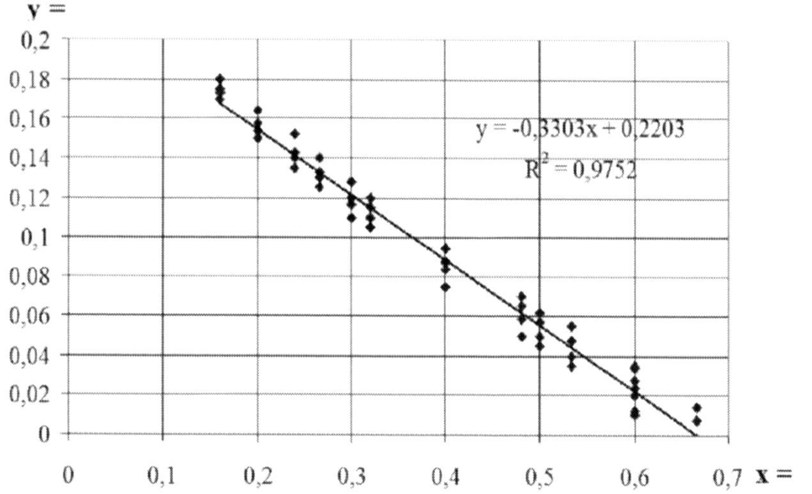

Figure 4. Results of numerical experiments and regression dependence of the relative height of roof collapse on the parameters of the problem

The regression equation allows us to express the absolute height of the roof collapse with reference to the rock tensile strength, depth and length of the production face in the following formula:

$$h = (-0,3303 \cdot \sigma_t / \gamma H + 0,2203) \cdot l. \tag{4}$$

In the resulting formula, the rock tensile strength can be replaced by hardness ratio, as at the ratio σ_t / σ_p = 1:10 the numerical value of tensile strength coincides with the hardness ratio.

The resulting expression for the height of collapsed rock can also be used to determine the initial step of roof caving, which in this case is equal to the length of the production face.

4 Conclusions

1. The numerical simulation of the strength state and the roof rock collapse at the initial stage of coal-face operations was made on the basis of the finite element method.

2. The dependence of the height of the initially caved roof layer on the length of the production face, roof strata strength and mining depth of coal seam was obtained.

3. It is proposed to use the suggested method of estimation of roof strata rock mechanics at the initial stage of longwall mining in the mines of JSC "SUEK-Kuzbass".

The authors express their gratitude to JSC "SUEK-Kuzbass" for the financial support of the given study.

References

1. Bieniawski Z. T., Int. J.Rock Mech. Min. Sci., **6**, 323 (1969)
2. Bieniawski Z., Mir, 254 (1990)
3. Zienkiewicz, O.C.; Taylor, R.L.; Zhu, J.Z. *The Finite Element Method: Its Basis and Fundamentals (*7th Edition*)*, (Butterworth-Heinemann, 2013)
4. B.H.G. Brady, E.T. Brown, *Rock Mechanics for underground mining* (Springer science, 2005)
5. J. Jaeger, N.G. Cook, R. Zimmerman, *Fundamentals of rock mechanics* (Wiley Blackwell, 2007)
6. M. Pastor, C. Tamaonini, *Numerical modeling in geomechanics* (Kogan Page science, 2004)
7. J.A. Hudson, J.P. Harrison, *Engineering rock mechanics. An introduction to the principles* (Pergamon press, 2000)
8. O. Fusao, *Computer methods and recent advances in geomechanics* (CRC Press, 2014)
9. M. Fremond, F. Maceri, *Mechanical modeling and computational issues in civil engineering* (Springer, 2005)
10. V.A. Gogolin, Y.A. Ryzhkov, Soviet Mining Science, **13**, 17 (1977)
11. V.M. Seryakov, J. Min. Sci., **45**, 420 (2009)
12. M.V. Kurlenya, V.E. Mirenkov, V.A. Shutov, J. Min. Sci. **50**, 1001 (2014)
13. V.A. Gogolin, Y.A. Ryzhkov, Soviet Mining Science **17**, 156 (1981)
14. Y.A. Ryzhkov, Y.V. Lesin, V.A. Gogolin, N.V. Karpenko, J. Min. Sci. **32**, 188 (1996)

The Study of Processes of Electrochemical Treatment of Soils at the Pilot Test Site

Sergey Prostov[1], *Eugene* Shabanov [1], *Alexandr* Shadrin[2]

[1]T. F. Gorbachev Kuzbass State Technical University, 28 Vesennaya st., Kemerovo, Russia
[2] Kemerovo State University, 650000, 6 Krasnaya st., Kemerovo, Russian Federation

Abstract. The techniques of field testing of controlled electrochemical cleaning of soil from oil pollution and the experimental setup are described. The results of engineering-geological surveys of groundmass artificially saturated with petroleum products (used oil, gasoline) are presented. The processes of formation of zones of drying, accumulation of petroleum products in electroosmotic transport, changes in the structure of soil as a result of the phase transformation and dissolution of petroleum products are shown in detail. The phenomenon was established of oil coagulation in the pores of the soil when subjected to electric treatment leading to an increase of particle size of sand and clay components of the soil and contributing to its deactivation. The efficiency of electrochemical treatment of tight sand and clay soils was proved experimentally. Experiments confirmed the effectiveness of the integrated electrophysical control of soil properties changes in the process of treatment, and in combination with the previously established correlation dependencies – the possibility of the average assessment of the contamination factor change.

1 Introduction

Intensification and globalization of human impact on the surrounding natural environment is a characteristic feature of our time, causing its negative consequence in terms of the environment pollution [1-4]. Petroleum products (fuel, oil and lubricants) [5] are among the most common pollutants of soil in Kuzbass.

Functionally, the methods of purification from oil pollution fall into four main types: physical, chemical, physicochemical [6, 7], and biochemical [8-10]. There are three main approaches to combating pollutants: immediate removal of oil by means of its extraction from the ground; inactivation (detoxification) of oil directly in the solid mass, localization of oil products by creating of a protective screen around the anomaly preventing the further spread of oil spills [11, 12].

When processing low permeability clayey soils, a very promising method of electrochemical cleaning is used based on the integrated impact by the active substance and electric current. The main physical processes in electro treatment are as follows:

- electrothermal stimulation resulting in the change of the pollutant phase state;
- dissolution of the contaminant into light fractions followed by evacuation.

* Corresponding author: psm.kem@mail.ru

When dispersive soil is exposed to electric current, its microstructure and its physical characteristics [13] change to a certain degree, including the conductive and dielectric properties [14, 15].

To identify patterns of changes of physical properties of soil in the electrochemical purification from oil pollutions, there were studies conducted in the KuzSTU laboratory which yielded the following main results:

- on a one-dimensional physical model, the phenomenon was established of the coagulation of oil in the soil pores during electro treatment leading to an increase in grain size of sand and clay components and correlated with the changes in the electrical resistance of the soil;
- the three-dimensional model confirmed that when electric current passed, it impacted the oil in the pores, and the oil moved in a solid cohesive state leading to an increase in the specific electrical resistance (SER) of the soil throughout the treatment area and to reduction in humidity.

In more detail the results of the laboratory studies are described in [16].

For industrial research of the processes occurring in the soil during the electrochemical cleaning from oil pollution the experimental plot of clay massif was prepared. The plan of the pilot area, the scheme of connection of electrodes and electric power installation are shown in Fig. 1.

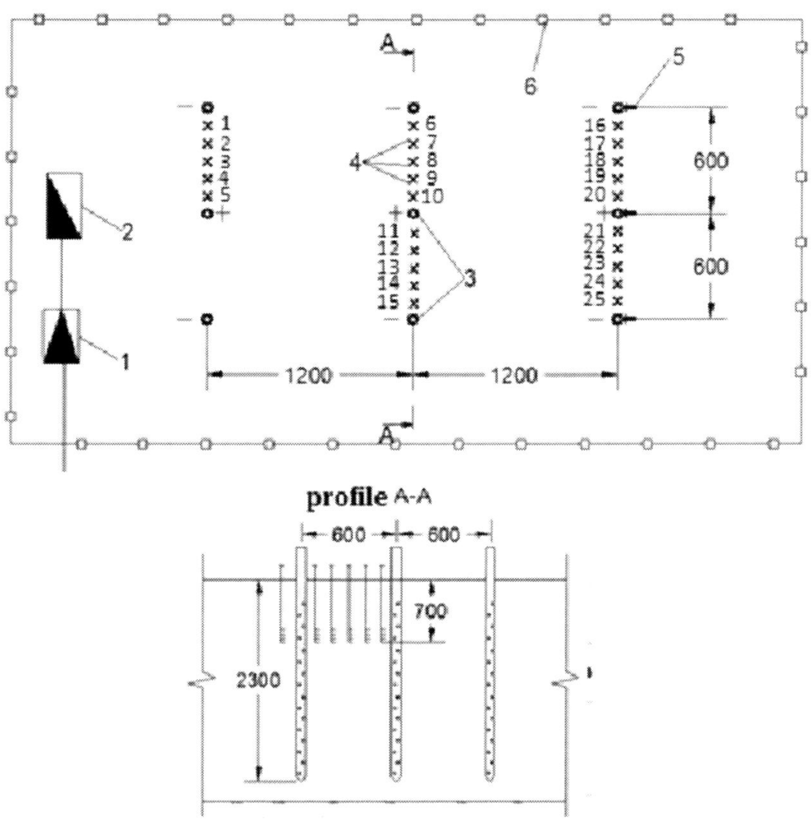

Fig.1. Diagram of the installation for field experiment: 1 – power transformer; 2 – rectifier-inverter; 3 – electrodes-injectors; 4 – microsensors of electrical resistance; 5 – feed of Hexane-n; 6 – temporary protective fencing; – electrodes-injectors; +, - polarity of the electrodes-injectors; x – SER microsensors

Near the place intended for cleaning there was installed an electric powertrain (power transformer TC-40, adjustable rectifier-inverter) and tanks for storage of polluting substances (waste oil and gasoline AI-80).

Direct current was supplied by cable type KG 4x25-0,66 from the installation to the electrodes. The range of the current intensity and voltages was from 0.1 up to 40 A and up to 360 V respectively. Perforated steel pipes with a diameter of 57 mm were used as electrodes-injectors. For electrochemical cleaning in series scheme of electrodes connection was used, with the polarity in the form: "cathode-anode-cathode". The distance between rows of electrodes was 1200 mm, and 600 mm between the opposite polarity electrodes.

The clay massif being in its natural state was artificially contaminated with oil for the experiment when the pollutant was injected in the blast holes with a 12 mm diameter to the depth of 700 mm (the cavity volume approximately 80 ml). The holes were placed every 80-100 mm uniformly in all directions. Oil contaminant from the cavities was absorbed into the soil, penetrating deeper and to the sides, thus having formed a contaminated area to a minimum depth of 2 metres, which corresponded to the parameters of soil contamination under natural conditions.

The main parameters of the experimental installation are the following:

- the total current consumption I·t – 3650 A·h;
- processing modes – electroosmotic mode for uncontaminated massif, electroosmotic mode for contaminated massif, electroosmotic with the dilution of the contaminant by the solvent "Hexane-n";
- polluting petroleum – gasoline A80; waste motor oil Shell Helix.

At all stages of the experimental research a continuous physical and technical control was carried out of the processes in the area of electro treatment including engineering-geological surveys and geophysical monitoring, both local (by SER microsensors) and integrated (electric sounding and ground penetrating radar). The information about the physical and technical methods of soil control is given in Table.1.

Table. 1. Methods of physical and technical control of processes in the treatment zone

Methods of control	**Equipment**	**Tasks to be solved, parameters to be defined**
Engineering-geological surveys	Field laboratory PL-2: a set of boxes to determine moisture, electronic balance, oven, a set of sieves	Determination of physical and mechanical properties and granulometric composition of the soil at different points in time
The SER microsensors	A logging tool KP-2 Sensors-microsondes	Monitoring of changes in the true resistivity of the soil in increments of 0.1-0.15 m
Electric sounding	A logging tool KP-2 rod electrodes	Determination of the effective electrical resistivity of the soil at various stages of electro treatment
GPR	GPR OKO-2	Diagnosing anomalies by the depth and in plan by the radargram

After the start of the clay massif electro treatment, measurements of all main characteristics were carried out twice a day, at the time of measurement the electric power unit was disconnected. At the end of electro treatment of the soil, sampling in the most characteristic areas (clean area, area contaminated with oil and area contaminated with gasoline) was performed, and physical and mechanical properties and granulometric composition of the soil were determined.

The total time of soil electrical treatment was over 168 hrs at current consumption per one pair of electrodes-injectors more than 600 A·h. The first area of clean clay was processed by a method of electro-osmosis without the addition of liquids in the electrodes in order to compare the obtained data with the results of treatment of contaminated clayey massif. The second and third areas contaminated with used oil and gasoline were treated by electric current without the use of active substance – solvent. The fourth and fifth areas were also contaminated with used oil and gasoline and treated by electric current and active substance – solvent Hexane. The solvent was supplied to the electrode-anode at the time point = 48 and 60 hrs in a volume of 4 liters at a time.

2 Results and discussion

To determine the physical properties of soils, a series of samples was selected in these areas prior to contamination of the clay massif, after its pollution by the oil products, and after completion of electrochemical treatment.

The results of the study of physical properties of soils before and after treatment of the massif are given in Table 2, and the granulometric composition of soil – in Table 3.

Table 2. Change of physical parameters of the soil as a result of electro treatment

Sample number	Characteristics (before contamination/before treatment/after treatment)		
	Moisture, %	Density in wet condition, g/sm^3	Density in dry condition, g/sm^3
# 1, clean clay	19.7/19.7/14.6	1.82/1.82/1.73	1.52/1.52/1.51
# 2, contamination with oil	19.7/20.1/17.1	1.82/1.80/1.71	1.52/1.5/1.46
# 3, contamination with gasoline	19.7/19.8/15.5	1.82/1.82/1.71	1.52/1.52/1.48
# 4, contamination with oil using solvent	19.7/20.1/15.7	1.82/1.91/1.79	1.52/1.59/1.55
# 5, contamination with gasoline using solvent	19.7/19.8/14.8	1.82/1.87/1.77	1.52/1.56/1,54

Table 3. Change of granulometric composition of soil due to the electro treatment

Sample number	The contents of fractions (in mm), % (before/after treatment)				
	2	0.5	0.25	0.1	<0.1
# 1, clean clay	2/2.2	17.6/18.5	15.5/17.4	18.8/21.2	46.1/40.7
# 2, contamination with oil	2/3	17.6/20.2	15.5/21.3	18.8/27.9	46.1/27.6
# 3, contamination with gasoline	2/2.5	17.6/19.7	15.5/18.5	18.8/24.4	46.1/34.9
# 4, contamination with oil, using solvent	2/1.8	17.6/17.8	15.5/16.6	18.8/29.4	46.1/34.4
# 5, contamination with gasoline using solvent	2/2.8	17.6/19.3	15.5/19.2	18.8/27.7	46.1/31

From the data of Table 2 it follows that in the soil both clean and contaminated with various petroleum products, after the electro treatment chemical binding of moisture and oil takes place, resulting in an overall decrease in soil moisture by 3 to 6 % and decrease in its density by 3 to 7% in both wet and dry conditions. At pollution of soil with petroleum products there was an increase in moisture, as the content of pore fluid at the experimental soil plot increased. In comparison with the results of laboratory studies, the changes in moisture and density are less significant, because of the influence of the atmosphere and surrounding rock with natural moisture.

In the granulometric composition of the soil the significant changes were identified:

- in the clean soil a small decrease in the content of fractions < 0.1 mm, and the increase of fractions of 0.5 mm, 0.25 mm and 0.1 mm were identified;

- in the soil polluted with used oil there is a significant increase in the number of particles fractions of 0.1 mm, a small increase in the number of particles of fractions 0.25 mm, 0.5 mm, and thus there is a significant decrease in particles fraction < 0.1 mm due to the process of oil coagulation, adhesion of small particles, with a likely transition of oil from a liquid to a solid state, which is considered less toxic and environmentally harmful, and in electrochemical processing with a solvent, this process was less intensive than in processing by electric current without solvent;

- in the gasoline contaminated soil there also occurs a significant increase in the number of particles fractions of 0.1 mm, a small increase in the number of particles of fractions 0.25 mm, 0.5 mm and a significant decrease in particles fraction < 0.1 mm as in treatment with solvent and without solvent; increase in the number of larger particles in the soil polluted with used oil is more intensive than when contaminated with gasoline, as used oil is more dense and viscous than gasoline, it coats the particles of the soil to a greater degree during pollution and electro processing.

To determine objective quantitative indicators of the condition of the soil massif in electrochemical processing of oil-contaminated soil, the measurements were carried out of the effective resistivity ρ_k by the method of electrical sounding from the earth's surface.

Graphically, the results of the measurements are presented in Fig.2.

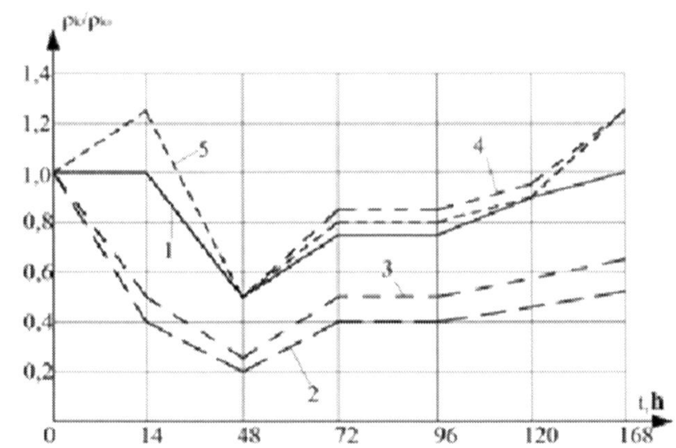

Fig. 2 The graph of changing of the relative effective electrical resistance in the electric sounding: 1 – clean clay massif; 2 – massif contaminated with used oil; 3 –massif contaminated with gasoline; 4 – massif contaminated with used oil processed with solvent, 5 – massif contaminated with gasoline processed with solvent.

In accordance with the theoretical background [17] outlined earlier, for transfer from the effective electrical resistance ρ_k to the contamination factor K_{cf} four dependencies were used: linear, logarithmic, parabolic and exponential. The equations shown in Table 4 were used to determine the coefficient of pollution in every moment of time measuring the effective electrical resistance ρ_k.

Table 4. The dependence of the rate of pollution K_{cf} on the relative change of the effective resistivity of the soil in the sample

Equation	**Constant *a* when polluted by:**		**Primary K_{cf0}, %**	**Final K_{cff}, %**	
	oil	**gasoline**		**oil**	**gasoline**
$k = a_1 \cdot \ln\left(\frac{\rho_k}{\rho_{k0}}\right)$	a_1=2.8	a_1=3,5	7	3,64	4,55
$k = a_2 \cdot \ln\left(\frac{\rho_k}{\rho_{k0}}\right)$	a_2=7,64	a_2=10,1		2	2,65
$k = a_3 \cdot \left(\frac{\rho_k}{\rho_{k0}}\right)^2$	a_3=1,12	a_3=1,75		1,9	2,96
$k = a_4 \cdot \exp\left(\frac{\rho_k}{\rho_{k0}}\right)$	a_4=0,575	a_4=0,95		2,11	3,49

The initial rate of pollution K_{cf0} (the ratio of volume of contaminant to volume of contaminated soil) was determined by the actual volume of the oil filled into the soil making K_{cf} = 7 %. Constants a_1-a_4 are included in the estimated equation.

Fig. 3 shows the change of coefficient of pollution in time. The graph shows that the rate of pollution over time decreases by the moment of processing time – t = 48 h, then increases until t = 72 h, and then almost does not change.

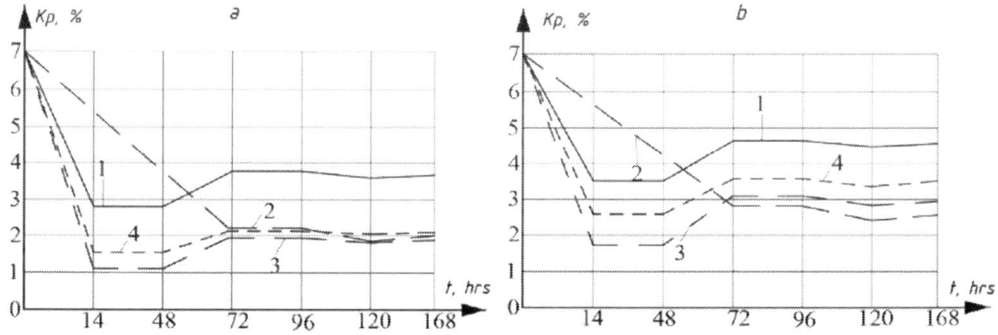

Fig. 3. Graph of changes in the concentration of used oil (a) and gasoline (b) in electro-processing depending on time t: 1 – linear relationship; 2 – logarithmic plot; 3 – parabolic dependence; 4 – exponential dependence.

The time of stabilization of value K_p is the criterion of the cessation of electro treatment, because the further electricity consumption does not lead to noticeable changes in the physical properties of soils.

3 Conclusion

Analysis of the results of engineering geological survey and geophysical soundings allowed us to make the following conclusions:

- when electric current passes and impacts oil products in the soil pores, they move in a solid cohesive state, and there is a decrease in moisture, which corresponds to the results of laboratory studies on one-dimensional and three-dimensional models, however, the ranges of these changes are lower, due to an additional supply of moisture from the atmosphere and surrounding soil with natural humidity;
- as a result of the electro thermal impact of direct current on the contaminated massif, the increase in the size of the particles and their adhesion take place, which leads to a change of granulometric composition of clay soil in the direction of increasing of the content of larger fractions and this increase is more intense in the contamination with more viscous oil products;
- in the processing of soil by electric current, the decrease in soil moisture and increase in the percentage of coarse particles lead to a decrease in soil density, both in wet and dry conditions;
- the process of phase conversion of oil-containing product in the soil pores due to electric and thermal coagulation occurs for any form of oil-pollutant, and when the solvent is used, the process of phase conversion is less intensive, however, this increases the volume of liquefied petroleum accumulated in the anode region;
- electrical sensing is provided by the integral control of changes of the soil properties during processing and in combination with the previously established correlation dependencies – the average assessment of changes in the rate of pollution. Electro treatment for 48-72 h reduces the degree of contamination of the soil by 2-3 times, while further dilution

of viscous oil by introducing into the electrodes of the solvent in the range of the ratio of pollution to 7% did not lead to noticeable positive effect.

References

1. Egorova K.V., Kayukova G.P., Romanov G.V., Naumova R.P, *Proceedings 7th International Conference 'Heavy Crude and Tar Sands'*, 1 (1998)
2. Krapivsky E.I., Nekuchaev V.O., Beljaev A.E., Charnetsky A.D., *Proceedings 67th European Association of Geoscientists and Engineers, EAGE Conference and Exhibition*, 2753 (2005)
3. Park S.-W., Kim K.-J., Baek K., Lee J.-Y., Yang J.-S., Separation Science and Technology, **45:12**, 1988 (2010)
4. Zhao L., Zhu N., Xie J., Oh K., Kimochi Y., *3rd International Conference on Bioinformatics and Biomedical Engineering, iCBBE 2009*, 51 (2009)
5. Abramov O.V., Abramov V.O., Myasnikov S.K., Mullakaev M.S., Theoretical Foundations of Chemical Engineering, **43:4**, 504(2009)
6. Korolev, V.A., Romanyukha O.V., *Book of Abstracts of 6th Symposium on Electrokinetic Remediation (EREM) 2007*, 119 (2007)
7. Archegova I. B., Khabibullina F. M., Shubakov A. A., Contemporary Problems of Ecology, 6, 548 (2012)
8. Trusei I. V., Ozerskii A. Yu., Ladygina V. P., Contemporary Problems of Ecology, **1**, 22 (2009)
9. Kolesnikov S. I., Zharkova M. G., Kazeev K. Sh., Russian Journal of Ecology, **3**, 157 (2014)
10. Efremova V. A., Dabakh E. V., Kondakova L. V., Contemporary Problems of Ecology, **5**, 561 (2013)
11. Korzhov Yu. V., Lapshina E. D., Khoroshev D. I. Contemporary Problems of Ecology, **3**, 292 (2010)
12. Vorobeichik E. L., Kozlov M. V., Russian Journal of Ecology, **2**, 89 (2012)
13. Seredina V. P., Sadykov M. E., Contemporary Problems of Ecology, **5**, 457 (2011)
14. Korolev, V. A., Romanyukha O .V., Abyzova A. M., Journal of Environmental Scince and Health. Part A: Toxic. Hazardous Substances and Environmental Engineering, **43:8**, 876 (2008)
15. Prostov S. M., Khyamyalyainen V. A., Bakhaeva S. P., Journal of Mining Science, **4**, 349 (2006)
16. E. Shabanov, S. Prostov, *Proceedings of the 8th Russian-Chinese Symposium "Coal in the 21st Century: Mining, Processing, Safety"*, 175 (2016)
17. S.M. Prostov, M.V. Gucal, E.A. Shabanov, *Taishan Academic Forum – Project on Mine Disaster Prevention and Control*, 433 (Atlantis Press, 2014)

The study of stress-strain state of stabilized layered soil foundations

Mikhail V. Sokolov[1], *Sergey M.* Prostov[1*], and *Viktor S. Zykov*[2]

[1]T.F. Gorbachev Kuzbass State Technical University, Vesennyaya St. 28, Kemerovo city, Russia
[2] VNIMI, Sovetckiy St. 63a, Kemerovo city, Russia

Abstract. Herein presented are the results of modeling and analysis of stress-strain state of layered inhomogeneous foundation soil when it is stabilised by injection to different depths. Produced qualitative and quantitative analysis of the components of the field of isolines of stresses, strains, stress concentration and the difference between the strain at the boundary of different elastic horizontal layers. Recommendations are given for the location of stabilised zones in relation to the border of different elastic layers. In particular, it found that stabilization of soil within the weak layer is inappropriate, since it practically provides no increase in the stability of the soil foundation, and when performing stabilisation of soil foundations, it is recommended to place the lower border of the stabilisation zone below the border of a stronger layer, at this the distribution of stresses and strains occurs more evenly, and load-bearing capacity of this layer is used to the maximum.

1 Introduction

In construction of mining buildings, civil and industrial buildings there is a frequent situation when soil foundations are made of two or more soils with different physical and mechanical properties while weak strata of the massif are located on top of more solid soil [1, 2]. The inhomogeneity of rock strata properties, in particular, its layered structure, considerably affects the distribution of stresses and strains [3-6]. The study of geomechanical aspects of the problem will improve the accuracy of prediction of soil foundations stability including foundations fixed by injection methods, as well as reduce the risks of technological accidents at the facilities of the mining and construction industries.

When solving such problems, the method of numerical simulation becomes increasingly common, its main provisions are described in several papers [7-10]. The concepts of the method are implemented in the framework of "Alterra" software for geotechnical calculations made by company "InzhStroyProekt". The basic principles of modeling and analysis of computer calculations are set out in the work [11].

The base model for the formation of a database and further analysis (Fig. 1) is implemented for the two-step strip reinforced concrete foundation loaded with a vertical concentrated force $P = 154.7$kN, the height $d_f = 3.0$m and the following mechanical parameters: modulus of deformation $E_f = 2000$MPa; Poisson's ratio $\nu_f = 0.15$; density $\rho_f = 2500$kg/m^3.

* Corresponding author: psm.kem@mail.ru

Soil massif consists of the upper (layer #1) and lower (layer #2) layers with specified physical and mechanical properties in accordance with examples of calculation [12-14] and SP 23.13330.2011 (Table 1), layers thickness $H_1 = H_2 = 10.5$m, the width of the model $B_m = 31.2$m.

Formation of the database for the analysis was carried out by setting the increment of physical and mechanical properties of the massif soils in the range of relations $E_2/E_1 = 1...5$ (Table 1) with various schemes of ground stabilisation by injection according to the recommendations adopted in the works [15, 16]:

- at location of zones of stabilisation within the boundaries of the weak layer (Fig.1,*a*);
- when stabilisation zones rest upon the border of layers (Fig.1,*b*);
- when the stabilisation zones are jammed in the strong layer (Fig.1,*c*).

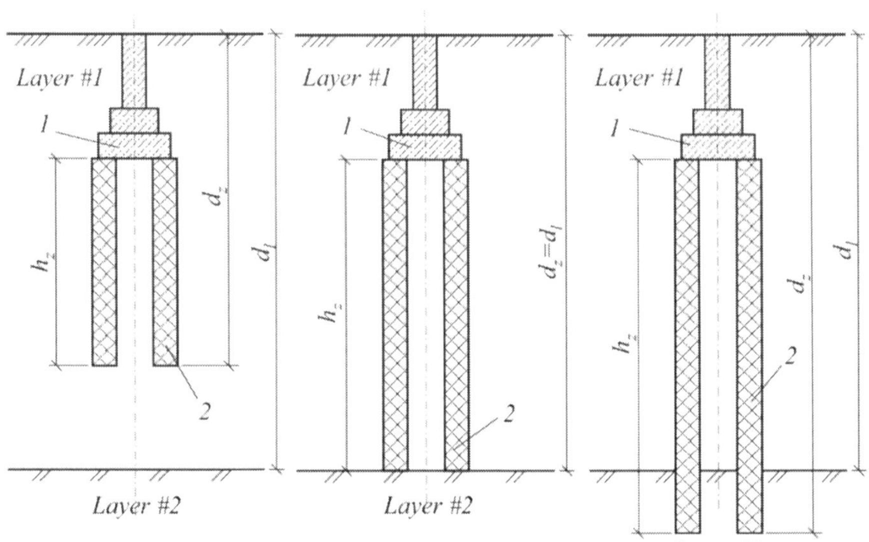

Fig. 1. Schemes of an arrangement of zones of stabilisation within the borders of the weak layer (*a*), when rested upon (*b*) and jammed in the strong layer (*c*): 1 - foundation; 2 - stabilisation zone

Table 1. Physical and mechanical properties of soils

Properties	Layer #1	Layer #2				
		relationship of E_2/E_1				
		1	2	3	4	5
Modulus of deformation E_i, MPa	5	5	10	15	20	25
Poisson's ratio ν_i	0.36	0.36	0.35	0.34	0.33	0.32
Average density ρ_i, kg/m^3	1750	1750	1800	1850	1900	1950
Angle of internal friction φ_i, degrees	17	17	18	19	20	21
Adhesion C_i, kPa	5	5	10	20	30	40

2 Results and discussion

Part of the results of simulation in the form of isolines fields of vertical stress σ_z is shown in Fig. 2.

Vertical stresses σ_z in the natural (non-bound) layered soil massif (Fig.2,a) are distributed symmetrically and have the area of concentration under the foundation, as well as characteristic changes on the border of layers. When forming the stabilization zones according

 DOI: 10.1051/e3sconf/20171501015

to the schemes (Fig. 1), additional areas of stress concentration are formed under the bases of these zones, as well as along their lateral surfaces.

In analyzing the character of the distribution of vertical stress σ_z from the depth the massif along the axis of symmetry of the foundation z (fig. 3) it was established that the formation of the field of vertical stress concentration is observed at z = 7.5…12.5m depending on the location of the stabilisation zones and relations E_2/E_1. In particular, the greatest effect from stabilisation is achieved at scheme #3 (Fig. 3,d), because in this case in the stabilised massif on the border of layers there are no surges of stress σ_z due to their increase in the weak layer in the range z = 7.0…10.0m, at the same time, the major part of the pressure is redistributed onto the strong layer.

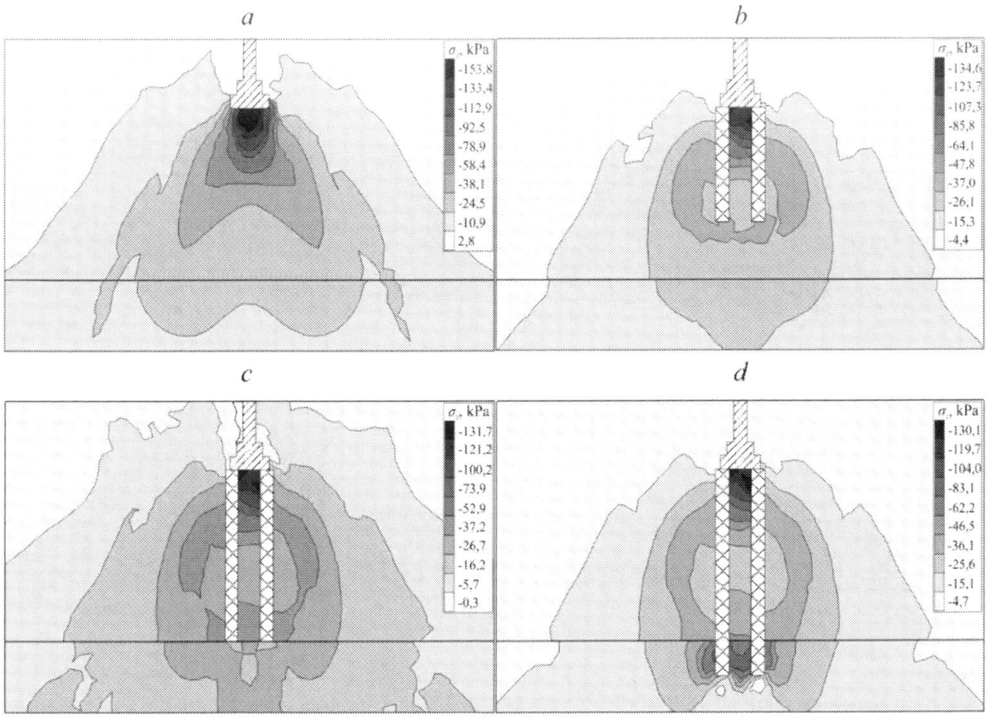

Fig. 2. Fields of isolines of vertical stress distribution in the natural soil foundation (*a*), at the location of zones of stabilisation within the borders of the weak layer (b), when resting (*c*) and jamming (*d*) in the strong layer

Dependences of vertical stresses σ_z on the depth of the model along the outer side surface of fixing zones z_s are shown in Fig. 4.

The graphs show that at the location of stabilization zones in the weak layer (scheme #1) the stresses decrease monotonically with increasing coordinate z_s in the absence of local concentrations on the border of the layers. When the stabilised zone rests on this border (scheme #2) and goes deeper in the strong layer (scheme #3), the stress state of massif is fundamentally different: stresses σ_z are more intensely redistributed to the lower layer.

For the numerical evaluation of the stress state of the soil massif, we introduce the integral criterion - stress concentration factor k_i, determined by the formula

$$k_i = \frac{\sigma_{\max}}{\sigma_{lost}},$$

where σ_{max} – the maximum stresses in the layer boundaries, kPa; σ_{lost} – stress in the pristine part of the layer under consideration (at the border of the model), kPa.

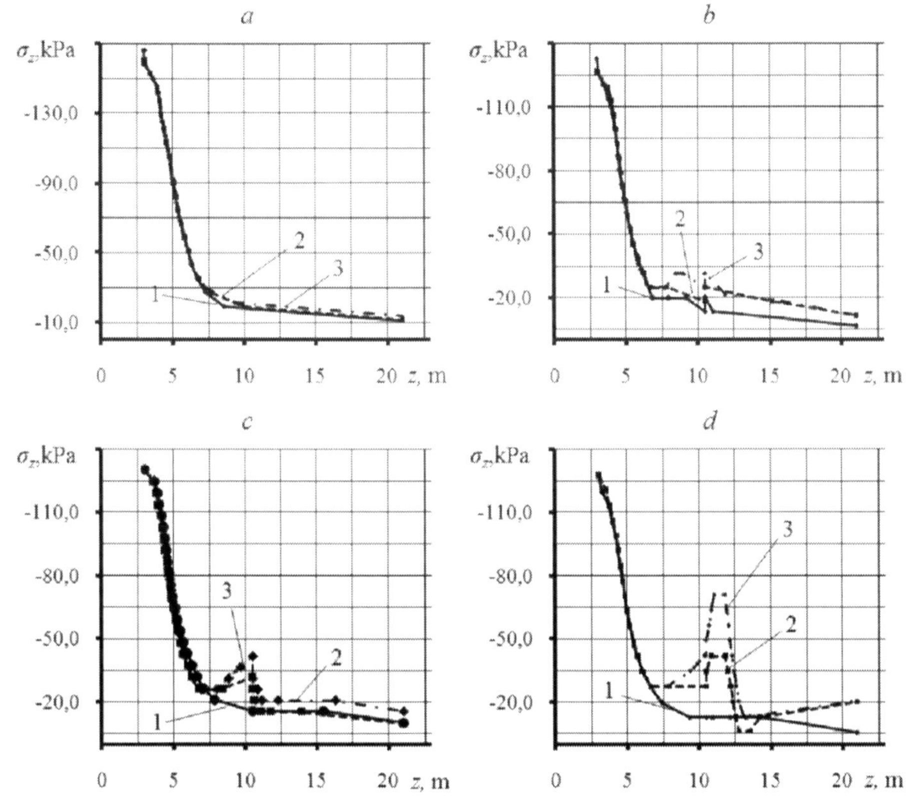

Fig. 3. The dependence of the stress σ_z on the depth along the axis of the model z in the natural (*a*) and stabilised massif as per schemes #1 (*b*), #2 (*c*) and #3 (*d*) and relation E_2/E_1 equal to: 1 – E_2/E_1 =1; 2 – E_2/E_1 =3; 3 – E_2/E_1 =5

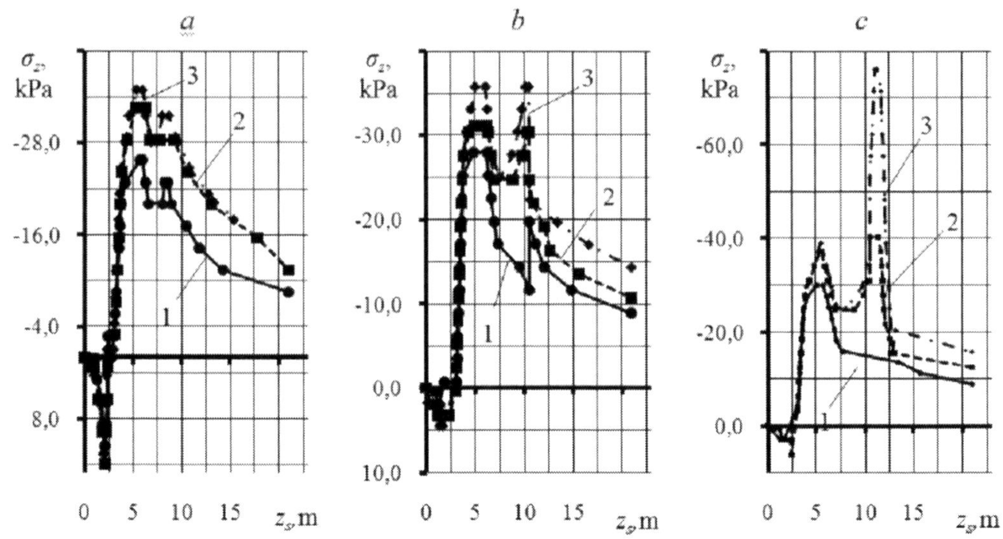

Fig. 4. Dependence of the stress σ_z on the depth of model z_s along the border of the stabilization zone when it is stabilized as per schemes #1, #2 and #3, and relation of properties E_2/E_1, equal to: 1 – E_2/E_1 =1; 2 – E_2/E_1 =3; 3 – E_2/E_1 =5

The dependences of relation of the stress concentration factors k_1/k_2 in the weak and strong layers on relation E_2/E_1 are shown in Fig. 5.

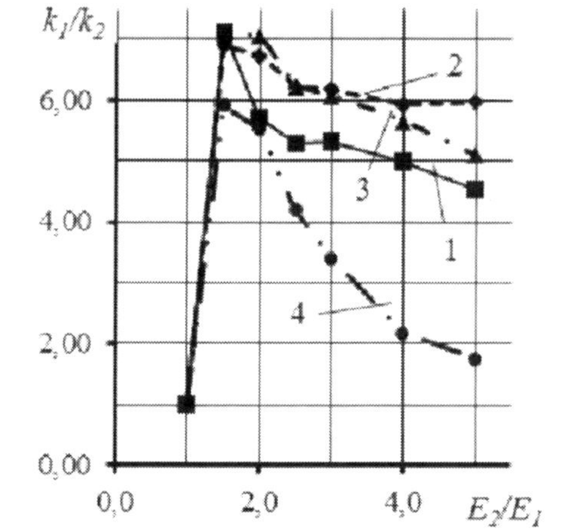

Fig. 5. Dependencies of relations k_1/k_2 on the relation E_2/E_1: 1 - natural massif; 2 – when stabilized as per scheme #1; 3 – scheme #2; 4 – scheme #3

According to the presented data, when relation E_2/E_1 increases in the range of $E_2/E_1 = 1.0...2.5$, there occurs redistribution of stresses between the layers with formation of the area of maximum stresses. At stabilization as per the schemes #1 and #2, there is some smoothing of the graphs in the range of $E_2/E_1 = 2.0...5.0$. The greatest effect is achieved at stabilization as per scheme #3, where there is a sharp decline in k_1/k_2 to values close to 1. Thus, the location of the base of the stabilisation zones within the boundaries of the strong layer contributes to a more favorable distribution of stresses in the layered soil massif.

The distribution pattern of vertical deformations ε_z in the soil massif is shown by the fields of isolines shown in Fig. 6.

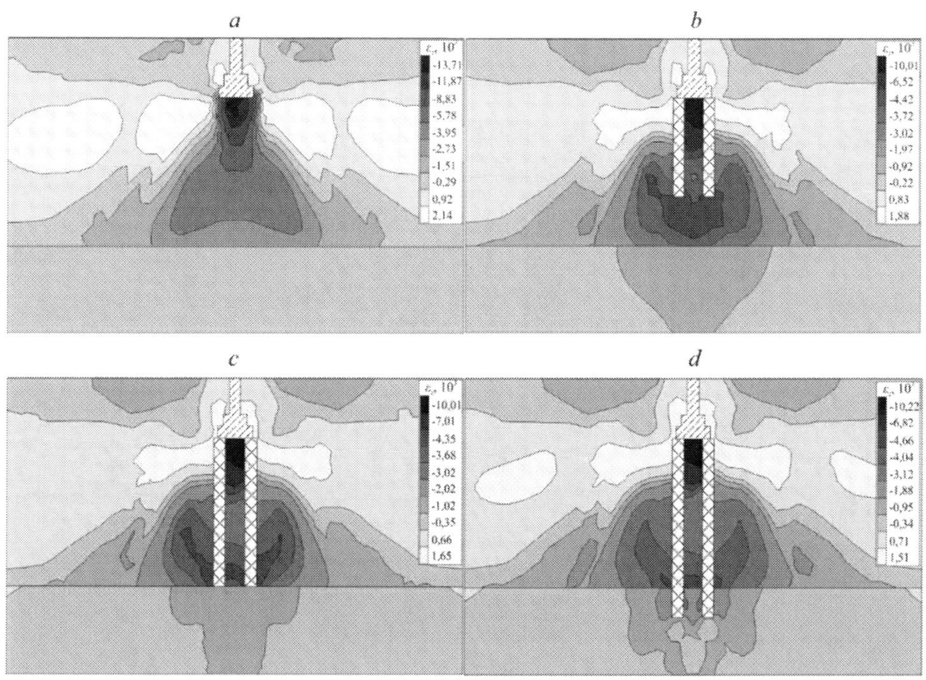

Fig. 6. Isoline fields of distribution of vertical deformations in the natural soil foundation (*a*), at the location of stabilization zones within the borders of the weak layer (*b*), at resting upon (*c*) and jamming (*d*) in the strong layer

Distribution of deformations ε_z occurs symmetrically, the areas of the maximum values stress are formed in specific parts of the massif: under the foundation at a depth of z = 3.0...6.0m 3.5-4.5m wide, and when stabilized - in inter-areas space; under the lower ends of the stabilisation zones at an average width 5.5-6.5m, in some cases 3.2-4.0m.

Along the axis z_s by the side face of the stabilisation zone (Fig. 7) there are significant lateral deformations of the stabilisation zones, since at the range of the depth z_s = 3.5-7.5m (in some cases up to 10.5m) a change in the character of the deformation into compression is observed, as well as significant increase of ε_z.

Fig. 8 shows the dependence of the deformation difference at the border of layers $\Delta = \varepsilon_{z2} - \varepsilon_{z1}$ on the relation E_2/E_1.

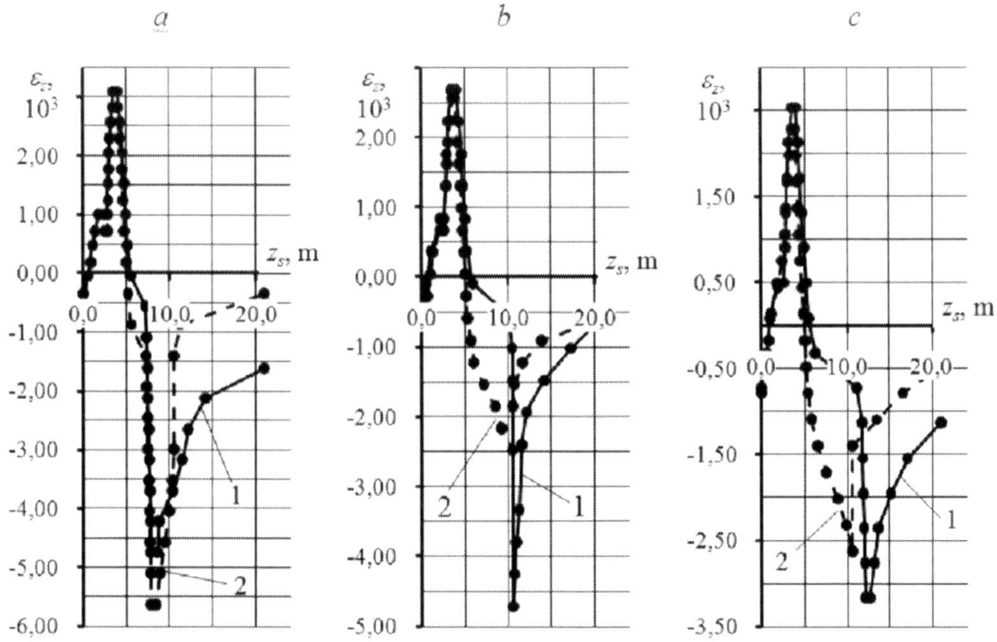

Fig. 7. Dependence of the vertical elastic deformations ε_z on the depth of the model z_s in the stabilised soil foundation according to schemes #1 (*a*), #2 (*b*) and #3 (*c*), at relation E_2/E_1, equal to: 1 – E_2/E_1 =1; 2 – E_2/E_1 =5

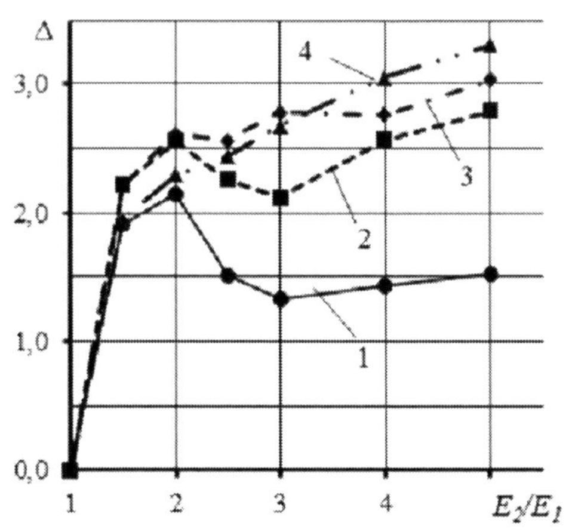

Fig. 8. Dependence of changes in the difference of the vertical deformation Δ on the border of layers on the relation E_2/E_1: 1 - natural massif; 2 – when stabilized as per schemes #1; 3 – schemes #2; 4 – scheme #3

The presented curves show that the highest values Δ are seen at stabilization as per scheme #3, at the same time the value Δ increases practically monotonously, without areas of local maximum values, in contrast to the other stabilization schemes. The lowest level of Δ is recorded in the natural massif, however, in the interval of relations E_2/E_1 =1.5...2.5 maximum values exceeding the value Δ = 1.5 at the ratio E_2/E_1 =5 are observed.

3 Conclusion

As a result of the analysis of the stress-strain state of the layered stabilised soil massif, the following has been found:

- stabilization of soil within the weak layer is inappropriate, since it practically provides no increase in the stability of the soil foundation;
- when performing stabilization of soil foundations, it is recommended to place the lower border of the stabilization zone below the border of a stronger layer, at this the distribution of stresses and strains occurs more evenly, and load-bearing capacity of this layer is used to the maximum.

References

1. S. V. Usanov, V. I. Ruchkin, O. D. Zheltysheva, Journal of Mining Science, **51**, 724 (2015)
2. O. B. Bocharov, V. Ya. Rudyak, A. V. Seryakov, Journal of Mining Science, **50**, 235 (2014)
3. A. B. Ponomarev, E. N. Sychkina, Soil Mechanics and Foundation Engineering, **51**, 111 (2014)
4. V. M. Seryakov, Journal of Mining Science, **45**, 420 (2009)
5. Z. G. Ter-Martirosyan, A. Yu. Mirnyi. Soil Mechanics and Foundation Engineering. **50**, 223 (2014)
6. V. I. Sheinin, E. P. Sarana, V. N. Soboleva, D. V. Gurevich, L. B. Katsnel'son, Soil Mechanics and Foundation Engineering, **53**, 304 (2016)
7. V. E. Mirenkov, A. A. Krasnovsky. Journal of Mining Science, **49**, 567 (2013)
8. S. G. Bezvolev. Soil Mechanics and Foundation Engineering, **45**, 77 (2008)
9. V. V. Orekhov. Soil Mechanics and Foundation Engineering. **53**, 1 (2016)
10. V.M. Ulitskii, M.A. Luchkin, Soil Mechanics and Foundation Engineering. **43**, 197 (2006)
11. Sergei M. Prostov, Mikhail V. Sokolov and Andrey V. Pokatilov. International Journal of Applied Engineering Research. **10**, 45297 (2015)
12. A. A. Tyrymov. Journal of Mining Science. **48**, 812 (2012)
13. Mikhail Sokolov, Sergey Prostov. Advances in Engineering Research. **92**, 206 (2016)
14. Sergei M. Prostov, Mikhail V. Sokolov. *Taishan Academic Forum - Project on Mine Disaster Prevention and Control,* 350 (2014)
15. M. N. Ibragimov. Soil Mechanics and Foundation Engineering. **46**, 17 (2009)
16. S. M. Prostov, N. Yu. Nikulin. Journal of Mining Science. **51**, 908 (2015)

Justification of parameters and technology of retaining prism filling to eliminate landslide

Valery Balakhnin[1,*], *Olga* Veretennikova[1], *Roman* Pobegaylo[1], and *Ekaterina* Mezina[1]

[1]Prokopyevsk Mining and Design Institute, 650036 Kemerovo, Lenina Avenue 90/5, Russian Federation

Abstract. The process of dumping on the surface mining is one of the most problematic in terms of industrial safety. The problem is compounded by the fact that the external dumps are often pilled on unprepared base, composed of clay. Therefore, such a situation occurs when a set of external factors leads to considerable deformations of the dumps. In particular, a few years ago on one of Kuzbass open pits there was a major landslide. According to the studies made, the reasons for the landslide were irregularities in the process of stacking, intensive snow melting and continuous heavy rains. All this together led to the deformation of the slope of the lower tiers of the outer dump and to the landslides. This article describes the measures of eliminating the danger zone, designed to prevent further landslides and to ensure safe mining works on external dump. The parameters of grounded prism are discussed and the technology of its piling is offered.

1 Materials and Methods

The formation of dump was carried out in two stages along the ravine's slope with the dip angle of 4-6°. The lower tier of the dump is mostly formed with Quaternary sediments – the loam. During the formation of external dump of overburden some technological errors were made. It was found that the lower tier of the external waste dump was formed with the rocks of Quaternary sediments due to lack of bedrock, which did not provide the desired safety factor during the period of intensive snow melting and long-lasted strong rains. It led to the deformation with landslides from the north-east slope of the lower tier. On the front of the dump the deformation processes were observed, expressed in the form of sliding rock mass on the contact of the slope with a dump's base (plantar landslide). Landslide body was formed at an angle close to the angle of the dump base (6-10°). From the foregoing we concluded that the main causes of deformation processes are the following:

1) The reduction of structural characteristics and loss of load bearing capacity by the dump's base as a consequence. The main reason for the loss of load bearing capacity by dump's base rock is the increase in its water-content under the influence of groundwater, with unloading of the ravine's thalweg [1-3].

2) Dip angle of the base (1-5 °) congruent to the slope angle and insufficient mechanical strength characteristics of the rock of the dump's base composed of soft Quaternary sediments [4-5].

* Corresponding author: v.v.balakhnin@pgpi.su

2 Results and Discussion

The criteria for assessing the slopes' sustainability are complying with the conditions of equilibrium of driving and resisting forces, expressed by the G.L. Fisenko formula (1):

$$n = \frac{\sum F_{rf}}{\sum F_{df}},$$

where n – the safety factor; $\sum F_{rf}$ – resisting forces; $\sum F_{df}$ – driving forces.

It is known that for assessing the stability of dumps the safety factor should be at least 1.2. At lower value of the safety factor the landslides may occur when the dump is standing without updating more than 1 year. It is possible to reduce the impact of adverse factors and to ensure overburden dump's slopes stability with implementing organizational and technical measures including drainage of the dump's base and loading the lower tier by retaining prism. Calculations have shown that in this case the safety factor is $n \geq 1.2$.

Figure 1 shows the general view of the retaining prism formation along the entire length of the landslide, which at the same time works as drainage trench. Because there is the urgent necessity in rain and melting snow water removing, the retaining prism has been piled of pre-distructed bedrock with high filtration rate. To calculate the parameters of this prism the cross sections of the landslide and the underlying surface were built. After that driving and resisting forces and retaining prism parameters were defined.

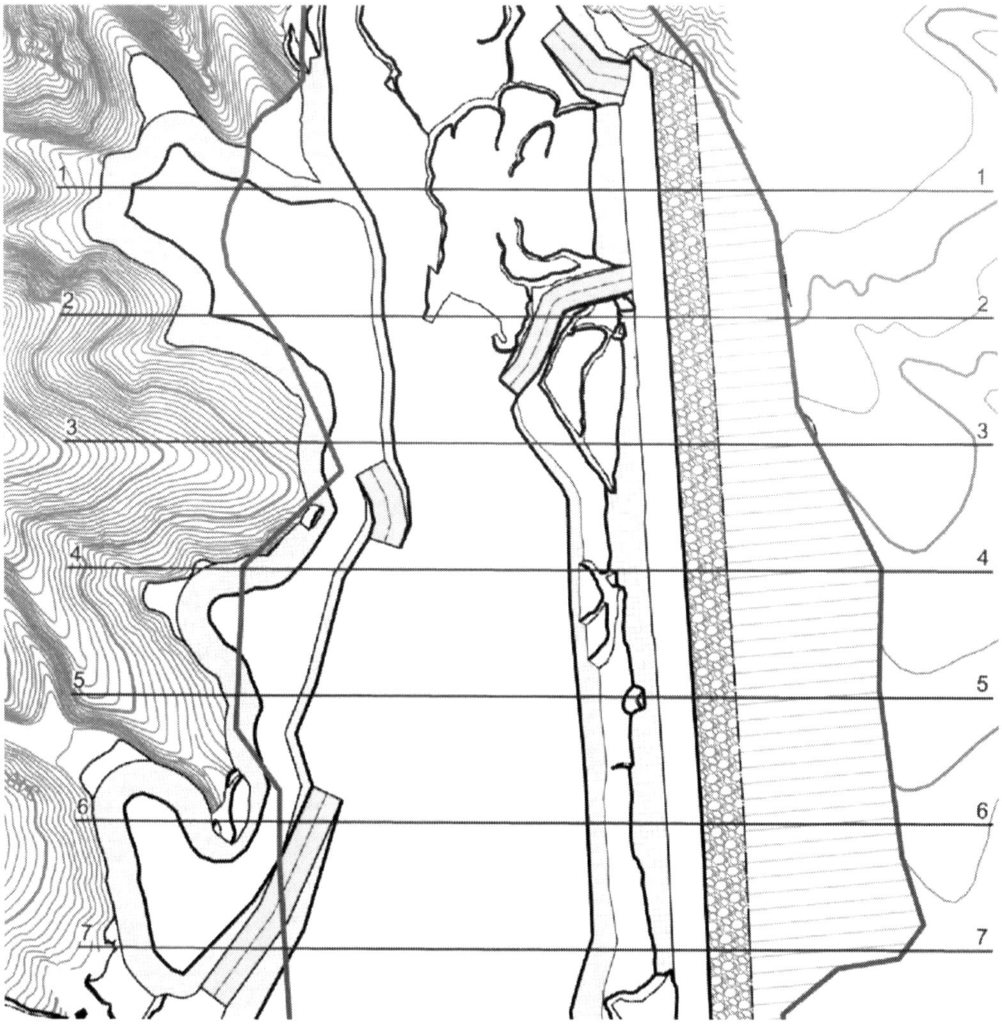

Fig. 1. The scheme of retaining prism formation

Fig. 2 and 3 show examples of cross-sectional areas of the landslide with the calculated parameters of the prism. For the use of hard rock for retaining prism constructing first it is necessary to load the landslide body in vehicles. For this purpose, dragline excavators and dump trucks with carrying are used. The conduit represents a working trench which is conducted by two layers.

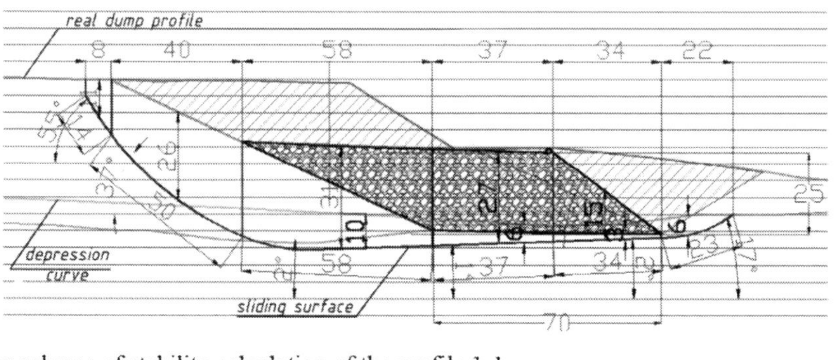

Fig. 2. The scheme of stability calculation of the profile 1-1

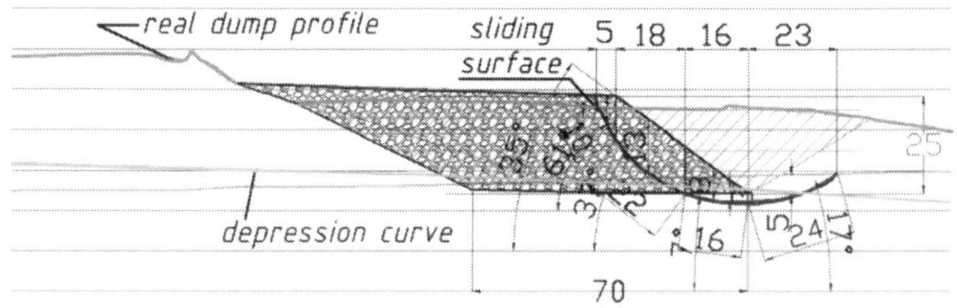

Fig. 3. The scheme of stability calculation of the profile 2-2

The technological schemes of working the first and the second layers are presented on Fig. 4 and 5. On the figures: R_{di}, R_{du} – respectively, digging and dumping dragline's radius.

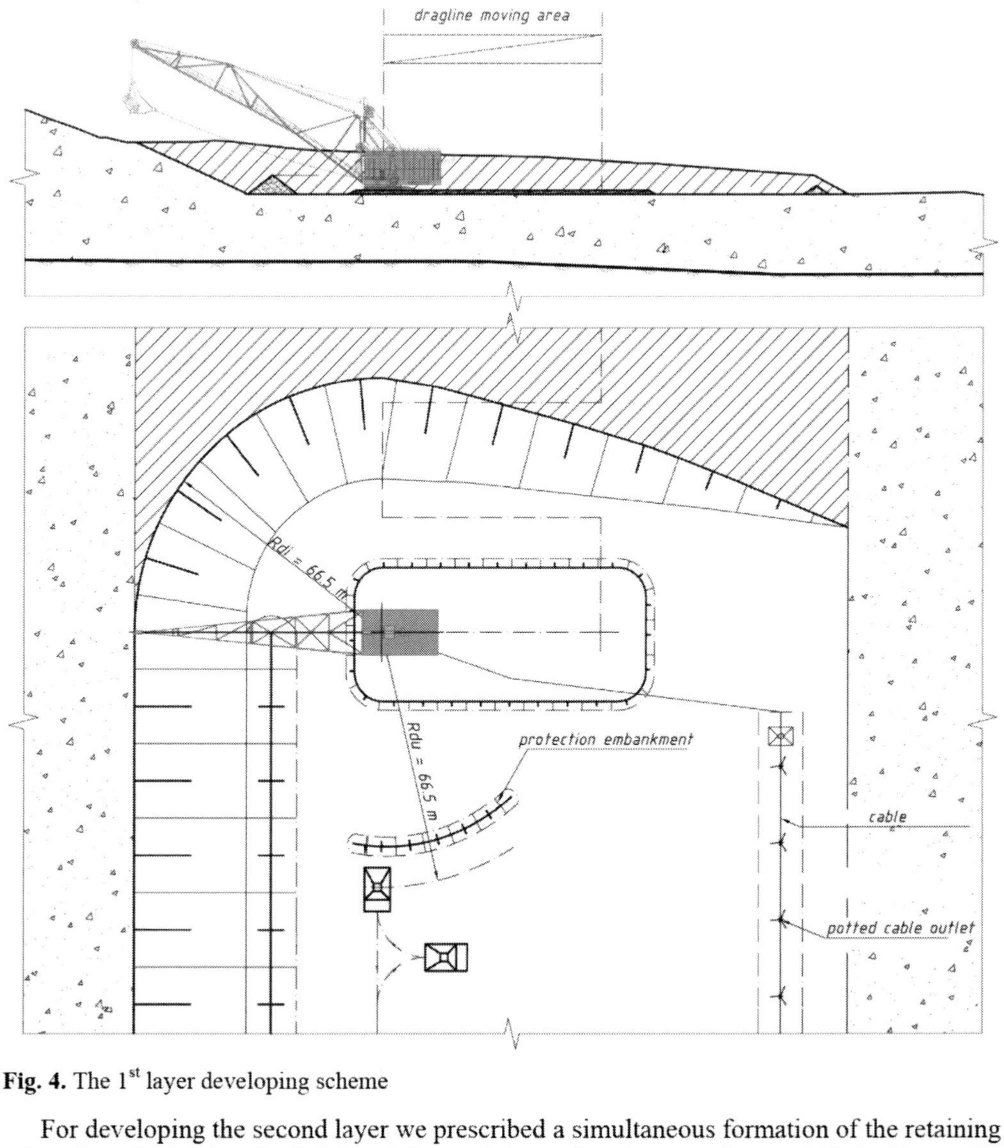

Fig. 4. The 1st layer developing scheme

For developing the second layer we prescribed a simultaneous formation of the retaining prism from bedrock. The height of the retaining prism is 30 m, base width is 70 m. The slope angle is adopted as 20°.

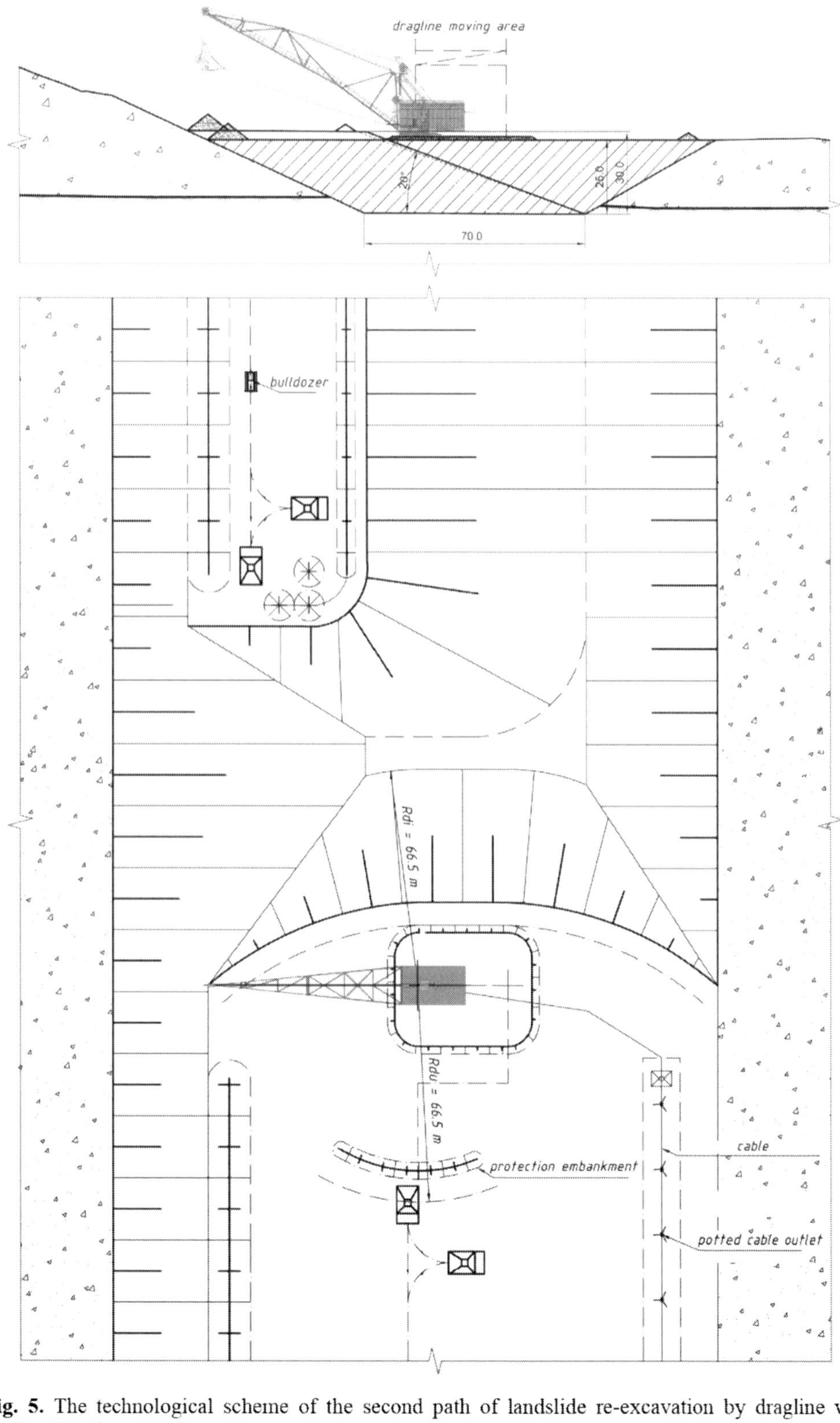

Fig. 5. The technological scheme of the second path of landslide re-excavation by dragline with loading into dump truck

3 Conclusion

The main reason for the deformation processes are the following:

- the structural characteristics reduction and, as a consequence, the loss of bearing capacity by the rocks of the bump's base which consist of 80% of Quaternary sediments (loams) due to increase in their water content under the influence of groundwater unloading of the ravine's thalweg.
- the angle of the slope, close to the dump's base dip angle (1-5 °) and soft mechanical characteristics of the base rocks composed of soft Quaternary sediments (mainly loam).

In the areas where landslides of overburden are in decay stage, their further loading (formation overlying dump's tier) without performing the necessary engineering measures (in particular, the drainage performance of the dump's base and piling the retaining prism from the bedrock), can trigger a relapse of overburden deformation processes and a violation of the stability of dump's slopes.

Made research allows developing complex technological schemes the usage of which ensures the landslide excavation with simultaneous formation of retaining prism. This will provide the necessary safety factor, as well as the safety of mining operations.

References

1. I. Vukotic, V. Kecojevic, W. Zhang, Q. Cai, Int. J. Min. Sci. Tech., **23(6)**, 901 (2013)
2. M.M. Bereznyak, A.V. Kalinin and V.G. Pronoza, Soviet Mining **6(6)**, 638 (1970)
3. B. Lokhanov, Y.A. Zakharov, M.M. Bereznyak, A.V. Kalinin, Sov. Min., **3(5)**, 523 (1967)
4. V. dos Santos, M. O. Medeiros, A. S. D. dos Anjos, C. A. Martínez-Huitle, D. R. da Silva, Chemical Engineering Transactions, **41**, 6 (2014)
5. R.Z. Velten, D.C. Lima, M.P.F. Fontes, C.A.B. Carvalho, Soils and Rocks, **32**, 11 (2012)

Use of Deep Peat-Processing Products for Hydrophobic Modification of Gypsum Binder

Oleg Misnikov[1*], *Valeriy* Ivanov[1]

[1]Tver State Technical University, A. Nikitin Street, 22, 170026, Tver, Russia

Abstract. The article deals with the issue of gypsum binder quality reduction during its storage and transportation. The study provides the main methods to protect gypsum from unauthorized exposure to moisture and water vapor. The author proposes hydrophobic modification as a perspective method for the preservation of gypsum activity and its water absorption reduction. The substantiation of cement hydrophobization with the bitumen released during peat thermolysis is provided. The author proposes to use this method in the technology of gypsum binder production. The basic idea is to combine the hydrophobization process with the calcination of calcium sulfate dehydrate. This is facilitated by temperature ranges used for dehydration of natural gypsum and the initial stage of thermal decomposition of the organic matter of peat. The author defined experimentally an optimal concentration of the organic component in gypsum binder. After adding 0.5-1% of the peat additive, the gypsum plaster preserved its grade strength and increased its storage time without caking and hydration, also under adverse conditions. The proposed method is adapted to the technological processes presently used in the production and doesn't require changing any equipment. The price of mineral raw materials and semi-finished products of peat are approximately equal which reduces the probability of increasing of the cost of hydrophobically modified gypsum binder.

1 Introduction

Gypsum binder (gypsum hemihydrate), used in the construction industry, is a fine powder obtained from natural gypsum dihydrate by heat treatment in the temperature range 140-190°C. Gypsum dihydrate refers to quick-setting and quick-hardening miner binding materials. It is widely used for plastering, manufacturing of gypsum concrete, gypsum building products, castings, forms, and as an additive to other binding materials [1].

One of the main issues of using gypsum binder is its activity reduction over time. This is due to its hygroscopicity. The intensity of the absorption of water from the atmosphere depends on the humidity and ambient temperature. This leads to a gradual transformation of gypsum hemihydrate into the dihydrate one during storage, and reduces its activity upon

* Corresponding author: oleg.misnikov@gmail.com

setting and hardening. The longest storage life is considered to be three months – during this period the activity of gypsum binder reduces on average by 30-50%. Moreover, the high water absorption of gypsum-based building materials results in low frost resistance. It does not allow them to be used for exterior works without additional treatment by traditional hydrophobic compounds [2, 3].

By now we know a number of methods for increasing water repellent properties of mineral dispersed materials [4]. According to their effect they can be divided into two main groups. The first group includes the additives that have inherent hydrophobic characteristics. They are obtained from animal and plant raw materials, as well as from some minerals and are called "hydrophobic additives". The second group, called "hydrophobizing additives", includes materials and chemical compounds that do not possess inherent hydrophobic properties but acquire them during their interaction with mineral binders.

There are several established methods for the protection from exposure to moisture and water vapor. For example, hygroscopic materials can be stored in dry places in an airtight plastic container. Another method is the processing of their particles with isolating coatings or their powdering with flowing agents. The use of this method prevents or hinders the access of water molecules to the protected particles. In this case an isolating protective film, liquid or solid flowing agent should possess a high water-repellant property – hydrophobicity.

Peat is one of the natural organic materials, which can be used as a raw material for obtaining hydrophobic coatings. Its known reserves in the Russian Federation are about 186 billion tons (37% of global reserves).

During drying, peat passes from hydrophilic to partially hydrophobic state, during which its water-absorption capacity reduces considerably. The natural hydrophobicity of peat can be decreased or increased by deep processing [5]. One of such scientific directions is thermal treatment, which causes a change in the group chemical composition of organic matter of peat [6]. The nature of such changes is determined (other conditions being equal) by magnitude and duration of thermal treatment of the material. When heated to temperatures above 150-180°C (in the absence or lack of oxygen) the peat begins to manifest the characteristic property of all biofuels – thermal instability. Its organic matter undergoes a complex transformation, the complex of which is called thermal decomposition, resulting in formation of solid, liquid and gaseous products [6]. It was established that by setting up certain parameters it is possible to obtain the maximum amount of thermal bitumen, which possess high hydrophobic characteristics. This gives the possibility to extract these products and further apply them as a coating film on an insulated material [7].

Thus, the aim of the study is to develop a method for hydrophobization of gypsum binder by organic components extracted from peat raw materials. The idea of the method consists in combining of two stages in the technological process of gypsum binder production: gypsum dehydrate calcination with simultaneous application of nano-films from peat bitumen onto the surface of its particles, which further isolates it from moisture.

2 Method used

The main physical and chemical process in the production of gypsum binding materials is the disengagement of chemically bound water from the crystal lattice. The dehydration of calcium sulfate dihydrate occurs during heating according to the following chemical reactions:

$$CaSO_4{\cdot}2\,H_2O = CaSO_4{\cdot}0{,}5\,H_2O + 1{,}5\,H_2O, \quad (1)$$

$$CaSO_4 \cdot 0{,}5\ H_2O = CaSO_4 + 0{,}5\ H_2O\ . \qquad (2)$$

The extracted products have different modifications depending on the reaction conditions [1-3]: calcium sulfate dihydrate ($CaSO_4 \cdot 2H_2O$), α-calcium sulfate hemihydrate (α-$CaSO_4 \cdot 0.5\ H_2O$), β -calcium sulfate hemihydrate (β-$CaSO_4 \cdot 0.5\ H_2O$), α -dehydrated calcium sulfate hemihydrate (α-$CaSO_4$), β-dehydrated calcium sulfate hemihydrate (β-$CaSO_4$), α-soluble anhydrite (α-$CaSO_4$), β-soluble anhydrite (β-$CaSO_4$) and insoluble anhydrite. Scheme and temperature ranges of the transition of calcium sulfate dihydrate to various modifications are shown in Figure 1.

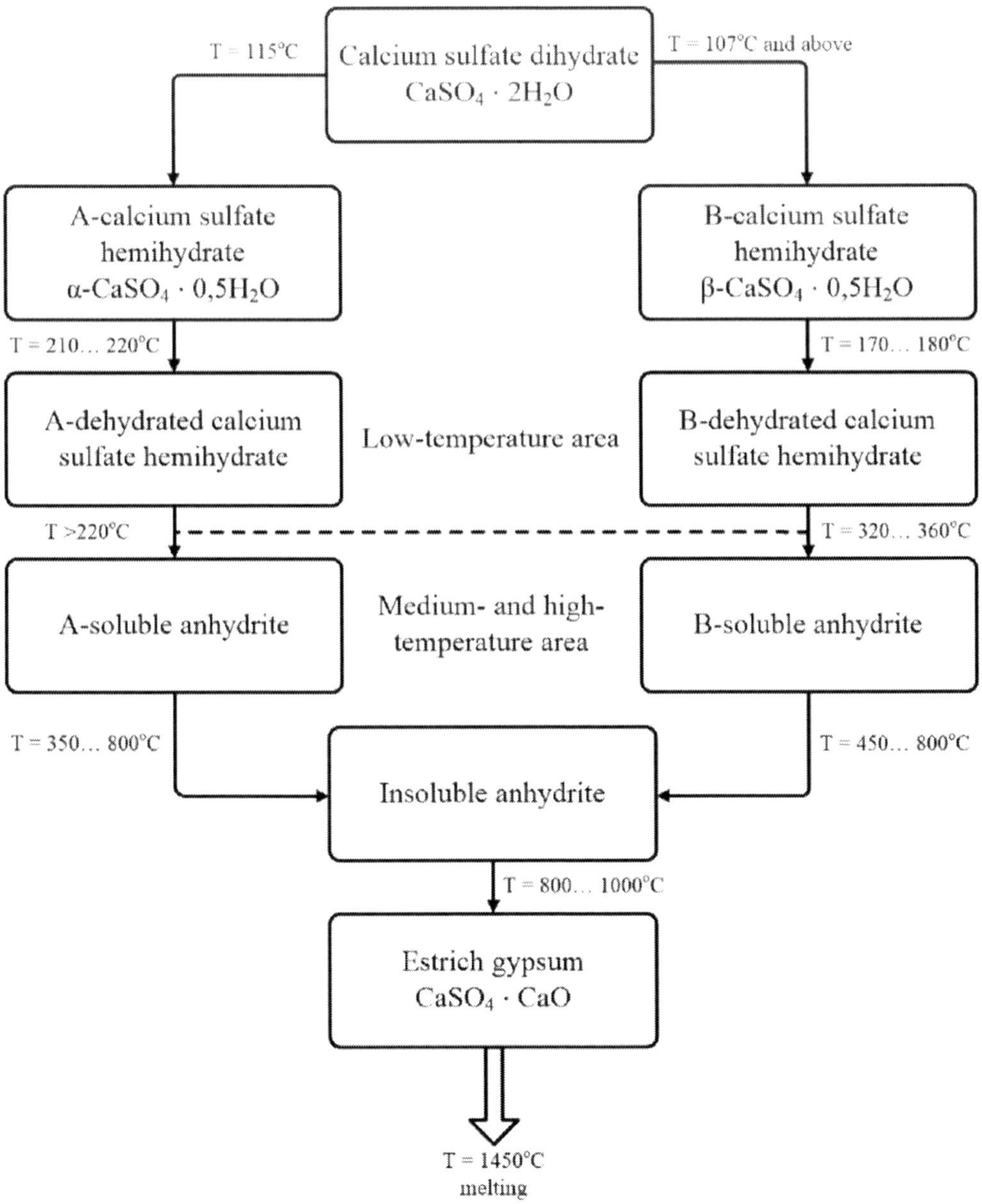

Fig. 1. Scheme of modification formation conditions of calcium sulfate in the anhydrous and the hydrated forms

In normal conditions gypsum dihydrate is a thermodynamically stable solid-phase system. The free energy of the system is minimal, which allows it to be in a balanced state.

During thermal exposure there occurs separation of water molecules. This process changes its balanced state. The major product in this case is gypsum hemihydrate, the specific modification (α- or β-modification) of which depends on the magnitude and duration of the heat treatment.

Hemihydrate calcium sulfate can be quite rarely found in pure form in nature. α-calcium sulphate hemihydrate, in its turn, is formed by thermal exposure (T = 97-115°C) to gypsum dihydrate in an aqueous medium, saturated steam, and in a number of salt solutions [1] in accordance with the mechanism (1). That is, it is necessary to create an environment in which liquid droplet moisture will be released from gypsum dihydrate. Later on, the gypsum will start to crystallize in large, dense or transparent prisms or needles.

Next gypsum modification (β-calcium sulfate hemihydrate) is formed by heating gypsum dihydrate at T = 100-160°C under normal conditions and under evaporation of moisture in the superheated steam state. This type of gypsum represents the smallest aggregates of not distinct crystals. Owing to such a process structure of the particles attains a developed inner surface. Specific surface of β-calcium sulfate hemihydrate is 2-2.5 times higher than the one of α-calcium sulfate hemihydrate.

The speed of setting of α-hemihydrate calcium sulfate when mixed with water is lower than the one of β-hemihydrate modification, which conversely has a higher hydration rate, since it has a larger specific surface of the particles [2].

With increasing treatment temperature on β-hemihydrate, up to 170-180°C, and on α-hemihydrate up to 200-210°C remaining crystallization moisture removes and they move to β- and α-dehydrated hemihydrates in accordance with the reaction (2). In this case the dehydration of the latter ones does not cause any visible structural changes.

Upon further temperature increase of β-dehydrated hemihydrate to 320-360°C, and of α-dehydrated hemihydrate to 220°C and above there is a change of their crystal lattices and they convert into β-and α-soluble anhydrites. The main differences of the obtained anhydrites from hemihydrate are in their higher water demand, quick setting and reduced strength [1]. A further thermal treatment in the range of 360 to 800°C on the above modifications of calcium sulphate results in complete rearrangement of the crystal lattice and, consequently, their conversion into insoluble anhydrite. This formation is very poorly soluble in water, and its solution almost does not set and harden.

In the next temperature range (T = 800-1000°C) insoluble anhydrite form decomposes into calcium oxide, oxygen, and sulfur dioxide. The product (Estrich-gypsum) obtained in these parameters includes insoluble anhydrite and small amounts of calcium oxide (2-3%).

In industry, the thermal treatment on the dihydrate gypsum is performed for 1-2 hours in special boilers with constant stirring. Generally, the maximum temperature of the material at the time of discharge from the boiler, depending on properties of raw materials and requirements to mineral binder, is in the range of 140-180°C. The process is divided into three stages. The first phase lasts about 20 minutes. During this time the temperature slowly increases from 80 to 120 ° C, which corresponds heating of the powder from the temperature at charging stage to the temperature of the beginning of intensive dehydration of gypsum. In the second stage the powder is kept for 30 minutes at the temperature of 120°C. This stage is characterized by intensive occurring of material dehydration with the removal of crystallization water. In the final third stage, the duration of which is 35-40 minutes, the temperature rapidly rises to 180°C. Dehydration intensity is lowered at the same time. After cooling, a ready-to-use material, semi-aquatic (building) gypsum, is produced.

The peculiarity of physical and chemical properties of raw peat is caused by its composition that includes a large number of different groups of chemical compounds [8]. However, most of them relate to hydrophilic components (humic and fulvic acids, cellulose, lignin) and only a minor part comprises hydrophobic compounds: bitumen consisting of waxes, paraffins, oils and resins [7].

Theoretical basis for the creation of hydrophobizing additives on the basis of peat is studies of Russian scientists [9-11] that identify changes in the chemical group composition of peat upon its thermal treatment. It is known that upon drying of natural and (or) artificial peat it undergoes irreversible changes that lead to a significant decrease in water absorption capacity [12]. Moreover, the nature of such change depends on the quantity and duration of thermal exposure to the organic matter.

Thus, it is determined that upon heating peat up to the temperatures of 200-500°C and higher, in the latter it is significantly increased the content of hydrophobic substances compared to the source (native) peat. However, the main objective of the research is to provide such a method which will utilize not only the final hydrophobic materials obtained from peat, but also intermediate liquid and gaseous substances released upon its thermal decomposition.

Based on the analysis of different methods for obtaining hydrophobic additives based on organic and mineral materials, there was developed a method for a water repellent processing of mineral binding powders by an organic peat-based additive [4].

The method of the hydrophobic modification of mineral dispersed materials by native and thermogenic bitumens, extracted from peat, was developed in relation to cement [7, 13]. Cement is mixed with a semi-finished peat additive with particle size of about 20-70 microns. Bog and mud peat with a medium decomposition level is used as an organic raw material. Preference is given to bituminous types of peat.

Further, heat treatment (activation) of the organo-mineral mixture is conducted at temperatures of 150-300°C. The concentration of peat component in cement enables to conduct the process even in the presence of oxygen in the air without the risk of ignition of the organic part. When treated thermally the organic matter of peat decomposes forming additional hydrophobic components. Water vapor and volatile matters remove themselves from the mixture at the initial stage of thermal decomposition. Hydrophobic pyrolysis products, primarily bitumen compounds, are sorbed in the form of thin films on the particles of mineral cement grains.

Image analysis of the Portland cement particles before and after thermal application of bitumen films on them allows to detect quality changes in distribution of chemical elements on the surface after the processing [4]. An extremely high carbon content, which is originally not included in the composition of cement clinker and gypsum, demonstrates presence of the organic film on the surface of the cement grains.

When implementing the developed method, hydrophobic coatings appear from the fluid resin products of peat organic matter pyrolysis on the cement particles and prevent them from wetting (Figure 2).

Fig. 2. Hydrophobicity test of the modified mineral binder: water droplets on the cement surface

The effectiveness of this hydrophobization method is high – we noted the increase of cement water-wetting time, regulated by GOST (Russian State Standard) No. 10178-85. When mixing the mortar, the mechanical action breaks continuity of the hydrophobic coating, which leads to its normal hydration and hardening.

3 Results & Discussions

Use of the method in the presented way in cement industry is extremely problematic due to a number of technological and economic issues. It requires adaptation to existing processes, or their minimal change. At the same time, the developed method can be well adapted to the process of production of gypsum binder. Basic parameters (temperature, time, presence of oxidation agents, etc.) of gypsum calcination technology [1, 2] meet the basic parameters of the hydrophobic modification process [7]. Experiments on calcination of gypsum dihydrate with peat hydrophobizers at T = 180-200°C completely confirm this assumption. Optimization of the quantitative composition of peat additives allowed getting hydrophobically modified gypsum binder with low sorption capacity for water vapor and high strength.

Simulation of experimental gypsum binder storage under 100% relative humidity indicates that at a concentration of 2.1% additive there is a decrease in water vapor absorption of the modified materials by 2 times (Figure 3).

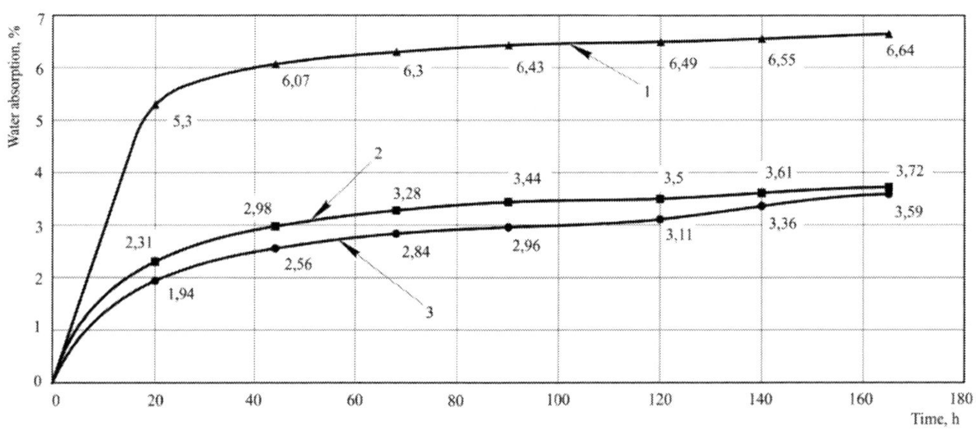

Fig. 3. Decrease in speed and capacity of water vapor sorption during storage of gypsum binder under 100% relative humidity: 1 – control sample; 2 and 3 with 1%- and 2%-additive, respectively

Moreover, increasing of the additive concentration does not result in a proportional decrease of water vapor absorption. This confirms the hypothesis that a protective barrier for the water molecules is a continuous bitumen film released during thermolysis of organic matter of peat and applied to the surface of the gypsum. Under these processing conditions the gypsum particles may keep per unit area of only the amount of bitumen which is determined by the energy of its interaction with the mineral surface. Most probably, the numerical value of the film thickness of the modified gypsum particles surface will be in the same range as that of the previously set for cement (11-22 nm) [13].

Peat is an inherently good organic binder which activity increases with the degree of mechanical dispersion. It is usually attributed to colloidal capillary-porous systems hardening due to coagulative structure formation upon drying [14, 15]. This feature is widely used in the production of various kinds of molded materials based on peat [16]: extruded fuel pellets or sod peat, peat-mineral fertilizers, insulation boards, spherical adsorbents [18] and other products. The mechanism of gypsum hardening is based, in its turn, on the reactions of hydration and occurs in three stages. That is, mineral binder and peat component do not chemically react with each other. If there are any chemical reactions, they may dominate and effect on the interaction of components only with relatively high concentrations of organic additives.

Proceeding from this it is possible to create building materials and constructions based on gypsum binder hydrophobically modified by peat additives the strength of which will not be lower than the one of control samples. However, it should be kept in mind that high strength properties can be achieved only by optimization of particles size, concentration of the hydrophobically modifying component and suppressing air entrapment [1]. In this case, there will be achieved a high hydrophobicity of the material that will not lower its strength.

For example, for Portland cement, use of mud-peat as a basic raw material for obtaining hydrophobic additives shows that there is a significant decrease in the strength of the samples only with the concentration of more than 2% of the mass of mineral binder. At the same time with 1%-content of an additive based on high-moor peat with the degree of decomposition R_t = 25% it is observed a slight increase of flexing and compression strength [4].

The chart in Figure 4 shows the dependence of relative strength, R_{rel}, on peat additive concentration, C

$$R_{rel} = R_{max} / R_i = f(C), \quad (3)$$

where R_{max} – maximum strength of the control sample (no additives) with uniaxial compression, MPa; R_i – experimental sample strength, MPa.

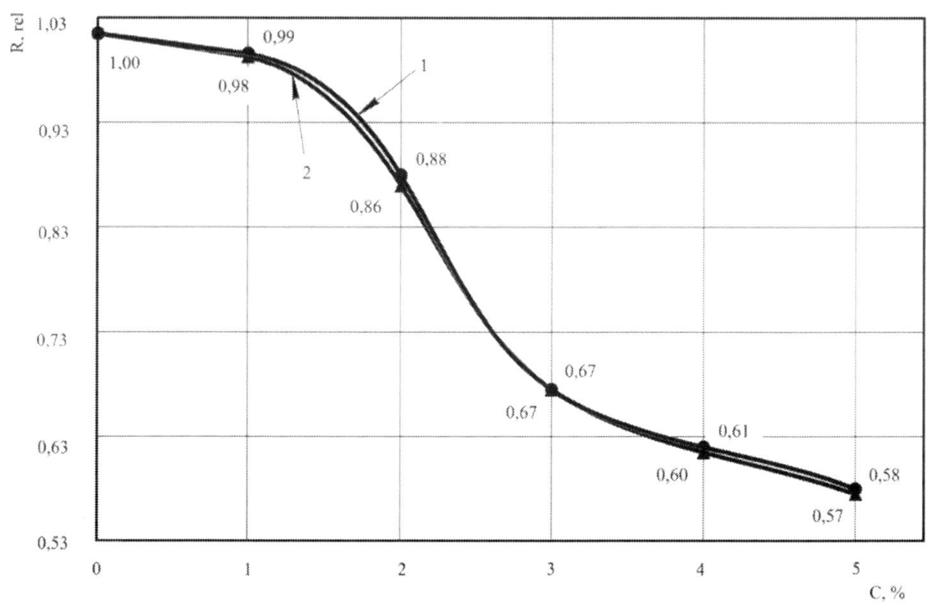

Fig. 4. Dependence of strength of gypsum binder experimental samples on concentration of organic additive based on peat: 1 – bog peat; 2 mud-peat

Dependence analysis (3) shows (Figure 4), that at a concentration of the peat hydrophobizing additive of 0.5-1% there is no critical decrease in strength characteristics as well as there is no increase in the setting time of gypsum samples. But at the same time, its storage time increases significantly without activity loss, also under adverse conditions. When the concentration of the organic component is of more than 1.5-2%, there is a decrease in the strength of gypsum, an expected negative factor. The strength reduces due to the decrease in the number of contact interactions in gypsum, caused by defects in its structure due to the presence of an organic component.

Despite high hydrophobicity of the obtained mineral binders, upon intensive mechanical stirring the hydrophobic bituminous films are «torn off» from the surface of the gypsum particles and are distributed throughout the volume of the mortar. Consequently, together with solid products of thermal degradation of organic matter of peat they create a "protective barrier" from moisture penetration into the sample. The effectiveness of this «barrier» depends on how evenly and orderly hydrophobic components are distributed in the solution.

Electron micrographs of cement stone [13] revealed fragmentary chaotic distribution of its organic components on the structural elements. Size of hydrophobic entities during zooming of individual sections ranged from 500 nm to 10 microns. However, even such a discrete distribution gives a sustainable hydrophobic effect on the surface of the hardened mortar – water does not penetrate into the sample. This does not occur even with the presence of open pores, which could potentially serve as moisture-conducting channels. Drops of water on the surface of the modified cement sample may be retained long enough. Moreover, wetting angle magnitude is substantially higher 90°, which indicates a high hydrophobization of the material surface [17].

It is expected that a secondary effect from the use of hydrophobically modified gypsum binder can be a significant reduction in moisture permeability and moisture capacity of solidified solution by making their structural elements water-repellent. For example, when experimental samples of hydrophobic cements were completely submerged into the water, moisture absorption speed and water absorption capacity decreased by 2-3 times in comparison with the control samples [4].

Preliminary studies revealed that the developed method can be successfully applied also for hydrophobization of various kinds of bulk mineral and organic materials: lime, sand, clay, nitrile butadiene rubbers and so on, which opens up new directions for its use in other industries. At present, the authors conducts successful experiments to expand the raw material base for obtaining hydrophobic modifying agents. Good results are shown by peat subjected to self-heat, organic sapropel, wood residues extracted from peat deposits, industrial wood waste, and especially boghead coal used for this purpose. This is due to the conditions of formation of sapropelite coals from water-weeds residues, rich in fats. Their characteristic feature is the extremely low hygroscopic property (2-5%). When boghead coals are extracted with the alcohol-benzene mixture they disengage up to 10-12% of organic compounds: fatty acids, anhydrides, ketones and hydrocarbons. Accordingly, the organic matter of sapropelites is a macromolecular compound having a linear-condensed structure of polyesters type and comprising interconnected relatively long methylene chains and cycles. Because of a large number of hydroxyl and carbonyl groups in boghead coals there are preconditions for the formation of hydrogen bonds. Furthermore, when evaluating the prospects of using this material in hydrophobization processes an important role should be given to the fact that its organic matter has a relatively low thermal stability.

4 Conclusions

Thus, the results of the study allow to develop a technology for the production of gypsum binder hydrophobically modified by peat additives. The proposed method is completely adapted to the processes currently used in the production and do not require any technological equipment changes. Its use in the proposed form would not lead to an increase in the modify product cost, as the price of the semi-finished hydrophobic peat product equals to the cost of the source raw gypsum.

References

1. L. Sycheva, Glass and Ceramics, **50**, 9-10, 440 (1994)

2. V. Korovyakov, Rossijskij Khimicheskij Zhurnal, **47**, 4, 18 (2003)

3. H. Fischer, B. Vtorov, ZKG International, **55**, 5, 92 (2002)

4. O. Misnikov, Mires and Peat, **18**, 1 (2016)

5. A. Afanas'ev, A. Boltushkin, Colloid Journal of the Russian Academy of Sciences: Kolloidnyi Zhurnal, **58**, 2, 139 (1996)

6. Y. Kosivtsov, K. Chalov, Y. Lugovoy, E. Sulman, A. Stepacheva, Chemical Engineering Transactions, **45**, 667 (2015)

7. O. Misnikov, Theoretical Foundations of Chemical Technology, **40**, 4, 423 (2006)

8. N. Gamayunov, B. Maslennikov, Eurasian Soil Science, **24**, 7, 32 (1992)

9. A. Afanas'ev, A. Efremov, Theoretical Foundations of Chemical Engineering, **45,** 1. 120 (2011)

10. Y. Navosha, S. Prokhorov, V. Strigutskiy, A. Tomson, Eurasian Soil Science, **24**, 5, 1 (1992)

11. K. Chalov, Y. Lugovoy, Y. Kosivtsov, M. Sulman, E. Sulman, V. Matveeva, A. Stepacheva, Bulletin of Chemical Reaction Engineering and Catalysis, **11**, 3, 330 (2016)

12. N. Gamayunov, V. Mironov, D. Stotland, Kh. Shebab, Colloid Journal of the Russian Academy of Sciences: Kolloidnyi Zhurnal, **56**, 3, 342 (1994)

13. O. Misnikov, Polymer Science. Series D, **7**, 3, 252 (2014)

14. A. Afanas'ev, S. Gamayunov, O. Misnikov, Colloid Journal, **61**, 3, 274 (1999).

15. N. Gamayunov, S. Gamayunov, Eurasian Soil Science, **38**, 3, 297 (2005)

16. P. Bel'kevich, L. Chistova, R., L. Rogach, M. Sobol, Solid Fuel Chemistry, **14**, 4, 125 (1980)

17. E. Bryuzgin, V. Klimov, O. Dvoretskaya, L. Man', A. Navrotskiy, I. Novakov, Russian Journal of Applied Chemistry, **87**, 8, 1119 (2014)

18. N. Gamayunov, B. Maslennikov, Y. Shul'man, Eurasian Soil Science, **24**, 4, 122 (1992)

Software for Automated Production Line of Peat Briquettes

Vladimir Lebedev[1], *Olga* Puhova[1]

[1]Tver State Technical University, A. Nikitin Street 22, 170026, Tver, Russia

Abstract. The article examines the software development of peat briquette production. Integrated peat development is one of the important tasks of peat industry. The importance of peat as local raw material increases with the increase in demand for various products. The production efficiency can be improved with new automated peat processing technologies based on advanced engineering. In view of market and technology requirements the peat industry development is hampered by slow development of automated peat processing and utilization technologies. The acceleration of technological progress can be achieved by designing automated lines of advanced peat processing. In accordance with this mission the development of technological process monitoring parameters is of great importance thanks to the possibility of end product physicochemical properties being predicted and managed to make efficient use of peat products for energy needs. The software with an intuitive interface for automated production line of peat briquettes has been developed. This provides means for managing and controlling the peat briquette production process.

1 Introduction

Russia has significant peat reserves. Management and integrated development of peat is one of the important tasks of peat industry. The increased demand for peat as a local fuel and raw material for various products leads to its energy value increase, with the development and improvement of peat production technologies as well as extensive use of peat and peat products taken in view. The significant issue of technological processes of production, processing and peat use [1-3] is peat structure formation in the process of heat-and-mass transfer. To increase the production efficiency the current peat processing and drying technologies [4-6] are to be improved and new technologies are to be sought around the results of modern research and engineering achievements.

In new market conditions and due to continuous changes in peat production technology requirements as well as wide peat variety the industry development is hindered by the lagging pace of the improvement and development of peat production, processing and use technologies. The acceleration of technological progress in the peat industry can be achieved with in-depth research and engineering solutions. Thus the problem of studying the change regularities of peat physical properties in its processing and drying becomes relevant in connection with solving the problems of predicting peat physicochemical properties and their controlled management to use peat products for power industry and other applications better and more efficiently.

The automated production line makes up a complex dynamic system which behavior in practice requires a mathematical description and a time consuming programming [7-8]. An automated technological process can monitor the parameters of peat briquette production process such as temperature, humidity, filling level, dust content, processing degree [9]. It can be done with a direct digital control. All the monitored parameters of a technological process are transmitted in full from sensors to a master industrial controller which computes control signals for actuators. The control issues are of special priority for mining technologies. This is primarily due to the following features of peat production:
– the complexity and high speed of technological processes
– explosion and fire hazards of processed substances [9-10]
– high temperatures and high pressure of the flow process [11]
– high sensitivity of a process number to technological mode violations [12].

It is necessary to take into account that not all technological parameters are available for direct and continuous measurement [13-16].

2 Results

The automated technological process of peat briquette production is a complete production cycle with peat stuff at the input and peat briquettes at the output (Figure 1).

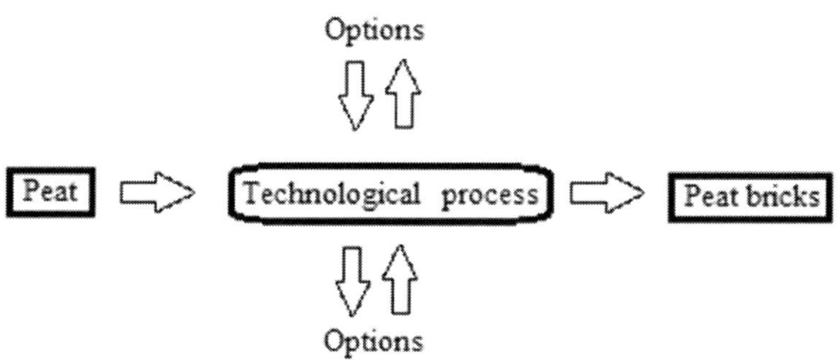

Fig. 1. The flow chart of peat briquette production

Briquetting is a process of turning milled peat into regular-shaped pieces. The technological process is divided into four stages which are interconnected (Figure 2).

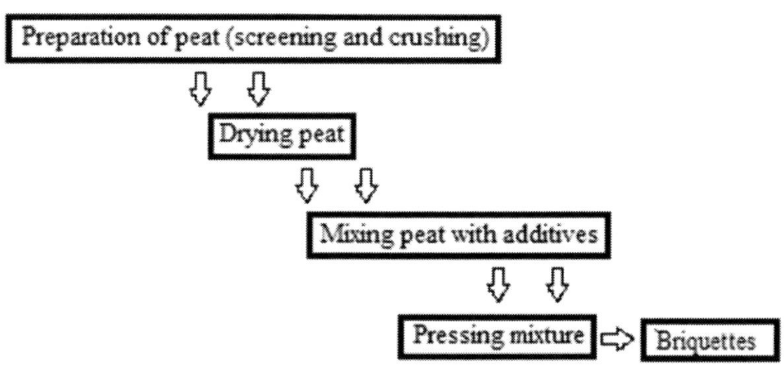

Fig. 2. Technological stages of the peat briquette production

Transferring equipment and the peat level in a charging bin (bunker) are controlled at the peat preparation stage.

Peat mass drying is one of the major stages where the end product quality is determined. The humidity and dispersiveness of dried peat, the discharge of peat being fed, the temperature in a drying chamber are measured and controlled at this stage.

The levels in dried peat bins, the performance of a scraper conveyor and a mixer are controlled at the mixing stage.

The total duration of drying is calculated by formula (1)

$$\tau = \tau_{in} A_d A_c A_h \qquad (1)$$

where τ_{in} is the initial duration of drying peat raw materials from initial humidity of 60% to 12% in the chambers; A_d, A_c, A_h are coefficients for drying modes A_d, intensity of circulation A_c, initial and final humidities A_h.

For A_d the following values are set:

- soft mode –1,70
- normal mode –1,00
- forced mode – 0,80.

The pressing stage provides for automatic alternation operations and control over performance press.

An automatic line of the peat briquette production is given in detail in the functional flow diagram of the technological process of the production in question (Figure 3).

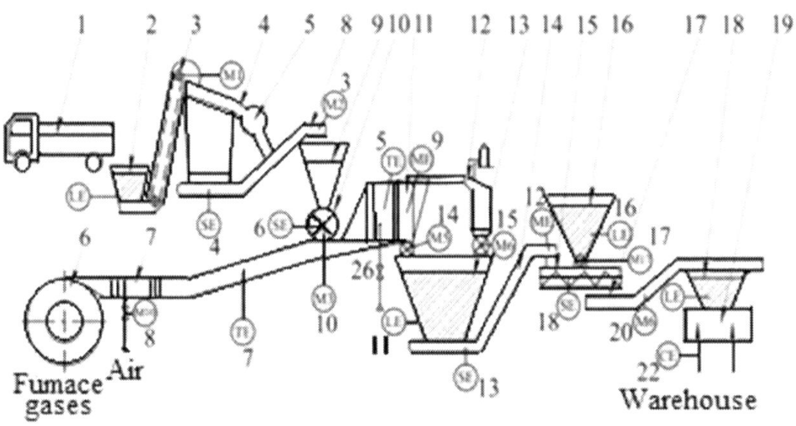

Fig. 3. Functional flow diagram of the automatic peat briquette production line

Initial peat is delivered to the shop by vehicles (1) and discharged into a charging bin (bunker) (2) over the bucket elevator (5). Raw material can be delivered in wagons or containers. Double screen (4) segregates factions of 10-20 mm out of a stream and then they are processed in hammer crusher (5), with the collected coarse rejects being sent to livestock farms. The developed peat is fed by a scraper conveyor (8) to a batch bin (9) of a continuous drying plant (11). The peat is dried up in hot air which is pumped by a forced fan (6) to the humidity value of no more than 25%, with the temperature in a heat generator (7) being 110-140°C. Peat from a dryer enters a storage bin (13), so do small fractions of peat deposited by drying agent in a cyclone (12). Dried peat from a storage bin (13) and additional components from a weighing hopper (16) are fed at regular intervals to a mixer (15) by a scraper conveyor (14). The developed mixture is delivered to a bunker (18) of a press (19) through a scraper conveyor (17).

Pressed briquettes come to carton-, plastic-, or paper-packaging line in baffle boxes (20). Finished products are delivered to a warehouse by electric cars.

Technical process is rather complex for software implementation [16, 17]. It contains a large number of control and monitoring elements. Software for the automatic line has an intuitive interface (Figure 4). The program interface consists of several panels and tabs with the established production configuration, the setting-up of technological process parameters, the schedule control and the description of graphic elements.

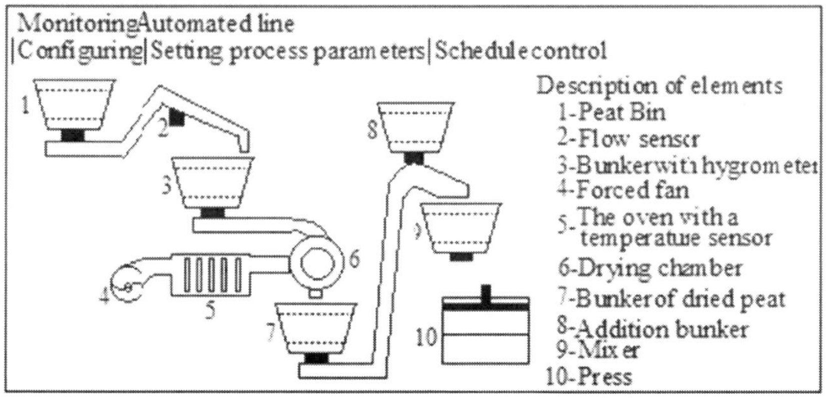

Fig 4. Interface of the control program for the automatic peat briquette production line

3 Discussion

To implement the automated line the following technological elements were chosen: a peat bin, a double screen, an addition bunker; a bunker with a hydrometer; a forced fan and a peat dryer, a bunker of dried peat, a mixer, a press. When required (if the level of peat in a charging bin is below minimum), the automatic sensor gives a signal to fill the bin up to the maximum permissible level. Double screen segregates factions of 10-20 mm and then they are fed to a hammer crusher for further processing. Ratio of large and small fractions is one of the options available for monitoring. Solids flow sensor (impeller) carries out the control. Peat humidity is measured in the bunker after crushing.

The dryer temperature is kept within the range of 110 to 140°C:

Humidity > 60 % – temperature grows

Humidity 25-60 % – temperature reaches 125 °C;

Humidity < 25 % – temperature drops.

Peat after drying is fed to the seventh bunker, with the humidity being < 25 % and the consistency being reduced to a normal state.

Addition bunker contains sawdust and plant raw materials.

To comply with the proportions, components are fed to a mixer when a bunker filling level is above the minimum level.

From the last bunker a peat mixture comes to the press. The press makes briquettes.

The control program for the automated peat briquette production process can set up and monitor the following parameters:

– bunker capacity (in litres)

– output stream (liters per second, movement speed)

– filling of a bunker to its minimum level (%, a starting signal of bunker filling)

– filling of a bunker to its maximum level (%, a termination signal of bunker filling up).

It is also possible to configure:

– the percentage of coarse fractions in peat raw materials (over 10 mm)

– the peat supply from a vehicle (peat is fed to the first bunker when the charging level is minimal)

– the supply of additives (additives are fed to the fourth bunker when the charging level is minimal).

The program interface changes in the course of technological process. It displays the levels of bunker filling, the work of a press (by a color-change indicator), the outgoing stream, and the feedstock supply. Changes in graphic display under different parameters of a technological process are given in Figure 5.

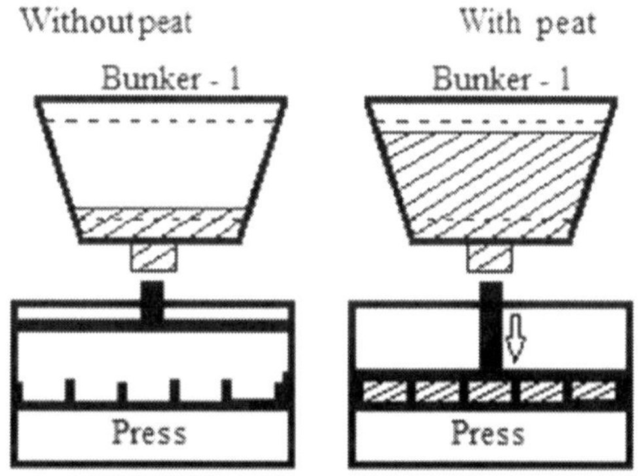

Fig. 5. Changes in a graphic display under different parameters of technological process

Software builds graphics, securing the values of different parameters of a technological process (Figure 6).

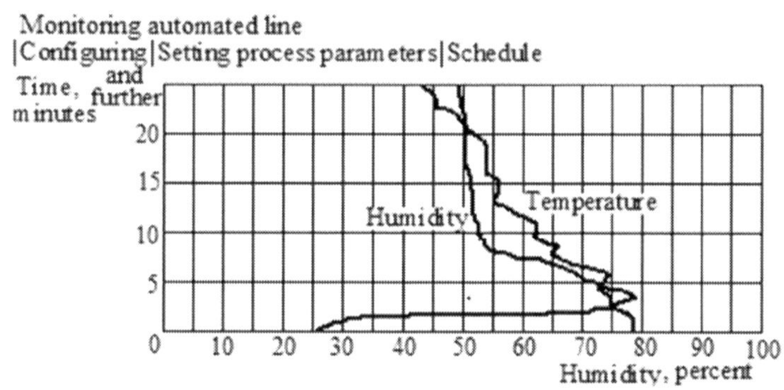

Fig. 6. Graphic display of the drying chamber temperature versus peat humidity

According to Fig.6, an operator watches the dashboard charts to monitor the current status of technological parameters of peat briquette production on a real-time basis. An operator can scale the image using the mouse wheel button. The horizontal axis of a line chart is marked from 0 to100, the vertical axis shows time. It displays up to five parameters simultaneously. There are altogether eight parameters:

– first five parameters show the filling of each of the five bunkers (from 0 to 100%)
– peat humidity (from 0 to 100%)
– percentage of coarse fractions
– dryer temperature (from 100 to 150°C).

The operator interprets the display parameters and produces configuration of the automated process control systems.

The main factors of increasing revenue by means of drying control are as follows:

– increase in efficiency / reduction of drying period
– improvement in the quality of peat briquettes
– saving of energy.

Currently the improved management of the drying process is connected with the performance improvement of a drying chamber. Energy saving, in its turn, is also achieved by drying chamber performance improvement.

4 Conclusion

The process control is connected with the technological advances in mining. The possibilities of automatic control and creating automated production lines bring about:

– automatic control of technological parameters and maintaining the optimum operation modes;

– automatic prediction of the technological process;

– production capacity increase, improvement in the quality and volume of peat products;

– cutting down the costs of raw materials and energy per product unit;

– security of technological process, reduction in number of operating regime disturbances;

– technological process reliability improvement (due to the reduction of downtime), lifetime extension of technological equipment;

– prevention of environmental pollution by industrial wastes.

The efficiency of an enterprise with the automatic peat briquette production line is determined by the quality of finished products and security of supply.

References

1. S.N. Gamayunov, O.S. Misnikov, Inzhenerno-Fizicheskii Zhurnal (IFZh), **71**, 233 (1998)
2. A.E. Afanas'iev, S.N. Gamayunov O.S. Misnikov, Colloid Journal: Kolloidnyi Zhurnal, **61**, 274 (1999)
3. O.S. Misnikov, A.E. Afanas'iev, Theoretical Foundations of Chemical Engineering, **37**, 582 (2003)
4. O.S. Misnikov, Theoretical Foundations of Chemical Engineering, **40**, 423 (2006)
5. O. Misnikov, Polymer Science. Series D, **7**, 252 (2014)
6. O.S. Misnikov, E.Yu. Chertkova, Eurasian Mining. Gornyi Zhurnal **1,** 63 (2014)
7. Abu-Abed F.N., Martynov D.V., Ivanova A.V., Dopira R.V., Kordyukov R.Y., ARPN Journal of Engineering and Applied Sciences, **11,** 9636 (2016)
8. Ivanov V.K., Palyukh B.V., Sotnikov A.N., Lobachevskii Journal of Mathematics, **37**, 244 (2016)
9. O.S. Misnikov, O.V. Dmitriev, E.Yu. Chertkova, Eurasian Mining. Gornyi Zhurnal, **2**, 30 (2015)
10. O.S. Misnikov, O.V. Dmitriev, V.I. Popov, and E.Yu. Chertkova, Polymer Science. Series D, **9**, 133 (2016)
11. O. Misnikov, A. Timofeev, O. Pukhova, Polymer Science. Series D, **8**, 66 (2015)
12. O. Misnikov, Mires and Peat, **18**, 1 (2016)
13. Burdo G., Paliukh B., Advances in Intelligent Systems and Computing, **451**, 229 (2016)
14. Palyukh B.V., Shprekher D.M., Bogatikov V.N., IJAER, **10**, 45724 (2015)
15. Palyukh B.V., Ivanov V.K., Egereva I.A., IJAER, **10**, 45736 (2015)
16. Shprekher D.M., Matveev Y.N., Bogatikov V.N. IJAER, **10**, 45703 (2015)
17. Matveev Y.N., Palyukh B.V., Stukalova N.A., Bogatikov V.N. IJAER, **10**, 45717 (2015)

Assessment process of concept for mining and its impact on the region

Michal Cehlár[1,*], *Juraj* Janočko[1], *Zuzana* Šimková[1], and *Tomas* Pavlik[1]

[1]Technical University of Kosice, Faculty of Mining, Ecology, Process Control and Geotechnologies, Letná 9, 04200 Kosice, Slovak republic

Abstract. Raw materials policy of the Slovak republic is currently in stage of the stagnation due to complex conception and outdated data (not updated from 2003). Nowadays there is an effort of the state to develop a completely new concept of the raw material policy using best available techniques in the given field. Paper is based on the study case of assessment of regional impacts of mining and processing the gold ore to public finance derived from real contitions of the Slovak republic region. Results achieved originates from the models for the assessment of the project of mining as well as from the model of environmental and social impact assessment. The main aim of the contribution is to increase the effectiveness in raw materials using with regard to environmental and social impact of the assessed region which could be implemented to the raw materials policy on the national level.

1 Introduction

The raw material policy of the Slovak republic reflects the all-society interest are essential for the sustainable functioning of modern societies [8]. Criticality of different metallic and mineral raw materials has gained attention in recent years. Consequently, the systematic evaluation of raw material criticality has been the subject of numerous studies in recent years [9]. Several working groups around the world have made criticality assessments for raw materials to analyze the driving impact factors for the instability [10]. Access and affordability of mineral raw materials are crucial for the proper functioning of the national economy. Sectors such as construction, chemicals, automotive, electronics, aerospace, mechanical engineering are all depended on access to raw materials.

The process of utilization of raw materials is currently governed by legislative standards that divide bearings into two categories – exclusive and non-exclusive bearings. Non-exclusive bearings are characterized by attributes property that is assigned to the land. Exclusive bearings are part of the mineral wealth of the state and under the Constitution they are the property of States, what is invariable. Effectiveness of the use of state property is interpreted through rationality. Since mineral resources are non-renewable, the state takes care of their economical use in order to ensure the conservation and rational utilization of

* Corresponding author: michal.cehlar@tuke.sk

mineral resources as a property of the Slovak Republic, together with respect for the principles of sustainable development.

From raw materials policy strategy follows that an important and necessary condition for the smooth development of the Slovak economy, as primary inputs in the production process are mineral resources. As it was mentioned, under Article 4 of the Constitution, the domestic mineral resources are owned by the State. Since it is non-renewable, they must be protected effectively. Achieve this objective is monitored by raw material policy, which defines the goals of the company in the use of domestic mineral resources in response to the long-term needs of economic and social development of the society with regard to the environmental aspects of sustainable development, also covering geological research and exploration and exploitation of proven reserves of minerals [5], [7].

Geological structure of Slovakia is due to geological diversity and rich mineral deposits an interesting site for the location of the project based on the exploration and subsequent exploitation of mineral resources. A change of geopolitical situation, application of new technologies in the industry resulting in different economy of mineral deposits resulted in current revival of mining activities after their recession during the last few decades in the Slovakia.

1.1 Raw materials policy of the Slovak republic

The raw material policy is a policy of how to dispose with raw materials. The strategic view has to be seen in the context of the importance of raw materials for the development of the country and its raw material security. A vital condition for the development of the country is the availability of raw materials that fulfill the complex requirements of the society including economical, environmental and social ones. The use of own resources is both the most economical and ignore this option would be denying ownership, which is included in the Constitution.

The main content of the raw material policy of the country is the analysis of domestic resources and determination of the rules for the protection and prudent use of mineral resources of the country according to the principles of sustainable development. Because the extraction of raw materials for different branches of industry has a positive effect on the economic growth of the country, the development of a proper raw material policy is of outmost interest of the Slovakia [8].

2 Model of optimal use of the resource base

Model of the optimal use of the resource base based on the above mentioned facts and selected economic instruments can be divided into three parts:

1 Model for the assessment of the project of geological exploration;
2 Model for the assessment of the project of mining;
3 Model for the assessment of the project of mining on the environment.

Model for the assessment of the project of geological exploration is an important phase of the initiation of the complex mining project. The whole model is a functional process of assessment of deposits while besides the natural - geological conditions it respects the laws of behaviour of the capital, businesses and public authorities, which are regulated by the relevant legislation.

The model makes possibility to assess whether it is possible to assemble a functional unit and thus evaluate the success of the geological project as a prerequisite of exploitability deposits. The role of regions, in this case, is given only by the possibility to approve or not the zoning plan. If there is no appreciation of the deposits, through the extraction and use of effects for the region, the summary of positive and negative effects for the region cannot be estimated. Thus, this model serves as an advance preparation for the evaluation of deposits, so it should have an irreplaceable role in mineral policy making.

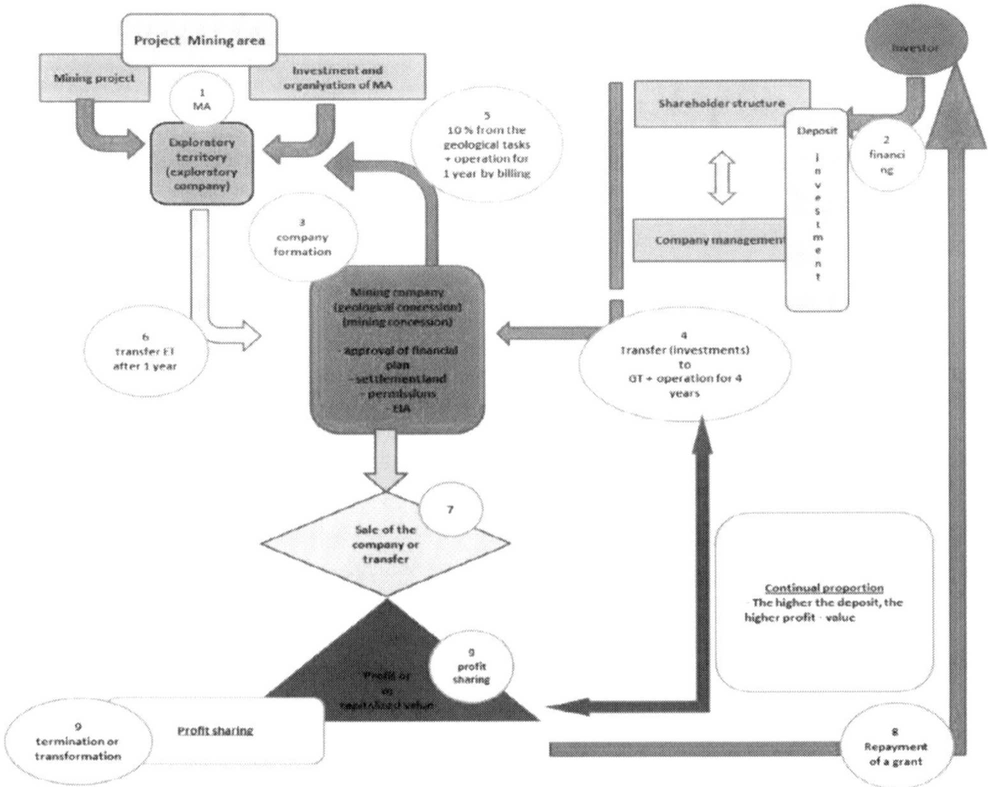

Fig. 1 Model for the assessment of the project of mining

Model for the assessment of the project of mining is an essential part of the evaluation of the optimal use of the resource. The assesment involves several parameters including regional policy for raw material utilization. The absence of this policy may cause in improper decisions, which often are in conflict from the perspective of sustainable development impacting other industries and the service sector. The role of the region, in this case, is again given only by the possibility to approve or not the zoning plan for using the deposits. A more effective tool for decision making is the fact that regional territorial plan is then binding for smaller spatial plans. This affects the representatives of the municipalities, which are parties in decision-making and allocation of mining licenses.

Then, managing conflicts of interest with contracts or sale of the land brings a strong regional charge. This task is not only indirect from the effective point of view, but there is already a recovery of the deposit by the extraction and use of effects for the region, which can be positive and negative. Thus, this model is essential for the evaluation of deposits and in the processing of raw materials policy of the region and should have an irreplaceable role. Both models, in combination with regional development policies materialised in the form of land-use decisions, are the basic structure for objective raw materials policy development of the region.

In contrast to the current state, the deposit is regarded as a resource, value that can only be assessed in relation to the project and its utilisation.

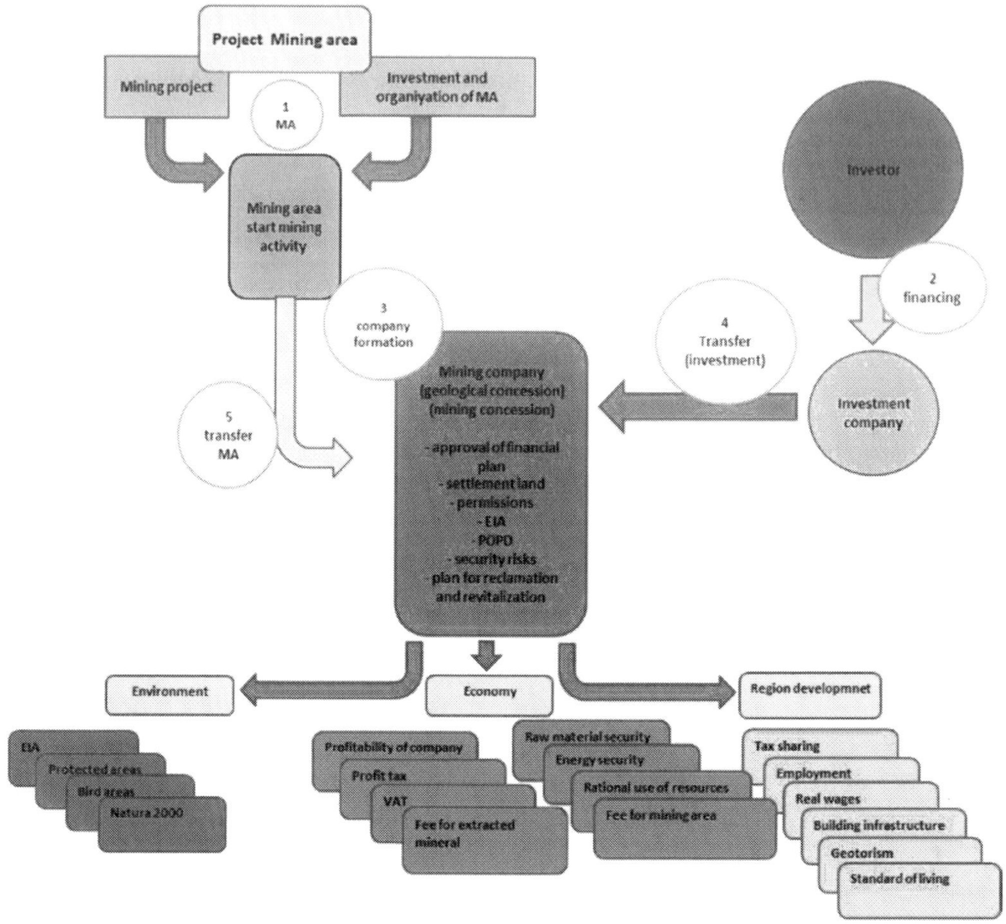

Fig. 2 Model for the assessment of the project of mining

The model for the assessment of the project of mining (Fig. 2) to the environment is closely linked to EIA (Environmental Impact Assessment). The purpose of EIA is to identify, describe and assess the direct and indirect impacts of the proposed mining on the environment, clarify and compare advantages and disadvantages of the proposed mining, including its variants, even compared to the zero option, identify measures to prevent pollution of the environment, mitigate pollution environment, or prevent damaging the environment, get a professional basis for the issuing of a mining permit under special regulations (Pavolová et al., 2012).

2.1 The concept of raw materials policy of Košice region

The concept of raw materials policy of Košice region is a tool for the rational use of mineral resources and for supporting the development of the region while respecting sustainable development with strict respect to the environmental protection. It means to find an optimal way of using, in which the miners are technically able to mine mineral deposits, technologically process material and economically realise all the works up to the location of raw materials in the market.

The concept of the raw material policy is a model that includes:

- legislative division of raw materials (exclusive + non-exclusive deposits),

· assessment factors of raw materials - legislative, economic, environmental, technological, social and mine-technological and autonomous skills that represent working group composed of representatives of organisations that have a real impact on the decision-making process for the authorization of exploration or mining activities.

· outputs of the concept of raw materials policy:

a) the environmental impacts of the resource base of the Košice region,

b) the economic impact of the use of the resource base of the Košice region,

c) social impacts of the resource base of the Košice region,

d) development of the Košice region in a wider context.

These outputs are the main arguments for their incorporation in the process of amending and approving the zoning region plan and then to the land use plans of individual towns and villages. The model of the conception of raw material policy is a process that should occur in the sequence as indicated in the Figure 3. The fundamental facts are competences.

The result of objectification is essentially the division of mineral deposits on the concept of usable and unusable. Usability and unusability depend on factors that are part of the evaluation of mineral deposits: price, production costs, investment costs, environmental and social requirements and resource availability by size and structure. It follows that this process needs to be updated, for what is the concept of model-making raw materials policy of the region prepared.

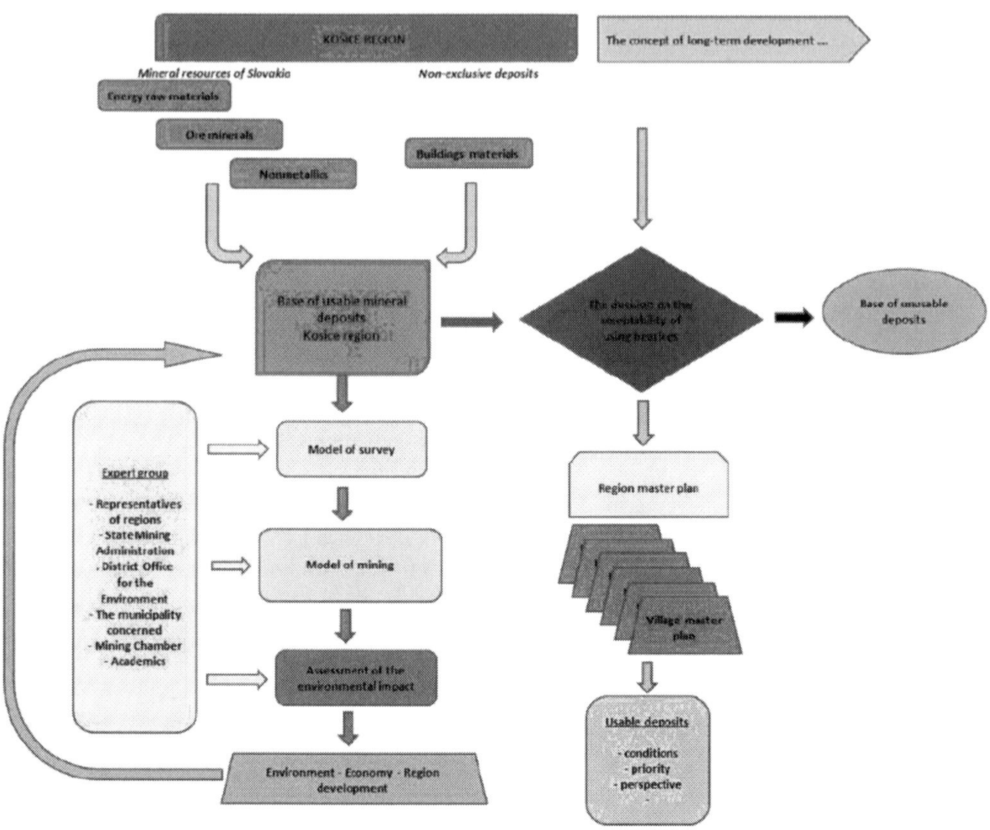

Fig. 3 The concept of raw materials policy Košice region

3 Verification of the model – assessment process

Assessment approach is based on three assumptions. The first is that the economic aspects are assessable through the evaluation model, which is given by the methodology of discounted cash flow. This methodology is officially used to assess the economic efficiency of the relative "value of bearing in the time period of preparing plan for mining development versus size and structure of the planned investment vehicles for the mining and mineral processing in the selected time period, taking into account the capacity of business and economic conditions - evaluation of bearing.

The second assumption are the environmental aspects which are assessable through the model of the impact assessment EIA extractive activities in accordance with Act No. 24/2006 Coll., where is also incorporated a public participation as an important aspect of democracy in decision-making regional governance.

Finally, the third assumption of the concepts is verification for which it was selected case study assessment of regional impacts of mining and processing of gold-bearing ore on public finances.

3.1 Economic and social assessment of the investment

Case study focuses on the analysis of macroeconomic parameters considered for the project in gold mining. It deals with the evaluation of benefits for state, regional and project implementers. The study aims to assess the possible implementation of the mining project by the verification of the model. Delivereably of the study case is an expert support for decision to allow or disallow the realization, respectively inclusion in a portfolio bearing deposits, which should be useful in terms of making the raw material policy of region. Mining project is a demanding not only because of expertise but also because of the the economic side. The basic element is particularly high initial investment and subsequent benefits, both social and economical for the state, region and project implementers for compliance with the high standards of environmental area, set by the legislation.

3.1.1 Geological background

The study area is built up by the Miocene volcanic extrusions having dome-like structure and composed of amfibolitic andesite. The gold mineralization that is the target of mining activity, is associated with quartzite and forms vein-type mineralization. The ore body has a vertical to subvertical position and widens toward depth.

The ore zones are characteristic by increased occurrence of Zn and Cu. The accessory minerals are magnetite, pyrite, chalkopyrite and molybdenite. The total geological reserves for the studied deposite were calculated as 140 200 000 tons of ore with mean content of gold 0.57 g/t.

Following is the geological structure shown on the map of the relevant area (Fig. 4). Overview of the free available reserves are shown in the Tab. 1 as well as free reserves not available shown in Tab. 2.

Tab. 1 Free available reserves [2]

Category of reserves	Number of blocks	Amount of reserves (thsousands of tons)	Mean quality (g/t)	Utilization
Z2	1	17 031	0.82	Production Au (bullion)
Z3	1	14 497	0.76	Production Au (bullion)
Z-2 + Z-3 total:		**31 528**	**0.792**	

Fig. 4. Geological structure of the site area of interest and the wider surroundings [2], [3]

Tab. 2 Free reserves not available [2]

Category of reserves	Number of blocks	Amount of reserves (thousands of tons)	Mean quality (g/t)
Z3	1	108 742	0.508

Overburden abover the reserve block represents totaly an amount of 892 155 m^3.

3.1.2 Social-economic parameters of the project

The situation in ore mining in Slovakia is in critical condition. Over the last five years there has been a tenfold decrease in ore mining and production of products.

Project of gold ore bearing has potential for development as the economic position of the region as well as improving the socio-economic status. At the same time launching the project would have to support production in metal processing and employment of staff in the sector.

The development of prices on the world markets especially in the field of precious metals indicates an increase in demand for these products what has an impact on the preparation and implementation of such projects.

During project implementation at mining area foresees the creation of 220 direct jobs - management positions of control commercial and economic matters, technical and economic, technicians and other staff. Implementation of the project will be given the potential to create more jobs and indirectly. These positions will mainly be related to the provision of services.

The clasification of employees will be implemented based on the organizational structure of the organization. Conducting the necessary geological and technical mining and other management and leadership activities require professional qualifications of staff.

Mining company can create employment in the standard working environment with the assessment of employees based on their qualifications, quality of work, and the degree of difficulty of the work respecting the legislative requirements in the field of labor law.

Estimated average amount of gross wages of employees is in the amount of € 1,600 / month.

Tab. 3 Ore production in Slovakia

Factory	**Ore mining (kt)**						**Production of concentrates (t, kg)**				
	2005	**2006**	**2007**	**2008**	**2009**		**2005**	**2006**	**2007**	**2008**	**2009**
Rudňany	29.1	62.1	25.9	20.6	2.,9	Fe					
						Cu					
						Hg					
						BaSo4	4.2	25.8	12.5	7.3	8.628
Nižná Slaná	603.5	660.4	640.3	443.8	0	Fe	258.5	310.6	348.5	180.6	
Slovenská banská Ltd. Hodruša Hámre	19.3	19.5	15	14.74	34.99	X*	0.4	0.4	0.4	0.465	0.741
Together	**651.9**	**742**	**681.2**	**479.14**	**64.89**	Fe	258.5	310.6	348.5	180.6	
						Cu	4.4	4.4	2.1		14.28
						Hg				6.26	
						BaSo4	4.2	25.8	12.5	7.3	8.628
						Au	108.6	83.7	92		346.11
						Ag	65.2	74.4	50.1	198.45	201.07
						Pb	23.2	15.1	11.4	104.86	62.514
						Zn	20.9	21	11.6	30.019	54.104

Total average annual income of the state formed by tax of employer, tax of employee and payroll tax represents the amount of 2,748 million € at 220 employees.

We need to take into account the telegraph arrival curve associated with job creation. It is believed that these average values representing the socio - economic parameters of the project will be achieved in the second year of the project.

Tab. 4 Overview of developments in the average monthly wage and the minimum wage [4]

Year	**The gross monthly minimum wage**			**Statistically recorded average monthly wage for the year (PM)**	**MM/ /PM**
	€/ month	**Valid from**	**According the regulation**	**€ monthly**	**%**
2016	405	1.1.2016	279/2015		
2015	380	1.1.2015	297/2014 Z.z.		
2014	352	1.1.2014	321/2013 Z. z.	858,00	41,03
2013	337.70	1.1.2013	326/2012 Z. z.	824,00	40,98
2012	327.20	1.1.2012	343/2011 Z. z.	805,00	40,65
2011	317.00	1.1.2011	408/2010 Z. z.	786,00	40,33
2010	307.70	1.1.2010	441/2009 Z. z.	769,00	40,01
2009	295.50	1.1.2009	422/2008 Z. z.	744,50	39,69
2008	268.87	1.2.2008	663/2007 Z. z.	723,03	37,19
2007	268.87	1.10.2007	450/2007 Z. z.	668,72	40,21
2006	252.27	1.10.2006	540/2006 Z. z.	622,75	40,51

3.1.3 Social-economic assessment of the project

For the evaluation of the project – gold-bearing ore as a means for assessing the socio-economic assessment of the project it is necessary to evaluate the overall economics of the project. It consists of projected investment costs, the fixed and variable costs, which are known at the moment of evaluation of the project, the anticipated yields and so on. In summary, this is a project evaluation in the mined gold ore. Determine the economic efficiency of financial investments of extraction and processing technology in the deposit is made by assessing the efficiency of investment. The overall parameters of the volume of extractable reserves in respect of the mining-technical conditions, environmental and economic conditions are listed below:

- The total volume of extracted raw materials 54 000 kt
- The total volume of ore: 23 000 kt
- Average annual extraction of 5 000 kt
- The average annual volume of extracted ore 2 130 kt
- The average annual yield of gold: 2 343 kg

The total amount of probable reserves of gold-bearing ore at the deposit is 23 million. ton and at this annual extraction represents bearing life about 10 years. To access the reserves of mineral raw materials for mining should be carried out preparatory work and uncovering. To increase the volume of reserves suitable for mining-technical application it is necessary to continue the geological survey to increase the exploration of deposits and reclassification from the list of non-balanced to balanced reserves. This changes the above contemplated amount of total mined raw materials and other parameters.

To start the project is required investment of € 45,536 mil. €. The successful implementation of the project it is needed to invest the financial resources in the total amount of 45,536 mil. € with the timeline of 2.5 years.

Annual direct (variable) and indirect (fixed) costs of production from level of the year 2010 are composed of the following items (calculated at an average rate of annual extraction of 5 mil. Tonnes per year):

- preparatory work consisting of: construction of transport routes,
- uncovering work : overburden works, quarry face, dewatering
- extraction of raw materials: primary and secondary rock disintegration, transport of raw material
- treatment and refining: mechanical treatment, chemical-technology treatment etc.

Taking into account the annual production is the direct production costs determined on the basis of the above at: € 2.35 /t mined of raw material. Mentioned direct production costs (€ 2.35 / t) does not contain the labor and financial costs as well as the depreciation. They are in the methodology calculated separately.

The average selling price per 1 oz of mined gold will be fixed by contract price, dependent of contracted quantity and quality of the product and also on the price of gold on world commodity market and exchange rate in € / USD. This study takes into account the price of gold based on the world stock exchanges values, where the date of 10.11.2010 the price is € 960 /oz in the amount of $ 1.4 /€. The average selling price is € 30.87 /g of Au. Determination of the final price will be the responsibility of the sales department and statutory company [2].

3.1.4 Economic assessment of the project

Overall layout drawing of funds in time is as follows: 100% in the first year (investment of own funds in the amount of 45.586 mil. €). Based on the data declared by the National Bank of Slovakia is done updating economic parameters. For the project of gold-bearing ore mined at the relevant site is a summary measure of updating the value of 6%.

In the case of "the payment of fees for extracted mineral", which regulated the Ministry of Economy in the range of 0-10% it is considered a limit of 5%, according to NVSR no. 50/2002 Coll. This value corresponds to the value of the mineral and its method of production.

Income tax is at 19%, consisting of a percentage level of 2011, the implementation period for the assessment of deposits for the purposes of exploitation, the period of execution of the decision.

Based on the evaluating of deposit, available reserves in terms of planned mining activities in the mining area referring to state the following facts:

1. Development of Cash Flow (CF) - In the first two years shows a negative value (investment and to modernize and increase the capacity of technology), then in the third year and the following years (with respect to the total volume of extraction and currently proven reserves), reaching a positive value. The actual investment in the project mining and processing of gold ore at those sites is negligible compared to the turnover. Investments (45 mil. €) needed to provide property rights to movable and immovable property, to obtain the mining area to build the plant and the related financing operations are considered from its own resources. An increase is noticeable in the 6th and 7th year of the project. This sharp upward trend is also provable for seven years horizon, but this is conditional upon subsequent geological survey, which for the purposes of geological law must result in approvals of new, workable reserves. The result is the ability to project self-financing and thus the operate the mining project from its own resources.

Present value (PV), without taking into account the time of updating reaches after expiration the period related to the 2010 during the extraction will be carried out according to the plans submitted is at level of 201 608 221 - €.

2. Value of the deposit (Net Profit Value) - defined by the referred exploration area, with a time horizon of assessment to 10 years, capacity of 5 million ton of minerals utilized by this project, with regard of conditions of 2010 and planned finish of investment plan; determined on 120 171 320 - €.

4 Conclusion

The issue of exploitation of mineral resources of Slovakia is fully reflected in accordance with applicable legislation in its management. This treatment is documentable through processes that are embodied in raw materials policy. Although this level of law is prepared, it contains only general declarations that without institutional instruments can not be effectively used and implemented. Nevertheless gives a framework which is binding and inspiring.

Taking over decision-making powers and brought with it the responsibility for decision-making. Just as it is the case in connection with the decision of the zoning plan to be adopted at the level of municipal governments and also municipalities and cities. Deposit utilization and consent to it is fully within the competence of the higher territorial unit and in approving the amended land use plan and the village or town for the same reason and in addition from the relevant municipality or city is subject to the parties in the selection and approval of the mining and mining area.

Global importance of raw materials is obvious interpretation of their impact on the world economy, the national economy and regional economy. Cost of raw material availability was reflected even through the availability of energy resources, which has occurred in the recent past and the impact on the economy across Europe were apparent.

Raw material security is all about availability and our own raw material base is more important than is apparent today. The European Commission started works on mineral policy in last years. In 2008 launched the EU Raw Materials Initiative in order to ensure security of raw material supply for its economic growth [11]. European Union can not be restricted or disfavoured by aspect of inefficient or incorrect decisions of Member States. Especially countries that do not have a direct impact on its raw material base and keep it deformed often by tendentious views on raw material basis and its use.

The decision-making model that combines information, competence and strategy is of crucial importance to maintain the ability to decide on their property. Decide competently, professionally, responsibly and at the same time take into account in this process specifics deposit (its size, quality, geotechnical conditions of mining, market position, location, complexity of technology, mining method, number of employees, their qualifications, risks, management of mining and processing of of raw material, etc.), the specifics of the region (flora and fauna, protected areas, industrial activity, age and educational structure of population, employment, architecture, recreation, development plans, etc.), specifications interests of the State as owner (raw material security, Non-renewability of the source, the strategic position of source income of the state budget, etc.) and the specifics of miners (know-how, management skills, marketing links, financial opportunities, market share, etc).

The model provides the opportunity to work with known facts or contrary extract facts that are required, for example, the future potential miners. Success of application depends only on the willingness to take a decision to address the region's raw materials at conception, conception process raw materials policy of the region and apply it effectively in decision-making.

References

1. Pavolová, H., Cehlár, M., Soušek, R., *Vplyv antropogénnych činností na kvalitu životného prostredia* (Institut Jana Pernera, 2012)
2. AMC Consultants (UK) Limited: Biely Vrch Scoping Study. Eastern Meditteranean Resources Slovakia, s.r.o. 2010.
3. Hlavný banský úrad. (2016). Evidencia dobývacích priestorov: Obvodný banský úrad v Košiciach. Retrieved February 15, 2016 from http://www.hbu.sk
4. Institute of Informatics and Statistics (2016). Retrieved February 15, 2016 from http://www.infostat.sk
5. Aktualizácia surovinovej politiky Slovenskej republiky pre oblasť nerastných surovín. MH SR, Uznesenie vlády č. 661/1995 aktualizovaný materiál a legislatíva k 31.12. 2003.
6. Cavender, B.: *Determination of the optimum lifetime of a mining project using discounted cash-flow and option pricing techniques* (Mining Engineering, 1992)
7. The mineral policy and joining of European Union. Retrieved March 15, 2017 from http://library.envifor.eu:8282/greenstone/collect/PDF/SK_000927.pdf
8. Puzder M., Koščová M., *XVIII International Coal Preparation Congress*, 457, 2016
9. Glöser-Chahoud S., Espinoza L, Walz R., Faulstich M., Resources, **5**, 45 (2015)
10. Achzet B., Helbig Ch., Resources Policy, **38(4)**, 435 (2013)
11. Kulczycka J., Kudelko J., Wirth H. Role and aim of mineral raw materials policy presented in national strategic documents. Retrieved March 15, 2017 from https://www.researchgate.net/publication/302193446_Role_and_aim_of_mineral_raw_materials_policy_presented_in_national_strategic_documents

Experimental Study of Methane Hydrates in Coal

Vyacheslav Smirnov[1]*, *Valery* Dyrdin[1], *Tatyana* Kim[1], *Andrey* Manakov[2] and *Aleksey* Khoreshok[3]

[1]T.F. Gorbachev Kuzbass State Technical University, Department of physics, 650000 Kemerovo, 28 Vesennaya st., Russian Federation
[2]Nikolaev Institute of Inorganic Chemistry SB RAS, laboratory of clathrate compounds, 630090 Novosibirsk, Ac. Lavrentiev Ave., 3, Russian Federation
[3]T.F. Gorbachev Kuzbass State Technical University, Department of mining machines and systems, 650000 Kemerovo, 28 Vesennaya st., Russian Federation

Abstract. The possibility of gas hydrate formation in porous space of coal has been studied. The experiments conducted have proven the possibility of methane gas hydrate formation in moist coal. It has been demonstrated that the decomposition points of methane gas hydrates in coal are near to the phase equilibrium curve for bulk methane hydrate. Only part of water absorbed by coal can be involved in the methane gas hydrate formation. With the increase in gas pressure increases the amount of gas hydrate formed in natural coal. For formation of hydrates at a positive temperature, the pressure in the system has to be at least 2 MPa. At the same time the speed of formation and decomposition of gas hydrates in coal is big enough.

1 Introduction

In the process of metamorphism of coal matter about 300 m^3 gaseous substances are formed per each ton of coal [1-4], generally they include carbon dioxide, methane, and nitrogen. The reservoir pressure of methane, on average, linearly depends on depth and for different coal seams it is in the range $(0.4 \div 1)\gamma H$, where γH is the rock pressure of undisturbed coal seam, $\gamma = 25\ 000\ \text{Н}/\text{м}^3$ is the average specific weight of rocks, and H (m) is a seam depth from a daylight surface. Respectively, for the seams lying at a depth from 400 to 600 meters, gas pressure can reach 4÷15 MPa. In spite of the fact that coal bed methane is a valuable raw material for industrial use, in modern conditions of mining technology methane is frequently considered as an accessory and dangerous product of mining [4], especially in case of underground mining. In fact, almost all researchers [2-6], carrying out estimates of amount of the methane in a gaseous state absorbed on a surface and adsorbed in volume of solid substance, come to the conclusion that the actual content of methane in

* Corresponding author: smirnovvg@mail.ru

natural coals often surpasses their estimates. Therefore, it is necessary to look for other forms of methane location in coal mass. The assumption was made of a possibility of finding methane in a form of gas hydrates, which are formed from coal bed methane and natural moisture under suitable pressure-temperature conditions [7-9]. Gas hydrates are clathrate compounds in which molecules of gas-hydrating agent are included in cavities of the polyhedral framework constructed of water molecules. One volume of methane hydrate can contain about 170 volumes of gas [10-14]. The lowest homologues of hydrocarbons (C1-C4), carbon dioxide, nitrogen, oxygen, etc. can act as a hydrating agent. The existence of hydrates needs increased pressure and, as a rule, below room temperature. In most cases temperature of decomposition of hydrates increases with growth of pressure. So, equilibrium pressures of hydrate of pure methane at 0 and 10 °C are up to 2.51 and 6.95 MPa respectively [10]. In this paper the results of pilot studies of formation and decomposition of methane hydrate in pore space of natural coal are presented. The main objective of research was to define pressure-temperature conditions under which the formation and decomposition of gas hydrates in pore space of natural coal occurs, and also to find out amount of water, capable of being involved in the formation of gas hydrates.

2 Experimental unit

The basic component of the used experimental unit is the cylindrical high pressure autoclave made of stainless steel and furnished with the gate valve, the pressure transformer and the thermocouple incorporated inside. The autoclave is located in the programmable cryostat. Before the experiment, the coal sample is loaded into the autoclave, free volume is washed out by methane several times for removal of air and the working pressure of gas is created. For experiments pure methane and carbon dioxide were used. In both cases the amount of impurity in each of these gases was less than 0.01%. The thermocouple is installed in the geometric centre of the cylindrical autoclave. Gas pressure in the autoclave, temperature inside the autoclave, and also temperature of the thermal liquid washing the autoclave outside are measured by means of the pressure transducer and temperature transducers. Data from the transducers are registered every 20 seconds by the computer.

Before loading in the autoclave, the coal preparation had been carried out. The selected natural coal was dried up in the vacuum furnace at a temperature of 110 °C within 4 hours till complete loss of moisture. Then the sample of coal was located in a desiccator with the constant air humidity, its humidity was measured in accordance with the change of mass of coal. To maintain the constant humidity in the desiccator, a Petri dish with the distilled water or saturated salt solution was put inside. Within several days (5 and more) the mass of coal reached the constant value, the equilibrium humidity of coal without contact with liquid water was reached. The sizes of particles of the coal taken for experiments were in range from 0.5 to 2.0 mm.

3 Results of the experiments and discussion

3.1 Isotherms of water sorption by coal sample

The formation of gas hydrates in porous space of coal depends on the form and the amount of moisture present in coal. Before carrying out experiments the water sorption isotherm of the K grade coal sample taken from outburst-prone seam XXVII of mine "Berezovskaya" in Kemerovo district was studied. Dependence of the established humidity of coal on relative air humidity in which the coal was kept is given in fig. 1 – the sorption isotherm at 22 °C. By its form the received curve is similar to those found in literature [16-18].

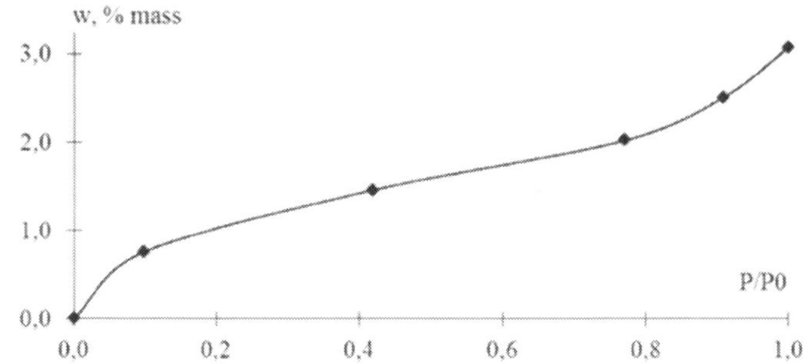

Fig. 1. Water sorption isotherms of K grade coal at temperature 295K.

3.2 Experiments on formation and decomposition of gas hydrates in coal

For carrying out experiments the prepared coal of the known humidity was located in the autoclave in which the working pressure of gas was created. On the programmable thermostat the program of linear decrease of temperature was set to study formation of gas hydrates when cooling. To study the decomposition of gas hydrates, the program of linear temperature increase was set on the thermostat. In the absence of phase transformations, the linear change of temperature led to linear change of pressure in the autoclave. During the formation and decomposition of gas hydrates the step on a pressure curve occurred. Experimental data allowed calculating the volume of the methane in a gaseous state and therefore the amount of the methane involved in the formation of gas hydrates. On the basis of the amount of the methane participating in hydrate formation and the formula of hydrate $CH_4 \cdot 6H_2O$ it is possible to calculate percent of the moisture Wg participating in hydrate formation. The P-T chart of change of pressure and temperature in the autoclave containing gas and moisture coal in the course of the experiment under code 208 is given in Fig. 2. Gas emission peaks are clearly visible: positive peak when heating and negative peak when cooling of system. The arrows indicate the direction of the temperature change.

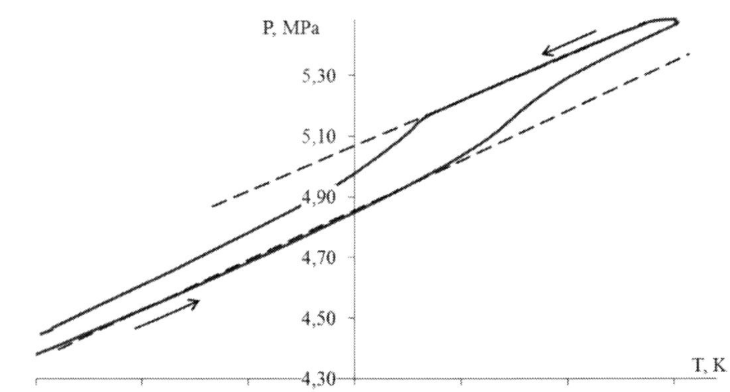

Fig. 2. P-T chart for experiment 208

To detect the trends of formation and decomposition of gas hydrates in natural coal a series of experiments was conducted. Usually (except 207, 208) experiments began at a temperature of +20 °C, then a linear decrease of temperature to –25 °C occurred, the hold for 60 minutes and again linear temperature increase with a constant speed. Decrease speed θ1 and increase speed θ2 temperatures for each experiment, and also pressure and temperature of the centre of the bend arising at formation and decomposition of gas hydrates are specified in Table 1. In Table 1 the humidity of the coal W0 loaded into the autoclave, and also Wg – the amount of water which is involved in formation of gas hydrates converted into percent from the mass of coal is specified.

Table 1. Experimental points of formation and decomposition of gas hydrates, and amount of water involved in the formation of gas hydrates

Code	w_0, % mass	Formation upon cooling				Decomposition upon heating			
		Θ_1, mK/s	P, MPa	T, K	Wg, %	Θ_2, mK/s	P, MPa	T, K	Wg, %
207	2.67	-	-	-	-	5.56	5.07	280.1	0.97
208	2.67	5.56	5.14	275.9	0.79	5.56	5.13	279.8	0.67
625	2.81	2.22	4.48	272.7	0.71	1.04	3.78	275.5	0.84
627	2.81	6.25	2.38	271.9	0.47	1.11	2.16	268.4	0.69
629	2.81	6.25	5.30	276.6	0.84	2.08	5.27	278.4	0.93
709	2.95	2.08	4.16	273.2	0.57	0.83	4.17	276.9	0.86
711	2.95	6.25	2.31	268.5	0.60	2.50	2.33	269.4	0.69
716	2.42	4.17	5.34	271.4	0.42	1.04	5.54	278.9	0.67
717	2.42	-	-	-	-	2.50	2.46	269.5	0.43

In Figure 3 the experimental points of decomposition and formation of gas hydrates in natural coal are given. The zone of stable existence of methane hydrates is above and more to the left of the phase equilibrium curve drawn as a continuous line. As the moment of phase transition the centre of the bent has been accepted, i.e. the section with the biggest inclination on P-T chart. Taking into account the extent of phase transition it is possible to say that all points of decomposition lie near the phase equilibrium curve of methane hydrate. The formation of gas hydrates happens when the system cools down on 3÷5 K in reference to the equilibrium curve of phase transition. The received results can be explained by the influence of porous structure of coal and they can be compared to the data obtained by other researchers [19].

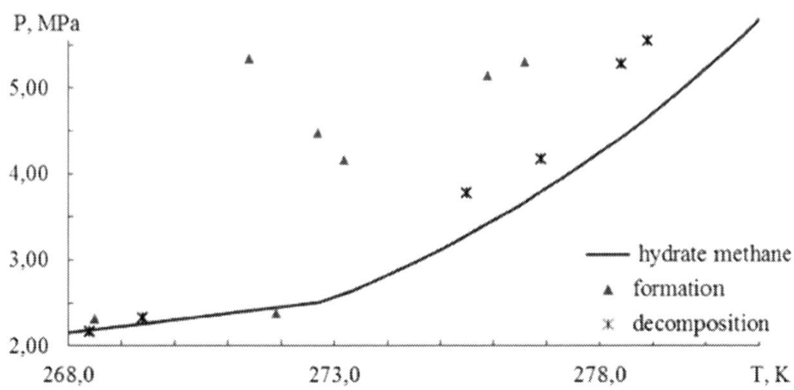

Fig. 3. The experimental points of decomposition of hydrates in natural coal.

The analysis of experimental data (Table 1) has shown that the amount of the moisture involved in hydrate formation is influenced by pressure of methane and humidity of coal. For construction of graphs in Fig. 4 and Fig. 5 the implemented experiments were arranged according to values of close humidity and close gas pressure. It can be seen that for each group of experiments with increase in the humidity of coal the amount of water, capable to be involved in hydrate formation monotonously increases. On the basis of the type of dependence, it is possible to assume that for pressure of 2.2÷2.5 MP the moisture participating in hydrate formation exceeds a threshold in 1.7%, and for pressure of 5÷6 MP this threshold is 1.3% of absorbed moisture, it decreases at increase of gas pressure. This size of humidity (about 1.5%) approximately corresponds to plateau level on a sorption isotherm of water (Fig. 1).

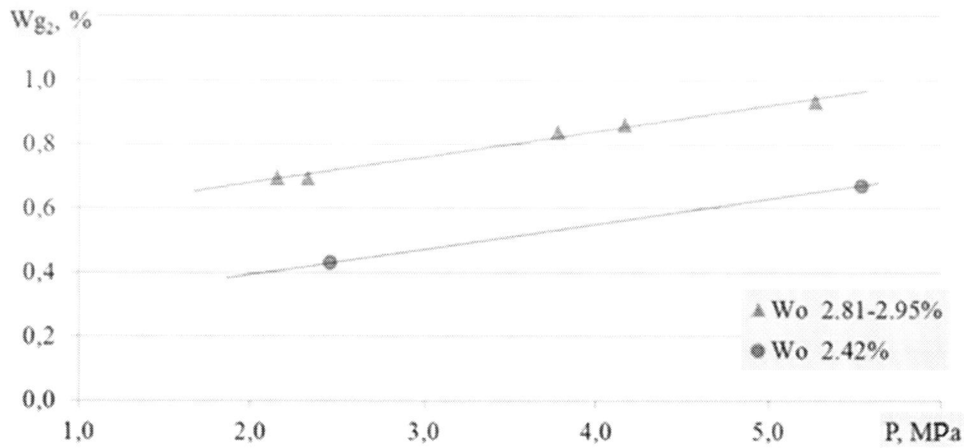

Fig. 4. The dependence of moisture involved in hydrate formation Wg on pressure of methane.

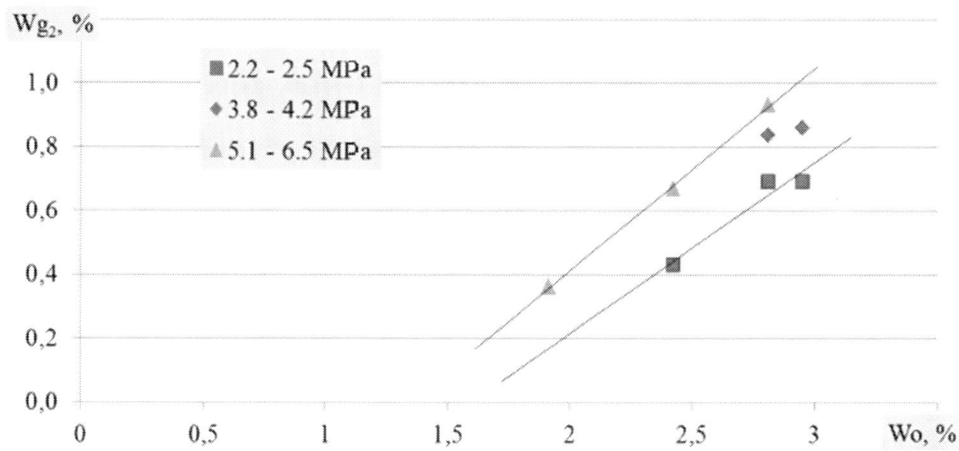

Fig. 5. The dependence of moisture involving in hydrate formation Wg on total moisture of coal Wo

4 Conclusions

Experiments show that in porous structure of natural coal, at the humidity corresponding to the term "air - dried coal" and suitable pressure and temperature, methane hydrates can be formed. Pressure and temperature conditions of formation and decomposition of methane hydrates in coal are close to the phase equilibrium curve. The amount of water involved in formation of gas hydrates always is less than humidity of coal. With increase in pressure of methane the amount of the hydrate formed in the taken coal sample increases. For formation of hydrates at a positive temperature, the pressure in the system has to be not less than 2 MPa. At the same time, the speed of formation and decomposition of gas hydrates in coal is big enough.

This work was supported by Russian Foundation for Basic Research grant 15-05-02240.

References

1. Van Krevelen DW. *Coal: typology, physics, chemistry, constitution* (Amsterdam, London, New-York, Tokyo, 1993).
2. Fengde Zhou, Guangqing Yao, Stephen Tyson, International Journal of Coal Geology **146**, 14 (2015)
3. A. D. Alekseev and E. P. Feldman, Technical Physics Letters, **34**, 609 (2008)
4. C.Özen Karacan, Felicia A. Ruiz, Michael Cote, Sally Phipps, International journal of coal geology **86**, 121 (2011)
5. Weishauptova Z, Medek J., Fuel, **77(l/2)**, 71 (1998)
6. Alexeev AD, Vasylenko TA, Ul'yanova EV, Solid State Commun, **130,** 669 (2004)
7. Dyrdin VV, Smirnov VG, Shepeleva SA, J Min Sci., **49(6)**, 908 (2013)
8. Yan L, Chen G, Pang W, Liu J, J Phys Chem B, **109**, 6025 (2005),
9. Hao S-Q, Kim S, Qin Y, Fu X-H., Ind Eng Chem Res, **52**, 18492 (2013)
10. Sloan ED, Koh CA., *Clathrate hydrates of natural gases. 3rd ed. Boca Raton* (London, New-York, CRC Press, 2008).
11. Milkov AV., Earth Sci Rev, **66(3–4)**, 183 (2004)
12. Makogon YF, Holditch SA, Makogon TY., J Petrol Sci Eng, **56**, 14 (2007)
13. Vyacheslav G. Smirnov, Andrey Yu. Manakov, Elissa A. Ukraintseva, Galina V. Villevald, Tamara D. Karpova, Valeriy V. Dyrdin, Sergey Yu. Lyrshchikov, Zinfer R. Ismagilov, Irina S. Terekhova, Andrey G. Ogienko, Fuel, **166**, 188 (2016)
14. Kurihara M, Funatsu K, Ouchi H, Sato A, Yasuda M, Yamamoto K et al, *Proceedings of the 7th international conference on gas hydrates (ICGH 2011)* (Edinburgh, Scotland, United Kingdom, 2011).
15. Zimny T, Finqueneisel G, Cossurutto L, Weber JV., J Colloid Interface Sci., **285**, 56 (2005)
16. Charriere D, Behra P., J Colloid Interface Sci., **344**, 460 (2010)
17. Kaji R, Maranaka Y, Otsuka K, Hishinuma Y., Fuel, **65,** 288 (1986)
18. Aladko EY, Dyadin YA, Fenelonov VB, Larionov EG, Manakov AY, Mel'gunov MS, et al., J Phys Chem B, **110**, 19717 (2006)

Expert evaluation of innovation projects of mining enterprises on the basis of methods of system analysis and fuzzy logics

Alexander Pimonov[1,2*], *Elena* Raevskaya[1], *and Tatyana* Sarapulova[1]

[1] T.F Gorbachev Kuzbass State Technical University, Kemerovo, Russia
[2] Institute of Economics and Industrial Engineering, Siberian Branch of the Russian Academy of Sciences, Novosibirsk, Russia

Abstract. This paper presents the multipurpose approach to evaluation of research and innovation projects based on the method of analysis of hierarchies and fuzzy logics for the mining industry. The approach, implemented as part of a decision support system, can reduce the degree of subjectivity during examinations by taking into account both quantitative and qualitative characteristics of the compared innovative alternatives; it does not depend on specific conditions of examination and allows engagement of experts of various fields of knowledge. The system includes the mechanism of coordination of several experts' views. Using of fuzzy logics allows evaluating the qualitative characteristics of innovations in the form of formalized logical conclusions.

1 Introduction

The strategy of innovative development of the Russian Federation for the period until 2020 and the long-term program for the development of the Russian coal industry for the period until 2030 indicates the need to develop the innovative potential of coal-mining enterprises. In 2016, Kuzbass produced 227.4 million tons of coal, which is 5.4% more than in 2015. At the same time, according to expert estimates, the total area of disturbed lands covers already about 100 thousand hectares, and the area of uncultivated land has sharply increased over the past ten years. Coal mining companies require sound scientific support, new technological solutions are needed. In this regard, the actual task is to evaluate the potential of innovative projects [1].

The infrastructure of coal-mining enterprises is based on the integrated functioning of structural units in various fields of activity. Managers constantly face multidirectional tasks, decisions on which cannot be made using the same factors and models (comparison of enterprises by the level of investment attractiveness, evaluation of the effectiveness of mines, open-pit mines, coal processing plants, comparison of the competitiveness of coal grades and products of processing, comparison of the ecological state of the areas of mining operations, comparison of the technical level of mining machines and equipment, selection and evaluation of engineering and technical workers in the coal industry) [2]. The persons

* Corresponding author: pag_vt@kuzstu.ru

making decision on these tasks periodically face the need for an integrated assessment of the potential of innovative projects. When carrying out such an examination, a large number of functional indicators [3, 4] (mining-geological, technological, technical, economic, ecological, social) must be taken into account, which may have a different nature - quantitative and qualitative. Thus, an important task is the formation of methods and algorithms for evaluating innovation projects of mining enterprises which allow taking into account the maximum possible number of factors associated with the implementation of innovation with minimal time, labor costs and a low proportion of subjectivity.

2 Expert evaluation of innovation projects

2.1 Existing techniques of innovation projects selection and evaluation

In the Russian Federation, a system of financial support for enterprises and research teams involved in the innovative development of a particular industry is widely developed. On the territory of the Russian Federation there are various systems for competitive selection of projects and their funding: federal scientific and technical programs based on government contracts; support of initiative projects in the field of humanities and social sciences by the Russian Humanitarian Scientific Foundation (RGNF) and the Russian Foundation for Basic Research (RFFI); Support of young innovators by the Foundation for Promoting the Development of Small Forms of Enterprises in the Scientific and Technical Sphere (UMNIK), and others. As a rule, in such competitions the evaluation of innovations is carried out only on the basis of the score-expert method [5, 6], and in this connection quite often the decisions made are subjective in nature.

There are three main approaches to the examination of innovative projects, which are based on various methods: *descriptive, comparing the positions "before" and "after", and comparative*. Pros and cons of various methods result in their combined application by expert structures. But at present, there is no universal formalized methodology that would allow a qualitative and effective evaluation of projects. In addition, the analysis of existing methods of innovation management shows that there is no unified approach to the formation of the composition of the criteria for assessing their commercial potential.

2.2 Specifics of evaluation of innovation projects

Since an innovation project by its the content is an investment project aimed at modernization of any type of goods, works or services, in Russia, expert questionnaires developed on the basis of "Methodological recommendations for assessing the effectiveness of investment projects and their selection for financing" (approved by a joint decision of the State Construction Committee, the Ministry of Economy, the Ministry of Finance and the State Committee for Industry of Russia on March 31, 1994, No. 7 - 12/47) are used, as a rule, to assess innovations. At the same time, the main indicators of the effectiveness of the project are commercial efficiency, budget efficiency and economic efficiency. However, innovation projects have certain specificity in terms of assessing their efficiency, in connection with which it is necessary to take into account their difference from investment projects: the need to create objects of intellectual property, increased costs during introduction of new equipment; specific and sufficiently significant risks; use of specific forms of financing (budget, venture, etc.). Therefore, it would be wrong to fully project the methods of investment projects evaluation on the system for assessing the effectiveness of innovation. Thus, there is a need to develop an expert system for assessing the innovative potential of an enterprise, which, under uncertainty, would allow an unlimited number of criteria to be used for evaluation in comparing available alternatives, and in addition would make it possible to assess on the basis of criteria having a different nature.

It should be noted that for the evaluation of innovation projects it is advisable to use not only numerical methods of data processing [7, 8], but also to carry out a purely qualitative assessment of the situation on the basis of logical conclusions, presenting the obtained quantitative values of variables as some linguistic parameters. The advantages of using this approach include the possibility of using the expert's experience. As one of the tools for supporting decision-making in a condition of uncertainty, one can use the apparatus of the theory of fuzzy sets. The basis of the theory of fuzzy sets is the production of fuzzy logical inferences, that is, the derivation of conclusions in the form of a fuzzy set, which corresponds to the current input values using a fuzzy knowledge base and fuzzy operations [9, 10].

3 Expert evaluation of innovations using the methods of system analysis and fuzzy logics

The authors are working to create an information support system for decision-making (Figure 1) on the basis of methods of system analysis and elements of fuzzy logic.

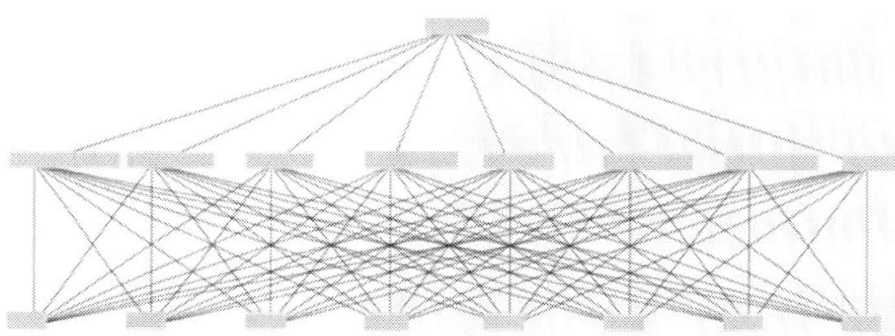

Fig. 1. Information system interface

As a result of the study, the database of characteristics and criteria for project evaluation was determined. As statistical data for filling the test information base, the project evaluation criteria were used which had been used earlier by experts at the following scientific events: competition of the Foundation for Assistance to Small Innovative Enterprises in Science and Technology - UMNIK (2015, Kemerovo), "RFFI. Regional competition SIBERIA" (RFFI - Kuzbass) (2015, Kemerovo). The results of the research showed that in most cases for the preliminary selection of projects by expert structures the following six groups of indicators were used in different combinations.

1) *Scientific and technical level of development*. This group of indicators includes evaluation of the commercial potential of development, patent protection, the relevance of the idea, the authors' awareness of the current state of the problem in this field of activity, the volume and nature of the product market.

2) *Economic efficiency of the project*: the possibility of commercialization of the results of operations, the availability of a business plan for the implementation of the project, the compliance of the volume of declared investments with the objectives of the project, the time to reach the break-even point, the projected average annual profit, the net present value, the internal rate of return, the payback period, presence of distribution risks, potential customers, increase in labor productivity.

3) *Production criteria*: the need in equipment and raw materials, the need in staff, the project's compliance with available fixed assets, the availability of a pilot sample, the degree of development of project documentation, the possibility of approbation of the results of work.

4) *Social significance*: compliance of the project with priority areas of the Russian economy development, possibility of creating new jobs, the use of labor of social categories of citizens, improving the quality of labor, developing social infrastructure, improving the level of labor safety, compliance with environmental standards and safety standards.

5) *Authors' team*: the author's initiative, assessment of capacities of the project leader and its team, scientific contribution of each team member in the problem.

6) *Project demonstration*: assessment of the quality of the prepared presentation of the results of the work / project, the extent to which the author was aware of the issues related to the project during its defense.

The methodology was developed for a comparative expertise of innovative projects, which is based on obtaining a numerical assessment of the potential of innovation projects by defining a linguistic variable for each of the above comparison criteria.

The proposed methodology using the method of analysis of hierarchies and the fuzzy logics apparatus consists of 4 stages.

Stage 1. Elimination of apparent unpromising projects.

Stage 2. Application of the method of analysis of hierarchies (MAI) T. Saati. The method consists in decomposing the problem into simpler parts and stepwise setting the priorities of the components using paired comparisons (Figure 2).

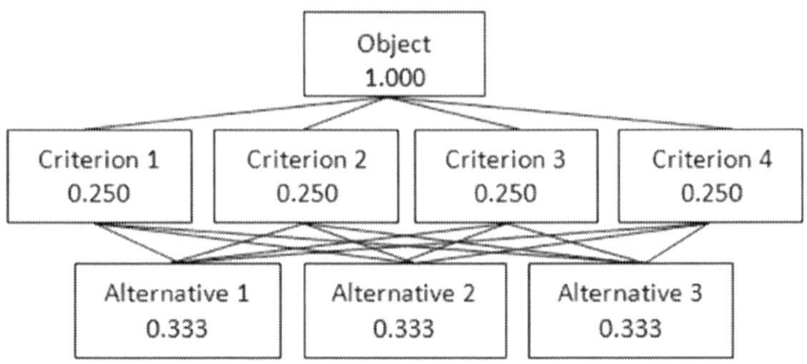

Figure 2. Hierarchy problem presentation

The method of analyzing hierarchies involves implementation of 11 stages [11, 12]:

1) Determination of the list of experts involved in the procedure of evaluating alternatives and taking decisions;
2) Decomposition of the problem in the task hierarchy;
3) Identification of the task solution evaluation criteria;
4) Construction of matrix of paired comparisons of criteria (priorities are set to each criterion based on opinion of each expert);
5) Construction of matrix of pared comparisons of alternatives with consideration of the set level of criteria importance for each decision-maker;
6) Calculation of priorities;
7) Synthesis of priorities;
8) Conformance testing;
9) Calculation of the global vector of priorities by all the levels of the hierarchy;
10) alternate selection of each of the experts for the examination in this project;
11) calculation of the final priority vector taking into account the opinion of all experts who participated in the assessment.

Stage 3. Consideration of criteria as fuzzy sets, which are given on universal sets of variants with the help of membership function in the form of triangular or trapezoidal fuzzy numbers.

Stage 4. Ranking of variants on the basis of intersection of fuzzy set-criteria, which correspond to Bellman-Zadeh scheme known in decision-making theory. When assessing indicators, experts set the lower estimates - pessimistic and the upper estimates – optimistic. Further processing of fuzzy formulated opinions of experts is proposed to be carried out by dephasing of the obtained fuzzy partial estimates and their further processing in a de-phased form [13-15].

Thus, the task of decision making under the given technique is as follows.

There is $V = \{v_1, v_2, ..., v_n\}$ – a set of innovative projects that are subject to multicriteria analysis;

$C = \{c_1, c_2, ..., c_m\}$ – a set of quantitative and qualitative criteria by which alternatives are evaluated;

$B = \{b_1, b_2, ..., b_k\}$ – competence of estimates k of experts conducting experts' assessment.

The problem is to arrange the elements of the set V by the criteria from the set C, taking into account the competence B of the expert estimates. For example, in the role of qualitative criteria for comparing alternatives, we will use the relevance of the idea (K1), the possibility of commercialization (K2), the stage of project elaboration (K3), the expected effective demand for products based on the use of the project (K4); evaluation of the authors' team (K5); scientific novelty (K6). In accordance with the above list of criteria, one can distinguish linguistic variables Xij, where i is the number of the alternative; j is the criterion number. We define the range for all linguistic variables as Ui = [0, 1] for i = 1, 2, 3, 4; J = 1, 2, 3, 4, 5, 6.

The sets of values of linguistic variables (term sets) for each of the criteria are formulated as follows:

$T(X_{i1})$ = the idea is not new + the project has some unique features that create unobvious technological or operational advantages + a significant part of the development is new + the proposed idea is absolutely new.

$T(X_{i2})$ = there are no prospects for the commercial implementation of the product + the market analysis is superficial, the prospects are not well grounded or are questionable + the market analysis is detailed, but the rationale for the commercial feasibility is questionable + the market analysis is detailed, the rationale for the commercial implementation is clear, the rationale for project risks and measures to reduce them don't admit of doubt;

$T(X_{i3})$ = the project is not worked out + the project is not well developed + the project is medium-developed + the project is sufficiently developed + the project is fully developed;

$T(X_{i4})$ = extremely low + low + average + high + extremely high;

$T(X_{i5})$= great experience of the authors of the project, there is a scientific capacity for the project + there is a scientific capacity for the project, but there is no experience with the authors + there is experience of research with the authors, but there is no scientific capacity for the project + there is no scientific experience with the authors of the project and scientific capacity.

Further, the following stages of the examination are expected:

- Fuzzy variables of each term set are used as qualitative estimates of alternatives according to one of the criteria;
- To obtain evaluation of the alternatives under review, the method of expert evaluation is applied and survey of the expert group is conducted;
- Using average estimates of alternatives for each of the criteria, a matrix of average estimates is formed;
- In view of the fact that the criteria have different degrees of importance, their coefficients of relative importance are determined using the hierarchy analysis method, then

the Eigen values of the matrix are calculated and the coefficients of the relative importance of the criteria are calculated;

○ A fuzzy set of estimates of the criteria given by experts is modified by raising to a power corresponding to the coefficient of relative importance of the criterion;

○ Calculation is made of set D needed for selection of the preferred alternative based on the Savage criterion (max (min)). To do this, select the maximum value of the indicator from the minimum for each alternative. The alternative thus obtained is given preference in selecting a project for implementation [16].

Thus, the decision-making procedure takes a clear formalized form, excluding possible subjectivity when taking into account the views of various experts.

4 Conclusion

The developed methods and algorithms were implemented as part of a decision support system that allows to take into account an unlimited number of criteria for evaluating available alternatives, to obtain an estimate based on criteria having a different nature - quantitative and qualitative by using the scales of relative comparisons of T. Saati, and it also takes into account the opinions of several experts who participate in evaluation of projects.

The use of the software reduces intellectual and time costs in the examination of available alternative scientific developments in deciding on implementation of innovation regardless of the specific field of application, the nature of the indicators, the complexity of the project being evaluated and the number of experts that will make decision making comfortable, technological, and, what is most important, effective.

The publication is prepared in the framework of research project № 16-32-00062 supported by RGNF

References

1. Yu.A. Fridman, G.N. Rechko, E.Yu. Loginova, Journal of Mining Sci, **51**, 924 (2015)
2. Yu.A. Friedman, E.Yu. Loginova, G.N. Rechko, Nauka i studia, **3**, 925 (2016)
3. Yu.A. Friedman, G.N. Rechko, *4th International Conference on Science and Technology « Science and Society» ISPC 2014*, 39 (London, 25-26 June 2014)
4. Yu.A. Fridman, G.N. Rechko, A.G. Pimonov, Competitive Advantages and Innovation in Regional Economies, **2**, 206 (2012)
5. A.A. Zakharova, IOP Conf. Series: Materials Science and Engineering, **127**, 1 (2016)
6. A.A. Zakharova, IOP Conf. Ser.: Mater. Sci. Eng, **91** (2015)
7. B. Kosko, Neural Networks and Fuzzy Systems: A Dynamical Systems Approach to Machine Intelligence, 452 (Englewood Cliffs, NJ: Prentice Hall, 1991)
8. A.A. Zakharova, Applied Mechanics and Materials, **379**, 207 (2013)
9. D. Dubois, H. Prade, Fuzzy Sets and Systems: Theory and Applications, **394** (1980)
10. B. Kosko, Int. J. Intell. Syst., Fuzzy knowledge combination, **1**, 293 (1986)
11. T.L. Saaty, *Theory and Applications of the Analytic Network Process* (RWS Publications, 2005)
12. T.L. Saaty, *The Analytic Hierarchy Process* (McGraw-Hill, New York, 1980)
13. L.A. Zadeh, Fuzzy sets, **8**, 338 (1965)
14. L.A. Zadeh, *Aspectson Network and System Theory*, 209 (Winston, N.Y., 1971)
15. M. Sugeno, Fuzzy Measure and Fuzzy Integra, **8**, 218 (1972)
16. M. Sugeno, Fuzzy Decision Making Problems, **6**, 85 (1975)

Gas hydrates in coal seams and their impact on gas-dynamic processes in underground mining

Tatyana Kim[1,*], *Valery* Dyrdin[1], *Vyacheslav* Smirnov[1] and *Valery* Nesterov[2]

[1]T.F. Gorbachev Kuzbass State Technical University, Department of physics, 650000 Kemerovo, 28 Vesennaya st., Russian Federation
[2]T.F. Gorbachev Kuzbass State Technical University, Department of mining machines and systems, 650000 Kemerovo, 28 Vesennaya st., Russian Federation

Abstract. This paper deals with the development and analysis of the method of predicting gas-dynamic phenomena due to the gas hydrates in coal seams. With the help of the method of mathematical modelling the rate of dissociation of gas hydrates depending on the size of particles was determined as well as the correlation between the speed of moving the boundary line of the dissociation of gas hydrates in the pore space of coal and that of extending of the wave of crushing. The regularity of gas pressure distribution in a coal seam marginal zone after dissociation of gas hydrate depending on its saturation capacity and taking into account the changing coefficient of permeability of this zone was stated. It was also found that this kind of gas-dynamic processes in the presence of gas hydrates in the pore space of coals depends on critical value of permeability of a coal seam marginal zone.

1 Introduction

The main objective of this research is to determine the safe working conditions of Kuzbass coal miners namely to develop the method of predicting gas-dynamic phenomena (spontaneous gas emission and sudden coal and gas outburst) due to the gas hydrates, dissociation of which produces extra gas volume making major contribution to the gas balance of the bottom-hole zone of a coal seam thus increasing prediction accuracy. In coal seams the natural gas can be found in different states: free, retained and in the form of a solid coal gaseous solution of the natural gas like gas hydrates [1–4]. Phase transitions from a solid phase into a free one and vice versa are possible providing that thermodynamic parameters are bigger or smaller than the equilibrium ones. Gas hydrates are inclusion compounds where methane molecules are embedded into a water molecules frame. When the dissociation of gas hydrates occurs, extra gas volume is released influencing the formation of different kinds of gas-dynamic phenomena. Present methods of predicting the gas-dynamic phenomena for example, spectro-acoustic method of outburst prediction, do not take into account the release of extra gas volumes due to the dissociation of gas

* Corresponding author: tanyakim@list.ru

hydrates as well as the increasing of the gas pressure which is an active force of the gas-dynamic phenomena development.

2 Materials and Methods

From works [5-9] it follows that in coal seams under certain thermobaric conditions and humidity at some distance from the mine face there can be the areas containing a solid coal gaseous solution.

If there is a sharp stress drop caused by the mine face getting into a zone of increased rock pressure, or into a zone of a coal seam with low strength properties, in a thin coal seam in advance of the face under the action of the tensile stress there could appear the flaws in the pores in a plane orthogonal to a direction of the least squeezing that is parallel to the surface of the mine face. The extending of small flaws causes a drop of gas pressure to a state below the equilibrium. This provokes the beginning of two processes: gas desorption from the surface of macro pores and flaws, and also a dissociation of gas hydrates which lead to the release of extra gas volumes, the increase of the gas pressure in a marginal zone and the formation of gas-dynamic phenomena.

3 Results and Discussion

The possibility of methane gas hydrates formation in the coal under the low humidity has been tested at the Nikolaev Institute of Inorganic Chemistry (Russian Academy of Sciences) in the laboratory of clathrate compounds. It has been found that gas hydrates are formed in a coal matrix providing the thermobaric parameters determine the state of system above an equilibrium curve, and the rate of moisture is 1.5 % or more. In this case the solid coal gaseous solution is formed in pore channels and evenly distributed in the form of crystals to the entire volume of coal providing the given parameters. So, it has been experimentally proved that the formation of solid natural gas solutions in coal pores can occur at moisture level characteristic for the coal seams dangerous because of the gas dynamic phenomena. To define the possible influence of solid coal gaseous solutions on extra gas emission from the solid block we set a task to determine the dissociation rate of gas hydrates depending on their sizes and allocation in a coal matrix. Drilling and maintenance of the deep oil and gas holes shows that the formation of gas hydrates in solid mass at their bottom leads to stopping of the process [10-14], as the speed of their decomposition is connected with the lowering the temperature of the environment. In this regard, we determine the gas hydrate dissociation rate in a coal seam, that is we determine how fast the decomposition into gas and water of a gas hydrate particle with the size corresponding to the size of pores in the structure of coal takes place. Let us take T – current temperature of coal; $T_0 = 278K$ – the temperature at which the decomposition of a gas hydrate takes place; r_2 –radius of a gas hydrate particle.

According with the Fourier equation, the heat flow from coal in solid to a gas hydrate particle equals

$$dQ_1 = \chi \cdot \frac{\partial T}{\partial r} \cdot 4\pi r^2 dt\,, \tag{1}$$

where χ – thermal conductivity of coal;

The heat necessary for gas hydrate dissociation equals:

$$dQ_2 = \lambda \cdot dm\,, \tag{2}$$

where $\lambda = 533 \cdot 10^3$ J/kg – specific heat of gas hydrate decomposition, $dm = 4\rho_1 \pi r_2^2\, dr_2$ – the mass of decomposing gas hydrate; $\rho_1 = 840$ kg/m3 – density of methane gas hydrate.

Equaling the expressions (1) and (2), we get

$$\frac{dr_2}{dt} = \frac{\chi}{\lambda\rho_1} \cdot \frac{r_1^2}{r_2^2} \frac{\partial T}{\partial r_1}, \tag{3}$$

where $\frac{dr_2}{dt} = v$ – rate of gas hydrate particle dissociation in the block of coal mass.

It was experimentally stated that in the process of gas hydrate particles decomposition temperature T recovers for the fractions of seconds thanks to the heat influx from the surrounding rock mass.

Gas hydrate particles in the coal mass have structure of Type-I which is close to the spherical structure.

Taking into consideration initial and marginal conditions for the chosen model, we have the temperature of the coal mass at the initial time equal: $T(r_1,0) = T_1$, and the temperature of a gas hydrate particle at the beginning of decomposition also equal T_1, i.e. $T(r_2,0)\big|_{r_2=r_1} = T_1$. With the help of mathematical modelling we got the dependence of rate of gas hydrate particles dissociation from their sizes (fig.1.) by reference to the characteristics of methane gas hydrate and coal [15, 16].

The rate of a dissociation of gas hydrates nonlinearly depends on sizes of dissociating particles and if the diameter is 4 nanometers, it is equal to 28 nanometers/ms while with the increase of the size of gas hydrate particles the rate of a dissociation decreases. It is significant that the module of velocity of a hydrate dissociation in pore space of coal considerably surpasses the velocity of a crushing wave (A. S. Khristianovich). As the natural moisture content of coal seams, dangerous and endangered by gas-dynamic phenomena, is not more than 1.5-2.0%, so gas hydrate particles are at a certain distance from each other, and their dissociation rate depends on their size only, i.e. the size of the pore channels, where they are formed.

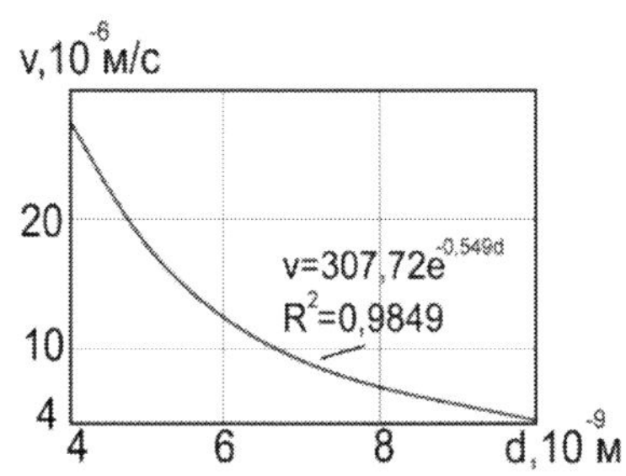

Fig. 1. Dependence of rate of gas hydrate particles dissociation from their sizes d –size of a gas hydrate particle

The dissociation of clathrate hydrates in the solid rock at the initial time occurs near the boundary of the inelastic and elastic deformations and eventually moves deep into the rock mass. The rate of moving the dissociation boundary line we find from the solution of the differential equation by A. M. Maximov:

$$k_1 \cdot (\frac{\partial p_1}{dx})|_{x=\xi} - k_2 \cdot (\frac{\partial p_2}{\partial x})|_{x=\xi} = \frac{W\mu\ p_0}{p_p} \cdot \frac{d\xi}{dt}, \quad (4)$$

And the current pressures p_1, p_2 in the areas of inelastic and elastic deformations (areas 1 and 2) we define from the equation of methane moving in a layer of one-dimensional isothermal flow, where p_0 – atmospheric pressure, Pa; μ – dynamic viscosity of the gas, Pa·s; ς – boundary line coordinate of hydrate decomposition, m;

W – the content of gas (hydrated state) depending on the porosity and moisture of the coal seam in a unit volume of the porous environment; k_1, k_2 – permeability coefficients in the areas 1 and 2, darcy; p_p – hydrate decomposition pressure at seam temperature, Pa.

The content of a gas (hydrate state) is the following:

$W = n(x) = n_0(1-e^{-bx}) \cdot V_0^1$, where $n(x)$ – function of particles quantity of the distance, V_0^1 – Volume of one particle of a gas hydrate, $n_0 \cdot V_0^1 = n_0 \cdot \frac{4}{3} \cdot \pi \cdot r^3$;given that $r = \sqrt{\frac{S_{пов}}{4\pi}}$, and Wcan be found from the formula: $W = \frac{n_0 S}{3V}\sqrt{\frac{S}{4\pi}}(1-e^{-bx})$, where n_0- concentration of gas hydrate particles.

The given task was solved for the seam "Vladimirovsky" of "Severnaya" mine and was reported by the authors in. Finding from (4) $v = \frac{d\xi}{dt}$ we determine the rate of movement of gas hydrate dissociation boundary for different decomposition pressures characteristic for coal seams with different thermodynamic parameters. The results of calculation are given in fig.2.

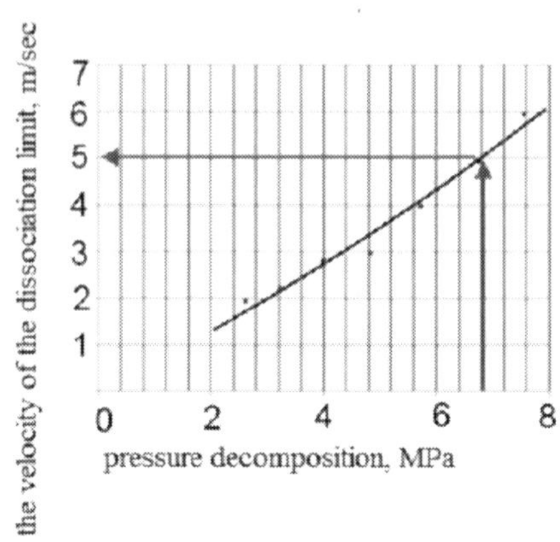

Fig. 2. Rate of movement of gas hydrate dissociation boundary for different decomposition pressures

Analyzing the found dependence we can see that the rate of movement of gas hydrate dissociation boundary for decomposition pressure equal $6.8 \cdot 10^6$ Pa is 5 m/s and is comparable to the speed of extension of a crushing wave calculated according to the method of A. S. Hristianovich which is up to 5.95 m/s for the given requirements. Thus, the crushing wave movement is accompanied by the decomposition of gas hydrates and the release of natural gas extra volume which accounts for the possible participation of gas hydrates in formation of the gas dynamic phenomena.

The distribution of gas pressure in a gas hydrates decomposition ahead of a working face depending on its saturation capacity and taking into account the changing coefficient of permeability is also of great interest.

In solving this task, we suppose that at the maximum point of the zone of bearing pressure, we have the minimum permeability of rock mass. Out of the zone of mine entry influence the coefficient of permeability takes constant value k_∞.

The other boundary conditions for the permeability coefficient are:

a) permeability at a development face $k(x)\big|_{x=0} = k_0$, i.e. the beginning of the x-axis coincides with a development face of a mine working and $x > 0$ – on the depth of rock mass;

b) $k_3(x)\big|_{x=\xi} = k_4(x)\big|_{x=\xi} = k$, where $x = \xi$ - distance from the mine face to the beginning of the zone, where the gas hydrates dissociation takes place, i.e. it is assumed that the dissociation of clathrate hydrates begins at the boundary of the elastic (x_3) and inelastic (x_4) deformations.

Methane movement in a coal seam is presented by the Bussinesk differential equation for the one-dimensional isothermal flow gained by L. S. Lejbenzon:

$$\frac{\partial p}{\partial t} = \frac{k}{2m\mu_g} \cdot \frac{\partial^2 p^2}{\partial x^2}, \tag{5}$$

m – porosity coefficient of environment; p – pressure of gas; μ_g – dynamic viscosity of gas; k – coefficient of permeability. First of all, we define the number of moles v of the gas present in hydrated state in a coal seam:

$$v = \frac{V_{coal}(1-\varepsilon)\rho_{hydrate}}{\mu_{water} n}\beta, \text{ where } \beta = \frac{W}{1-\varepsilon}\frac{\rho_w}{\rho_h} \text{ – saturation capacity of hydrate;}$$

Water density – ρ_w = 1000 kg/m^3; $\varepsilon = 0.126$ – natural mass of gas in a mass unit of gas hydrate; density of hydrate of rock gas ρ_h = 883 kg/m^3; W – natural moisture of coal; μ_{water} – molar mass of water; n – quantity of molecules of water on one molecule of gas in each mesh of gas hydrate of Type-1.

As an example, we have calculated the distribution of gas pressure after the beginning of the dissociation process for the "Vladimirovskiy" seam of "Severnaya" mine. The results are given in Fig. 3.

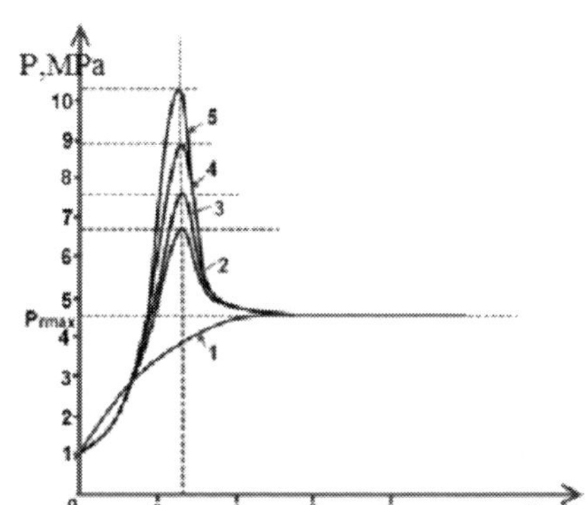

Fig. 3. The distribution of gas pressure after the beginning of the gas hydrates dissociation process and changing permeability of a boundary zone of a coal seam: 1 - the initial distribution of the gas pressure; 2 - hydrate saturation at 1.9%; 3 - hydrate saturation at 2.6%; 4 - hydrate saturation at 3.9%; 5 - hydrate saturation at 5.2%; P_{max} – gas pressure out of the zone of mine entry influence

The extra volume of the free gas depends on extent of the area containing gas hydrates, i.e. on its hydrate saturation. Thus, when the inherent humidity of a seam is from 1.5 to 2.5%, additional gas pressure in the zone of gas hydrates decomposition increases by 2-2.5 times directly affecting the increase of gas pressure ahead of a working face and can cause spontaneous gas emission and sudden coal and gas outburst.

To determine the type of gas-dynamic phenomena (spontaneous gas emission or gas hazard), we calculate the hazard ratio of emission gas in the reservoir R^n by the method of prof. V.I. Murashev, taking into account the pressure rise due to the dissociation of gas hydrates of methane, and by the transition point of the curve through zero we define critical $k_{кр}$ for the given seam.

To define the permeability coefficient in practice, we assume that in the marginal zone of a coal seam ahead of a working face it varies according to the law:

$$k(x) = k_0 e^{-b \cdot \sigma(x)}, \quad (6)$$

where k_0 - coefficient of permeability of coal in the initial state, m^2 (determined in the laboratory conditions by the method of V.V. Hodot); b - a constant which depends on the compression properties of the coal seam; $\sigma(x)$ - tension at the point considered.

It is necessary to know the coefficient of permeability of rock mass at the boundary line of inelastic and elastic deformations where the clathrate hydrates dissociation begins. The distance to the boundary line is determined from the inflection point of the experimental culm yield curve in drilling wells and the stresses at this point are found in accordance with their linear approximation in the zone of maximum stress state. For this purpose we determine the stresses at the bottom and at the peak of the bearing pressure according with the well-known formulas of VNIMI.

For the values of the permeability coefficient k at the interface between elastic and inelastic deformations exceeding the critical value for the given coal seam, gas content production occurs (since there is an intensive gas filtering to the mine face) and for $k < k_{kr}$ - outburst situation is formed.

As an example, for the seam "Vladimirovskiy" of the mine "Severnaya" it was calculated that the practical value of the permeability coefficient at l = 3.5 m is 3 10^{-15}m^2, and the critical value for these conditions is 10/10^{-15} m^2, i.e.

k_{pr} (3.5 = $3 \cdot 10^{-15}$ m^2 < $k_{кр}$=$10 \cdot 10^{-15}$ m^2, so this section is prone to spontaneous outbursts of coal and gas (Figure 4).

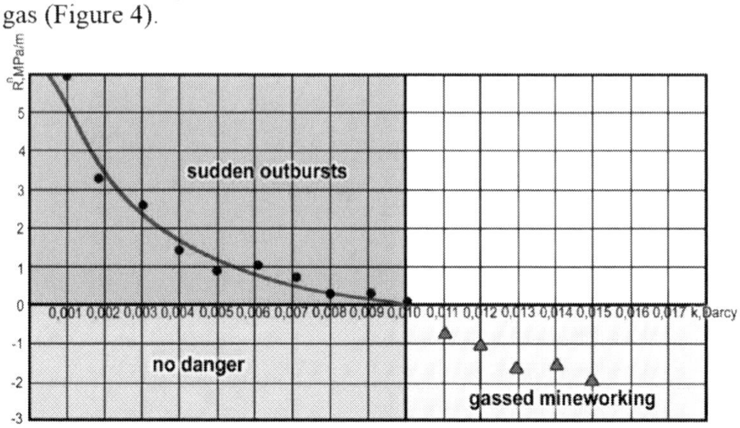

Fig. 4. Dependence of the gas dynamic phenomena types on the rock mass permeability at the interface between elastic and inelastic deformation

4 Conclusions

This technique allows us to identify areas in coal seams prone to gas-dynamic phenomena taking into account the possibility of methane gas hydrates decomposition and to improve the reliability of the current forecast of dangerous coal and gas outbursts by taking into account the dissociation of gas hydrates, which are the major contributors in the gas balance in the process of gas-dynamic phenomena formation.

References

1. Yu.A. Dyadin, E.G. Larionov, A.Y. Manakov, E.Y. Aladko, Mendeleev Communications, **209** (1999)
2. Melnikov V.P., Nesterov A.N., Podenko L.S., Reshetnikov A.M., Shalamov V.V., Chemical Engineering Science, **71**, 573 (2012)
3. Yao SuPing, Chinese Science Bulletin, **25**, 2706 (2011)
4. Sloan, E. D. *Clathrate Hydrates of Natural Gases – 3rd ed.,* (CRC Press, 2008)
5. V.V. Dyrdin, V.G. Smirnov, S.A. Shepeleva, Journal of Mining Sci, **49**, 908 (2013)
6. V.G. Smirnov, A.Y. Manakov, E.A. Ukraintseva, V.V. Dyrdin, I.S. Terekhova, A.G. Ogienko, Fuel, **166**, 188 (2016)
7. Clarkson, C. R., R. M. Bustin, Fuel, **78**, 1333 (1999)
8. Christophe Dicharry, Pascal Gayet, Gerard Marion, Alain Graciaa and Anatoliy N. Nesterov, J. Phys. Chem., **109**, 17205 (2005)
9. Lijun Yan, Guangjin Chen, Weixin Pang, J. Phys. Chem., **109**, 6025 (2005)
10. Liu, J, Jiang X, Huang X, et al. Fuel, **89**, 3884 (2010)
11. Skoczylas N., Arch Min Sci, **57**, 861 (2012)
12. Sobczyk J., Fuel, **90**, 1018 (2011)
13. Yi J, Akkutlu IY, Karacan CÖ, Clarkson CR., Int J Coal Geol, **77**, 137 (2009)
14. Jianming, Yu Bufan, Wang Youan, al. *The technology hand book of coal and gas outburst control* (China University of Mining and Technology Press, 2006)
15. Lama, R. D., Australia, Report GSM21 (1981)
16. State Administration of Quality and Technical Supervision, State Administration of Coal Mine Safety. Prevention of Coal and Gas Outburst Provision (Beijing: ChinaCoal Industry Publishing House, 2009)

Environmental and Economic Efficiency of Comprehensive Technology of Sulfur Oxides, Nitrogen Oxides and Mercury Removal from Flue Gases

Svetlana Grigashkina[1*], *Tatyana* Galanina[1], *Vladimir* Mikhailov [1], *Tatyana* Koroleva[1], *Evgeniya* Trush [2]

[1]T.F. Gorbachev Kuzbass State Technical University, 28 Vesennyaya st., 650000, Kemerovo, Russia
[2]Tomsk State University of Architecture and Building, 2 Solyanaya Sq., 634003, Tomsk, Russia

Abstract. The article reviews the issues of environmental and economic efficiency of the comprehensive technology of removing sulphur and nitrogen oxides and mercury from flue gases. The purpose of the study is to justify the choice of the methods for the analysis of environmental and economic effectiveness of the designed air protection activities and to perform a number of calculations. The article analyses main processing methods for reducing the concentration of sulfur and nitrogen oxides in the flue gases from the combustion of coal fuels, thus generating maximum ecological and economic benefits. The main result of the work is development of the adapted system of technical, environmental and economic indicators characterizing the high efficiency of the proposed technical solutions. The results of the study can be recommended to heating enterprises in support of effective environment-friendly management decisions justifying implementation of appropriate flue gas cleaning technologies.

1 Introduction

The problem of air pollution is the most relevant for such countries as Russia and China with a high concentration of thermal power plants. This situation is largely due to the use of coal-fired CHP (combined heat and power), having a high degree of versatility, on the one hand, and, on the other hand, such a disadvantage, as high physical and moral wear of the equipment, as well as significant environmental pollution. Currently increasing production growth rates and the tightening of environmental legislation [1, 2, 3] generate new requirements for the equipment used in the thermal power stations, in particular, special attention is paid to employed boiler plants which must comply with the new requirements related to the provision of permissible values of emission of some combustion products of solid, liquid and gaseous fuels. Fuel combustion at CHP is linked to generation of combustion products containing fly ash, sulfur dioxide, nitrogen oxides, and other products of incomplete combustion. In general, in the Russian cities the average of the maximum measured concentrations of impurities other than sulfur dioxide and nitrogen oxide exceed 1 MAC, the highest level of air pollution occurs in cities located on the territory of the Siberian and Ural

*Corresponding author: grigashkina_si@mail.ru

Federal Districts.

A similar situation exists in China, where the main cause of pollution of the atmosphere is a low level of energy product treatment. Another problem is also connected to moral and physical wear and tear of the equipment of enterprises. Furthermore, coal produced in China has a high sulfur content. Cities located in the basins poorly permeable to winds are particularly vulnerable to air pollution (Chongqing, Wuhan, Beijing, Nanjing and others) [4, 5].

The reviewed situation characterizes the significant environmental problems requiring immediate solutions:

- Reduction of emissions amount;
- Introduction of new environmentally clean technologies, including the best available techniques (BAT);
- Minimization of the impact of the anthropogenic complex on the environment;
- Restoration of the natural balance.

The scientific novelty of the work lies in the theoretical and practical justification of environmental and economic efficiency of the comprehensive technology developed by a group of scientists from KuzSTU for effective removal of sulfur oxides, nitrogen and mercury from flue gases. The practical result of the work relates to the possibility of its implementation at coal-fired CHP and district power plants of the Kemerovo region and other regions which adversely affect the environment.

2 Results and Discussion

The analysis of scientific, technical and patent literature has shown that in order to reduce the content of sulfur and nitrogen oxides in the flue gases from the combustion of coal fuels three main technological methods are used [6, 7, 8, 9, 10, 11, 12, 13]:

- Reduction of the concentration of sulfur in the original coal coming to power and steam stations;
- Reduction of the concentration of sulfur oxides formed during combustion of coal fuel containing sulfur by introducing Sulphur Capture Agents (SCA) in the combustion zone;
- Reduction of the concentration of sulfur and nitrogen oxides in flue gases by means of purification of the latter using a number of known methods, most effective of which are methods of sorption and catalytic purification of flue gases.

Despite the fact that all the techniques described are effective methods, each of them used alone does not achieve complete cleaning. KuzSTU scientists have developed a new technology based on a comprehensive efficient cleaning of coal-fired thermal power plants flue gases of sulfur oxides, nitrogen oxides, and mercury.

This technology has the following advantages:

- Possibility of burning both coal and coal-water fuel;
- Possibility of differentiating the ways of coal fuel combustion (in a swirl burner, in a compact bed, in a fluidized bed);
- Introducing SCA to the combustion zone as part of the fuel or separately;
- Removing harmful impurities from the flue gases as part of the furnace and in the flue gas and smoke path;
- Performing a comprehensive removal of flue gases by several methods:

1) desulphurization at the coal fuel combustion stage, including SCA feed to the fuel during its preparation for combustion;
2) SCA feed directly in the burner in accordance with the individual scheme;
3) catalytic purification of waste gases from sulfur and nitrogen oxides;
4) absorption cleaning of flue gases from sulfur oxides, nitrogen oxides and mercury.

This comprehensive technology allows an increase of the cleaning efficiency up to 90% and significantly reduces the burden on the environment.

In order to assess the negative impact, the ecological and economic parameters [14-17] are used, of which the economic damage being the most crucial. The economic damage of air pollution (E_{dap}) is determined by the following formula:

$$E_{dap} = \gamma_s \cdot R_{ind} \cdot C_{esa} \cdot Mr, \text{ roubles,} \quad (1)$$

where γ_s – the specific economic damage from air pollution of one specific tonne of pollutants equivalent, 47.5 RUB./sp.t.; R_{ind} – indexation rate; C_{esa} – coefficient of ecological situation for atmospheric air (it is differentiated by environmental elements and by regions, for Kemerovo region it is 1.44); M_r - reduced mass of the annual emission of harmful substances into the atmosphere, sp.t./year. M_r is calculated by the formula:

$$M_r = \sum_{i=1}^{n} m_i \cdot A_i, \tag{2}$$

where m_i – actual mass of i- air pollutant, t; i – type of pollutant; n – total quantity of pollutants; A_i – indicator of relative hazard of i-th pollutant (sp.t./t) that is determined by the formula 3:

$$A_i = 1/ADM_{ci}, \tag{3}$$

where ADM_{ci} – average daily maximal allowable concentration of i-th air pollutant, mg/m^3.

The simplified calculation of economic efficiency of environment protection activities (without time factor) is made by the formula:

$$R = \Delta P - C_r, \tag{4}$$

where R – net economic effect from implementation of an environment protection activity, RUB.; ΔP – economic result of environment protection activities, RUB.; C_r – reduced costs of implementation of the environment protection activity, RUB.

$$\Delta P = P_d + AI, \tag{5}$$

where P_d – prevented damage, RUB.; AI – additional income (increased income based on improved technical and economic indicators, including decrease in pollution charge), RUB.;

P_{da} is calculated by the upgraded formula 1:

$$P_{da} = \gamma_s \cdot R_{ind} \cdot C_{esa} \cdot \Delta M_r, \text{ roubles} \tag{6}$$

where P_{da} – prevented economic damage from air pollution, RUB.; ΔM_r – change in the reduced mass of annual pollutant emission, sp.t./year, that is determined by the formula:

$$\Delta M_r = \sum_{i=1}^{n} \Delta m_i \cdot A_i, \tag{7}$$

where Δm_i – change (decrease) in the actual mass of i- air pollutant, t, that is calculated by the formula:

$$\Delta m_i = m_{i1} - m_{i2}, \tag{8}$$

where m_{i1}, m_{i2} – actual masses of i- air pollutant, respectively, before and after implementation of an environment protection activity, t.

C_r is determined by the formula:

$$C_r = CO_c + CI \cdot E_{pti}, \tag{9}$$

where CO_c – current and operating costs of environment protection activity implementation, RUB., CI – capital investment in the environment protection activity, RUB.; E_{pti} – profit-to-investment ratio which is recommended to be taken equal to 0.15.

To estimate the prevented economic damage as a result of implementation of the comprehensive flue gas cleaning system, Table 1 shows the calculation of the change in the reduced mass of air pollution for power plants with capacity of 75 MW and 325 MW. For ease of notational convenience, the following abbreviations are taken: SCA - cleaning method using Sulphur capture agent; CC - catalytic cleaning; AC - absorption cleaning.

Based on the data presented in Table 1, the prevented economic loss was calculated using formula 6 comprising 184,150 thousand RUB. for 75 MW power plant, and 797,974 thousand RUB. for 325 MW power plant.

Table 2 shows the calculation of the economic benefit from implementation of the comprehensive environment protection activity on flue gas cleaning.

Table 2 shows the high value of the potential additional income obtained as a result of calculation of pollution charge reduction (over 2 million RUB. for 75 MW power plant and about 10 million RUB. for a 325 MW power plant).

Based on the data presented in Table 2, the economic effect of the environment protection activity on flue gas cleaning was calculated by formulae 4 and 9 for power plants of various capacities shown in Table 3.

Table 1. Calculation of changes in the reduced mass of air pollution for power plants of various capacities

Indicators	**75 MW**			**325 MW**		
	SCA	**CC**	**AC**	**SCA**	**CC**	**AC**
Pollutant	SO_2	SO_2	SO_2	SO_2	SO_2	SO_2
		NO_2	Hg		NO_2	Hg
m_{i1}, t/year	52500	15.750	1575	227.500	68,250	6825
		143	-		620	-
m_{i2}, t/year	15750	1575	78,75	68.250	6825	341.25
		14.3	-		62	-
Δm_i, t/year	36750	14.175	1496.25	159.250	61.425	6483.75
		128.7	0.00625		558	0.00625
ADM_{ci}, mg/m^3	0,05	0.05	0.05	0.05	0.05	0.05
		0.04	0.0003		0.04	0.0003
A_i, sp.t/t	20	20	20	20	20	20
		25	3333.3		25	3333.3
ΔM_i, Sp.t./year	735000	283500	29.925	3185000	1228500	129.675
		3217.5	20.8		13.950	20.8
ΔM, total, Sp.t./year	1051663.3			4557145.8		

Table 2. Calculation of the economic benefit from implementation of environment protection activity

Heating capacity of power plant	P_d, thous. RUB.	*AI*, thous. RUB.	ΔP, thous. RUB.
75 MW	184,150	2297	186,447
325 MW	797,974	9959	807,933

Table 3. Calculation of the economic effect of the environment protection activity on flue gas cleaning

Heating capacity of power plant	ΔP, thous. RUB.	*CI*, thous. RUB.	C, thous. RUB.	C_r, thous. RUB.	*R*, thous.RUB.
75 MW	186.447	119.032	137.695	155.549	30.898
325 MW	807.933	419.117	437.023	499.891	308.042

Table 3 shows that the net economic effect from the developed comprehensive technology ensuring decrease in the concentration of sulphur oxides, nitrogen oxides and mercury in the gas emissions of coal-fired CHPs of 75 MW and 325 MW amounts, respectively, to 30.898 thousand RUB. and 308.042 thousand RUB.

3 Conclusion

The study of environmental and economic efficiency of the comprehensive purification technology for thermal power plants flue gases led to the following conclusions:

- The environmental degradation in China and Russia caused by air pollution by thermal power plants was analysed;
- The priority environmental problems requiring immediate solutions were identified;

- The scientific, technical and patent literature was analysed in order to study the technology aimed at reducing the concentration of sulfur oxides, nitrogen oxides and mercury in the flue gases from combustion of coal fuel;
- The main advantages of the proposed treatment technology were formulated;
- The techniques for calculation of basic ecological and economic indicators and for assessment of the cost-effectiveness of environmental protection measures were selected;
- The calculation of basic ecological and economic indicators characterizing the ecological and economic efficiency of the proposed measures on flue gas cleaning was performed;
- The use of the proposed three-stage technology ensuring the maximum degree of flue gas purification for similar objects of heating power industry was justified.

References

1. E.V. Ryumina, A.M. Anikina, Studies on Russian Economic Development **18**, 5 (2007)
2. V.M. Tumin, A.G. Koryakov, E.P. Nikiforova, World Applied Sciences Journal **25**, 6 (2013)
3. A.I. Shinkevich, A.A. Lubnina, A.G. Koryakov, V.G. Mikhailov, E.L. Vodolazhskaya, International Review of Management and Marketing **6**, 2 (2016)
4. http://www.iiav.org/archives_icsv_last/2014_icsv21/content/papers/papers/full_paper_600_20140314144048495.pdf.
5. http://english.mep.gov.cn/News_service/news_release/201602/t20160219_330460.htm
6. Z.O. Siagi, M. Mbrawa, R&D Journal of the SOuth African Institution of Mechanical Engineering **1** (2008)
7. I.M. Zasypkin, V.I. Murko, V.I. Fedyaev, M.P. Baranova, Therm. Science **16**, 4 (2012)
8. N.V. Shikina, S.R. Khairulin, N.A. Rudina, T.N. Teryaeva, E.S. Mikhailova, Z.R. Ismagilov, Eurasian Chemico-Technological Journal **17**, 2 (2015)
9. Z.R. Ismagilov, M.A. Kerzhentsev, S.A. Yashnik, S.R. Khairulin et al., Eurasian Chemico-Technological Journal **17**, 2 (2015)
10. V.I. Murko, V.I. Fedyaev, V.I. Karpenok, I.M. Zasypkin, Yu.A. Senchurova, A. Riesterer, Thermal Science **19**, 1 (2015)
11. Z.R. Ismagilov, N.V. Shikina, N.A. Rudina, V.A. Ushakov, R.R. Potokina, T.N. Teryaeva, Solid Fuel Chemistry **49**, 4 (2015)
12. V.I. Murko, V.I. Karpenok, Yu.A. Senchurova, V.A. Khyamyalyainen, O.V. Tailakov, *Coal in the 21st Century: Mining, Processing and Safety* (2016)
13. N.V. Shikina, O.V. Tailakov, Z.R. Ismagilov, *Coal in the 21st Century: Mining, Processing and Safety* (2016)
14. T.V. Galanina, M.I. Baumgarten, V.G. Mikhailov, T.G. Koroleva, G.S. Mikhailov, IOP Conf. Series: Earth and Environmental Science **50**, 012030 (2017)
15. S.A. Zhironkin, A.A. Khoreshok, M.A. Tyulenev, G.A. Barysheva, M.C. Hellmer, IOP Conf. Ser.: Mater. Sci. Eng., **142**, 012127 (2016)
16. V.G. Mikhailov, G.S. Mikhailov, A.G. Koryakov, Journal of Mining Science **51**, 5 (2015)
17. N.N. Golofastova, V.G. Mikhailov, T.V. Galanina, Coal in the 21st Century: Mining, Processing and Safety (2016)

Utilization prospects for coal mine methane (CMM) in Kuzbass

Oleg Tailakov[1,2*], *Denis* Zastrelov[1], *Vitaliy* Tailakov[3], *Maxim* Makeev[1], *Peet* Soot[4]

[1] Institute of Coal of Federal Research Center SB RAS, 650065, Kemerovo, Leningradsky ave., 10, Russia
[2] T.F. Gorbachev Kuzbass State Technical University, 650000, Kemerovo, Vesennya str., 28, Russia
[3] JSC "Uglemetan Service", 650000, Kemerovo, Tereshkova str., 40, Russia
[4] CMM Energy LLC, 333 S. State St. # V-133, 97034, Lake Oswego, Oregon USA

Abstract. Coal mine methane (CMM) is a considerable, relatively unused, resource in Russia. With the second largest proven coal reserves in the world, behind the USA, Russia has a tremendous coalbed methane (CBM) resource. The coal mine methane (CMM) emissions which are created from the CBM resource by way of coal mining yield a significant fraction of the anthropogenic greenhouse gas emissions in the country. Developing CMM utilization projects will mitigate at least some of these emissions. Having evaluated the CMM resources at various underground coal mines in Kuzbass, it has been determined that it is economically viable to apply a variety of CMM utilization technologies in Russia. CMM use for electricity production was found to be quite capital intensive. Although economically viable, this option was the least attractive of the four technologies which were considered. Thermal energy production from boilers and delivering liquefied CMM at vehicle fueling stations were of intermediate potential. The most attractive option is the fueling of vehicles with compressed CMM. This required the lowest capital investment and had the highest paybacks, including a very rapid 2 ½ year payback.

1 Introduction

Russia has the second largest proven reserves of coal in the world, behind only the USA (BP, 2014). However, in coal production they are number six (6) in the world, having produced 165 million tons of coal in 2013. The coal reserves provide a large source for coalbed methane (CBM) and the production still yields a significant amount of coal mine methane (CMM) from underground coal mines. This CMM is the source of a significant fraction of the anthropogenic greenhouse gas emissions in this country. Developing CMM utilization projects will mitigate at least some of these emissions.

* Corresponding author: oleg2579@gmail.com

The Kuznetsk Basin (also known as Kuzbass) is not the largest coal basin in Russia, but is the location of the most underground coal mine production. Hence, it is the source of most of the CMM production in Russia. The CBM resource in Kuzbass is 13 trillion m^3 (Gazprom, 2014). About 20% of these CBM resources are at mineable depths down to 600 meters. Another 40% are at depths down to 1200 meters, with the balance of this defined resource found within 1800 meters of the surface. There are likely additional resources in deeper horizons, but those have not been defined. The rest of Russia, outside Kuzbass, holds much more CBM (reportedly over 70 trillion m^3), but it is not located near major population centers like Kuzbass and has not been explored as much. The Kuzbass CBM resource is larger, 13.1 trillion m^3 vs. 8.4 trillion m^3 than the resource in the San Juan Basin in the USA (Potential Gas Committee). And, the San Juan Basin is the source of the largest production of CBM in the USA. The low permeability of the coal in Kuzbass will not allow as prolific CBM production from Kuzbass as from the San Juan, but the resource is still there.

This article considers the CMM resources in Kuzbass, taking into account the various methane sources in underground coal mines. Those include the coal seam being mined as well as the coal seams overlying and underlying the active seam during and after mining. Four technologies were considered for the utilization of the CMM:

- Electric power generation;
- Thermal energy production in boilers;
- Fueling of vehicles with compressed CMM; and
- Fueling of vehicles with liquefied CMM.

The economic viability of each of these options were evaluated with an economic analysis which calculated net present value (NPV), internal rate of return (IRR), investment payback period in years and profitability index (PI).

It views evaluation approaches of coalbed methane resources developed in the area, taking into account the peculiarities of its supply from coal seams during and after mining. It was given an estimation of economic efficiency of methane utilization in generating stations, boilers and gas stations, fueled by compressed and liquefied coal mine methane.

2 CMM Resources

The coal mines in Kuzbass yield the CBM in the coal in the various forms of coal mine methane (CMM): drained gas, ventilation air methane (VAM), and abandoned mine methane (AMM). The drained gas is often the only form of coal mine methane that is considered by developers, but the other forms of CMM also exist and are a potential resource for utilization.

Kuzbass coal mine degasification systems bring 100-200 million m^3 of drained methane to the surface annually. This methane comprises up to 80% of the methane/air mixture that is produced. One can use this methane/air mixture in a variety of applications, including production of electricity and/or heat, for motor fuel and for chemical products. The use of degassing methane in these volumes, for example, would allow the production of 30-60 MW of electricity. To evaluate the possibility and effectiveness of developing CMM projects, a complex approach was considered, which consisted of assessing the CMM resources and performing an economic analysis of alternative utilization options.

Mining experience at Kuzbass coal mines show that degassing boreholes drilled from the surface into the goaf ("gob" in American terminology) are the most stable sources of gas with the required concentration. According to applicable regulations, the methane cannot be less than 25% of the drained gas for transportation.

The goaf is that zone which is formed from the collapse of the strata overlying the coal when the mined seam has been extracted with the longwall mining machine. Since this zone is not isolated from the active mining area, the mine is ventilated in such a manner as to minimize the movement of the methane from the goaf to the active area and degassing boreholes are drilled into the goaf from the surface.

In the underground coal mines, the goaf methane consists of methane entering that zone during the unloading of the coal seam massif and which has migrated into that zone from the underlying and overlying coal seams, and from the remaining coal mass formed from remaining broken coal pillars. Given this mechanism, it is important to begin the understanding of the CMM resource by studying the formation of the goaf. In general, the amount of coal mass in the goaf is described by the mathematical formula in Equation 1.

$$V_i^{seam} = \sum_{j=1}^{n} \iint_{S_j} m(x, y) dxdy, \tag{1}$$

where $j = \overline{1,n}$; $i = \overline{1,k}$; n – number of pillars in the i-th seam; k – number of coal seams; $m(x,y)$ – thickness of the i-th coal seam.

Figure 1 shows a 3-D distribution of the density of methane resources in one of the Kuzbass coal mines within the developed coal seam.

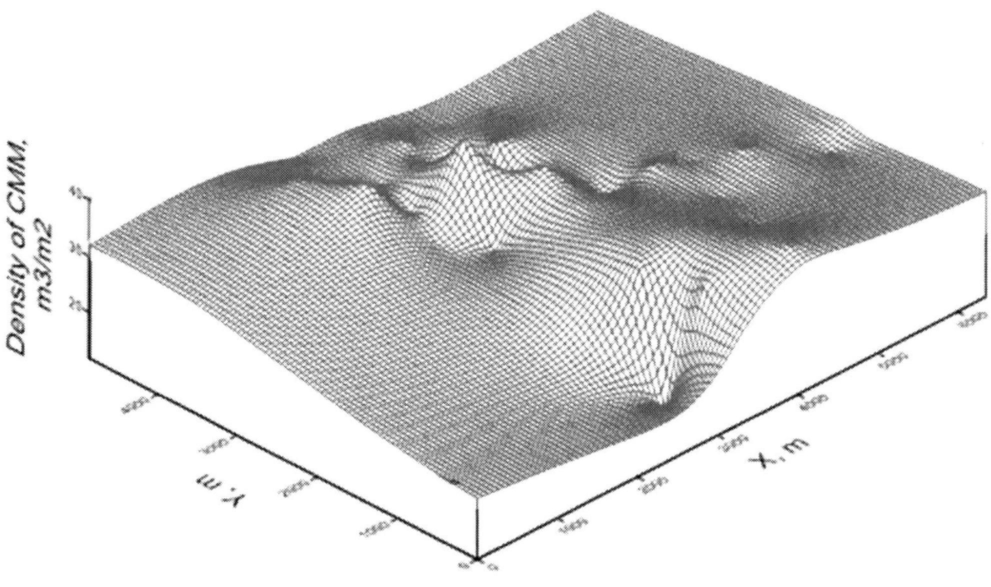

Fig. 1. 3-D representation of the density of coalbed methane resources in a Kuzbass underground mine

When drilling boreholes in the dome of the goaf, the concentration of methane-air mixture is achieved, sufficient for its use as efficient fuel.

Having information on the potential volume of CMM in these man-made reservoirs (goafs), which is based on data regarding methane in the goaf, provides the basis for evaluating the economic viability of CMM utilization projects – which is described below.

3 CMM Utilization Options

Technologically, the most effective options for methane utilization are its use in generating stations, boilers, and gas stations working on compressed and liquefied CMM. Several projects on CMM utilization were fulfilled in Kuzbass. For example, installation of the modular boiler station of 0.7 MW capacity for thermal power generation at the Krasnogorskaya mine in Prokopyevsk; commissioning the gas generating unit of 0.9 MW capacity for electric power generation at the Kirova mine in Leninsk-Kuznetsky. Also, the project of coalmine methane utilization in power generator with 0.4 MW electrical capacity is implemented on the Komsomolets mine. The technology of simultaneously use of two sources of air-methane: low (less than 3.5%) and high (over 25%) concentration of methane was successfully tested (Tailakov et al., 2004, Tailakov et al., 2014, Backhaus et al., 2009).

4 Economic Analysis

The main factors affecting the economic viability of CMM utilization projects are technological, environmental (the reduction of greenhouse gases in the atmosphere), social (the creation of additional jobs), financial, and marketing. If there is an increase in energy prices the investment attractiveness of such projects will also increase. Income sources of such methane projects are the revenue from selling the generated electrical and thermal energy, the savings on payments to the government for methane emissions into the atmosphere, and/or the savings over traditional fuels because of the use of CMM as a fuel for vehicles. Capital costs of the methane utilization project have many elements, including the following:

– defining the costs for development of the concept and technical and economic substantiation of the project (typically referred to as the front-end engineering and design – FEED);

– project documentation, examination and approval of the project with the state regulatory agencies, the acquisition of basic and auxiliary equipment for methane utilization, area preparation, construction of foundations, pipelines, installation of valves, transformer station (typically referred to as the engineering, procurement and construction - EPC), and commissioning activities.

The amount of investment required for the construction of a power generating station (IC engines) with a total power capacity of 30 MW is estimated at 1174 million rubles.

Economic analysis of such projects shows that the amortization of CMM power generating station investments are the largest component of operating costs and estimated as 81% from the overall cost of the project while the next largest component is materials estimated as up to 14.5 % from the overall cost (see Fig. 2).

This confirms the high capital intensity of the power generating option from among the various CMM utilization technologies.

The same applies to CMM utilization in gas boilers, where the amortization charges are again the largest element of operating costs – at 76.6%.

And for vehicle fueling stations supplying liquefied methane amortization charges in the proportion of 76.2% of the operating costs are the largest single cost element. This technology option has the materials and components comprising 20.6% of operating costs, which is the largest share for this cost element of any of the evaluated CMM utilization technologies. This latter point is explained by the high costs of maintenance of complex technological equipment for the liquefaction of CMM. The power output of vehicle fueling stations supplying liquefied methane is 10.4 million liters per year.

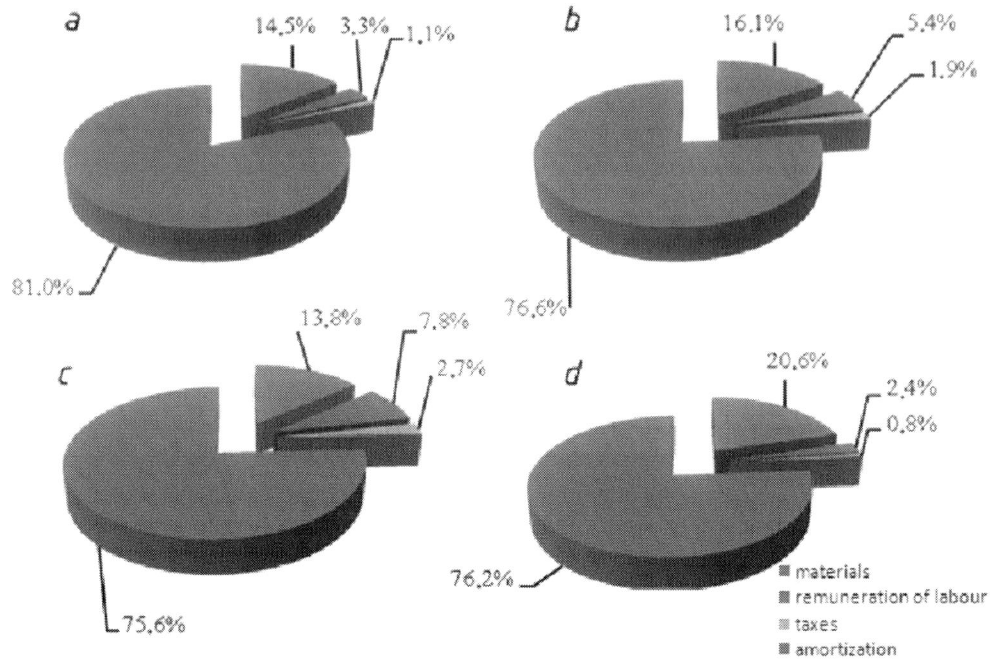

Fig. 2. The structure of the operating costs for CMM utilization projects: a) – 30 MW power generating station, b) – 30 MW gas boiler station, c) – gas station on compressed methane, and d) – gas station on liquefied methane

The project of construction of gas stations, in which compressed mine methane is used to fuel vehicles, is characterized by the lowest amortization charges (75.6%) in the structure of the operating costs of the project. The power output of vehicle fueling compressed CMM station is 1.971 million m^3 of fuel per year.

Having defined the capital and operating costs, various economic analysis results were calculated for each of the technology options. The economic viability results are presented in the Table 1.

Table 1. Economic evaluation of the CMM utilization projects

Index	Index Value	
	Generating Station (30 MW)	**Gas Boiler (30 MW)**
Investment costs (mln RUR)	1174	284
Payback period	3.9	2.8
Net present value (NPV) for 5 years (mln RUR)	267	187
IRR (%)	8.6	21.8
Profitability Index (PI)	1.23	1.66

Table 2. Economic viability evaluation for CMM utilization for compressed and liquefied methane as a fuel for vehicles

Index	Index Value	
	Compressed Natural Gas (CNG) Station	**Liquefied Natural Gas (LNG) Station**
Investment costs (mln RUR)	81	460
Payback period	2.5	2.7

Net present value (NPV) for 5 years (mln RUR)	67	326
IRR (%)	26.5	23.3
Profitability Index (PI)	1.82	.,71

The CMM utilization project using a boiler is considerably more attractive for potential investors when compared to the construction of the gas power generator, since the payback period of a modular boiler is less (2.8 years vs. 3.9 years) and its implementation requires a significantly lower amount of investment (284 mln RUR compared to 1174 mln RUR).

The project where gas stations provide compressed CMM as a fuel is much more attractive than gas stations that deliver liquefied CMM because the LNG stations would require over 5 times more capital expenditure (Table 2). The payback period of the project, involving the use of technology to compress the gas, is 2.5 year, with investment costs of 81 mln RUR. Net present value over 5 years is 67 million rubles, the internal rate of return – 26.5%, profitability of investments (PI) – 1.82.

5 Conclusions

Further development of the CMM utilization projects in Kuzbass will contribute to the rational use of energy resources that are suitable for thermal and electric energy production; will reduce greenhouse gas (methane) emissions into the atmosphere and create new job places. Also, the use of coal mine methane will increase the energy efficiency of coal enterprises. According to the produced economic analysis the project of converting the vehicles on fueling by the compressed CMM is the most attractive option for the CMM utilization in Kuzbass, but its implementation will require existence of the high concentrated methane (not less than 80% CH_4 in the air methane blend) source.

References

1. BP Statistical Review of World Energy, 2014.
2. Gazprom, 2014, Russian CBM Resources, http://www.gazprom.com/, internet site accessed 11/06/14.
3. Potential Gas Committee, 2006, Russian Forecasted CBM Resources.
4. O.V. Tailakov, D.V. Islamov, D.N. Zastrelov, V.O. Tailakov *The 7th International Conference on Greenhouse Gas Control Technologies* (Vancouver, Canada, 2004).
5. O.V. Tailakov, D.N. Zastrelov, E.A. Utkaev, A.I. Smyslov, A.N. Kormin, *Taishan Academic Forum – Project on mine Disaster Prevention and Control*, 450, 2014.
6. C. Backhaus, V.A. Bezpflyug, E.V. Mazanik, C. Hoppe, Coal, **11**, 50 (2009).

The Experience of Using Innovative Artificial Filter Arrays on South Kuzbass Open Pit: Case Study

Maxim Tyulenev[1,*], *Yury* Lesin[1], *Ekaterina* Tyuleneva[1] and *Elena* Murko[1]

[1]T. F. Gorbachev Kuzbass State Technical University, 650000, 28 Vesennyaya st., Kemerovo, Russian Federation

Abstract. The article describes the detailed example of placement of artificial filter arrays in one of Kuzbass open pits. It is proved that the use of the existing rock dump as an artificial filter array allows cleaning the quarry wastewater to the required parameters. Therefore, for water purification, it is possible to pile a new array as well as to use the existing dump, which will reduce the operating costs and the land consumption of open pit mining operations. Also, a brief methodology for determining the parameters of the artificial filter array is given in the article, taking into account the volume of contaminated water, the degree of its contamination and the location of the filter. The proposed complex of measures significantly improves the geo-ecological situation of mining region.

1 Introduction

Main geoecological problems of Kemerovo region are:

- every year the water bodies of region receive about 2 billion m^3 of quarry water;
- more than 500 million m^3 of water coal enterprises discharge without any treatment;
- aggregate weight of pollutants reaches more than 600 thousand tons;
- the main polluting substances in quarry waste water are suspended solids (finely-dispersed rock and coal particles);
- the concentration of suspended solids in quarry water is up to 250 mg / l, in some cases 2000-3000 mg / l, which exceeds maximum permissible concentrations (MPC) in many times.

Therefore the solution of these problems must take into account the following factors:

- continuous toughening of requirements for quality of waste water discharge;
- the use of holding and clarifying ponds does not provide quarry waste water purification according to MPC;
- existing treatment facilities occupy considerable areas;
- catchments basins are distantly located from each other and change location in the process of mining.

* Corresponding author: tma.geolog@kuzstu.ru

2 Material and method

Calculation of the water intake consists in checking the ability of the drainage pipe of a given length and diameter to pass the entire flow of water being purifyed. Preliminarily we take the drainage pipe length $L_t = 5$ m and the diameter $2r_t = 0.5$ m.

Considering the filtration flow around the drainage pipe as flat-radial, and neglecting the gravitation, the differential equation characterizing the water consumption at a distance r from the pipe can be written as:

$$q = -2 * \pi * r * k \frac{dh}{dr} \tag{1}$$

where: q – the flow of liquid per 1 m of the length of the drainage pipe ; $q = Q / L_n$, Q is the productivity of purifying facilities; L_n – the reduced length of the drainage pipe, considering its walls water-permeable, m; k - the rate of filtration; h - the value of the liquid head at a distance r from the pipe.

The solution of the differential equation is reduced to the separation of variables with subsequent integration. As a result, we obtain the following formula:

$$q = \frac{2 * \pi * k * H}{\ln \frac{R}{r}} \tag{2}$$

where: H – the amount of pressure the of liquid in the drainage pipe; r_t – the radius of drenage pipe, m; R – the radius of the drainage trench filling is equal to the distance from the center of the pipe to the walls of the trench, m.

Taking the value of q from formula (2), the reduced length of the drainage pipe is:

$$L_n = Q/q \tag{3}$$

The calculated pipe lengh L_{tr} in this case can be calcelated as:

$$L_{tr} = k_n \cdot L_n \tag{4}$$

where: k_n – the rate considered the presence of perforations.

The coefficient k_n can be found as the ratio of the surface area of the drainage pipe to the total area of the perforations.

When the capacity of water treatment plants Q is 500 m^3/day, the filtration rate is 0.01 m/s and the pipe diameter $2r_t$=0.5 m, we find that $L_m = 5$ m $> L_{tr}$. Consequently, the given dimensions of the drainage pipe and the trench ensure the passage of all water entering the purifying.

3 Results and Discussion

The choice of the location of the water intake and the calculation of water filtration in the dump massif must be besed on the mutual location of open pit office and the rock dump, the configuration of the dump and the relief of the underlying surface. The optimal location is near the point indicated on open pit mining plan (Fig. 1).

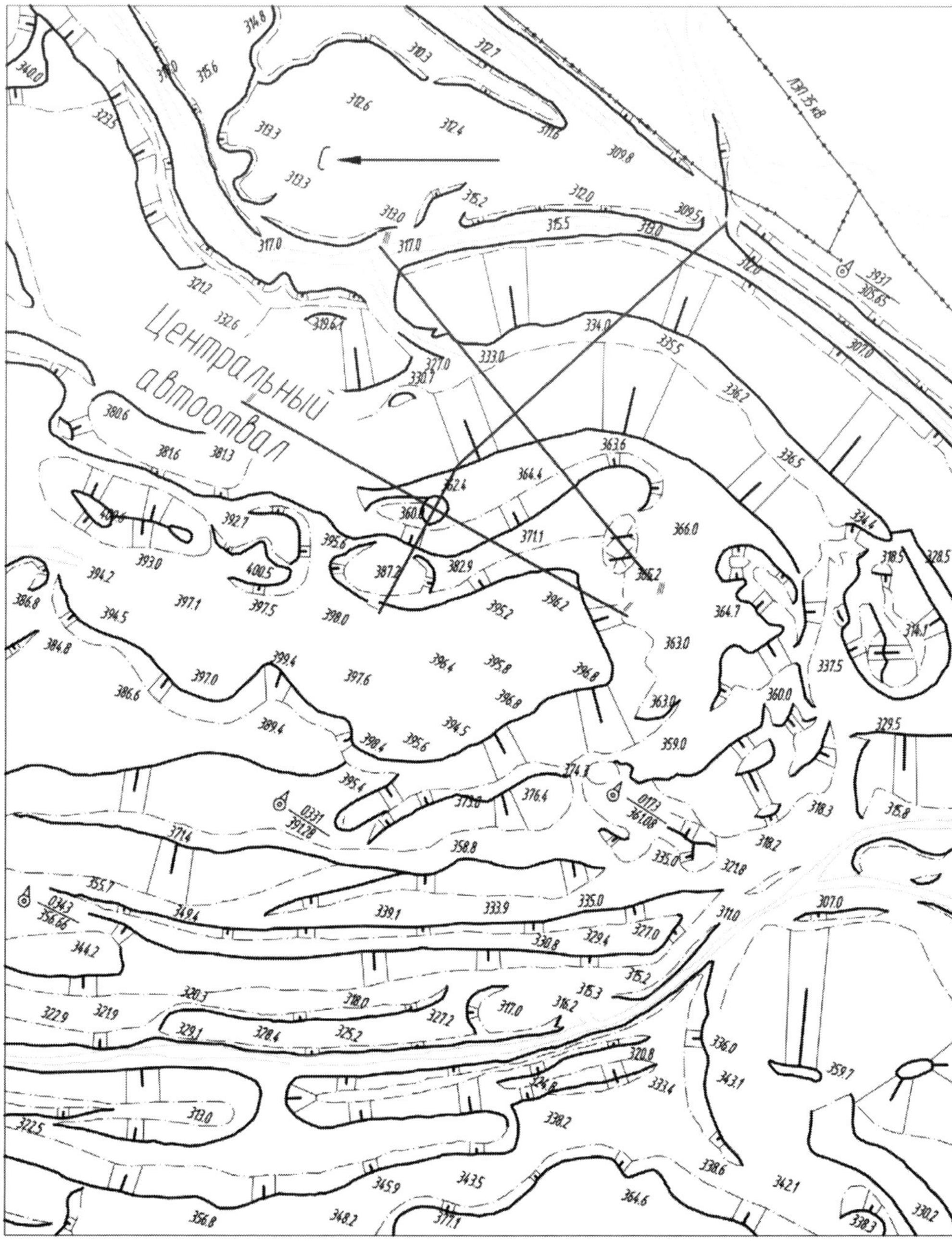

Fig. 1. The location of the water intake

According to the plan of the underlying surface (Fig. 2), we determined the likely direction of the water filtered in the dump, and found the point of seepage of purified water from the dump. The probable direction of water movement in the dump is shown in Fig. 3 by the dotted line. The place of water seepage is located as the point of intersection of line I-I with the line of the lower contour of the slope of the first tier of the dump.

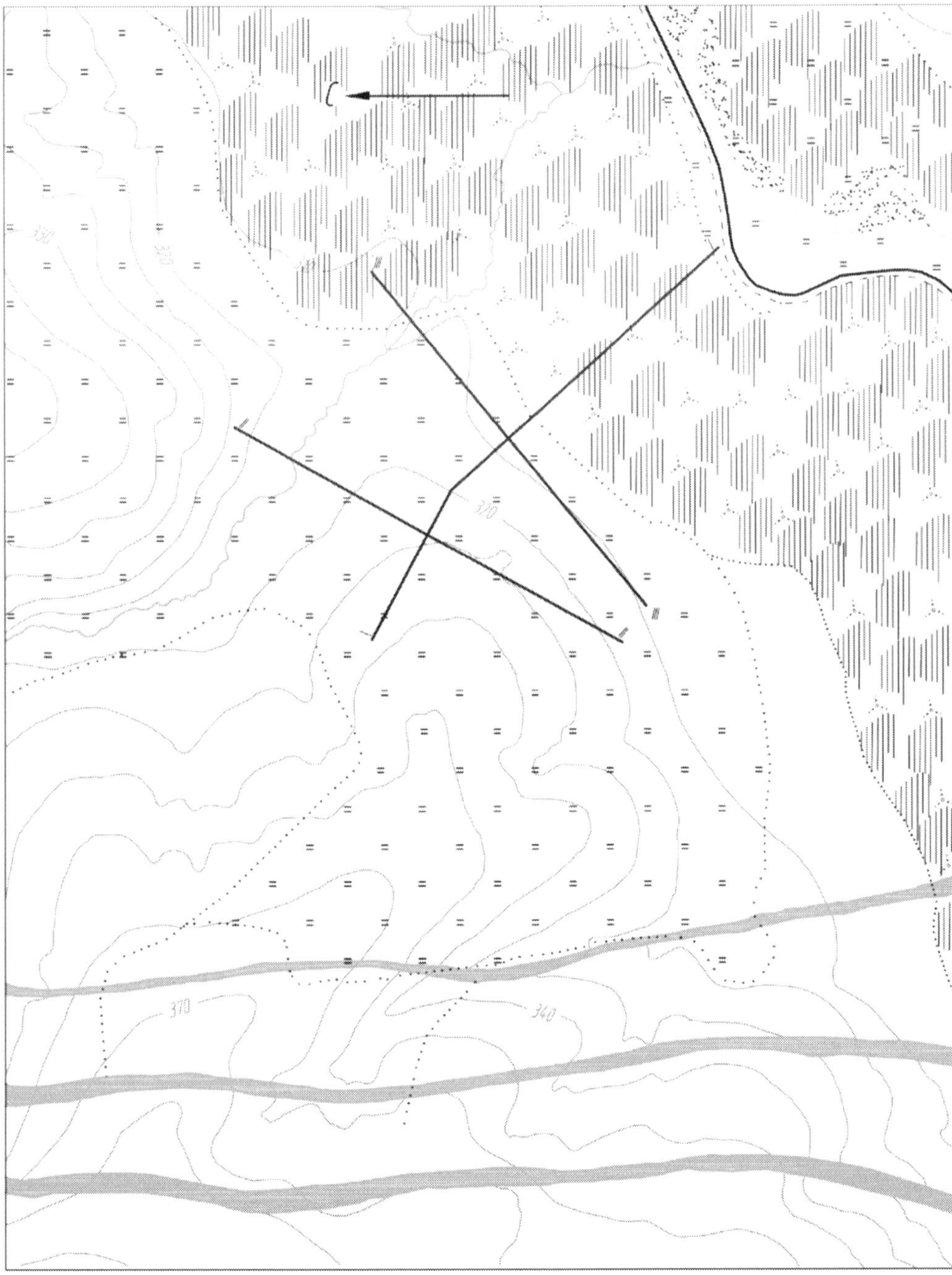

Fig. 2. The plan of the underlying surface

The longitudinal and transverse profiles of the rock dumps are shown in Fig. 3. The scale of the plans was adopted 1: 5000, profiles 1: 2000.

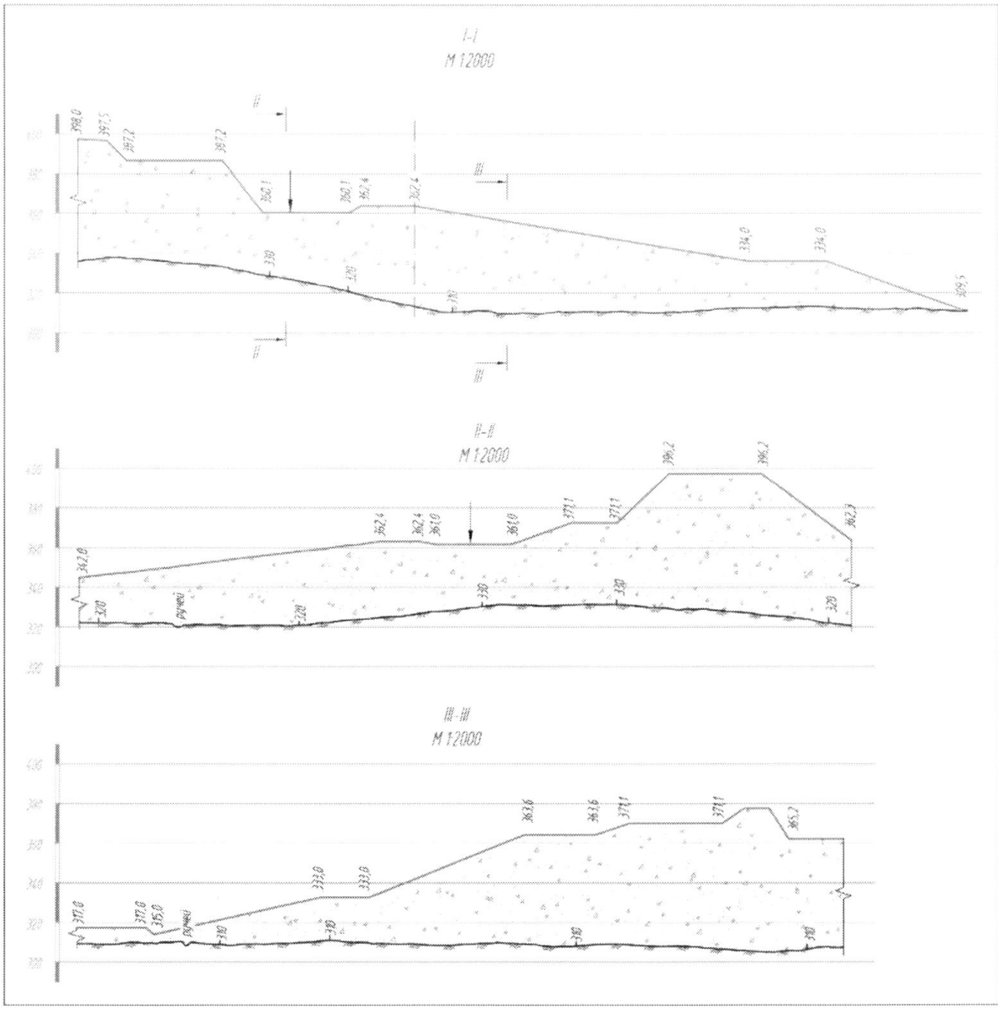

Fig. 3. The longitudinal and transverse profiles of the rock dumps

The distance between water entry and exit points of the dumps is 340 m.

When filtering contaminated water in lump rock massifs, sediments are suspended in the pores and the water is purified. This is due to the process of natural penetration or artificial insertion of small particles into rock's fractures – colmatation process [1-6].

It was found that the following dependence between the concentration of suspended solids C and the length of water filtration in the dump L there can be expressed as (2):

$$C = C_0 \exp(-\eta L) \tag{5}$$

where: C_o – initial consentration of suspended solids;

η - filtering index.

The filtering index η depends on the size of the components of the filtering array, its petrographic composition, water filtration rate and other factors.

For rock dumps and other arrays from bedrock rocks of Kuzbass open pit it was found within 0.005-0.05.

Calculations using formula (5) showed that with an initial suspended matter concentration of 380 mg/l, the filtration length L = 340 m and a filtering index of 0.035, the concentration of suspended solids in purified water does not exceed 1 mg/l.

In this case the filtration calculation of the rock dump for the passage of all water entering the treatment is not needed, because the calculation of the water intake is performed.

4 Conclusion

Thus, for the chosen location of the water intake on the dump and the distance between the entry and exit points of water from the dump (equal to 340 m), the necessary degree of wastewater treatment from polluting suspended solids is provided.

References

1. S.L. Arnon, P. Marx, K.E. Searcy, A.I. Packman, Hydrol. Proc., **24**, 108 (2010)
2. T. Esakkimuthu, D. Sivakumar, S. Akila, Poll. Res., **33**, 567 (2014)
3. S.K. Gupta, K. Ramesh, S. Sameer, Poll. Res., **34**, 111 (2015)
4. J.I. Jones, J.F. Murphy, A.L. Collins, D.A. Sear, P.S. Naden, P.D. Armitage, Res. Appl., **9**, 67 (2011)
5. D.L. Karwan, J.E. Saiers, Water Resour. Res., **48**, 15 (2012)
6. M.M. Sharma, Y.C. Yortsos, AlChE J., **33**, 1636 (1987)

The results of air treatment process modeling at the location of the air curtain in the air suppliers and ventilation shafts

Aleksandr Nikolaev[1*], *Nikolaj* Alymenko[1], *Anton* Kamenskih[2], *and Viktor* Nikolaev[1]

[1]Perm National Research Polytechnic University, 29 Komsomolsky av., Perm, 614990, Russia
[2]Mining Institute of the Ural Branch of the Russian Academy of Sciences (78a Sibirskaia st., Perm, 614007, Russia

Abstract. In the existing shaft air heater installations (AHI), that heat the air for air suppliers in cold seasons, a heater channel is used. Some parts of the air from the heater go to the channel, other parts are sucked through a pithead by the general shaft pressure drawdown formed by the main ventilation installation (MVI). When this happens, a mix of two air flows leads to a shaft heat regime violation that can break pressurization of intertubular sealers. The problem of energy saving while airing underground mining enterprises is also very important. The proposed solution of both tasks due to the application of an air curtain is described in the article. In cold seasons the air treatment process should be used and it is offered to place an air curtain in the air suppliers shaft above the place of interface of the calorifer channel to a trunk in order to avoid an infiltration (suction) of air through the pithead. It's recommended to use an air curtain in a ventilation shaft because it reduces external air leaks thereby improving energy efficiency of the MVI work. During the mathematical modeling of ventilation and air preparation process (in SolidWorks Flowsimulation software package) it was found out that the use of the air curtain in the air supply shaft can increase the efficiency of the AHI, and reduce the electricity consumption for ventilation in the ventilation shaft.

1 Introduction

In cold weather, the air supplied to the mine shafts must be heated to a temperature of not less than +2 ° C, the temperature of this value should correspond to the entire volume of air entering the barrel throughout its cross section. However, conducted in [1-3] studies have shown that these rules are not respected in practice. This is due to the fact that the air supply divide in the two airflow supplied: hot air heater installation in the shaft (AHI) supplied by air heater channel and sucked through the pithead by mine depression. Consequently, in practice, the air supplied from the AHI is heated to temperatures much higher than required, resulting in the barrel is often observed a significant «overheating» air. In the air supply shaft portions are present, the temperature which is slightly above

* Corresponding author: nikolaev0811@mail.ru

value of +2 ° C, which indicates the low efficiency of the existing process control method of air handling and proximity to the temperature disturbance in the shaft.

This situation leads to disruption of the pressurization of intertubular sealers, resulting in there is a threat to the integrity of the trunk lining. The presence in the air supply shaft with different temperatures flows becomes the cause of them so-called «airlock» that prevent air entering on them and cause a malfunction of ventilation modes [1].

Thus, temperature rise air in the feed mine to excess results in increased energy costs, work spent on the AHI, i.e. to a waste of energy.

Due to the fact that, according to [4-6] from 30 to 50% of the total spent on underground mining enterprise power are consumed for ventilation, another important challenge for the production is the need to improve energy efficiency work main ventilation installation (MVI).

The situation described above, in addition increase to energy cost overruns at the MVI and the deterioration of production safety conditions, causing an associated problem - the emergence of between mine shafts negative general natural draft [7-12], ie when injected into the warm air of air supply shaft tends to rise upwards, and cooled in the mine (mine) - goes down. This problem is particularly acute for the shallow (up to 500 m or less) underground mining.

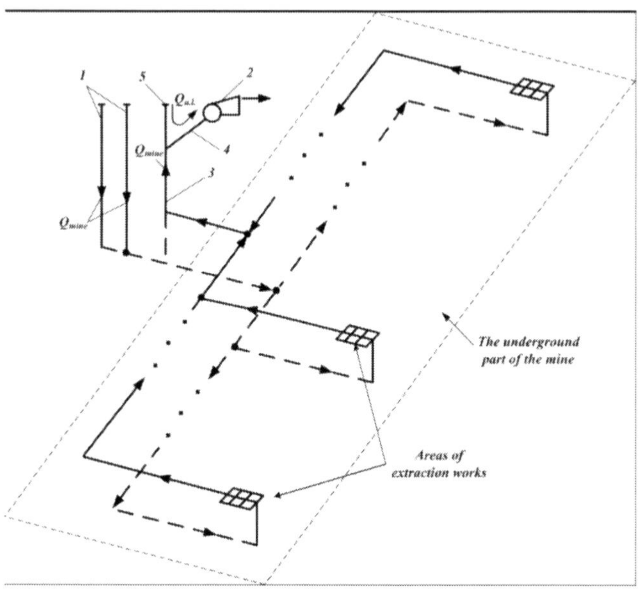

Fig.1. A simplified diagram of the mine operating on the suction method of ventilation

Additionally, inherent to the suction method of ventilation problem – the emergence of outside air leaks [13], for the following reasons. The air in the mine (Q_{mine}) is supplied through air suppliers shafts 1 by the underpressure created by the MVI 2, located on the ventilation shaft 3 (Fig. 1). In addition to the air sucked into the mine, with a similar pattern of ventilation in the channel of MVI 4 is also sucked into the outside air through the mouth of the ventilation shaft 5, which is mixed with the general stream will again be released into the atmosphere. Therefore, in addition to useful work (fresh air injection), the MVI will make useless work due to external air suction (external leakage) $Q_{a.l.}$. The quantity of leakage (for example potash mines) according to [13, 14] ranges from 9 to 51%, ie there are mines, where the efficiency of the MVI work is only 49%. In this regard, reduction of external (surface) leakage is one of the main directions in the complex of measures to reduce costs in the mining ventilation underground mining enterprise.

2 Results

In order to improve air handling efficiency and reduce electricity costs spent on the work of the MVI, the present study the authors proposed in the air suppliers and ventilation shafts to establish an air curtain (Fig. 2). The air supplier shaft air curtain should ensure infiltration reduction (choke) of the outer cold air through the pithead. In this case, the AHI work will require less energy costs, as well as the normal air mixing process in the shaft. The air curtain in the ventilation shaft is needed to reduce external air leaks [13, 14].

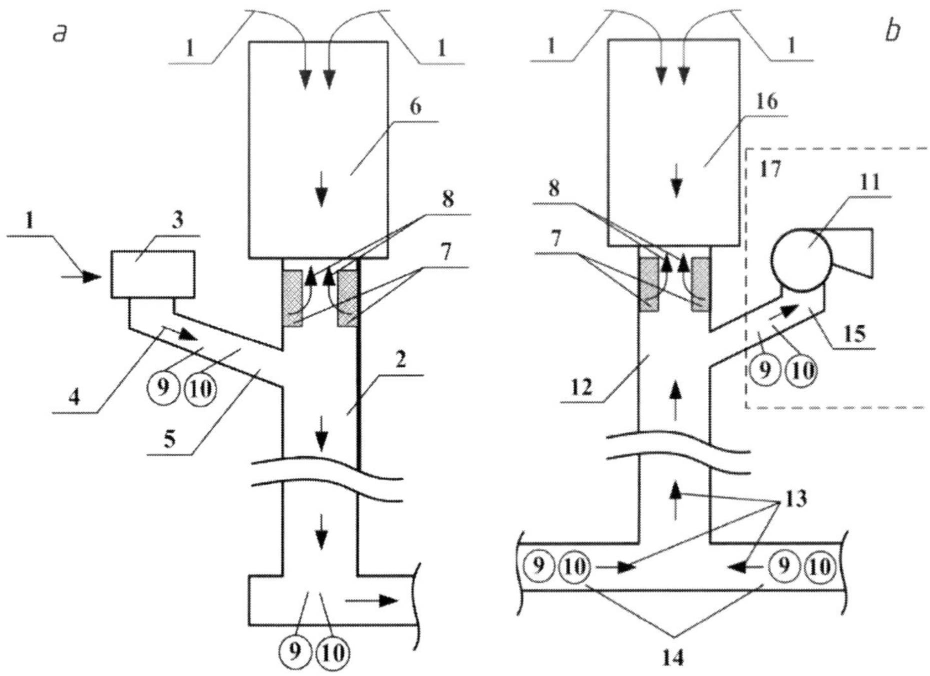

Fig. 2. Implementation of ventilation and air handling proposed method: *a* – air curtain in the air supplier shaft; *b* – air curtain in ventilation shaft. 1 – outside air (cold); 2 – air supplier shaft; 3 – AHI; 4 – the heated air; 5 – air heater hot-air channel; 6 – pithead on the air supplier shaft; 7 – air curtain; 8 – the air output from the air curtain; 9 – air flow sensor; 10 – a temperature sensor, a pressure (or density meter); 11 – MVI; 12 – ventilation shaft; 13 – outgoing air for ventilation shaft; 14 – the main ventilation shaft production; 15 – channel of MVI; 16 – pithead on the ventilation shaft; 17 – superficial complex MVI

In Fig. 3 shows the layout and operation of the air curtain at the mouth of the shaft. With fans 1 of the air curtain, installed on shelves on one of tiers reinforcement stem 7 shall intake air through the ducts 3 in the place indicated by 4. The jets of the air curtain is ejected at a speed V_r from the dispenser 2, at an acute angle α meet the air flow external air leakage. After bending, jet air curtain cover the cross section of the shaft, thereby increasing the aerodynamic resistance of the shaft section from zero mark to interface with the channel. Dispensers air devices are installed on the shelves of one of the reinforcement of the barrel 7 tiers with fans 1. The place of installation of the air curtain overlaps part of the shaft section of a hollow flooring. The central execution is set 5 longitudinal bulkhead limiting the impact of the jets of the veil at each other. To improve the effectiveness of the air curtain, and set the partition 6 on the perimeter of the flow cross section of the shaft.

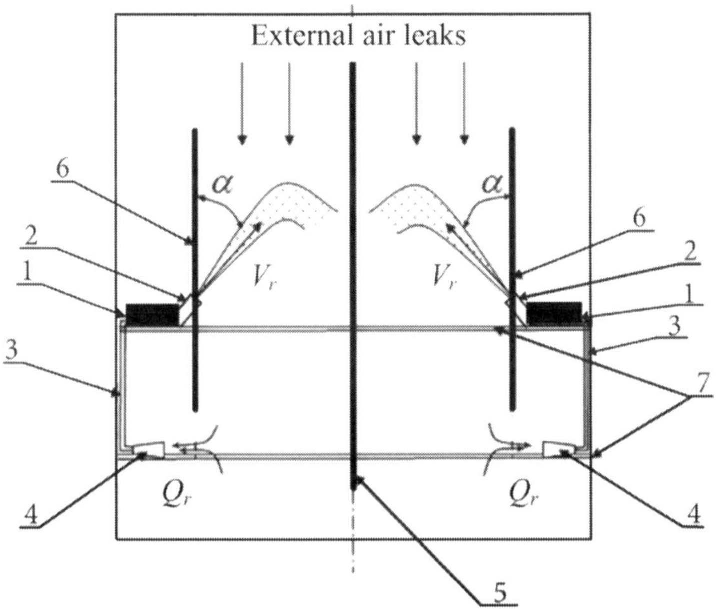

Fig. 3. The layout and operation of the air curtain in the mouth of the shaft: 1 – fan air curtain; 2 – a box for the issuance of air; 3 – pipes (ducts); 4 – the air intake; 5 – longitudinal bulkhead; 6 – longitudinal walls (on the perimeter of the section to light); 7 – tier shaft reinforcement

Engineering practice is based on experimental studies that when proper planning usually provides high reliability of experimental studies. A significant drawback in the present conditions of these studies – the high cost and time required for preparation and conduct of such studies. Therefore, an effective measure is the use of parametric calculations (computational studies or numerical simulations). They are accurately simulate physical phenomena that determine the characteristics of the system, with experimental studies needed to test this accuracy. Calculations should be performed using the software, ensure the implementation of the entire set of necessary calculations in a reasonable amount of time with sufficient accuracy.

For a detailed analysis taking into account all the geometric features of the shaft and the proposed construction of the air curtain in the present study carried out a three-dimensional modeling and numerical simulation using computational fluid dynamics methods. Mathematical modeling conducted in the software package SolidWorks Flowsimulation in the supercomputer (system «PGU-Tesla») [15].

The test model is considered a variant design and placement of the air curtain in the mouth of the ventilation shaft. Fig. 4 shows a model of the placement of the air curtain in the mouth of the ventilation shaft, which is constructed according to the geometrical dimensions of the barrel, and a scheme for allocating the air curtain in his mouth (Fig. 3).

3 Discussion

Research carried out on the three-dimensional mathematical model with different numbers of cells (the number of computational cells was brought to 3 million, thanks to high-performance computing system «PGU-Tesla»), which made it possible to obtain reliable results. Based on the results of the study, the calculations must also be carried out on a computational grid with the number of cells at least 3 million, as in this case the results obtained with the smallest deviation from the mean.

In Fig. 5 shows the results of modeling air handling method of the currently accepted (Fig. 5, *a*) and by using an air curtain (Fig. 5, *b*).

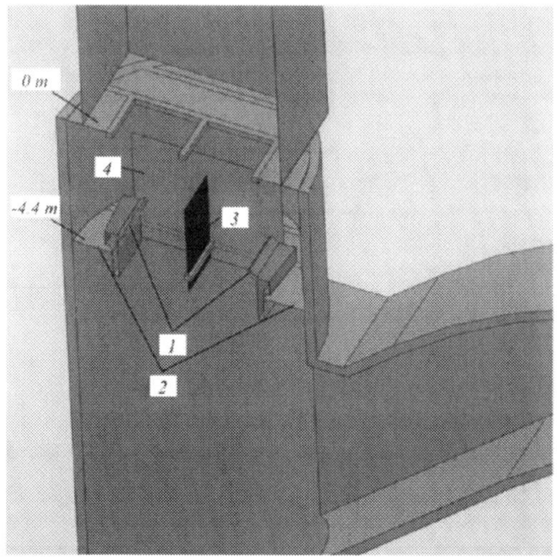

Fig. 4. The air curtain in the mouth of the shaft: 1 – a box for the issuance of air (fans air curtain); 2 – shelves; 3 – longitudinal bulkhead; 4 – transverse partition

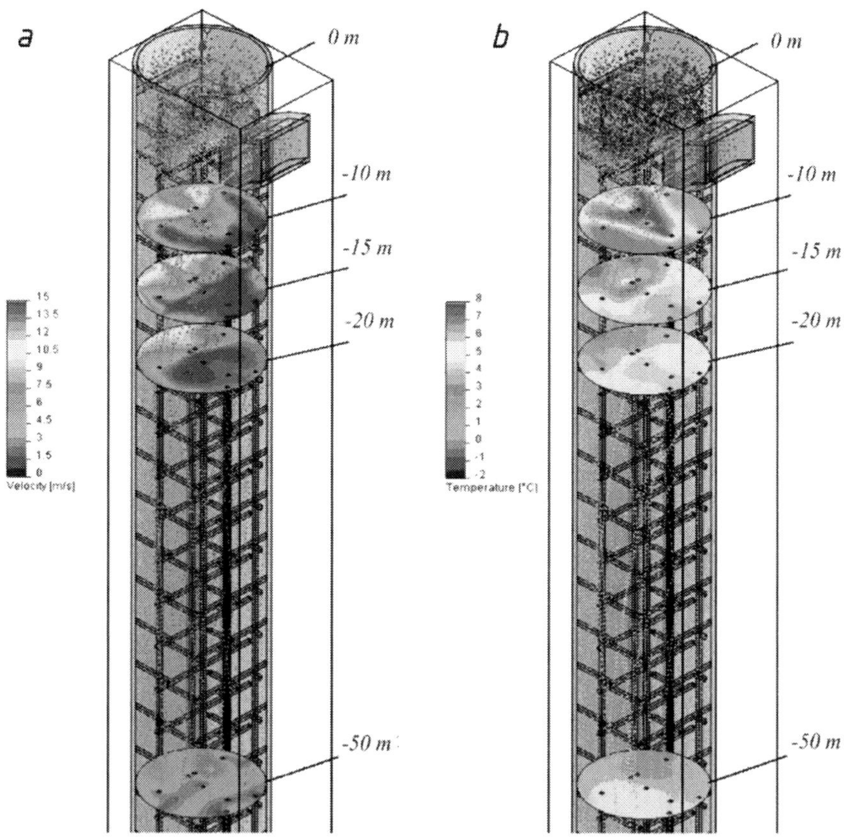

Fig. 5. Air temperature distribution over the cross section of air supplier shaft at different depths by mixing cold and hot air using an air curtain installed in the mouth of the barrel: 20 ^{0}C cold air (20 m^3/s), a warm air +8 °C (147 m^3/s): *a* – without air curtain in the air supplier shaft; *b* – using an air curtain in the air supplier shaft

As seen from the figure, as compared with the inventive method has used a number of advantages:

Firstly, in the air supplier shaft are no zones with temperature close to (or even more below) + 2 ° C, e.g. obsolescence threat violation the pressurization of intertubular sealers. At the same time work on the AHI will need a smaller amount of energy, ie, reduce the cost of air handling. Compared with the amount of energy spent on the work of the AHI, the electricity costs spent on the work of the air curtain disproportionately small. Moreover, due to the fact in the shaft will not be necessary to heat the air to a higher temperature, between the trunks decrease negative (or increase positive) mine natural draft, resulting in the MVI will decrease energy consumption at times more than what is required in the operation of the air curtain.

Secondly, already at around -20 m thermal distribution on the cross section substantially uniform air supplier shaft, i.e. the likelihood of «airlock» it is virtually absent.

During the process of mathematical modeling of the air curtain in the ventilation shaft (Fig. 6), it was found that when the external air leaks accepted conditions will decrease from 60.43 m^3/s to 26.27 m^3/s, ie, approximately 43.5%, and the work of the MVI instead 443.80 m^3/s need to 409.64 m^3/s.

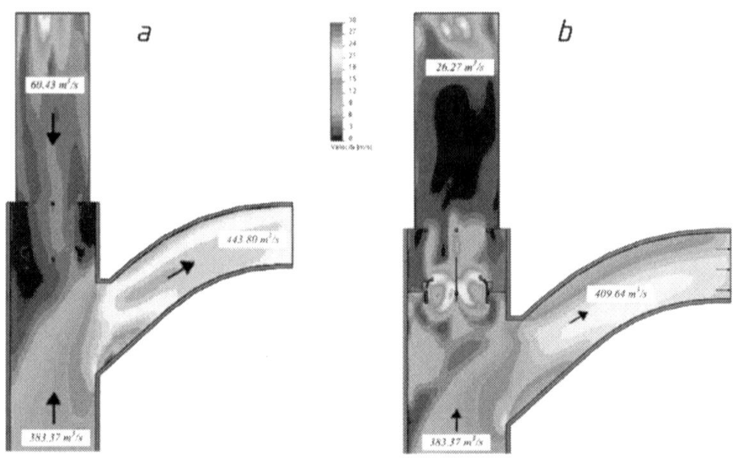

Fig. 6. The result of mathematical modeling of air distribution of air flow in ventilation shaft: *a* – without air curtain in ventilation shaft; *b* – using an air curtain in the ventilation shaft

In the action of the air curtain aerodynamic resistance of the shaft portion of the zero mark to interface with ventilation shaft increases, ie, increases aerodynamic drag of external air leaks. At the same time the aerodynamic resistance of the mine remains unchanged, and the aerodynamic resistance of the ventilation network, which operates the fan increases. Operating point MVI moves the aerodynamic characteristics of the fan in the region of lower pressure, causing by changing the installation angle of the blades of the axial guide vane changes the work of the MVI on a more economical mode. At the same amount of fresh air entering the underground mining company remains unchanged, and the consumption of electricity is reduced MVI.

4 Conclusion

Thus, in the course of mathematical modeling established that the proposed method of using an air curtain in the air supplier and ventilation shafts will prevent the negative and harmful processes associated with the uneven distribution of temperature field on the shaft cross section, overruns of energy to work AHI, and will save electricity, spent on the work of the MVI.

References

1. N.I. Alymenko, A.A. Kamenskikh, A.V. Nikolaev, A.I. Petrov, Eur. Min. **26**, 45 (2016).
2. K. Hanjalic, B.E. Launder, J. Fluid Mech., **52**, 609, 1972.
3. M.J. McPherson, G. Robinson, Mine vent. South Africa, **33**, 145, 1980.
4. B.K. Belle, *North American Mine vent. Symp.*, 427, 2008.
5. J.J.L. Du Plessis, W.M. Marx, and C. Nell, The Journal of The Southern African Instit. of Min. and Metall., **114**, 1033, 2014.
6. G. Cheng, M. Qi, J. Zhang, W. Wang, and Y. Cheng, Procedia Eng., **45**, 311, 2012.
7. W.E. Bruce, *Natural draft: its measurement and modeling in underground mine ventilation systems* (US: Dept. of Labor, Mine Safety and Health Administr., 1986).
8. P.F. Linden, Annual Review of Fluid Mechanics, **31**, 201, 1999.
9. Jianwei Cheng, Yan Wu, Haiming Xu, Jin Liu, Yekang Yang, Huangjun Deng, Yi Wang, Tunneling and Underground Space Technology, **45**, 166, 2015.
10. Van Ulden, A. Holtslag, J. Clim. Appl. Meteorol., **24**, 1196, 1985.
11. N.I. Alymenko, A.V. Nikolaev, Journal of Mining Science, **47**, 636 (2011).
12. G. B. Lyal'kina, A.V. Nikolaev, Journal of Mining Science, **51**, 342 (2015).
13. Y. Verma, Mining Engineering (Gr. Brit.), **140**, 47, 1980.
14. P. Berry, in *World Mining Congress*, New Delhi, **3**, 1, 1984.
15. R. E. Barr, D. Juricic, T. J. Krueger, *Engineering and Computer Graphics Workbook Using Solidworks 2015* (SDC Publications, 2015).

Mitigating Against Conflicts in the Kenyan Mining Cycle: Identification of Gaps in the Participation and Recourse for Rights Holders (Civil Society & Community)

Seroni Anyona[1,*], *Bernard* Rop[1]

[1]Jomo Kenyatta University of Agriculture & Technology, Mining, Materials & Petroleum Engineering Department, P.O. Box 62000 – 00200, Nairobi, Kenya

Abstract. One of the major obstacles to sustainable development of the mining sector in Africa is conflicts. These conflicts emanate from various sources. However one of the key sources of misunderstanding is lack of communication between the rights holders (community and duty bearers (government and corporate) and this breakdown of communication is due to poor or nonexistent mechanisms of involvement of affected communities in decision making and ignorance. This study sought to examine the Kenyan Mining cycle and identify procedural and systemic gaps that if not addressed could trigger conflict in the emerging mining sector and suggest ways of mitigating them. Key among the greatest barriers to communal participation is ignorance regarding mining activities and also the language of instruction. Most rural people are uneducated and yet a large number of communication and notices are in English. Poor management of expectations is also a critical source of misunderstanding and hence conflict. Poor communal organization and lack of some sort of recognized authority give room for disorder and hence poor channels of communication.

1 Introduction

The management of extractive industries is one of the most critical challenges facing many resource-dependent developing countries today. Rather than stimulating broad-based economic development, reliance on resource extraction has tended to concentrate wealth and power in the hands of a few, exacerbate corruption and inequalities, lead to environmental degradation and pollution, while doing little to reduce poverty, economic disparities and generate employment. Worse still, in many countries extractive resources have fuelled violent conflicts. The recent announcement of oil discovery and other minerals has propelled Kenya as a new player in the global market for hydrocarbons and valued minerals. The International Monetary Fund (IMF) projects that oil production in Kenya is

* Corresponding author: sanyona@jkuat.ac.ke

expected to start in two to three years from now, giving the country time to prepare to manage its endowment to achieve its development goals as stipulated in the 2030 Vision.

Kenya is well endowed with bulk mineral resources, some of which are already being exploited by private companies and some are yet to be prospected and exploited.

2 Material and Method

Kenya enacted a new law in 2016 to reinvigorate the mining sector. The new Act seeks to provide clear guidance on mining activities in Kenya. It introduces amongst other things legalization of Artisanal Miners and Transparency and Accountability through use of the Online Mining Cadastre portal for licensing and for management of mineral rights and permits. It also introduces Community Development Agreements, mandatory for all holders of large scale mining rights, and sharing of royalties among the national government, the county governments and the local communities.

The Act is intended to streamline the country's mining sector and open the gates for its development as well as ensure environmental conservation and sustainable development in the mining sector. It also seeks to address the key gaps in the Mining Act of 1940, and align the sector to the latest global trends such as value-addition and the use of technology to spur investor interest. The Cabinet Secretary for Mining has termed the Act as one of the most "progressive mining laws in the continent." (Oraro & Co. Advocates).

From a human rights based perspective the new Act introduces major changes in how communities (rights holders) will be engaged in the mining cycle and licensing process.

2.1 Considered Factors

2.1.1 Community Development Agreement

The Act makes it mandatory for a holder of a large scale mining licence to enter into an agreement with the community where the mining operations will be carried out. This agreement is known as the Community Development Agreement

2.1.2 Employment and Training

The Act seeks to ensure employment opportunities are created for Kenyans, ensure skills transfer and capacity building for the citizens. Section 46 thereof provides that each mineral right holder is required to submit to the Cabinet Secretary, a program detailing how it shall recruit and train Kenyans. Furthermore, the holder of a mineral right is required to give preference to members of the community and Kenyan citizens when it comes to employment and only engage non-citizen technical experts in accordance with such local standards for registration as may be prescribed in the relevant law.

2.1.3 Local Equity Participation

The Act protects local participation with provisions for equity participation in large mining operations. The Act also provides for the prioritization of local procurement of goods, services and workforce.

2.1.4 Land Use, Water Rights Laws and Pollution

A holder of a mineral right is also required to comply with the prevailing laws on water rights and use the land in accordance with the terms of the permit of licence. In doing so, the holder of the permit or licence will ensure that the land is used in a sustainable manner. Section 179 of the Act which deals with land use requires the holder of a permit or mining licence to ensure the sustainable use of land through restoration of abandoned mines and quarries; that the seepage of toxic waste into streams, rivers, lakes and wetlands is avoided and that disposal of any toxic waste is done in the approved areas only; that blasting and all works that cause massive vibration are properly carried out and muffled to keep such vibrations and blasts to reasonable and permissible levels in conformity with the Environmental Management and Coordination Act, (Cap. 387); and that upon completion of prospecting or mining, the land in question shall be restored to its original status or to an acceptable and reasonable condition as close as possible to its original state.

2.1.5 Health and Safety

The Act has provisions to ensure the health and safety of persons working at the mines. It provides that the holder of a mineral right will comply with the Occupational Health and Safety Act, 2007 with regards to the safety of workers and mine operations.

2.1.6 Revenue Sharing

The Act makes provision for the sharing of the royalties that are payable under the Act. Section 183(5) provides that the National Government will be entitled to 70%, the County Government 20% and the community where the mining operations occur will be entitled to 10% of the revenue. This provision has been viewed as being probably one of the most progressive provisions in the Act, if not the environmental provisions. It embodies the spirit of the Constitution that requires equitable distribution of revenue among the national and county governments and inclusion, more so, financial inclusion for the people of Kenya.

2.1.7 Dispute Resolution

Section 154 of the Act provides for dispute resolution. It provides that any dispute arising as a result of a mineral right issued under the Act, may be determined by the Cabinet Secretary, through a mediation or arbitration process as may be agreed upon by the disputing parties or as may be stated in an agreement; or through a court of competent jurisdiction. Section 155 sets out the disputes that may be determined by the Cabinet Secretary.

2.1.8 International Best Practice on Mining

The Act generally prescribes for the carrying out of prospecting and mining activities in accordance with international best practice. Below is a discussion on some of the international best practices on mining.

2.1.9 Sustainable Development in Mining

Sustainable development is at the heart of international best practice on mining. There is no universally adopted definition of sustainability in the mining industry. However, there are five (5) main pillars that embody sustainable mining practices. These are Economy, Safety,

Resource efficiency, Environment and Community. A brief description of each dimension follows.

2.1.10 Environmental Management

Environmental assessment is a process that will ensure the mineral right holder considers the potential environmental effects before they can begin the mining process. It is important that the impact to the environment is managed through specialized low impact exploration practices, water management, pollution prevention and control and the rehabilitation of mines after closure. The Act has several very progressive provisions on environmental management, which include the requirement for a person to obtain an impact assessment licence before being granted a mining licence. The Act also requires the holder of a mineral right to comply with the prevailing laws on water and use the land in a sustainable manner.

The Act has enhanced the provisions on the environment, health and safety. Section 176(1) states that a mineral right or other licence or permit granted under the Act shall not exempt a person from complying with any law concerning the protection of the environment. Section 176(2) further provides that a mining licence shall not be granted to a person under this Act unless the person has obtained an Environmental Impact Assessment licence, social heritage assessment and the environmental management plan has been approved. After completion of the prospecting or mining, the mineral right holder will ensure that the land in question is restored to its original status or to an acceptable or reasonable condition as close as possible to its original state.

3 Results and Discussion

For the management of the mining sector, the most significant human rights impacts, risks, and opportunities arise through the use of their products and services instead of their operations. This sector whose products and services present risks and/or opportunities to users or other rights holders, consideration of service (product)-level HRIAs is critical. A service (product)-level assessment identifies human rights impacts of products that can be used for beneficial purposes but can also be misused in a way that infringes on human rights(Institute for Human Rights and Business and Global Business Initiative, "State of Play: The Corporate Responsibility to Respect Human Rights in Business Relationships

4 Conclusions

After a thorough examination of the Kenyan mining cycle a number of gaps have been identified that if sealed will enable the sector to sustainably contribute to the development of the country. The major procedural challenges that inhibit the full participation of affected communities in decision making include:

- notices are published in the media which most n people may not have access to; English language used is not familiar to the larger section of the uneducated population
- Experience shows that the public participation in ESIA's is limited and normally skewed towards duty bearers
- Over centralizing decision making power in the position of the Cabinet secretary may lead to abuse of power.
- Ignorance on the part of the public regarding the mining sector and mining cycle may lead to exploitation and conflicts
- Community could be involved at the time the holder decides to surrender part of the prospecting licenced area that does not show potential for mineralization

• the environmental protection bond is paid once in the entire life of the mining cycle – This may not be sufficient to reclaim the mined areas after many years of mining

• The ESIA and ESMP are very technical documents for the uneducated population to comprehend

• No clear guidelines on compensation for land and other assets- the actual value of the land is left to concerned parties to decide

• Most local people are unskilled in complex negotiations dealing with compensation and resettlement with large multinational companies. This can lead to possible exploitation of unexposed communities

• Lack of understanding regarding the true value of land and dynamics and mining economics could lead to unrealistic expectations that often result in conflicts

• No clear regulations to identify and weed out land speculators

• No well elaborated dispute resolution system

• No clear and specific emphasis on safety for employees at the work place. This is hidden under the ESIA report which may not be standardized nor closely evaluated

• Solely depending on a statement, of attestation of no mining Act Offence Penal conviction of the applicant may not be sufficient to effectively capture the background criminal record or transparency check of the applicant

• Under these regulations the affected community or the public in general has no access to the monthly, quarterly and annual reports on mine development and mineral production

• There is no clear definition of the community where the mining operation occurs or the organizational structure to which 10% of the loyalties payable shall be entrusted. (Should it be based on clan or tribe or based on distance from the mining operation?)

• While there is a litany of regulations on mining there is little focus in Mineral Processing leaving a yawning gap that could be abused

References

1. Khalid Salim – *Mining Act, 2014 Of Kenya-An Overview*
2. Evans Khamala Masachi - *Implications Of The Old Mining Law Cap. 306 And The Proposed Mining Act* - Presentation December 2014
3. Bawa Ibrahim – Minerals Commission, Managing Africa's Resources Sustainably: Challenges of Small Scale Mining in Ghana.
4. Bridges, C.R. Gems & Gemology, **Summer**, 290 (1974)
5. Butler, J., *Gender Trouble: Feminism and the Subversion of Identity* (Routledge, 1990)
6. Çagatay N., *Trade, Gender and Poverty, UN Development Programme*, (2001)
7. Dreschler, B., *Small Scale Mining and Sustainable Development within the SADC Region* (Mining, Minerals and Sustainable Development, IIED, London 2002)
8. http://commdev.org/files/1798_file_asm_southern_Taita Taveta County (TTC).pdf.
9. Gueye D. Mining, Minerals and Sustainable Development, **73** (2001)
10. Hentschel T., Hruschel F. and Priester M., *Artisanal and Small Scale Mining: Challenges and Opportunities* (International Institute for Environment and Development and WBCD, London 2003)
11. Hinton, J., *CASM, an integrated Review for Development Planning* (2003)
12. Hinton J.J. et al., Journal of Cleaner production **11:99**, 115 (2003)
13. Hinton J., Veiga M.M. and Veiga A.T.C., Journal of Cleaner Production, **11**, 99 (2003)
14. IGF, (2010) A mining policy framework; Mining and Sustainable Development

Modeling of Power Consumption of the Mining Equipment Using "The Probabilistic Automata Method"

Alla Zakharova[1], *Veniamin* Kashirskikh[1*], *Irina* Lobur[1], *Nadezhda* Shauleva[1] and *Valeriy* Borovtsov[1]

[1] T.F. Gorbachev Kuzbass State Technical University, 28, Vesennaya st., 650000, Kemerovo, Russia

Abstract. The method of power consumption forecasting based on hierarchy of the objects forming system is considered. One of perspective methods of numerical modeling of casual processes, i.e. direct generation of the studied process realization by means of the Monte Carlo method is involved in the solution of this task. The mining equipment belongs to systems with final number of states, and the separate realization of their evolution process is enabled by method of "probabilistic automata method" which generates transitions between conditions of the system by some rules set in advance. The parameters which form the basis for drawing up a matrix of probabilities of transitions when modeling a running cycle of the shearer using this method are determined on the basis of statistical data. The regularities of power consumption by separate units, subsystems and the enterprise in general established by means of the offered model can be applied to the solution of a problem of increase in efficiency of the electric power use both at a design stage, and to conditions of normal operation of the enterprise.

1 Introduction

The power economy of the coal-mining enterprise is a difficult system. Its functioning is defined by the nature of interrelations between its elements, and also external and internal factors a part of which has casual character. At the same time, one of the main tasks arising is forecasting of electric energy costs for production of one product. To solve this task, it is required to know statistical characteristics of electric energy consumed by both the mining equipment (ME), and the received final product [1-6].

To forecast electricity consumption, it is necessary to solve an important question, namely, the record of casual impacts on the units of the technological chain as casual impacts on separate units form dispersions of system characteristics in advance up a hierarchy of the objects forming system (the coal enterprise).

* Corresponding author: kvg.ea@kuzstu.ru

2 Theoretical basics

The expected scheme of modeling of power consumption by the coal enterprise can be constructed by the principle of the "ascending stream of models" [7] based on the hierarchy of the objects forming system the essence of which is as follows. At the beginning, the behavior of objects of the lowermost level of hierarchy "units" is modeled (for example, some types of mining equipment: shearers, conveyors, belt elevators, pumps of the main mine drainage, etc.) at the set input characteristics of casual influences. The lower level of objects is characterized by the fact that its elements can't be separated into smaller components. Objects of the lowermost level - "units" - experience casual influences.

Output characteristics of "units" are input characteristics for conditions of the following level of hierarchy - "subsystems" (the set of mining equipment united by a certain feature, for example the mining equipment of certain mining districts).

The input parameters for modeling of objects of the following hierarchical level are received by modeling output characteristics of all objects of the hierarchical level second from below on the input parameters received on the previous step. The described process of advance up hierarchy continues until behavior of the "system" (the coal enterprise), the object of the highest hierarchical level, is simulated.

Creation of the expected scheme described above requires establishment of regularities of power expense formation in the form of the mathematical models considering influence of casual influences that demands attraction of the mathematical theory of casual processes describing transitions of system between its states under the influence of casual influences [8, 9]. For creation of mathematical model of a casual process it is necessary to define the number of the system's conditions and probability of system's transitions between its states. Further it is possible to simulate trajectories of casual process, the sequence of system's transitions between its states.

One of the most perspective methods of numerical modeling of casual processes is the method of direct generation of the studied process by means of the Monte Carlo method [10-15] based on generation of random numbers. For systems with final number of states to which ME belongs the separate realization of their process evolution can be carried out by one of perspective methods of direct modeling of dynamics of stochastic systems - a method of "probabilistic automata" [16].

"The probabilistic automata" is the system with final number of states $E_1, E_2, E_3, \dots, E_{B1}$ the transitions between which are described by probabilities of transition P_{ij}. "The probabilistic automata" generates casual transitions by means of the rule defined in advance and based on generation of random numbers. This rule also defines the structure of "the probabilistic automata". Modeling of dynamics of the system by means of "the probabilistic automata" is realization of the casual process described by Chapman-Kolmogorov's equation for the total probability of transition for any number of actions.

$$\Pr\{E_i \to E_j \text{ actions from k to k+m}\} = \sum_{\{n_1\}} P_{in_1}(k) P_{n_1 n_2}(k+1) .. P_{n_{m-1}}(k+m).$$

With set the initial condition of the system, the sensor of random numbers defines into what admissible states the system has to pass, and the system is transferred to the following state. On the following steps of the considered temporary interval, the same procedure of determination of probability of transition and transfer of the system to the corresponding state is carried out.

As a result of the described process of casual transitions during all considered time interval, concrete realization of casual process turns out – the casual sequence of states in which the system stayed in simulated process of evolution

$$s_0, s_1, s_2, s_3, \dots, s_N,$$

where s_k- number of the state in which the system was on step with number k.

Based on the received sequence of states, temporary dependence of any measured size connected with evolution of system can be constructed.

After the task of the system conditions, the matrix of probabilities of transitions is defined

$$P_{ij}(dt) = \begin{cases} 1 - \lambda_i(t)dt, i = j; \\ a_{ij}(t)dt, i \neq j, \end{cases} \quad (1)$$

where value $\lambda_i(t)dt$ makes sense to probability of exit of the system from condition E_i for time interval $(t, t+dt)$. The parameters characterizing the conditions of the system and the matrix of probabilities of transitions form a set of input characteristics of the model. Further calculation of output parameters and the analysis of results is made.

3 Application

For modeling the work of the coal-mining enterprise unit with use of the considered method as an example the operating cycle of shearer KGS-445 has been chosen as the shearer is the main technological machine performing loading to all units and the mine in general.

Now extensive material about determination of average duration of starting-up T_{inc}, duration of operating time tp and duration of operating cycle of shearer tц during shift [7]. Duration of starting-up T_{inc} of shearers with the modes of frequent start-up is determined by the formula

$$T_{inc} = \frac{(nt_s + t_w)}{t_c},$$

where n - number of shearer motors; t_s – duration of the motor starting; t_w - work duration under loading; $t_c = nt_s + t_w + t_0$ cycle duration; t_0– pause time. Start-up duration for faces with daily output to 1200 t makes 0.5 – 0.68; with output of 2000 - 2500 t – 0.7 – 0.81; with output above 2500 t – 0.85.

Distribution of the loading levels during an operating cycle is shown in Fig. 1. During time t_{min} the first motor (lower drum) starts and runs idling then during time $t_{min.av}$ the motor of the top drum starts and runs idling. Further, after short time of load increase (it makes 3 – 5 s. therefore it is neglected), the shearer runs with some loading which histogram of distribution is presented on Fig. 2. It is convenient to use this histogram for modeling.

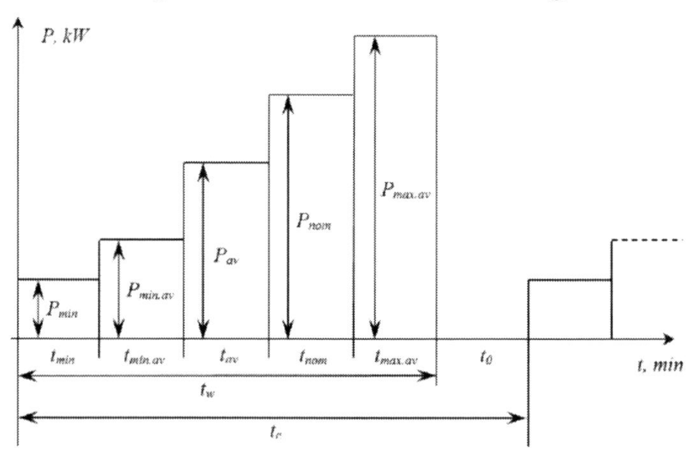

Fig. 1. Distribution of shearer capacity levels during operating cycle

During a production run which lasts from 25 to 80 *min*. it is possible to identify five levels of shearer loading: minimum P_{min}; minimum average $P_{min.av}$; average P_{av}; nominal P_{nom}; the maximum average $P_{max.av}$.

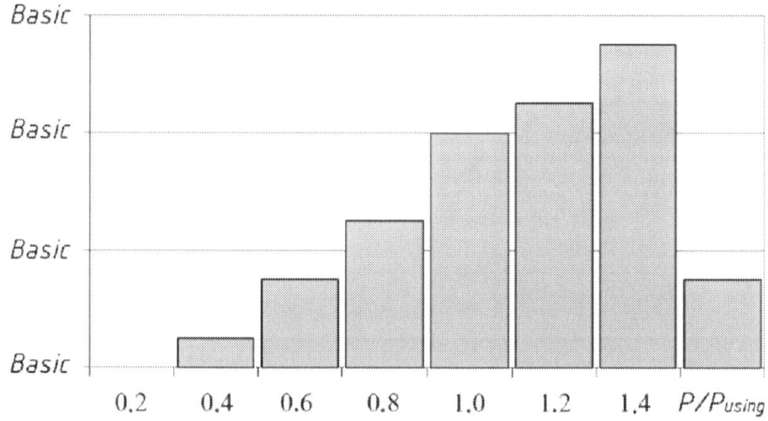

Fig. 2. Histogram of loading distribution of the shearer

The analysis of time observations of production at faces has shown:
- duration of the shearer stops for the intra face and extra face technical reasons has gamma distribution with density

$$f(t) = \frac{\lambda^{\eta}}{\Gamma(\eta)} t^{\eta-1} e^{-\lambda t}$$

where, λ, η – parameters of elementary distributions; $\Gamma(\eta)$ – gamma function;
- duration of stops for the extra face organizational reasons also has gamma distribution, and on intra face ones – exponential with density

$$f(t) = \lambda e^{-\lambda t}$$

- duration of stops because of mining-and-geological violations is described by exponential distribution;

- the probability of all types of stops is distributed by Poisson's law:

$$P(m) = \frac{\lambda^{m}}{m!} e^{-\lambda}$$

Operating time of shearers under loading has the exponential law of distribution, and duration of the cycle is described by Weybull-Gnedenko's distribution:

$$f(t_c) = \frac{b}{a}\left(\frac{t_c}{a}\right)^{b-1} \exp\left[-\left(\frac{t_c}{a}\right)^{b}\right]$$

where *a, b* – distribution parameters. Observations have shown that the output per face of 1500 t/day, the cycle duration with confidential probability p=0.9 is in the range of 40 to 50 min., and at the output up to 2500 t/day – up to 78 min. The faces working with output of 1.0 million tons/year are characterized by long cycle up to 135 - 180 min. Having the presented statistical data, it is possible to calculate probabilities of continuous work and a stop of the shearer, and also probability of continuous work in minutes will make no more t_w and duration of stops won't exceed t_o.

In the considered model during a running cycle of the shearer KGS-345 six temporary periods described in Table 1 are allocated, and amount of the consumed electric power is submitted through the total rated power of cutting motors $P_{nom.m}$ and rated power of the drive motor $P_{nom.d}$. To estimate probabilities of transition between the considered conditions of the shearer we will use results of researches [7] reporting us that the average duration of stops for various reasons makes 8÷28 min, and probabilities of shearer stops lie in the range of 0,04 – 0,52.

Table 1. Designation of a shearer condition

Period	Description	Power consumption	Designation of condition
t_0	Motors are switched off	Absent	E0
t_{min}	Motors of the lower drum idle	$P_{min}=(0,12\div0,16)\cdot P_{nom.m}$	E1
$t_{min.av}$	Motors of the lower and top drum idle	$P_{min.av}=(0,26\div0,35)\cdot P_{nom.m}$	E2
t_{av}	The shearer works with average output	$P_{av}=(0,7\div0,8)\cdot P_{nom.m}+0,8\cdot P_{nom.d}$	E3
t_{nom}	The combine works with a rated load	$P_{nom} = P_{nom.m} + P_{nom.d}$	E4
$t_{max.av}$	The shearer works with the maximum load	$P_{max.av}=(1,06\div1,36)\cdot P_{nom.m}+P_{nom.d}$	E5

4 Conclusion

The parameters of the running cycle shown in the form of the graph in Fig. 3 have been determined on the estimates of probabilities of transition and calculation of states parameters. Tops of the graph correspond to conditions of the system, arrows are the transitions from one state to another (where α – transition speed from one state to another).

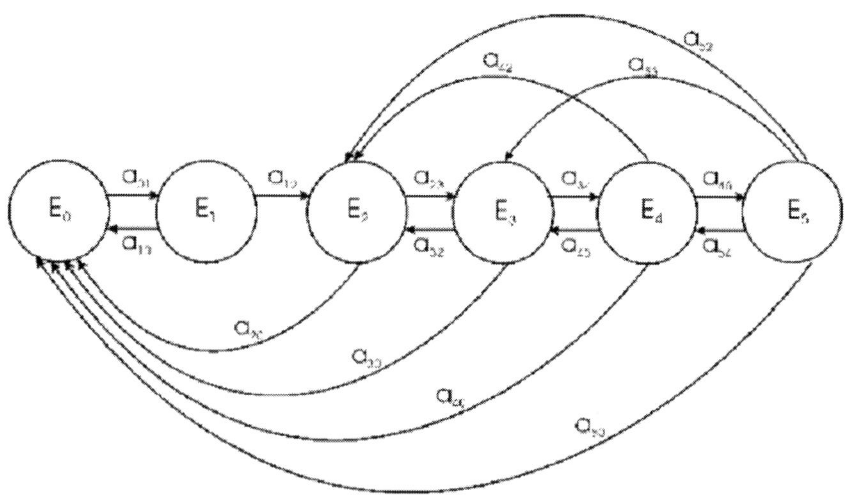

Figure 3. State graph of the "shearer" system

These data are the basis for drawing up the matrix of probabilities of transitions (1) corresponding to the graph by means of which input parameters for modeling of a running cycle of the shearer KGS-445 can be created.

Established by means of the offered model, the regularities of power consumption of certain units, subsystems and the enterprise in general can be applied to the solution of improving efficiency of the electric power use both at a design stage, and in the conditions of normal operation of the enterprise by optimizing the ME choice.

References

1. D. W. Bunn, E. D. Farmer, *Comparative models for electrical load forecasting* (Wiley, New York, 1985)
2. D. W. Bunn, E. R. Larsen, *Systems Modelling for Energy Policy* (Wiley, New York, 1997)
3. M. Munasinghe, P. Meier, *Energy Policy Analysis and Modelling* (Cambridge University Press, Cambridge, 1993)
4. J. Taylor, L. de Menezes, P. McSharry, International Journal of Forecasting, **22**, 16 (2006)
5. R. Weron, *Modeling and Forecasting Electricity Loads and Prices: A Statistical Approach* (Wiley, Chichester, 2006)
6. J. Taylor, International Journal of Forecasting, **24**, 645 (2008)
7. A. G. Zakharova, I. A. Lobur, N. M. Shauleva, V. A. Borovtsov, *Coal in the 21st Century: Mining, Processing and Safety*, 198 (Atlantis Press, 2016)
8. N. R. Draper, H. Smith. *Applied Regression Analysis* (John Wiley & Sons, New York, 1998)
9. K. Dougherty, *Introduction to econometrics* (Oxford university press, New York Oxford, 2001)
10. J. C. Bogdanoff and F.Kozin. *Probabilistic Models of Cumulative Damage* (John Wiley & Sons, New York, 1985)
11. D. W. Heerman, *Computer Simulations Methods in Theoretical Physics* (Springer-Verlag, Berlin, 1986)
12. L. Yu. Barash, L. N. Shchur, Computer Physics Communications. **185**, 1343 (2014)
13. L. Yu. Barash, L. N. Shchur, Computer Physics Communications. **184**, 2367 (2013)
14. B. Stickler, E. Schachinger, *Basic concepts in computational physics* (Springer International Publishing, Switzerland, 2016)
15. M. Newman, *Computational physics* (CreateSpace Independent Publishing Platform, USA, 2012)
16. H. Gould, J. Tobochnik, W. Christian, *An Introduction to Computer Simulation Methods. Applications to Physical System* (Addison-Wesley, 2006)

Application of Machine Learning for Dragline Failure Prediction

Amir Taghizadeh[1], *Nuray* Demirel[1*]

[1]Mining Engineering Department, Middle East Technical University, Üniversiteler Mah., Dumlupınar Bulvarı, No: 1, 06800 Çankaya, Ankara, Turkey

Abstract. Overburden stripping in open cast coal mines is extensively carried out by walking draglines. Draglines' unavailability and unexpected failures result in delayed productions and increased maintenance and operating costs. Therefore, achieving high availability of draglines plays a crucial role for increasing economic feasibility of mining projects. Applications of methodologies which can forecast the failure type of dragline based on the available failure data not only help to reduce the maintenance and operating costs but also increase the availability and the production rate. In this study, Machine Learning approaches have been applied for data which has been gathered from an operating coal mine in Turkey. The study methodology consists of three algorithms as: i) implementation of K-Nearest Neighbors, ii) implementation of Multi-Layer Perceptron, and iii) implementation of Radial Basis Function. The algorithms have been utilized for predicting the draglines' failure types. In this sense, the input data, which are mean time-to-failure, and the output data, failure types, have been fed to the algorithms. The regression analysis of methodologies have been compared and showed the K- Nearest Neighbors has a higher rate of regression which is around 70 percent. Thus, the K-Nearest Neighbor algorithm can be applied in order to preventive components replacement which causes to minimized preventive and corrective cost parameters. The accurate prediction of failure type, indeed, causes to optimized number of inspections. The novelty of this study is application of machine learning approaches in draglines' reliability subject for first time.

1 Introduction

Draglines work in a harsh working environment and prone to various failures as they operate. Lack of proper maintenance and investigation may increase the rate of failures and unavailability (Ebeling, 2010). A failure takes a place in a subsystem of a dragline halts the whole machine out of operation. Therefore, proper failure analysis and predicting the failures in advance may help in optimization of dragline preventive maintenance and eventually increase the availability and reliability of the system.

* Corresponding author: ndemirel@metu.edu.tr

The availability definition is equal with performability; moreover, quality, reliability, and maintainability should be considered together to assess the preformability. In order to achieve high performability of the system, appropriate and effective preventive and corrective actions should be taken on time. Corrective maintenance has to be carried out to recover system to its functional state. On the contrary side, preventive maintenance is utilized for forecasting failure(s) and its types as well. In this case, a predictor, which has been modeled based on the preventive step, benefits to repair, replace or make investigation for a component in a system. Since the preventive action can diminish the maintenance cost, which has 40 to 50 percent of the equipment operating cost in a mine (Forsman et al, 1992), appropriate prediction of failure types led to lower not only equipment operating cost but also total operation cost of a mine (i.e. maintenance cost is around to 20 to 35 percent of overall operating cost of a mine (Unger *et al.*, 1994).

In the mining industry like others, the systems breakdown has its effects particularly on production rate, time losses, and safety issue. Based on literature survey, the draglines, as high-capacity complex machines, are bottlenecks of mines, so availability of the mentioned excavator are vital. In this sense, reliability measurement of a dragline should be taken into consideration. Hence, various types of methods have been utilized in order to assess the reliability of a system such as: Effects and Criticality Analysis (FMECA), Fault Tree Analysis (FTA), Failure Modes, Markov Analysis (MA), Reliability Block Diagram (RBD). Since all the named methods are based on conventional linear algebra or statistical algorithms, the utilization of them in a domain which has high volume of data is impractical.

In order to handle the mentioned objective, machine learning approaches which use for data with high volume as well as high frequency are utilized. In this study, machine learning approaches (i.e. Multi-Layer Perceptron, Radial Basis Function, and K-Nearest Neighbor) have been implemented on a dataset which was shaped by gathered data from a dragline currently operating in a coal mine, Turkey. The utilized methods which are categorized as classification algorithms have one feature and 1,303 observations. There are six different failure types were identified in the dataset. Table 1 tabulates the causes of dragline failures and failure numbers associated with each type. The input data are Time to Failure (TTF) and the output data are called types of breakdowns. Table 2 presents sample datasets used as input for the algorithm.

Table 1. Causes and number of dragline failures

ID of Failure Type	Failure Type	Failure Number
1	mechanical failures	945
2	electrical failures	276
3	energy related failures	53
4	major revision	5
5	maintenance	14
6	power cut	7

Table 2. Sample dataset

Mean Time to Failure (Hour)	ID of Failure Type
1.6597	1
3.5625	1
14.4167	2
5.5625	6
0.5938	1

2 Application of Machine Learning Algorithms for Dragline Failure Prediction

The assessment of draglines' reliability aims to achieve high availability which is a crucial task in the mining industry. Based on idea mentions practical engineering system, which is not like a closed-form relation, encompasses various types of components, so the elaborated calculation is required. In this sense, Machine Learning (ML) approaches, which can model the complex system, can be utilized. Modeling the whole system in respect to assess the reliability in the case of quantitative data by means of ML is not only effective but also operational and logical action in case of big data management.

ML approaches have been implemented in the recent studies (Hurtado *et al.* (2001), Xu *et al.* (2002), Hurtado (2004), Zhiwei (2008), Gomes *et al.* (2004), Schueremans *et al.* (2005), Roco *et al.* (2002), Chua (2002)). Since the abilities of parallel processing with respect to learning, generalizing to classify, and organizing data is the main reason for using Artificial Neural Network (ANN) in this study. ANN models have been categorized to two groups: supervised training and unsupervised training. The structure of ANN, also, can divided to two parts feedback and feedforward architecture (i.e. in some cases combination of two architectures might be used). In this study, Multi-Layer Perceptron (MLP) and Radial Basis Function (RBF) algorithms as methods of ANN, which are supervised training methods, have been utilized (i.e. ANN is an approach of ML). MLP and RBF have been implemented on data which were acquisitioned from an operating dragline in a coal mine in 23 years. The arrays of pre-processing methods have been carried out on the dataset, such as: i) data cleaning, ii) data integration, and iii) data transformation. In order to tackle the missing values, which is located in data cleaning phase, data have been eliminated. The idea of MLP is in simplest format is close meaning of logistic regression, which contains one hidden layer and only a single neuron (Figure 1).

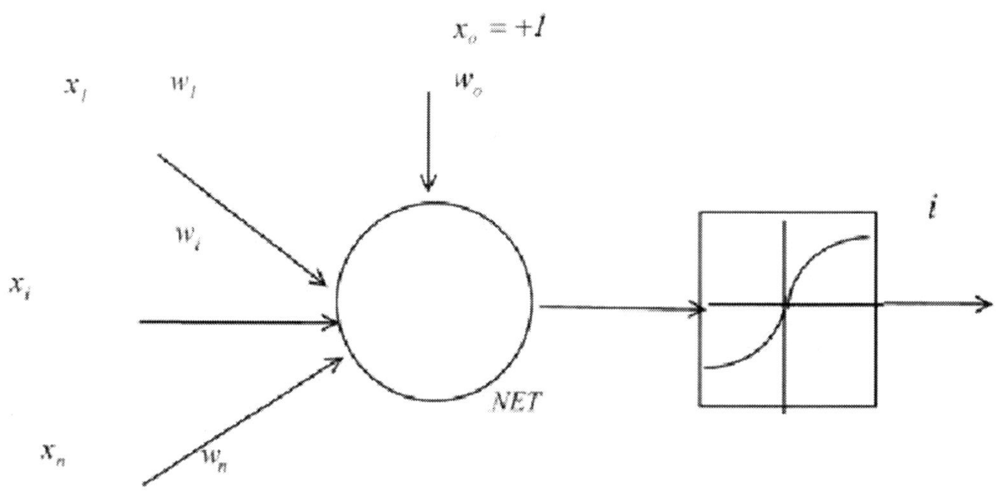

Fig. 1. Single neuron MLP

A single neuron unit was fed by input vector(s); then some calculation which is called transformation is carried out on inputs to reach output vector(s). Transformation was carried out in two stages, i) sum up of all inputs (x_i) and their corresponding weights (w_i), so the scalar is going to be calculated and ii) the output(s) based on the scalar and an activation function (i.e. an activation function can be linear or non-linear) is generated. A single processing unit makes connections with other processing units, so a Neural Network is shaped. In a network, it is necessary for weights to be adjusted. In order for weights adjusting, Gradient Descent method as an optimizer of weights was applied in this study.

An input layer and an output layer get their data from a network. In shaping a Neural Network, it is necessary to determine the number of hidden layer as well as the number of hidden units. Since a model is more complex, more hidden layer is required; moreover, the number of hidden units is determined by trial and error in practice (Fig 2).

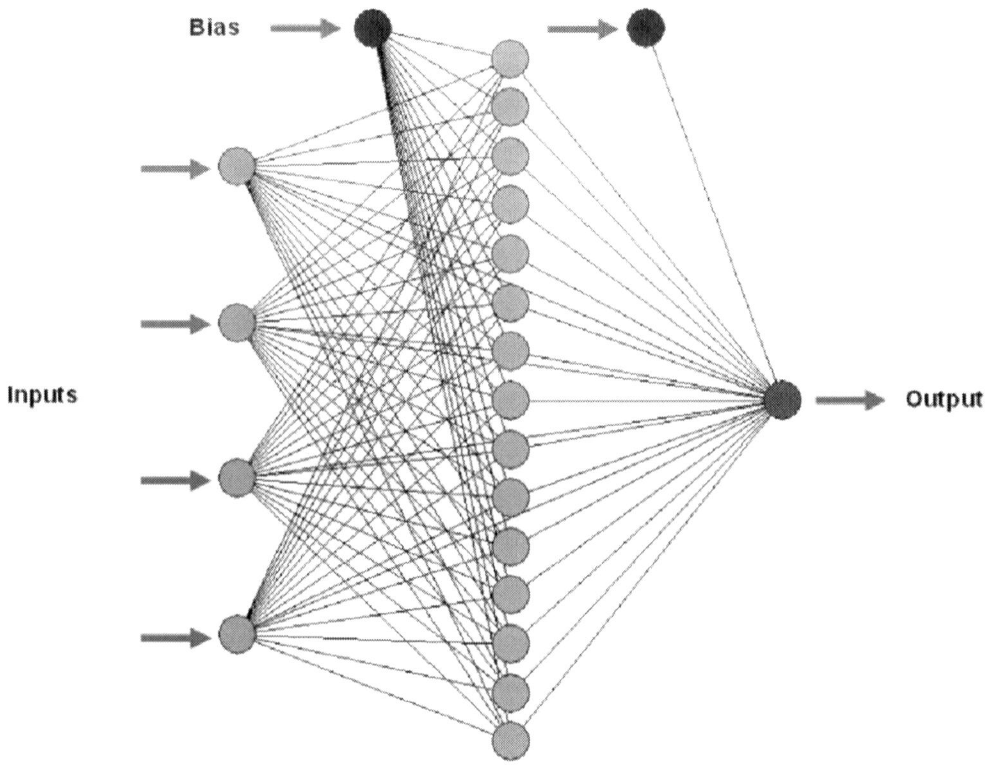

Fig. 2. Schematic view of multi-layer perceptron network

In this study, MLP with four hidden layers and five processing neuron in each layer has been utilized. The regression analysis of MLP is 37 percent; moreover, the Root Mean Square Error (RMSE) for MLP is 0.79. In the case of high RMSE and low regression analysis of MLP, RBF has been implemented on dataset. In RBF, input vector(s) feed to a Neural Network and instead of weight which taken in Perceptron neurons, the Euclidean Norm is calculated (i.e. the Euclidean Norm is distance between input vector(s) and center vector). The multiplication of Euclidean result and a threshold (b) is feed to non-linear function (i.e. the radial basis function (G)) (Fig 3). The calculation is shown in equation below:

$$a = (\|x - t\|) \bullet b \tag{1}$$

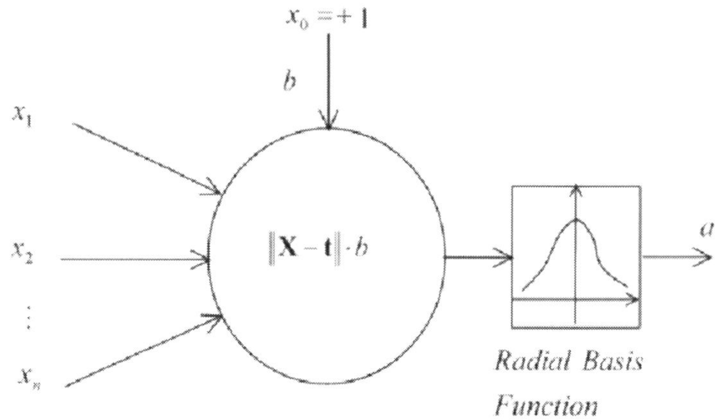

Fig. 3. Single radial basis function

When number of single radial basis function connected each other, the Neural Network of radial basis function shaped (Fig. 4).

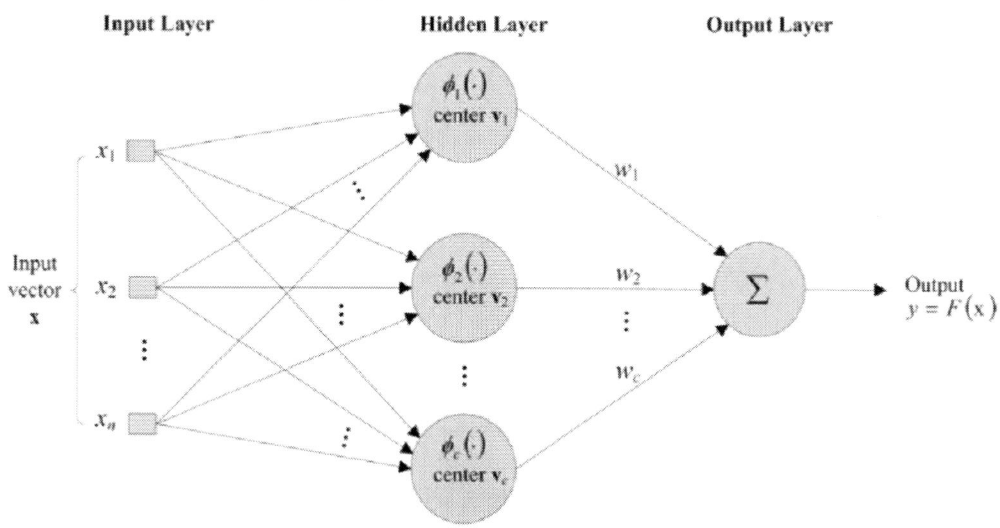

Fig. 4. Schematic view of radial base function network.

The regression analysis for RBF is 18 percent and RMSE is 0.74. Since MLP could not achieve the desired result, nor did RBF, K- Nearest neighbor algorithm has been utilized in order to achieve a result with higher regression analysis.

K-Nearest Neighbor (KNN) is non-parametric algorithm which is called lazy algorithm. Since it does not get any assumptions on the data distribution, it is categorized as non-parametric algorithm. In this sense, the application of KNN in the practical data, which does not have any typical assumptions, can be profitable. While KNN does not utilize any training data in order to do generalization, it is called Lazy algorithm. In the KNN algorithm, decision is made based on whole of training dataset. Therefore, KNN requires more time as well as more memory in order to make a decision. Since, data is in a feature space, the notion of distance exist, which can be Euclidean distance or any common used distance measuring algorithm. In the KNN algorithm, every data in feature space has label, in the simplest trend number of classes are to two. However, KNN works with uninformed number of classes. Based on the distance matrix the number of neighbor (K) is calculated. In this study, KNN has been applied for the dataset, and therefore, it yields regression analysis 72 percent. Table 3 presents the utilized machine learning algorithms and respective regression analysis results.

Table 3. Utilized ML algorithm and results

Algorithm	Regression Analysis (%)
Multi-Layer Perceptron	37
Radial Basis Function	18
K-Nearest Neighbour	72

3 Conclusions

Draglines, as a single operating machine in the coal mines, play a crucial role. In this sense, reliability of draglines should be taken into consideration, since it breakdown of the mentioned mining machinery causes losses in time and profit as well as higher direct and indirect costs. In this study, three types of machine learning approaches have been utilized in order to predict the failure type of a dragline and to make a predictive model of breakdown, so higher rate of regression analysis can fulfill the better prediction with respect to prognostic process in condition-based maintenance. Mean time to failure and failure type as the input vector and the output vector has been used, respectively. Three algorithms have been implemented on the dataset. As a result, K-nearest neighbor algorithm has a better result in contrast to other utilized algorithm. This study showed that machine learning approaches can effectively be utilized to predict the failure type of a dragline. As a conclusion, preventive maintenance policies should adopt machine learning algorithms to achieve optimized maintenance programs which yields lower maintenance costs, fewer machine failures, less repair downtime, reduce small parts inventory, longer life of dragline, increase of its production and improved operator safety.

References

1. K. S. Chua, Pattern Recognition Letters, **24**, 75 (2003)
2. C.E. Ebeling, *An introduction to reliability and maintainability engineering*. (Waveland Press Inc, 2010)
3. B. Forsmann and U. Kumar, Journal of Mines, Metals and Fuels, **40**, 267 (1992)
4. H. M. Gomes and A. M. Awruch, Structural Safety. **26**, 49 (2004)
5. J. E. Hurtado, Structural Safety. **26**. 271 (2004)
6. J. E. Hurtado and D. A. Computer Methods in mechanics and Engineering. **191**, 113 (2001)
7. C. M. Rocco and J. A. Moreno, Reliability Engineering and System Safety. **76**, 237 (2002)
8. L. Schueremans and D. V. Gemert, Structural Safety. **27**, 246 (2005)
9. R. L. Unger and K. Conway, *Impact of Maintainability Design on Injury Rates and Maintenance Costs for Underground Mining Equipment.* In R. H. Peters, Special Publication Report, 140 (1994)
10. G. Zhiwei and B. Guangchen, Chinese Journal of Aeronautics. **22**, 160 (2009)

Kinematic Parameters Of Rotary Transmission With Hydraulic Cylinders

Mikhail Blaschuk[1,2*], *Anton* Dronov[1], *Alexandr* Koperchuk[1], *Roman* Chernukhin[1], and *Victoria* Litvinenko[1]

[1]Yurga Institute of Technology of Tomsk Polytechnic University, 652055, Kemerovo region, Yurga, Leningradskaya str. 26, Russia
[2]Institute of Coal of the Siberian Branch of the RAS, 650000, Kemerovo, ave. Sovetski, 18, Russia

Abstract. The issue of designing drives, which provide low frequency (max. 1 rotation per minute) rotation with a big moment (min 1 MN m) of large technical bodies utilized in restricted spaces, is a complex and contradictory one. The drives of geokhod propeller, rotor actuators of tunneling machines with overload protection, as well as actuators of other machinery meeting aforementioned requirements are examples of such machines. The paper considers mathematical model developed by the authors which determines the relation of design factors of transmission tooled with hydraulic cylinders to kinematic parameters of output element movement. The paper also provides description of methods to determine pumping unit efficiency for rotary transmission tooled with hydraulic cylinders.

1 Introduction

Conventional drives with mechanic transmission and electric or hydraulic motors are difficult to utilize because of insufficient mass-overall parameters [1-9]. It is also a problem to use rather perspective wave gears with interposing rolling elements because of their technical complexity and technological imperfection [10,11]. An alternative is suggested: a swinging mechanism tooled with hydraulic cylinders placed on chord of a circle and linear – rotary motion converter designed as a free-wheel clutch mechanism [12–14]. As all hydraulic cylinders are cyclic reciprocating hydraulic motors, the continuity of output element rotation is provided with out of phase operating hydraulic cylinders in most mechanisms, that is, one half of cylinders makes a power stroke and the other one – an idle stroke [15,16]. Mechanisms with more hydraulic cylinders making a power stroke than an idle stroke are considered to be more power efficient. For this purpose hydraulic cylinders are to be placed in different phases of extension (Fig. 1, 2).

The authors have already considered issues focused on determination of energy-power parameters of this type of transmissions. We have suggested a mathematical model to calculate a turning moment made by transmission according to accepted dimensions of geokhod, necessary clearance of internal space, and conditions of mine working [2, 3].

* Corresponding author: mby.tpu@gmail.com

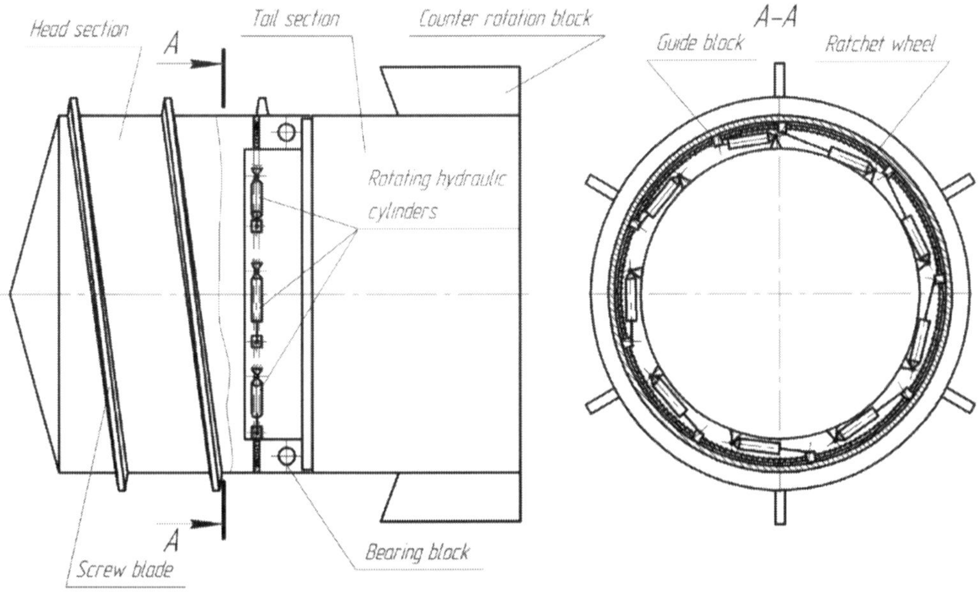

Fig. 1. Outlined design of constant bottom-supply geokhod transmission tooled with two groups of hydraulic cylinders in different extension phases in one plane.

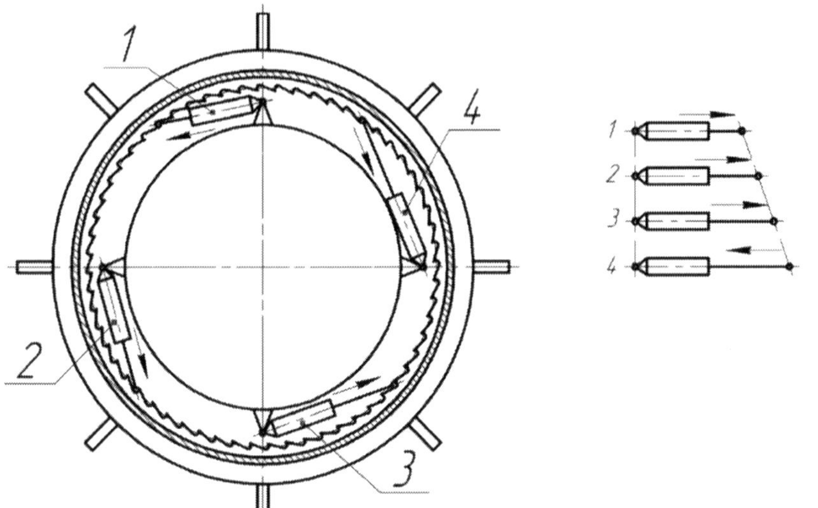

Fig. 2. Explanatory design of operating transmission with hydraulic cylinders in different phases

It is stated in paper that the moment made by transmission tooled with hydraulic cylinders will change periodically because of alternating lever of force exertion by hydraulic cylinders. We have also studied the effect of geometrical parameters of transmission and the number of hydraulic cylinders on the value of moment inequality. Another major influence on the geometric parameters of the transmission will have the technological features of manufacture geokhod housing [17–19].

2 The mathematical model

This paper is suggested to focus on determining kinematic parameters of rotary transmission tooled with hydraulic cylinders, in particular – on angular rate and output element angular rotation (head section with propeller) specific for a stroke of hydraulic cylinder rod, as well as on relation of these parameters to design factors of transmission and geometrical parameters of geokhod.

If a hydraulic cylinder rod is extended by the value of a stroke , the section is turned by an angle , which can be determined from the triangle ABO (Fig. 3)

$$\varphi_P = \gamma - \gamma_0, \tag{1}$$

where γ – central angle between supports of the hydraulic cylinder in the end position of rod extension, that is at $L_{W\max} = L_0 + L_m$;

γ_0 – central angle between supports of hydraulic cylinders in the start position of rod extension, that is at $L_{W\min} = L_0$.

$$\gamma = \arccos \frac{\dfrac{D_{PL.HC}^2 + D_{PL.ROD}^2}{2} - 2\cdot(L_0 + L_W)^2}{D_{PL.HC}\cdot D_{PL.ROD}}, \tag{2}$$

$$\gamma_0 = \arccos \frac{\dfrac{D_{PL.HC}^2 + D_{PL.ROD}^2}{2} - 2L_0^{\ 2}}{D_{PL.HC}\cdot D_{PL.ROD}}, \tag{3}$$

where $D_{PL.HC}$ – diameter of a circle formed by placed hydraulic cylinder trunnions on the shell ring of the tail section, m; $D_{PL.ROD}$ – diameter of a circle of rod trunnion rotation on the head section, m; L_W – spacing between body trunnions and hydraulic cylinder rod, provided that rod extension is maximum, m; L_0 – spacing between body trunnions and folded up hydraulic cylinder (minimal extension), m.

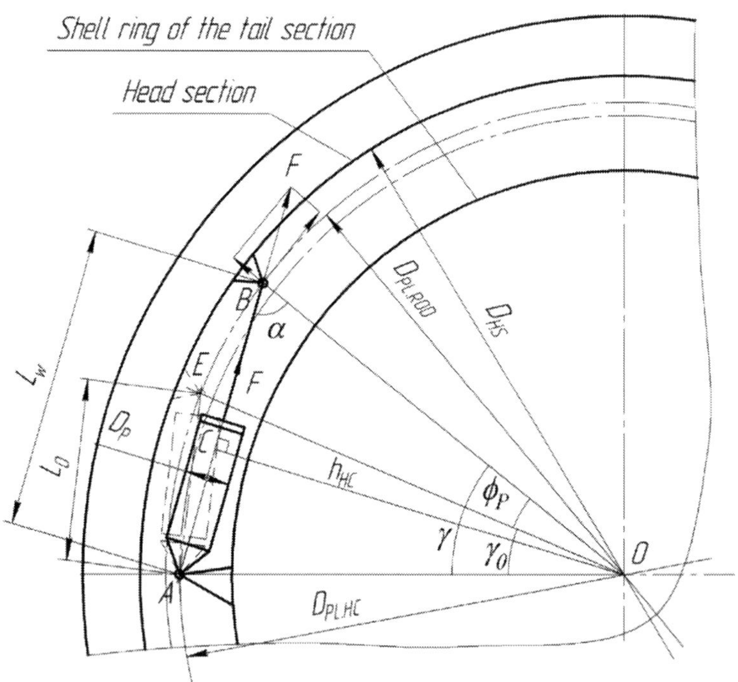

Fig. 3. Explanatory design to determine the angle of rotation of the head section of geokhod

After substitution of (Eq. 2) and (Eq. 3) into (1) we obtain

$$\varphi_{II} = \arccos \frac{\frac{D_{PL.HC}^2 + D_{PL.ROD}^2}{2} - 2\cdot(L_0 + L_m)^2}{D_{PL.HC}\cdot D_{PL.ROD}} - \arccos \frac{\frac{D_{PL.HC}^2 + D_{PL.ROD}^2}{2} - 2L_0^{\,2}}{D_{PL.HC}\cdot D_{PL.ROD}}, \quad (4)$$

If body trunnions and hydraulic cylinder rods are placed on the same circle, that is, $D_{PL.ROD} = D_{PL.HC} = D_{HC}$ the angle of rotation φ_P is determined by the expression

$$\begin{aligned}\varphi_P &= \arccos \frac{D_{HC}^2 - 2\cdot(L_0 + L_m)^2}{D_{HC}^2} - \arccos \frac{D_{HC}^2 - 2L_0^{\,2}}{D_{HC}^2} = \\ &= \arccos\left(1 - \frac{2\cdot(L_0 + L_m)^2}{D_{HC}^2}\right) - \arccos\left(1 - \frac{2L_0^{\,2}}{D_{HC}^2}\right).\end{aligned} \quad (5)$$

3 Determination of the angular rate of head section rotation

If consumption of hydraulic fluid passing into the head end of a hydraulic cylinder is constant $Q(t) = const$, piston and rod will be extended at a constant rate υ, determined from the expression

$$\upsilon = \frac{Q}{S_P}, \quad (6)$$

where $S_P = \pi \cdot D_P^{\,2}/4$ – area of a hydraulic cylinder piston, m^2; D_P – diameter of a hydraulic cylinder piston, m;

Extension of hydraulic cylinder rod (Fig. 3) from the start position is determined according to the formula

$$\delta(t) = \upsilon t = \frac{Q}{S_P}\cdot t. \quad (7)$$

The angle of rotation of the head section is $\varphi_P(t)$, as time variable t varies from L_0 to $L_0 + L_m$ when rod is extended, as it is stated in (Eq. 5)

$$\varphi_P(t) = \arccos \frac{\frac{D_{PL.HC}^2 + D_{PL.ROD}^2}{2} - 2\cdot(L_0 + \delta(t))^2}{D_{PL.HC}\cdot D_{PL.ROD}} - \gamma_0, \quad (8)$$

where $\gamma_0 = const$ – angle, determined according to the formula (Eq. 3);

The angular rate of section rotation $\omega(t)$ is formulated in the expression

$$\omega(t) = \frac{d\varphi_P(t)}{dt}, \quad (9)$$

After differentiating the expression (Eq. 8) with respect to time we obtain

$$\omega(t) = \frac{4\upsilon\cdot(L_0 + \upsilon t)}{D_{PL.HC}\cdot D_{PL.ROD}\cdot\sqrt{1 - \left(\frac{\frac{D_{PL.HC}^2 + D_{PL.ROD}^2}{2} - 2\cdot(L_0 + \upsilon t)^2}{D_{PL.HC}\cdot D_{PL.ROD}}\right)^2}}, \quad (10)$$

After differentiating the expression (Eq. 4) by analogy we obtain

$$\omega(t) = \frac{4\upsilon \cdot (L_0 + \upsilon t)}{D_{HC}^2 \cdot \sqrt{1 - \left(1 - \frac{2 \cdot (L_0 + \upsilon t)^2}{D_{HC}^2}\right)^2}}, \tag{11}$$

In terms of the expressions (Eq. 10) and (Eq. 11), the angular rate of head section rotation $\omega(t)$, as well as the turning moment made by transmission will change according to hydraulic cylinder rod extension (Fig. 4). Provided that hydraulic fluid passing into the head end $Q(t) = const$ the rate of piston movement is constant too $\upsilon(t) = const$, but the circular rate υ_R will change and determine variation of angular rate of head section rotation.

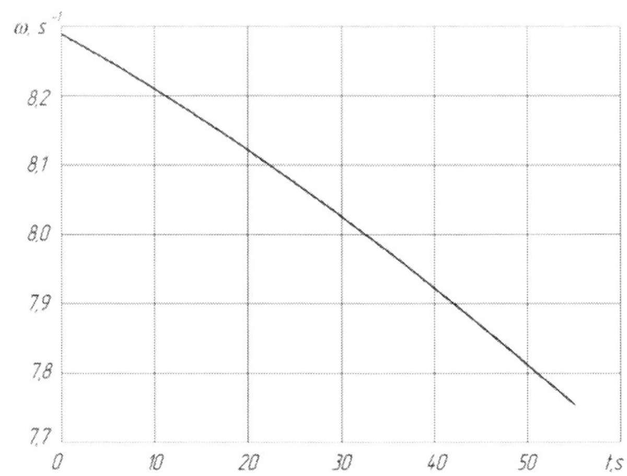

Fig. 4. The variation graph of angular rate of head section rotation during the time t of rod extension

It is evident that to obtain a uniform angular rate ($\omega(t)$ =const) during the period of cylinder rod extension the rate of extension $\upsilon(t)$ is to vary according to a certain law. After expressing the rate $\upsilon(t)$ in terms of (Eq. 11) we have

$$\upsilon(t) = \frac{\sqrt{4D_{HC}^2 - 4L_0^2 + (D_{HC} \cdot \omega \cdot t)^2} - L_0 \cdot \omega \cdot t}{4 + \omega^2 \cdot t^2} \cdot \omega \tag{12}$$

As it is stated in the known expression, the circular rate of head section rotation will be calculated as the product of rotation radius – R and angular rate – ω

$$\upsilon_R = R \cdot \omega, \tag{13}$$

A circular rate will be determine by the rate of hydraulic cylinder rod extension and angle α between the vector of the rod rate υ and angular rate υ_R (tangent drawn to the circle of rotation)

$$\upsilon_R = \upsilon \cdot \cos\alpha, \tag{14}$$

After expressing υ from (Eq. 14) and substituting it for υ_R in the expression (Eq. 13) we have

$$\upsilon = \frac{R}{\cos\alpha} \cdot \omega, \tag{15}$$

In the expression (Eq. 15) the first multiplier is a ratio

$$\frac{R}{\cos\alpha} = \frac{\sqrt{4D_{HC}^2 - 4L_0^2 + (D_{HC} \cdot \omega \cdot t)^2} - L_0 \cdot \omega \cdot t}{4 + \omega^2 \cdot t^2}, \tag{16}$$

In addition, it is to say, that in expressions (Eq. 12) and (Eq. 16) the product $\omega \cdot t$ is an angle of rotation, that is, $\varphi_P = \omega \cdot t$.

We can obtain the necessary variation of rod extension rate if we change the volume of hydraulic fluid passing into the head end of the cylinder per time unit, that is, by changing the consumption according to the expression $Q = \upsilon \cdot S_{\Pi}$.

$$Q(t) = \frac{\sqrt{4D_{HC}^2 - 4L_0^2 + (D_{HC} \cdot \omega \cdot t)^2} - L_0 \cdot \omega \cdot t}{4 + \omega^2 \cdot t^2} \cdot \omega \cdot \frac{\pi \cdot D_P^2}{4} \tag{17}$$

Figure 5 demonstrates the graph of changing hydraulic fluid consumption for one hydraulic cylinder

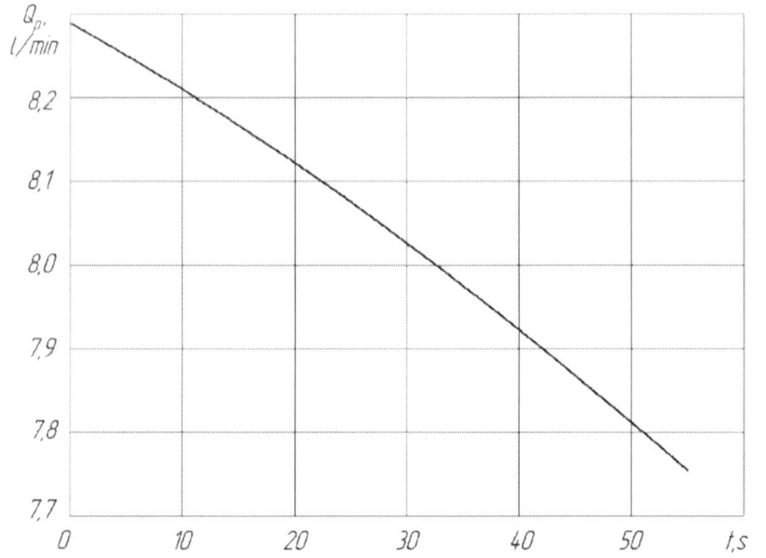

Fig. 5. The graph of changing hydraulic fluid consumption in the head end of the hydraulic cylinder during the time of rod extension t

Therefore, the angular rate of rotation of transmission output element (head section of geokhod) is constant provided that hydraulic fluid consumption in each cylinder varies according to the set variable (Eq. 17). Volumetric and throttling regulating can be utilized to provide the consumption of hydraulic fluid according to the set variable [15]. Volumetric feeder with stepping motor drive system is an alternative.

4 Conclusions

1. The angular rate of output element rotation changes periodically in rotary transmission tooled with hydraulic cylinders placed on chords, provided that hydraulic fluid consumption in hydraulic cylinders is constant. It is caused by changing attitude position of hydraulic cylinder rods (similar to turning moment irregularity). Additional research is required to evaluate kinematic irregularity.

2. Hydraulic fluid consumption is to vary according to a certain law in each hydraulic cylinder in order to provide regular rotation of output element in rotary transmissions tooled with hydraulic cylinders placed on chords. It is possible to achieve quite simple – with the use of volumetric feeders, proportional control, and other throttling regulating methods.

References

1. V.V. Aksenov, V.Y. Beglyakov, A.A. Kazantsev, I.V. Doroshenko, IOP Conf. Ser. Mater. Sci. Eng., **127**, 012031 (2016)
2. M.Y. Blaschuk, A.A. Dronov, S.S. Ganovichev, IOP Conf. Ser. Mater. Sci. Eng.,. **142**, 012128 (2016)
3. M.Y. Blaschuk, A.A. Dronov, S.S. Ganovichev, IOP Conf. Ser. Mater. Sci. Eng., **127**, 012033 (2016)
4. A.B. Efremenkov, *6th International Forum on Strategic Technology (IFOST - 2011): Proceedings: in 2 vol. IEEE*, **1**, 348 (2011).
5. V.V. Aksenov, A.B. Efremenkov, V.Y. Beglyakov, Applied Mechanics and Materials, **379**, 16 (2013)
6. A.V. Koperchuk, A.V. Murin, V.V. Filonov, IOP Conf. Ser. Mater. Sci. Eng., **127**, 012040 (2016)
7. A.A. Khoreshok , L.E. Mametyev, A.Y. Borisov, A.V. Vorobyev, IOP Conf. Ser. Mater. Sci. Eng., **127**, 012039 (2016)
8. S.A. Zhironkin, A.A. Khoreshok, M.A. Tyulenev, G.A. Barysheva, M.C. Hellmer, IOP Conf. Ser. Mater. Sci. Eng., **142**, 012127 (2016)
9. V.Y. Begljakov, V.Y. Timofeev, M.V. Dokhnenko, Applied Mechanics and Materials, **682**, 282 (2014)
10. V.Y. Timofeev, V.V. Aksenov, J.I. Galjamova, Applied Mechanics and Materials, **682**, 246 (2014)
11. G.B. Buyalich, A.V. Anuchin, K.P. Serikov, IOP Conf. Ser. Mater. Sci. Eng., **127**, 012012 (2016)
12. R.V. Chernukhin, M.Y. Blaschuk, V.Yu. Blumenstein, P.A. Chazov, IOP Conf. Ser. Mater. Sci. Eng., **142**, 012003 (2016)
13. M.Y. Blashchuk, A.A. Kazantsev, R.V. Chernukhin, Applied Mechanics and Materials, **682**, 418 (2014)
14. G.D. Buyalich, K.G. Buyalich, V.V. Voyevodin, IOP Conf. Ser. Mater. Sci. Eng., **91**, 012087 (2015)
15. J. Kretschmann, A.B. Efremenkov, A.A. Khoreshok, IOP Conference Series: Earth and Environmental Science, **50**, 012024 (2017)
16. A.V. Walter, I.F. Borovikov, M.F. Savin, IOP Conf. Ser. Mater. Sci. Eng., **142**, 012129 (2016)
17. A.V. Walter, I.F.Borovikov, R.V. Chernukhin, S.S. Nozirzoda, IOP Conf. Ser. Mater. Sci. Eng., **127**, 012017 (2016)
18. D.V. Lychagin, A.A. Lasukov, A.V. Walter, D.A. Arkhipova, IOP Conf. Ser. Mater. Sci. Eng., **125**, 012030 (2016)

 DOI: 10.1051/e3sconf/20171503004

The results of cutting disks testing for rock destruction

Aleksey Khoreshok[1*], *Leonid* Kantovich[2], *Vladimir* Kuznetsov[1], *Elena* Preis[1], *Dilshad* Kuziev[2]

[1]T.F. Gorbachev Kuzbass State Technical University, 650000, Kemerovo, 28 Vesennya st., Russian Federation
[2]National Research Technical University "NITU "MISiS", 119991, Moscow, 4 Leninsky av., Russian Federation

Abstract. To determine the rational order of disk tools placement on the working body is necessary to know the maximum amount of rock, destroyed by the disk tool in benching cutting mode depending on the tool geometry parameters, physical and mechanical parameters of rocks. The article contains the definition of rational parameters of cutting disk tools as well as power and energy parameters of the destruction process by cutting disks and by executive body of the coal cutter. The rational geometric parameters of cutting discs are specified. It was found that each step of cutting with a minimum depth of penetration has its own maximum height of bench outcrop. The dependence of the volumes of large items destroyed by the disk tool on the cutting step height was determined. The existence of the cyclic alternation of destruction phases, regardless the fracture parameters, the height of the ledge outcrop, and tools like free cutting geometry were found. In contrast to the free cutting in benching mode of destruction two large fragments of rocks in one cycle were observed. Consequently, the cyclical nature of the destruction process in the benching mode will be characterized by two chips and crushing, and this cycling repeats throughout the destruction process with the same parameters of destruction.

1 Introduction

At present time studies of mining machinery and equipment increased greatly in coal basin areas of Russia. The reason of this is coal mining boosting in Kuzbass – the greatest Russian coal basin located in Western Siberia. Last decade coal mining in Kuzbass has increase by 6-10 million tons annually. The main part of Kuzbass coal is mined by open pit enterprises - 65%. And now Kuzbass open pits are the leaders in new mining equipment implementation and research contractors [1]. Many studies have focused on the destructive influence of cutting tools on rock arrays [2-5]. Some studies were devoted to the analysis of different aspects of cutting process [6-14].

* Corresponding author: haa.omit@kuzstu.ru

Another group of works is devoted to accuracy of measurements [15-18]. Destruction process modelling has become the topic of active research since the 2000th [19-24]. Some studies are devoted to applying the results of research of cutting process in different kinds of equipment [25-29]. The significant part of research made by Kuzbass scholars (Western Siberia, Russia) belongs to improving characteristics of cutting instruments for rock destroying in coal mines and pits [30-37].

However production of highly effective cutting disks for rock destroying tools of mining machinery is still important for Kuzbass coal industry.

2 Materials and methods

The estimated scheme of the influence model of distributed single effort on a quarter of space with outgrowth is shown in Fig. 1, where the resultant of the force distribution on a horizontal surface is defined as:

$$P=-\int_{0}^{0.1}\int_{-1}^{1}(z/l)\left(1-\cos\frac{2\pi z}{l}\right)dzdx, \quad (1)$$

where l – length of contact of the tool with the rock.

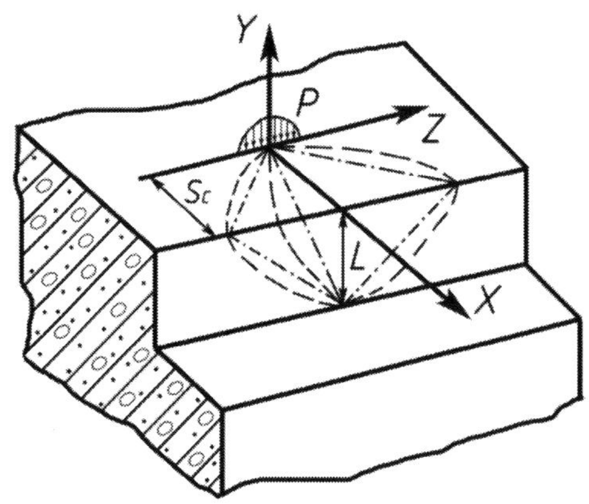

Fig. 1. The scheme of the disk tool and the rock interaction

Calculation of models was conducted at following input values:

$0.01 \le S_c \le 0.09$ m, $S_c \le L \le L^{\infty}$ with pitch $\Delta S_c = \Delta L = 0.005$ m, where L^{∞} – unlimited length of the exposed surface.

The fracture surface is described by the criterion:

$$F \le \frac{\sqrt{2}}{2}\sqrt{(\sigma_1-\sigma_2)^2+(\sigma_2-\sigma_3)^2+(\sigma_3-\sigma_1)^2}, \quad (2)$$

where σ_1, σ_2, σ_3 – primary stresses.

The theoretical results describing the volume of the large destroyed element by disk tool are best described by the formula of the ellipsoid.

In Fig. 1 bar and dot-dash lines show the fracture surface, built on the formulae of ellipsoid and the pyramid respectively.

As a result, conducted research showed that the step of cutting S_c – the distance from the line of action of the tool to the height of the projection influences on the distribution of the fracture surface plane X0Y.

According to the results of modeling, a few dependences were built to define the maximum amount of material destroyed disk tool V_{max} for cutting step S_c with unlimited height of projection's exposure (L).

For ellipsoid:

$$V_{el}(S_c)=\begin{cases}\frac{\pi}{3}S_c\left(-36+\frac{1080}{S_c+21}\right)\left(0.1S_c^2-0.19S_c+5.72\right),\text{where } 0.03\le S_c\le 0.09\,\text{m};\\ \frac{\pi}{3}S_c\left(\frac{1}{2}S_c^2+\frac{1}{2}S_c+3\right)\left(14.65-\frac{63.47}{S_c+4.33}\right),\text{where } 0\le S_c\le 0.03\,\text{m}.\end{cases} \quad (3)$$

For pyramid:

$$V_{pyr}(S_c)=\frac{V_{el}(S_c)}{\pi}. \quad (4)$$

Results of calculation of the dependence of the volume of large elements (V) destroyed by disk tool on the size of the cutting step S_c are shown in Fig. 2.

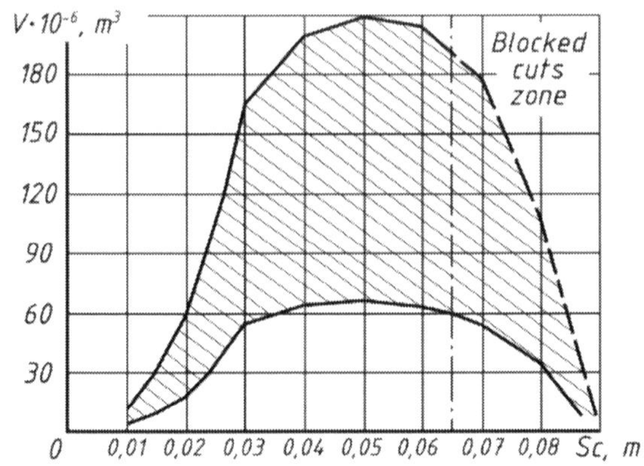

Fig. 2. Dependence of the volume of destroyed large element V and cutting step S_c

It is likely that the surface of large elements of real materials having structural defects should be between the fracture surface of the ellipsoid and pyramids. Here, the amount of real materials is shown by the shaded area.

These curves reflect a dome-like character of the distribution. Volumes of maximum large elements separated from the array by disk tool are located in the area of $0.04\,m\le S_c\le 0.06\,m$ and $0.072\,m\ge L_{\max}\ge 0.04\,m$. The maximum amount of large elements is achieved at $S_c=0.05\,m$ and $L_{\max}=0.055\,m$. Therefore if $0.555\le S_c/L_{\max}\le 1.5$ then maximum area of large elements (V_{max}) and maximum is achieved at the ratio of $S_c/L_{\max}=0.91$.

In order to confirm the results of simulations there were experimental studies conducted for destruction of sand-and-cement blocks with the following strength characteristics: compressive strength σ_c = 28.0 MPa, ultimate tensile strength σ_t = 4.0 MPa. The disk instruments had the following geometric parameters: diameter D = 0.16 and 0.18 m; wedge angle F = 30 and 35°. The geometrical dimensions of the projection and fracture parameters had the following values: cutting step S_c = 0.03, 0.04 and 0.05 m; penetration depth h = 0.005, 0.01 and 0.015 m; height of exposure L = 0.03, 0.04 and 0.05 m.

Experimental studies were carried out on a special stand with strain gauge head in the volume setting (*Px*, *Py*, *Pz*), developed in the laboratory of the department of mining machines and equipment of Kuzbass State Technical University [38-40].

As a result of experimental studies the existence of cyclical phase sequence of destruction was revealed, regardless of fracture parameters, the height of projection exposed and geometry of tools like free cutting.

Fig. 3 shows these phases of destruction depending on the different penetration depth (h) of a disk tool. In contrast to free-cutting, in projection mode of destruction we observed two large fragments of rocks in the same cycle. Consequently, the cyclical nature of destruction process in projection mode [41] is characterized by two fragments and common fragmentation, and all these cycles are repeated throughout the process of destruction with the same parameters of fracture.

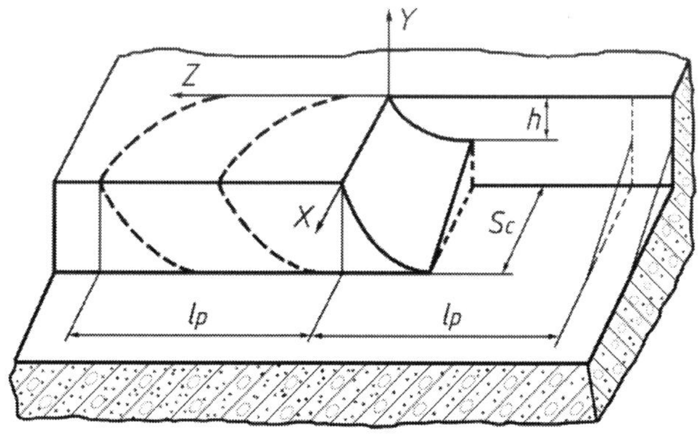

Fig. 3. Forms of fragments and the cycles of destruction with projection cutting mode

On the basis of conducted experiments it was established that phases of cycles don't depend on penetration depth (h) and have the same form of fragments. Also it was revealed that when changing diameter of the tool D, length of a phase of destruction of l_p remained constant.

As a result of calculations the following functional dependences of volume of the destroyed large elements of material (V) on a cutting step S_c and height of exposure of free surface (L) were calculated:

– for ellipsoid:

$$V_{el}(S_c,L)=\begin{cases}\dfrac{\pi}{6}S_cL\left(14.65-\dfrac{63.47}{S_c+4.33}\right),\text{where } 0.03\le S_c\le 0.09\,\text{m}\\ S_c\le L\le L_{\max}=\dfrac{1}{2}S_c^2+\dfrac{1}{2}S_c+3,\\ \dfrac{\pi}{6}S_cL\left(0.1S_c^2-0.19S_c+5.72\right),\text{where } 0.03\le S_c\le 0.09\,\text{m}\\ S_c\le L\le L_{\max}=-36+\dfrac{1080}{21+S_c}\end{cases} \quad (5)$$

– for pyramid:

$$V_{pyr}(S_c,L)=\frac{V_{el}(S_c,L)}{\pi}. \quad (6)$$

3 Results and discussion

Based on the functional relationship described above the chart (Fig. 4) has been plotted to show clearly the amount of damaged material, depending on the cutting step S_c, and the height of exposed surface (L). In Fig. 4 solid line denotes the amount of damaged material, calculated on the ellipsoid, and bar line – for the pyramid. The volume of real materials shown as shaded area. The dashed line represents the average volume of destroyed material. It can be seen that the most exposed surface L_{max} (for each step of cutting) destroyed by the maximum volume is observed for D = 0.04 m, 0.05 m and 0.06 m height to the exposed surface of L = 0.07 m; 0.055 m and 0.04 m, respectively. The greatest value of volume is observed at S_c = 0.05 m and L_{max} = 0.055 m, confirming previously given ratio S_c/L_{max} = 0.91 for V_{max}.

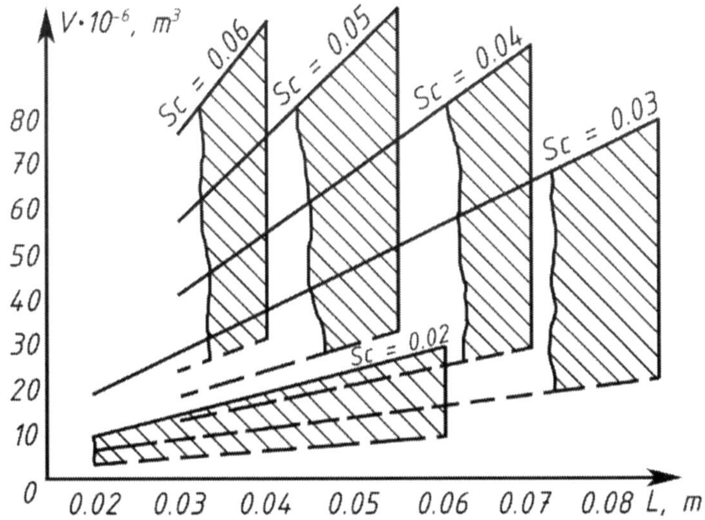

Fig. 4. Dependence of the volume of destroyed element (V) in one phase of cutting cycle on the height of exposure (L)

We modeled real zone of destruction for outgrowth system of development with working body for the ratio S_c/L_{max} = 0.91 (Fig. 5A), from which it follows that this ratio is characterized by significant area of intact material (indicated by the dashed lines), for further destruction of which the undue increase of energy consumption is required.

Therefore, to achieve the smallest area of intact material while maintaining the maximum possible amount of damaged material we modeled simulated modes and selected the parameters of destruction $S_c \geq 0.035$ m and L = 0.045 m (Fig. 5B).

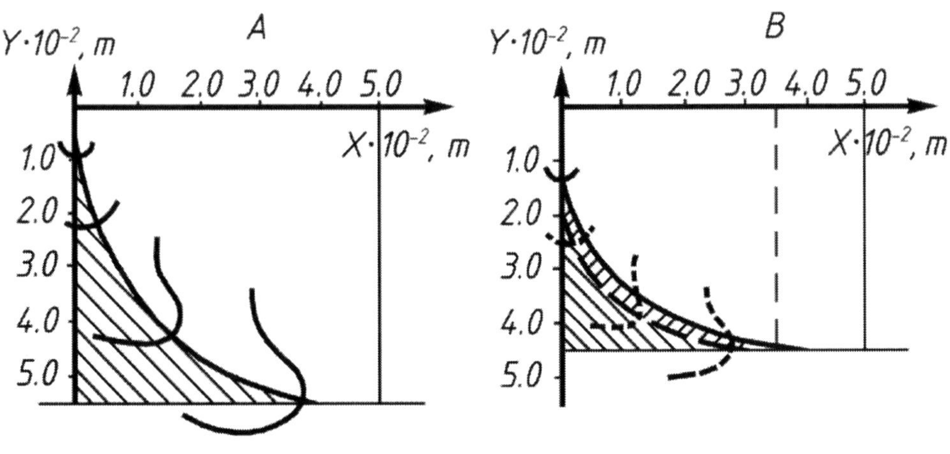

Fig. 5. The description of outgrowth disintegration in plane: A – with S_c = 0.05 m, L_{max} = 0.055 m; B – with $S_c \geq 0.035$ m, L_{max} = 0,045 m.

The results of theoretical and experimental studies of the volume of destruction of large elements (V) depending on depth of introduction (h) are given in fig. 6.

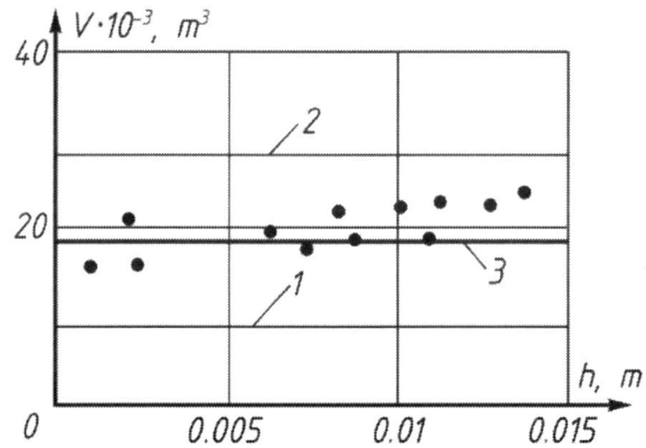

Fig. 6. The results of theoretical and experimental studies of the volume of large elements destruction (V) depending on depth of introduction h.

Horizontal lines in Fig. 6 designate theoretical dependences of volume of destroyed material (V) without deepening the disk tool. They are independent from depth of introduction of h character and are calculated for a pyramid (Fig.5 A,B, line 1), an ellipsoid ((Fig.5 A,B, line 2) and the average volume of the destroyed element ((Fig.5 A,B, line 3) $V_{av} = (V_{el} + V_{pyr})/2$. Results of experimental studies are designated by points depending on the depth of penetration of h.

4 Conclusions

Results of the maximum volumes of destroyed material (V_{max}) are located between the theoretical dependences calculated on formulas of an ellipsoid and a pyramid. Therefore, the offered theoretical dependences of creation of volumes of destroyed large elements (V) on influence of the disk tool without deepening are correct for depth of introduction of $h < 0.015$ m. At the destruction of solid rocks the disk tool penetrates into a depth h of 0.01 m. So, defined functional dependencies are valid for the destruction of solid rocks with outgrowth mode by disk tools.

References

1. M. Tyulenev, S. Zhironkin, O. Litvin, Pollution Research, **34**, 825 (2015)
2. G.E. Andreev, Min Sci and Tech **12**, 445 (1991)
3. Peide Sun, Min Sci and Tech **10**, 433 (1991)
4. Shixiong Zhang and Guangxu Tong, Min Sci and Tech **10,** 157 (1991)
5. Yuejin Li, Singh RN, Min Sci and Tech **12**, 279 (1991)
6. Lindholm US, Yeakley LM, Nagy A , Int J Rock Mech Min Sci **11** 181 (1974)
7. Atkinson T, Cassapi VB, Singh RN, Int J Min Geol Eng **4**, 151 (1986)
8. Kulatilake PHSW, Shou G, Huang TH, Morgan RM, Int J Rock Mech Min Sci Geomech Abstr **32**, 673 (1995)
9. Li HB, Zhao J, Li JR, Liu YQ, Zhou QC, Int J Rock Mech Min Sci **41**, 365 (2004)
10. Goktan RM, Gunes NA, Int J Rock Mech Min Sci **42**, 466 (2005)

11. C. Dogruoz, *Effect of pick blunting on cutting performance for weak-moderate rocks,* Ph.D. Thesis, Mining Engineering Department, Middle East Technical University, Turkey, 144 (2010)
12. RP Janeiro, HH Einstein, Int J Fract **164**, 83 (2010)
13. Zhi Cheng Tang et al., Rock Mech Rock Eng **49**, 1191 (2016)
14. Ö. Erdem, T. Güyagüler, N. Demirel, The Journal of The Southern African Institute of Mining and Metallurgy, **112**, 405 (2012)
15. Grasselli G, Wirth J, Egger P, Int J Rock Mech Min Sci **39**, 789 (2002)
16. Balci C, Bilgin N, *Comparison of small and full scale rock cutting tests to select mechanized excavation machines* (Istanbul Technical University, Mining Engineering Department 2005)
17. Tatone BS, Grasselli G, Rock Mech Rock Eng **46**, 657 (2013)
18. Tyulenev MA, Lesin Yu, Vik S and Zhironkin S, *Proceedings of the 8th Russian-Chinese Symposium "Coal in the 21st Century"*, 333 (2016)
19. Li SC, Cheng YM, Theoret Appl Fract Mech **44**, 234 (2005)
20. Thiagarajan G, J Eng Mech Asce **133**, 514 (2007)
21. Resende R, Lamas LN, Lemos JV, Calcada R, Rock Mech Rock Eng **43**, 741 (2010)
22. Ning Y, Yang J, An XM, Ma GW, Comput Geotech **38**, 40 (2011)
23. Engin IC, Bayram F, Yasitli NE, Rock Mech Rock Eng **16**, 755 (2013)
24. Dogruoz C, Bolukbasi N, Rostami J, Acar C, Rock Mech Rock Eng **49**, 213 (2016)
25. Brady BHG, Brown ET, Chapama, *Rock mechanics for underground mining, 2nd edn.* (London, UK 1993)
26. Yagiz S, *Assessment of brittleness using rock strength and density with punch penetration test, tunnelling and underground space technology* (Elsevier, Amsterdam, 2009)
27. Copur H, Balci C, Tumac D, Bilgin N, Int J Rock Mech Min Sci **48**, 269 (2011)
28. Khoreshok A A, Zhironkin S A and Tyulenev M A et al., IOP Conference Series: Materials Science and Engineering **142**, 012122 (2016)
29. Kovalev V, Gerike B, Khoreshok A, Gerike P, *Taishan Academic Forum – Project on Mine Disaster Prevention and Control*, 184 (2014)
30. Aksenov VV, Khoreshok AA, Beglyakov VYu, App Mech and Mat **379**, 20 (2013)
31. Lekontsev Yu, Sazhin P, Temiryaeva O, Khoreshok AA., J of Min Sci **49**, 757 (2013)
32. Khoreshok AA., Ugol' **7**, 67 (2002)
33. Ryzhkov YA, Gogolin VA, and Karpenko NV, Journal of Mining Science **28**, 6 (1992)
34. Ryzhkov YA, Lesin YV, Gogolin VA, Karpenko NV, J of Min Sci., 32, 188 (1996)
35. Tyulenev M A, Khoreshok A A, Garina E A, Danilov S and Zhironkin S, *Proceedings of the 8th Russian-Chinese Symposium "Coal in the 21st Century: Mining, Processing, Safety"*, 111 (2016)
36. Zhironkin SA, Khoreshok AA, Tyulenev MA et al., IOP Conference Series: Materials Science and Engineering **142** 012127 (2016)
37. Zhironkin SA, Khoreshok AA, Tyulenev MA, Barysheva GA, Hellmer MC, IOP Conference Series: Materials Science and Engineering **142**, 012127 (2016)
38. Tyulenev MA, Zhironkin SA, Garina EA, International Journal of Mining and Mineral Engineering **7**, 363 (2016)

39. Tyulenev MA, Gvozdkova TN, Zhironkin SA, Garina EA, Geotechnical and Geological Engineering, **35**, 203 (2017)
40. Tyulenev M, Zhironkin S, Kolotov K and Garina E, Pollution Research, 35, 221 (2016)
41. Gölbasi, O.; Demirel, N. *Proceedings of the International Mining Congress and Exhibition of Turkey* (Antalya, Turkey, 14–17 April 2015)

Grinding efficiency improvement of hydraulic cylinders parts for mining equipment

Aleksandr Korotkov[1,*], *Vitaliy* Korotkov[1], *Leonid* Mametyev[2], *Lidia* Korotkova[1] and *Tatiana* Terjaeva[3]

[1]T.F. Gorbachev Kuzbass State Technical University, Department of Metal-cutting machines and tools, 28 Vesennya st., Kemerovo,Russian Federation, 650000
[2]T.F. Gorbachev Kuzbass State Technical University, Department of Mining machines and complexes, 28 Vesennya st., Kemerovo,Russian Federation, 650000
[3]T.F. Gorbachev Kuzbass State Technical University, Department of Coal, Engineering Plastics and Environmental Protection, 28 Vesennya st., Kemerovo,Russian Federation, 650000

Abstract. The aim of the article is to find out ways to improve parts treatment and components of mining equipment on the example of hydraulic cylinders parts, used as pillars for mine roof supports, and other actuator mechanisms. In the course of the research work methods of machine retaining devices design were used, the scientific approaches for the selection of progressive grinding schemes were applied; theoretical and practical experience in the design and production of new constructions of grinding tools was used. As a result of this work it became possible to create a progressive construction of a machine retaining device for grinding of large parts of hydraulic cylinders, to apply an effective scheme of rotary abrasive treatment, to create and implement new design of grinding tools by means of grains with controllable shape and orientation. Implementation of the results obtained in practice will improve the quality and performance of repairing and manufacturing of mining equipment.

1 Introduction

Hydraulic cylinders are widely used in mining industry as hydraulic pillars for mine roof supports, as well as actuators for different mining equipments. Hard working conditions of hydraulic cylinders (heavy loads, high humidity, coal dust, and coal and rock fragments as active abrasive elements) lead to their intensive wear, damage or malfunction.

As a result, a problem of repairing often occurs, reconstruction of individual parts of cylinders or their complete replacement. As half of the main components of the hydraulic cylinder is made with high accuracy and low surface roughness and it has a "mirrored" surface, so parts treatment is closely connected with grinding and with grinding tools.

Analysis of ways to enhance the effectiveness of the grinding process of essential parts of hydraulic cylinders shows that there are significant and vacant reserves of potential opportunities in this problem [1]. They can be divided into two groups – technological ways to improve grinding efficiency and instrumental ones. The first group includes the

* Corresponding author: korotkov.a.n@mail.ru

development of advanced methods of grinding and construction of machine tools for their implementation. The second group includes design, manufacture and application of new high-performance constructions of grinding tools.

2 Materials and Methods

When plunger of hydraulic cylinders is being grinding, having large dimension (length and diameter, respectively, over 1000 mm and 400 mm) and weight, there is a problem with their run-out during rotation in the centers of a machine cutting tool (turning or grinding tool). Elimination of these run-outs, reaching $\Delta = 5 \div 10$ mm per side by means of mobile or stationary lunettes, does not always produce the desired effect – outruns are reduced but are not completely eliminated. Such situation becomes dangerous when plungers (or rods) are treated when a grinding disk is used as a grinding tool and a work piece is treated by a scheme of cylindrical grinding. If there are great run-outs a disk may fail during operation due to irregular allowance and large cyclic loads, and a part can get waste of the grinding surface.

Removal of such negative consequences under grinding is achieved by implementing of quite costly measures – increasing the fixity of the machine construction (for example, through the purchase of a new machine), which is under repairing is not quite cost-effective, or it can be done due to the replacement of tools and grinding scheme [2, 3, 4, 5]. Thus often under run-outs of machine parts, hard grinding disk is replaced by a grinding belt or a flap grinding disk, and the cylindrical grinding scheme is changed into the scheme of belt and flap grinding [6,7]. However, in this case, there is a disadvantage in treatment performance and increased consumption of tools, which are rapidly worn.

As an alternative variant of treatment of mentioned machine parts the KuzSTU department of "Metal-cutting machines and tools" proposed grinding with the help of a developed machine tool, which is schematically shown in Fig. 1.

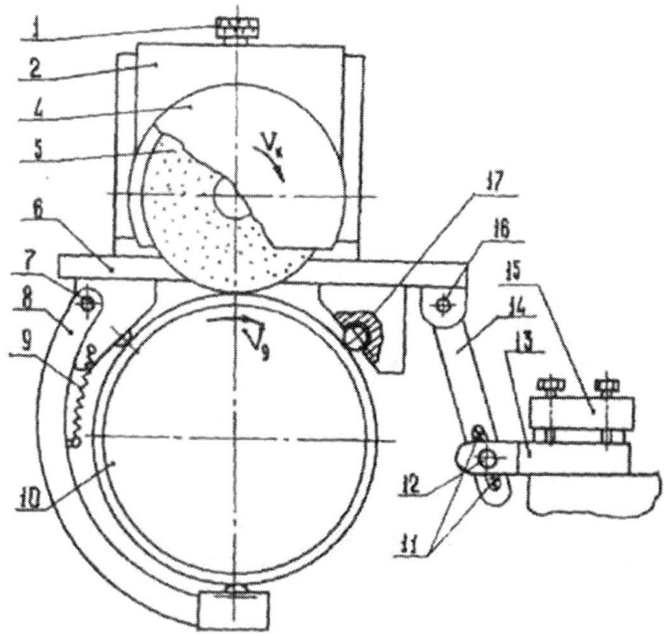

Fig 1. Machining device for grinding of hydraulic cylinders parts of large size

In the device by means of a micro screw 1 of vertical feed, a bed carriage 2 with attached to the carriage an electric motor 3 with a protective cap 4 and with a grinding disk 5, it is possible to set the vertical disk movement towards a platform 6, which by means of a pin 7, a bracket 8 and a spring 9 is fixed on a work piece 10. The platform together with the grinding disk can be moved towards a machine part by means of springs 11, a trunnion block 12, a pull-rod 13 and a crank arm14 by means of a cutter holder 15, fixed on a bed carriage of a turning (or grinding) machine.

Parts 11,12,13,14 provide flexible joint of the bed carriage and the cutter holder with the platform 6 and compensate their mutual movement towards to each other.

The platform 6 is install on the machine part on the four upper bearing parts 17 and is supported by two same bearing parts at the bottom.

The device operating principle is that the platform 6 with a grinding disk 5 and its electric drive are placed on the top of the machine part 10, and are held down by the bracket 8 and by means of the pull-rod 13 is connected with the cutter holder 15. The machine tool is turned on and the desired speed of part rotation is set up; then the grinding disk is driven to rotate, it is approached to the machine part by vertical feed and is plunged into the part to the desired cut depth. The longitudinal feed of a bed carriage of the machine tool is turned on and cylindrical grinding process of parts along its length is carried out.

For the device roll off from the treated part the special adapter is installed between the back center of the machine and the work piece. The diameter of the adapter corresponds to the diameter of the treated work piece, and its length is equal to the required length of overrun beyond the part edges.

The grinding disk is located behind the bearing parts 17, so it eliminates the possible violations of the treated surface by the trunnion ball 17.

Another progressive method of parts treatment of hydraulic cylinders is the introduction of rotational-abrasive grinding (circular plunge grinding) for cutting parts and grooving [8].

The scheme of such grinding method is shown in Fig. 2.

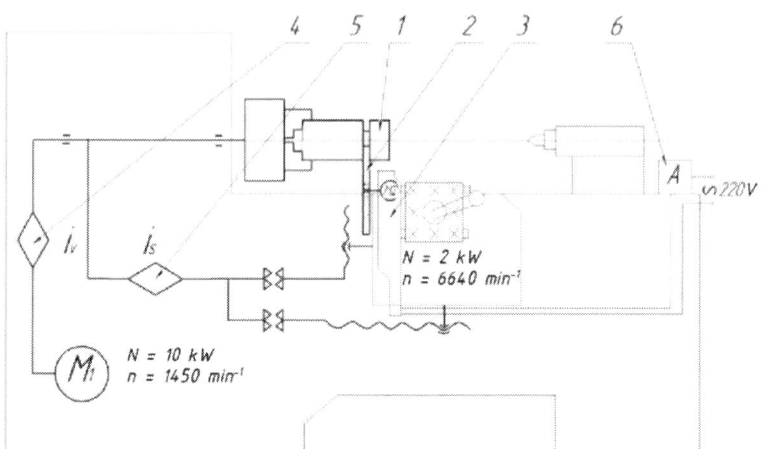

Fig 2. The scheme of rotational-abrasive grinding for cutting parts and grooving

To implement the method a turning machine is used, the model of which is chosen according to the size of work pieces.

The treated work piece 1 is fixed in the chuck of the machine tool and the treatment is done by a cutting grinding disk 2, which is installed on the main spindle of the angle grinder (AG) 3, which is fixed on the cutter holder of a machine tool. The part rotation frequency and the magnitude of cross-feed of a disk are regulated by the speed box 4 and feed box 5 of a turning machine. The ammeter 6, built-in AG chain, serves for registration of grinding effective power.

Instrumental improving of the treatment efficiency of parts of hydraulic cylinders can be considered the way, associated with the design, manufacturing and application of new advanced constructions of grinding tools, consisting of grains of certain form, and oriented in the tool body in a given direction.

The fact is that the standard grinding tools are made from grains of optional form, randomly placed in the tool body [9, 10]. That is why, only a small portion of such grains of the tools randomly acquires favorable geometry for cutting and cut metal. Most of the grains are not taking part in the overall micro cutting process or participates in it in a minimum volume, deforming and heating the metal. The share of active grains, ensuring the removal of metal, according to the data of various researchers, is only 10-15% of their potential possibilities [11]. Thus, there is a substantial reserve of unused grains possibilities, determined by their relatively low physical and mechanical properties, and also by their unfavorable and uncollected geometry of cutting parts.

The geometry of the grains is determined by two factors – their form and location on the working surface of a grinding tool. Grains form, due to the specificity of their manufacturing technologies [12, 13] by abrasive ingot crush and subsequent sizing of obtained abrasive mass into fractions with different particle size, varies within a wide range from isometric to plastic types. More accurate quantitative classification of grains form can be given by so-called form coefficient (C_f) equal to the ratio of diameters of inscribed and described spheres (in volume) or circulars (on the plane) into grains configuration. Developed the software that allows in automatic mode to determine the numerical values of the coefficients of the grains forms of various fractions according to the analysis of single grains in amount of at least 50-100 pieces.

To separate the original abrasive mass on the number of fractions with the grains of the same form the construction of the special vibratory separator was developed, which was patented and has a certificate for the invention of the Russian Federation (Patent 2248851 Russian Federation). To give a particular orientation to grains in a grinding disk, there was a method proposed, based on the usage of electric static effect, which is also patented (Patent 2369474 Russian Federation).

Thanks to the proposed methods to separate grains according to form and to give them a specific position in the tool body, it was possible to develop a technology to create new designs of grinding tools with controlled grains form and orientation. During the implementation of this technology the mixing machine was also developed to prepare the abrasive molding compounds. The remaining stages of the new technology don't fundamentally differ from the standard one and are completed by forming and heat treatment operations of grinding disks.

3 Results and Discussions

The device was developed in the way that allows grinding of a part of the hydraulic cylinder with high-quality and productively, while the cylinder is being manufactured or repaired and reconstructed. The device exploitation at the factory confirmed its effective functioning and the ability to perform the given tasks.

The principle included in the design of this device (the combination of parts and tools bases), allows under hard radial run-outs of samples to polish not only the plungers of hydraulic cylinders, but other similar parts. For example, axle trees and large diameter pipes, cylinder containers and others.

Rotational abrasive cutting and grooving has a number of advantages in comparison with conventional forms of treatment [14]. Among them there are:

- reduction of grinding temperature and, as a consequence, the elimination of burns on the parts due to their rotation and continuous displacement of the contact area and the part (each local area of the part manages to cool until the next contact with the tool);
- the tool is less intensively heated and worn;
- there are no flash and sharpened edges on the part after such treatment.

Besides axisymmetric cylindrical parts the work pieces of other sections (square, rectangular, elliptical, etc.) may be subjected to this kind of treatment.

New constructions of grinding tools with controlled form and orientation of the grains with enhanced effectiveness were developed. Comparative tests of this tools with standard ones demonstrated their advantages in a number of operational parameters [15, 16]. The utilization of these tools for grinding of parts of hydraulic cylinders increases the productivity and quality of their production.

Discussion of the results described above is reflected in the previous section. The results of the developments according to paragraph 2 suggest that a differentiated and well-reasoned approach to the choice of the form and orientation of grains in grinding disks gives a significant increase in their performance. In particular, in the schemes of Fig. 3 – Fig. 4 it is shown that the cutting capacity (Q_m) and surface roughness (R_a) is directly dependent on the form of the grains in a disk (C_f).

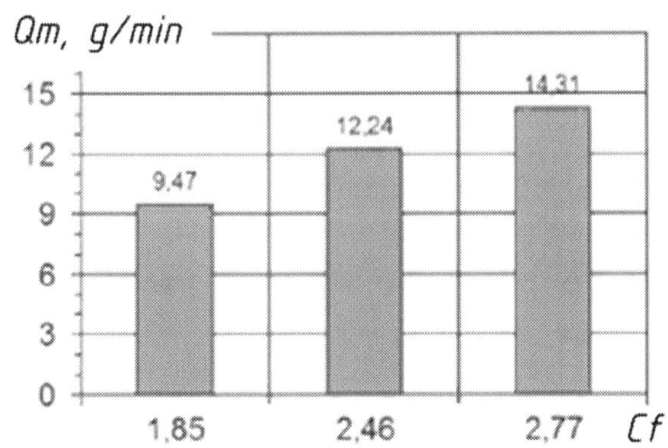

Fig 3. Cutting capability of disks with controlled form of grains

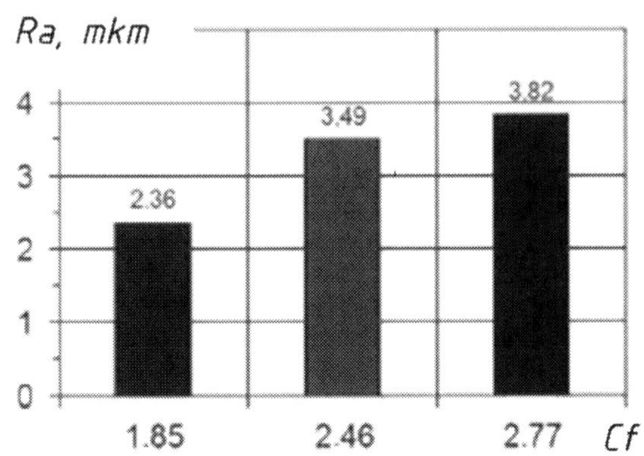

Fig 4. The roughness of the surface, treated by disks with controlled grains form

Specifically:

- with transition from isometric grains to grains of needled-plastic form the cutting capability of grinding disks, in dependence on the trade mark of abrasive and grinding scheme and condition increases in 1.51 times (Fig. 3);

- reduction ratio of grains form (transition to the isometric forms) can reduce the surface roughness in 1.6 times (Fig. 4);
- cutting capability of disks with grains oriented in the radial direction in 1.25-1.27 times higher than of conventional tools;
- application grains form with a tangential orientation in disks makes it possible to reduce the roughness of treated surfaces in 1.28-1.97 times.

4 Conclusion

The researches have confirmed the initial idea that parts of hydraulic cylinders used in mining engineering can be manufactured, repaired and reconstructed productively and efficiently by application of the proposed technological and instrumental approaches.

References

1. V. Korotkov, S. Petrushin, Applied Mechanics and Materials, **682** (2014)
2. A.Hirsch, *Werkzeugmaschinen. Grundlagen, Viewegs Fachbuecher der Technik* (2000)
3. *Taschenbuch Maschinenbau* (Verlag Technik Berlin, 1978)
4. *DIN 69 651 Werkzeugmaschinen fuer die Metallbearbeitung*, (Verlag Berlin, 1982)
5. *Werkzeuge. Fachkatalog* (Verlag Berlin, 1992)
6. A. Korotkov, D. Shatko, Applied Mechanics and Materials, **788** (2015)
7. W. Koenig, *Fertigungsverfahren, Band 1 bis 5* (VDI-Verlag, Duesseldorf, 1990)
8. V. Korotkov, IOP Conf. Series: Materials Science and Engineering, **91**, 012041 (2015)
9. V. Lukshin, Applied Mechanics and Materials, **682** (2014)
10. G. Lal, M. Shaw, Works of American union of engineers, **97** (1975).
11. V. Lukshin, A. Barsuk, R. Fazleev, IOP Conf. Series: Materials Science and Engineering **91**, 012047 (2015)
12. A. Korotkov, A. Hirsch, IDR, **2** (2002)
13. W. Degner, H. Lutze, E. Smejkal, *Spanende Formung: Theorie, Berechnung, Richtwerte* (Carl Hanser Verlag Muenchen Wien, 1993)
14. *Verein Deutscher Werkzeugmaschinenfabriken e. V. (Hrsg.): Werkzeugmaschinen Bezugsquellenverzeichnis*, (Ausgabe, Frankfurt am Main, 1997)
15. *Development Methods and Application Experience of Parallel Kinematics* (Verlag Wissenschaftliche Scripten, 2002)
16. *Handbuch der Metallbearbeitung* (Verlag Europa-Lehrmittel, 2000)

The Smart Grid using in the Kuzbass open-pit coalmine

Irina Semykina[1,*] *Andrey* Evstratov[1], and *Gennady* Lebedev[2]

[1]T.F. Gorbachev Kuzbass State Technical University, Department of Electric Drive and Automation, 650000, 28 Vesennyaya Street, Kemerovo, Russia
[2]T.F. Gorbachev Kuzbass State Technical University, Department of Mining and Power Industries, 650000, 28 Vesennyaya Street, Kemerovo, Russia

Abstract. The Smart Grid system become to apply all over the world. It provides significant efficiency increasing of power supply networks by supporting its balanced load. The key design features of each Smart Grid system depend on individual characteristics of particular power supply network which are determined by the structure of its electrical energy consumers. Relating to Kuzbass open-pit coalmines it is necessary to take into account the nonstationarity of its power supply scheme, the strict requirements of power supply reliability and the high capacity of consumers. This article is connected with the "Kedrovskiy" open-pit coalmine and describes its power supply scheme and the structure of its consumers. The downtimes and electrical equipment failures are analyzed and the connection between the number of emergency downtime and the number of working excavators is found out. The load of "Kedrovskiy" power distribution network is calculated for the strip mining operation. The results show power distribution network under consideration does not provide the effectiveness functioning and the implementation of Smart Grid is a good decision. There are the structure and design calculation of proposed Smart Grid in the article. It is finally depicted that Smart Grid" system decreases the downtimes of electric equipment and increases the power supply reliability.

1 Introduction

An electrical grid is a complicated system that consists of wires and transmission towers, cables, transformers and breakers, which are interconnected some way with each other and with electrical energy consumers. Traditionally power distribution network is supplied from a limited number of power supply sources, which are heat power station or diesel power station, but now power distribution network has tended to become a decentralized bidirectional intelligent grid of consumers and different kind of supply sources including renewable energy generation. Such distribution network is called Smart Grid [1].

Smart Grid system increases the power supply reliability due to automated analytics and raised controllability and observability of electric power grid. Smart Grid

* Corresponding author: siyu.eav@kuzstu.ru

controllability leads to not only raising of its efficiency but to decreasing of social environmental impact thank for using clean, renewable sources of energy like solar, wind and geothermal [2]. Implementation of Smart Grid, especially on high-voltage and heavily loaded power distribution network, demands new types of equipment, for example, alternating current to direct current interconnections or point-to-point bulk power transmission [3]. Some solutions were tried out in such a big project as UK's North Wes or Turkish Kayseri ve Civarı Elektrik Türk A.Ş. [4] or Sub-Saharan Africa [5].

Some category of electrical energy consumers has the special aspects of the power supply. For example, the above mentioned is intended to describe an open-pit coalmine. Modern coalmines are almost completely mechanized and the operations involve a set of technics specified to open-pit mines so that smart technology is going to interpenetrate there [6].

Mining machines like rope shovels and dragline are in need of power supply with a high capacity and according to the producing method, they may produce an increased load during critical periods. The average weekly load profile of any open-pit coalmine tell as the plant is not producing at full capacity which implies either maintenance is being carried out on the plant modules or it is stopped. The plant completely stops on some periods, thus resulting in a much lower baseload compared to when the plant is not producing at full capacity. The load is torn between 10th and 90th percentile that is a significant variation [7]. That is why it is important to apply the Smart Grid to open-pit coalmine, especially if the power system becomes more decentralized with higher renewable energy penetration to balance the grid.

Smart Grid also can report power usage across all of the consumers and its sensors on remote areas cause warning signs to flash at the central control center when problems arise. It has to be noted that miners already produce huge amounts of sensor data, potentially enabling them to implement in the Smart Grid. In additional, intelligent algorithms can be used to predict consumers cut out. Moreover, complex mining tasks such as geological modeling, on-the-day scheduling, and predictive maintenance are increasingly in the domain of smart statistical and optimization algorithms in Smart Grid [8]. Taking into consideration the open-pit coalmine specific, communication between sensors and control center may be organized with wireless technology [9]. Another part of Smart Grid system for open-pit coalmine may become the generators that use kinetic energy from transported gangue material [10] or supercapacitor for use in the rope shovels [11].

With a glance to all of the aforesaid, the main task in the creation of Smart Grid system for open-pit coalmine is an analysis of its consumer's structure and searching of Smart Grid system itself. So the aim of the research is an investigation of one of the open-pit coalmine supplied from the National Grid and a proposition of its Smart Grid.

2 Subject of research

According to the aim of research let us consider «Kedrovskiy» that is one of Kuzbass open-pit coalmine. The «Kedrovskiy» power supply has the combined scheme with 6 kV portable transmission lines that are direct-axis on the open-pit embankments and cross-axis at the embankments crossing. The main electrical energy consumers are excavators. They are connected to portable transmission lines by switching substations equipped with pressure vacuum breakers. The excavator's supply system has the isolated neutral. The number of connected excavators does not exceed the allowance which is specified by mining technology [12, 13]. The auxiliary electrical energy consumers are supplied by portable package transformer substations with 0.38 kV or 380/220 V voltage level and with solidly grounded neutral. These transformer substations have an overcurrent protection and short-circuit protection. It must be truth not only for Russia but for another country [14]

The «Kedrovskiy» emergency downtime book and electrical equipment failure book for the year 2014 got to the bottom based on which the paired regression analysis and the correlation analysis of power supply reliability were carried out. In the analysis, the influence of each factor was looked separately. The result is the multiple regression model:

$$N = 23.2 + 7.39L_p + 24.55N_e + 1.99N_a, \tag{1}$$

where N – the number of emergency downtime, L_p – the length of portable transmission lines, N_e – the number of connected excavators, N_a – the number of auxiliary electrical energy consumers.

The next step of research was multiple regression model verification. For that, the F-test was carried out. This test includes a comparison of determination coefficient R^2 with the critical value, which is determined by table [15], corresponded to the referred confidence level. If R^2 is less than the critical value then the multiple regression model is not verified and if not the model can be used.

In this research, the confident level was fixed at 5 % so the critical value of R^2 was 0.764. The obtained R^2 was 0.882 that means the multiple regression model of power supply reliability is verified. In the case of «Kedrovskiy», it goes to show the number of emergency downtime is mostly affected by the number of connected excavators.

3 Investigation of power distribution network

The next research stage was the investigation of power distribution network supplied the excavators. Let us consider Fig.1 that show the part of distribution network with two excavators and the decontamination station.

The load calculation for this electric scheme was carried out for the overburden-mining regime with specific energy consumption method. In the calculation, recommendations from handbook [16] were taken into account. The scheme load parameters are presented in Table 1.

The power distribution network under consideration was analyzed at the maximum load by computer simulation. The simulation model in relation to Fig. 1 contains the models of drive motors for «ESH 13/50» excavator, which is dragline, and for «EKG-10» excavator, which is rope shovel, the switching equipment models, and the distributed-parameters cable models.

Table 1. The parameter of «Kedrovskiy» load.

<table>
<tr><th colspan="3">Parameter</th><th>Value</th></tr>
<tr><td colspan="3">Voltage, kV</td><td>6</td></tr>
<tr><td colspan="3">Simultaneously operating consumers power, kW</td><td>17 854</td></tr>
<tr><td colspan="3">Equipment shift-working factor</td><td>0.85</td></tr>
<tr><td colspan="3">Demand factor</td><td>0.53</td></tr>
<tr><td colspan="2" rowspan="2">Power factor</td><td>cos φ</td><td>0.99</td></tr>
<tr><td>tg φ</td><td>0.14</td></tr>
<tr><td rowspan="4">Estimated maximum load</td><td colspan="2">active, kW</td><td>9 535</td></tr>
<tr><td rowspan="2">reactive, kVAr</td><td>lagging</td><td>1 356</td></tr>
<tr><td>leading</td><td>–</td></tr>
<tr><td colspan="2">full-power, kVA</td><td>9 631</td></tr>
</table>

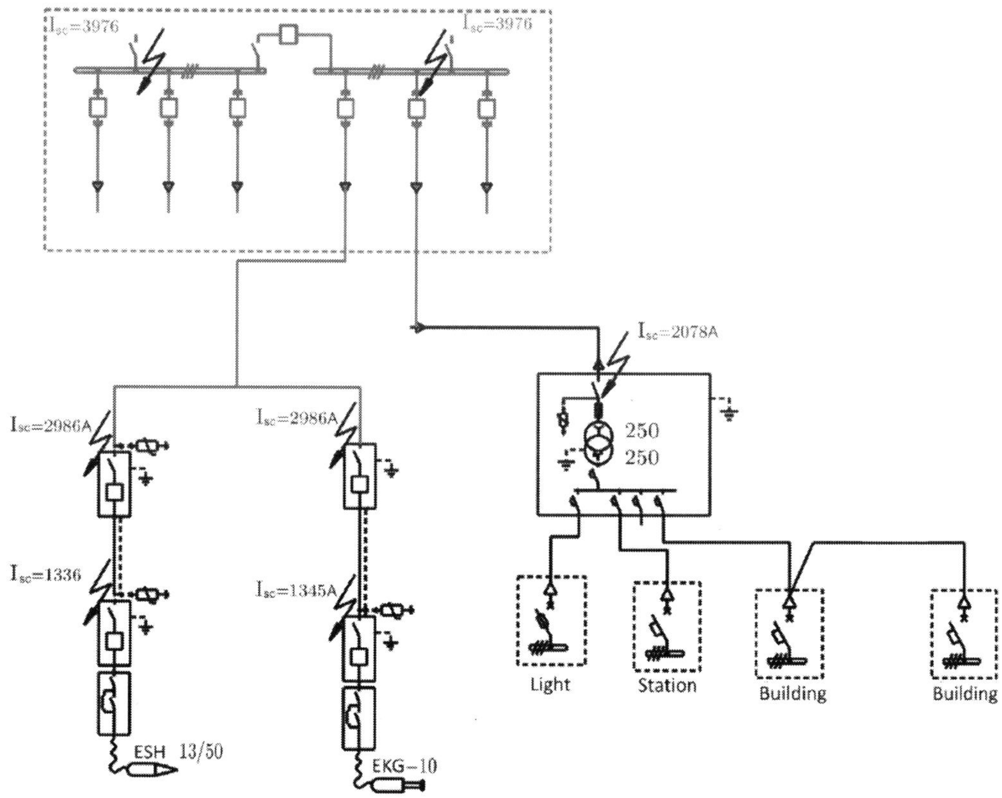

Fig. 1. The part of «Kedrovskiy» distribution network

Simulation results are shown in Fig. 2 where we can see the voltage at the excavator's motor terminal is 2.36 times smaller than the feeder voltage. It testifies the excavators cannot starting up because of this low voltage level leads to activation of the sustained start protection. All similar evidence have to be noted down at «Kedrovskiy» emergency downtime book.

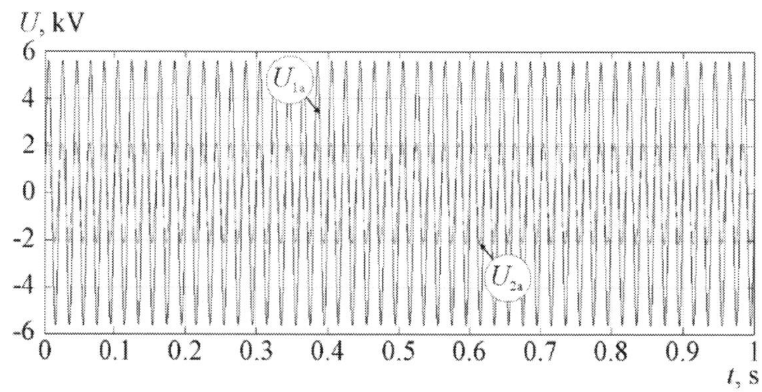

Fig. 2. The line-to-neutral voltage at the simultaneous starting-up of both excavators: U_{1a} – the feeder phase voltage; U_{2a} – the motor terminal phase voltage for «EKG-10» excavator.

The above mentioned confirms «Kedrovskiy» power distribution network does not provide all the way functioning of main open-pit coalmine equipment at the maximum load and coupled with the absence of a bright-line rule for auxiliary equipment using at peak usage times, this power distribution network need to modernization. The effective solution is the Smart Grid.

4 Analysis of Smart Grid and conclusion

The proposed Smart Grid system make compensation of voltage level at excavator starting up by using of accumulators and grid-controlled inverters with a capacity equal 40 % of startup power consumption total for all excavator's motor in the considered power distribution network. This Smart Grid system was simulated with discussed simulation model. The obtained result is presented in Fig. 3.

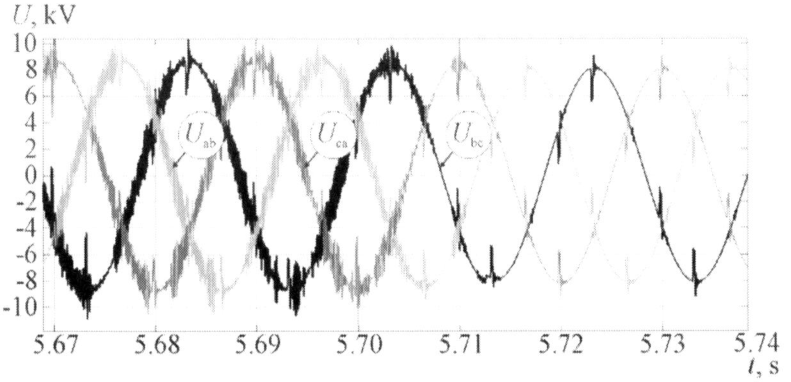

Fig. 3. The motor terminal line-to-line voltages for «EKG-10» excavator at the simultaneous starting-up of both excavators with Smart Grid.

Despite high-frequency harmonics, the voltage characteristics provide evidence the proposed Smart Grid system ensures power supply in «Kedrovskiy» power distribution network at the maximum load with high quality. In addition, it allows controlling of auxiliary equipment and provides it's using only at minimal load. So, Smart Grid system decreases the downtime of electric equipment and increases the power supply reliability thereby the open-pit coalmine partly becomes the self-supporting object.

Regardless of Smart Grid using requires high investment cost to reconstruction the power distribution network and it is too expensive to implement it at «Kedrovskiy», it should be noted Smart Grid system will be effective at the newly built Kuzbass open-pit coalmine.

References

1. C.W. Gellings *The Smart Grid: Enabling Energy Efficiency and Demand Response* (The Fairmont Press, 2009)
2. *The Smart Grid: an introduction*, prepared for the U.S. Department of Energy by Litos Strategic Communication (2008)
3. M. Claus, D. Retzmannl, D. Sörangr, K. Uecker, *17th Conference of the Electric Power Supply Industry* (Siemens AG, 2008)
4. Geospatial World, **5**, 16 (2015)
5. Donald I. Bleiwas *Estimates of Electricity Requirements for the Recovery of Mineral Commodities, with Examples Applied to Sub-Saharan Africa* (USGS National Minerals, 2011)
6. *100 Innovations in the Mining Industry* (The Mining Association of Canada, 2012)
7. Coenraad Benjamin Pretorius *Development of a Demand Response Programme for the Coal Mining Industry* (Stellenbosch University, 2016)

8. *Mining's next performance horizon: Capturing productivity gains from innovation*, Metals and Mining (McKinsey & Company, September 2015)
9. *Wireless communications for open-pit fleet management*, Network. The newsletter of ABB's Network Management business unit, **3** (2014)
10. S.A. Linkov et al., *Procedia Engineering*, **129** (2015)
11. Babak Parkhideh, Subhashish Bhattacharya, Joy Mazumdar, Walter Koellner, *IEEE Industry Applications Society Annual Meeting* (2008)
12. William A. Hustrulid, Mark Kuchta, Randall K. Martin *Open Pit Mine Planning and Design*, 1308 (CRC Press, 2013)
13. Kshirod Chandra Brahma *A Study on Application of Strategic Planning Models and Operations Research Techniques in Opencast Mining* (The National Institute of Technology, Rourkela, 2007)
14. *Electrical Protection and Earthing, Mine Safety Operations EES005* (NSW DPI Technical Reference, 2006)
15. Scott H. Irwin *The Multiple Regression Model: Joint Hypothesis Testing* (University of Illinois, 2006)
16. R.J. Thompson *Surface Strip Coal Mining Handbook* (South African Colliery Managers Association, 2005)

Multifunctional Testing Rig for Machinery Safety Equipment

Stefan Vöth[1,*] *Julian* Tschersich[1], *Tim* Schwartz[1]

[1]Technische Hochschule Georg Agricola, Centre for Drive and Lifting Technology ZAFT, Bochum, Germany

Abstract. At the Centre of Drive and Lifting Technology ZAFT of Technische Hochschule Georg Agricola a new testing device will be available. The device is suitable for testing drivetrain components like safety clutches and hoist safety components like snag protection devices. In focus are continuous tests with regard to fatigue as well as transient tests with regard to switching characteristics. The article gives information on requirements on the testing rig out of the main purpose and out of the given environment. Furthermore concept, main data and engineering design of this equipment are demonstrated. Aspects and first results of the mechanical assembly are shown.

1 Projects at ZAFT

At the Centre for Drive and Lifting Technology ZAFT different projects related to drive and lifting technology are followed up. Examples are following:

Ship-to-shore cranes are vital elements of container port operations. They are used to load and to unload containers. One case of emergency happening in this context is a container getting stuck within the ship construction during lifting operation with higher speed (so called snag case). To limit the loads occurring to the hoist and the structure of the crane in this situation special safety devices are used (Snag protection devices). The development of concepts for those systems is part of activities at ZAFT [1].

Mining machinery like crushers is exposed to high variable loads. Temporarily the loads can exceed the loads permissible for the equipment. To prevent damage to the machinery safety couplings are implemented to the drivetrain. Those couplings have to meet certain properties to show the system behaviour required. Of interest are for example reaction times and stability of switching torques. The development of those safety couplings is part of activities at ZAFT [2].

2 Objectives

The testing rig is planned to serve different purposes.

First purpose is the testing of components in drive trains. It is intended to execute continuous and transient tests especially on safety clutches. Continuous tests are suitable for

* Corresponding author: stefan.voeth@thga.de

example to check durability of the components. Transient tests are carried out to reveal the dynamic behaviour of them. The testing rig is rated for P=45kW n=291.4min^{-1}.

Second purpose is the testing of safety devices for hoists. For this, a hoist is system planned. The winch is to be constituted by using above-mentioned drivetrain with a rope drum unit instead of a specimen. The pulley is located in a pulley tower of maximum achievable height. Intention is a hoisting process with following maximum data:

- Acceleration $a=1.0m/s^2$
- Acceleration time $t_a=0.5s$
- Maximum speed $v_{max}=0.5m/s$
- Maximum hoisting height $h_{max}=3.0m$
- Total process time $t_{tot}=6.5s$.

3 Boundary conditions

3.1 Floor load

The testing rig will have a considerable mass. This will be of use due to the dynamic forces expected. The mass is acting as a filter and reducing dynamic loads to the ground. Looking on this the restricted floor pressure of p=12kN/m^2 had to be regarded. This sets the maximum weight of a test bed of A=1.5mx5.0m to m=9175kg, including all installations on the test bed. Similar considerations apply for the pulley tower. All plans had to be carried out with regard to this. The situation planned was discussed with and proofed by a static expert.

3.2 Safety regulations

According to BGDV6 of Deutsche Gesetzliche Unfallversicherung (DGUV) a safety clearance from rigid edges of surroundings to rigid edges of cranes of s=0.5m has to be kept [3]. This applies to the roof, to the walls and to any object below the crane if people presence is given in these zones. In the present case, the planned tower should be of maximum height. With a lowest rigid edge of the crab at l=4.73m (Fig. 1) above floor the maximum height of the tower must not exceed h_{max}=4.23m.

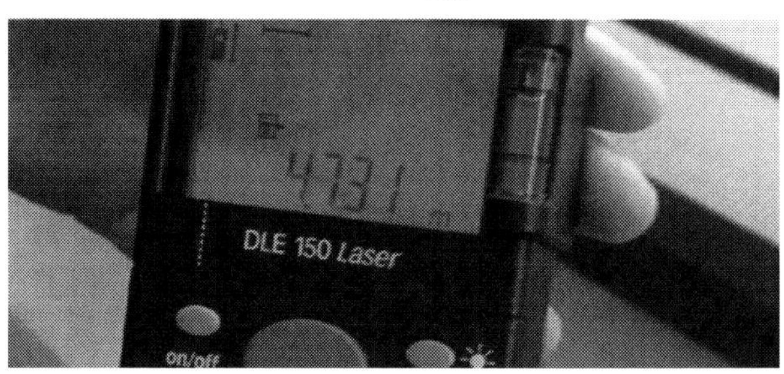

Fig. 1. Height of lowest rigid crab edge

3.3 Safety concept

The testing rig is an equipment planned and built for research purposes. Configurations are set up for different modes of operation: finishing operation, continuous operation, switching operation and hoisting operation. Each operation, that means research purpose, is planned for a period of up to a maximum of three years. Due to this the testing device is exempted from the machine directive. A declaration of conformity is not established. Risk analysis and operations instructions are created anyway [4]. A multi-stage concept is provided for obtaining safety.

4 Design

4.1 Mechanical

The mechanical design consists of a drive side, the specimen slot and the load side, all located on the test bed.

An asynchronous motor drives the testing rig via a frequency inverter with energy recovery. Behind the motor a disc brake is located. This brake consists of two identical brakes acting on one disc. By this a stepped braking torque may be realized. Behind the brake a gear for the transformation of the mechanical performance on a level of lower speed is located. Between motor and brake are two clutches located. A safety clutch for the protection of the drivetrain and an aligning coupling (Fig. 2).

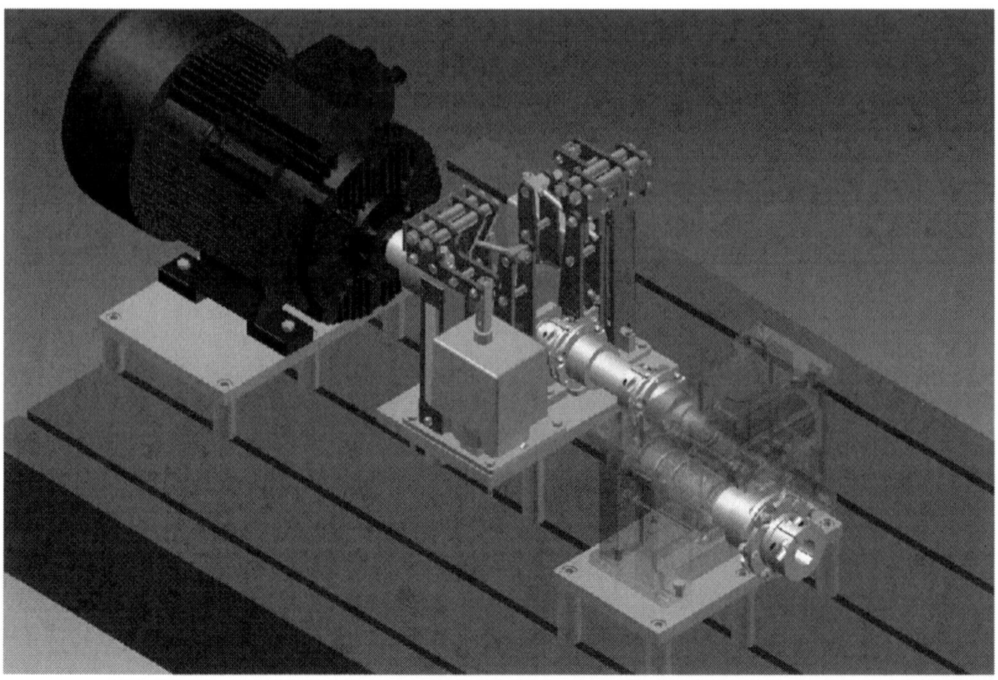

Fig. 2. Mechanical design of the drive side of testing rig

On the load side the performance is transformed back on a level of higher speed again in the first step. On this high-speed level are located two brakes. At first a magnet powder brake and behind a brake in form of an asynchronous motor driven via a frequency inverter with energy recovery.

The specimens are located in a slot between drive side and load side. They are mounted on a pallet which is adapted to the rig via defined interfaces, realized by aligning couplings. So preparation for new tests may be organized parallel to testing on the rig and exchange is realized in a safe manner and a short time (Fig. 3).

Next to the test rig is the pulley tower located (Fig. 4). In connection with a rope drum unit as specimen on the rig this completes to a full hoisting system. Maximum load is m=1t, reevings 1/1, 2/1 and 4/1 can be realized. The special design allows a quick change to part loads m=250kg, m=500kg and m=750kg as well as between the different reevings. This hoisting system is the basis for testing of safety devices for hoists.

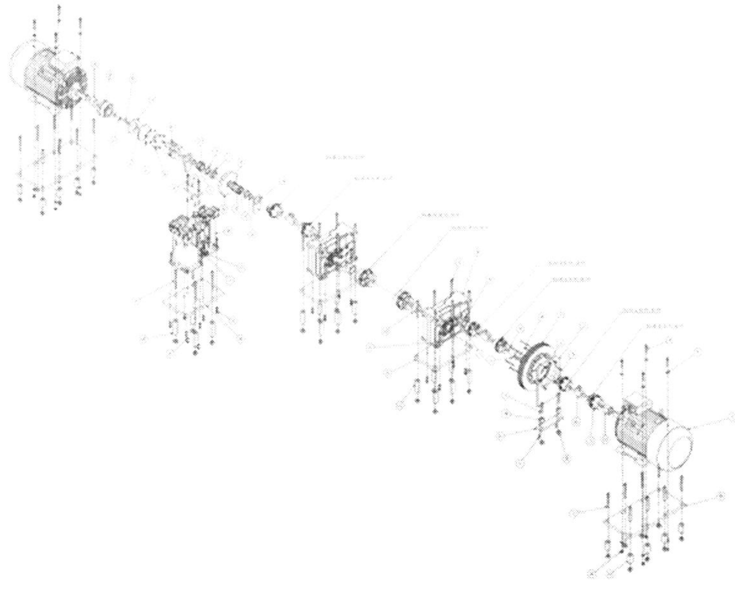

Fig. 3. Mechanical design of the drive side and the load side of testing rig

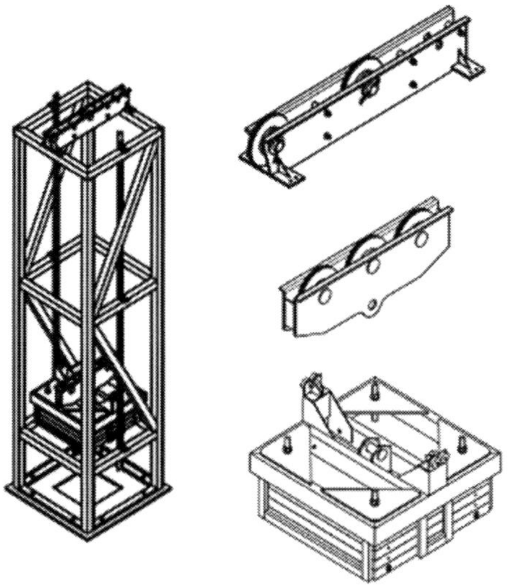

Fig. 4. Pulley tower (left) with top sheave (top), bottom sheave (middle) and load attachment device including load (bottom)

Elements of the multi-stage safety level are (Fig. 5):

Level 1 "Information": A corresponding information informs each person entering the hall about the presence of the testing device and main safety instructions.

Level 2 "Authorized persons": A list will be kept of the persons authorized to perform certain activities related to the test facility.

Level 3 "Walkway": All persons in the machine hall have to move exclusively on the marked walkway in the middle of the hall.

Level 4 "Ground marking": The yellow-black ground marking must not be passed.

Level 5 "Protection wall": In order to avoid gripping into the test device and to shield from the testing device protective wall of 1.5 m height is installed on the front edge of the test bed.

Level 6 "Local protection equipment": If tests are carried out on the rig, which systematically include the destruction of components, the potential destruction areas are to be isolated by means of the planned local protection device.

Level 7 "Light grid": If continuous tests are performed unattended on the test stand, the test bench area is monitored on all sides by a light grid. The interruption of the light grid leads to the immediate stop of the test stand.

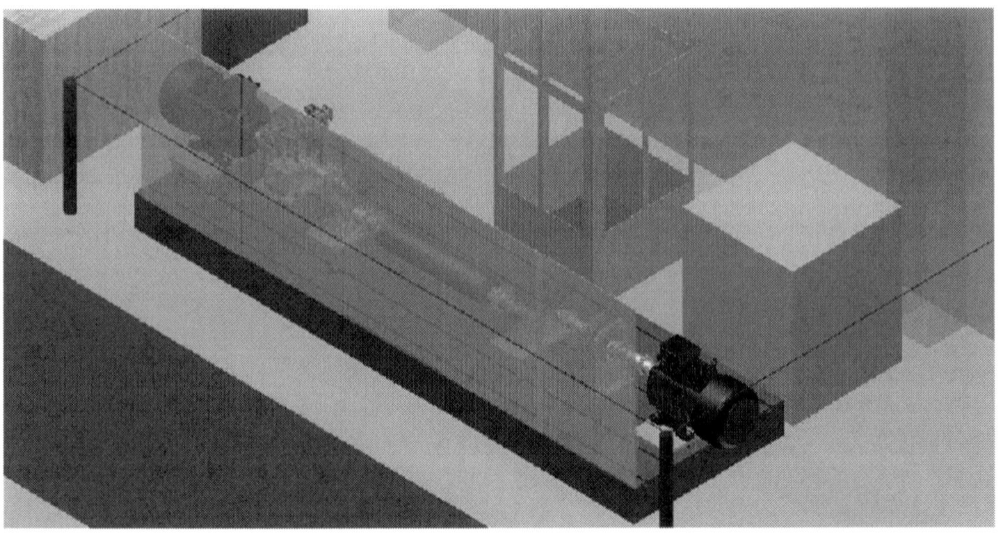

Fig. 5. Elements of the multi-stage safety level

4.2 Measurement and Control

The testing rig comprises three measuring shafts: The first one is located between brake and gearing on the drive side. Its main purpose is the measurement of motor actions, especially in the hoist modus. Two further measuring shafts are located on the load side, one between gearing and magnet powder brake and one between magnet powder brake and motor. Main task of these measuring devices is the collection of data related to the braking actions. All measuring shaft deliver a signal on the speed beside the torque signal as well.

The disc brakes are controlled in their position by switches. The brakes comprise, in addition to the motoric drive, hand levers for lifting. Switches control the position of these hand levers as well.

A load-measuring bolt is located at the fixation point of the rope at the top of the pulley tower.

Limit switches for the hoisting movement of the load are located in the upper and the lower position.

In case of continuous tests a light grid secures the test rig.

As testing programs change the alternation between different control systems takes place. Basically the different control systems consider the different nature of continuous testing and transient testing, both of which are executed on the rig.

DOI: 10.1051/e3sconf/20171503007

5 Transport and assembly

Transport of all components into the hall did not require special efforts. Trucks with the components could enter the hall via a gate with a large headroom. Unloading was done by a 10t bridge crane with a maximum hook height over floor of H=4.30m. With this unloading of the biggest part, the test bed with m=6670kg, was realised as well as other big components like the pulley tower structure.

As a first step of assembly, the test bed was adjusted. This sounds easier than it was done. The test bed is rather a flexible than a rigid body. Due to this, a certain strategy had to be applied to gain a plane and horizontal test bed surface. Afterwards further equipment was installed mechanically as supports, drive motor, safety clutch, alignment coupling, and motor side brakes (Fig. 6).

Fig. 6. Mechanical assembly of drive side

6 Outlook

Next steps to be carried out are: Mechanical installation of drive side, mechanical installation of load side, mechanical installation of pulley tower, installation of safety devices, electrical installation, definition of detailed requirements of control system, programming and installation of control system, commissioning via defined test scenarios.

References

1. S. Vöth, *Proceedings of the 17th ITI-Symposium, ITI*, 127 (Dresden, 2014)
2. S. Vöth, J. Sippel, Eigenschaften von drehmomentgesteuerten Schaltkupplungen nach dem Rastprinzip, Der Konstrukteur, Antreiben-Steuern-Bewegen, p. 50-52, Vereinigte Fachverlage, 2014
3. BG Holz und Metall (Hrsg.): Unfallverhütungsvorschrift, DGUV Vorschrift 52, Krane, 2013
4. Deutsche Gesetzliche Unfallversicherung (Hrsg.): BGI/GUV-I 5139, Information, herstellen und Betreiben von Geräten und Anlagen für Forschungszwecke, CE-Konformität und Betriebssicherheit, 2011

Development of the preventive maintenance system for belt conveyors reducers

Boris Gerike[1,2,*], *Ivan* Panachev[2], Eugene Kuzin[3]

[1]Federal Research Center of Coal and Coal Chemistry SB RAS, 650003, 10 Leningradsky av., Kemerovo, Russian Federation
[2]T.F. Gorbachev Kuzbass State Technical University, 650000, 28 Vesennyaya st., Kemerovo, Russian Federation
[3]Branch of T. F. Gorbachev Kuzbass State Technical University in Prokopievsk, 653039, Kemerovo region, 19a Nogradskaya str., Prokopievsk, Russia

Abstract. Heavy operating conditions of mining machines as well as the high level of dynamic loads lead to reduction of their service life. The quantitative estimation of the machine reliability by one of the feature – service life – has become widely distributed in all the branches of engineering. Technical diagnosis is one of the important methods of improving the reliability in operating conditions. The diagnostics sub-system should provide for: non-destructive inspection of a technical condition of objects, the definition of sudden and parametric failures of mining machines and their systems, the detection of gradual failures by predicting changes in the monitored parameters, a continuous and periodic technical inspection. The obtained results given in this article prove the possibility of creating a group of common diagnostic criteria suitable for assessing the technical state of reducers of mining machines and equipment, but also being a prerequisite for the effective short-term prediction of the parameters under study when developing adaptive mathematical models.

1 Introduction

All the parts of mining machines and equipment can be subdivided into the following groups:

1. The first group includes dipper teeth and cutting lips; bases and canopies of powered supports; augers; bulldozer bars, blades; crawler shoes, idler rollers, pins, bushings, sprockets of crawler undercarriage and other parts with the service life duration depending on abrasive wear;
2. The second group includes slotted and threaded parts, gear couplings, mounting seats for rolling bearings of shafts, machine tools, surfaces of gearing, etc. with the service life duration determined by mechanical wear of their surfaces;

* Corresponding author: gbl_42@mail.ru

3. The third group includes parts of internal combustion engines of dump trucks, bulldozers, scrapers, diesel locomotives, thermal rock fracturing and jet drilling rigs, etc. with their service life duration limited by molecular – chemical or corrosive mechanical wear;

4. The fourth group includes rolling and plain bearings, shock absorbers, springs, connector rods, connecting rod bolts, etc. with their surface life duration depending on metal fatigue strength.

2 Formulation of the Problem

At present, the coal mines operate a significant number of belt conveyors [1, 2], and the performance indicators of the entire coal industry of Kuzbass depend on their operating condition. In the nearest future, it is expected that power consumption and technical equipping of belt conveyors increase, as well as the performance and length of rock mass transportation, variable frequency drives are widely introduced.

The increasing volumes of coal production by integrated mechanized coal faces, together with an increase in production safety require the creation of reliable transport systems. High efficiency and trouble-free performance along with reduction of energy consumption present the main challenge faced by manufacturers of production lines of mine belt conveyors. Another no less important goal is reduction of the cost of their maintenance and repairs [3, 4]. To ensure reliable operation of a belt conveyor for as long as possible, it is required to find the causes of failure of the various constituent elements [5-7]. The analysis of downtime caused by failure of a belt conveyor reducer shows that it varies from 7.4% to 18.2% and accounts on average for 12%, with the average recovery time is from 24 to 48 hours. It should be noted that the most common failure is belt rupture (up to 50%), while the average time to fix this failure is from 1.5 to 2 hours. Hence, identification of the actual technical condition of mine belt conveyor reducers is really quite an urgent objective.

3 Results of Study

The vibration control method [8-10] proved to be useful in checking the technical condition of mechanical equipment. The vibration diagnostics is used to monitor the current state of equipment, to identify possible defects, to assess the remaining service and to define repair time and scope. The analysis of domestic and foreign experience of monitoring the technical condition of systems with rotational motion of power assemblies shows that for the detection of potential failures the most effective (77%) is control of the machinery condition by vibration parameters [11], and with the involvement of other methods of the functional diagnostics – like spectral oil analysis [12] and thermal control, the accuracy of recognition of the cause of the defect increases to 95%.

The functional diagnostics, in terms of safe operation of mining machinery, should play a key role in the field of research and development, production and quality control of the process equipment. And while in the coal mining and ore mining industries the forms of technical service of technological equipment based on its actual condition [1, 8] are becoming more common, in the manufacture of such equipment new forms of quality control are not yet implemented in spite of the introduction of quality standards GOST ISO 9000-2011. Defects that arise in the gearbox manufacturing process can be divided into errors in the manufacture of gearbox elements and errors in gearbox assembly.

A full analysis of the technical condition of the gearbox after assembly and trial running on the test bench will allow us to identify and localize manufacturing defects, but also to eliminate possible supply of poor quality products to the consumer. In addition, the findings could form the basis for development of automated quality control system.

The analysis of vibration control methods allows us to conclude that it is appropriate to apply the method of support masks for automation of control of products manufactured for the coal industry. This method is based on the fact that the defects formed as a result of assembly operations generate a vibration in certain frequency bands with a certain ratio of the values of controlled parameters. The method of support masks makes it possible to set the width of the frequency band, its position and the values of the evaluation criteria, which are compared with the current values in random order. Based on the analysis of the controlled parameter changes in the frequency band (the number of bands may vary from 6 to 30), the condition of the equipment is evaluated and forecasted [4].

Frequency ranges of the spectrum mask (band width) are usually taken based on the following conditions:

1. "High-energy" components of the spectrum accompanying the misbalance or misalignment – $(0,5...1,5)\times f_r$ and $(1,5...2,5)\times f_r$;
2. "Low-energy" components of vibrations accompanying the defects of rolling bearings – $(7,5...15,5)\times f_r$;
3. $(2,5...10,5)\times f_r$— total failure of the system stiffness;
4. The first medium frequency band $(3... 15)\times f_r$;
5. The second medium frequency band $(15...40)\times f_r$;
6. The first high frequency band 40×fr...20 kHz;
7. $(0,1...0,9)\times f_r$ – for detection of oil wedge defects of slide bearings;
8. $(n\pm1)\times f_r$ – for detection of damages in the components of couplings.

The use of modern automation technologies for control of equipment technical condition enables an individual approach to each manufactured mechanism in the evaluation of its technical condition and sets the thresholds of the initial operable and limit state. As an example, Figures 1 and 2 show the spectrum of the vibration signal and its averaged spectral mask in 1 checkpoint of gearbox RKC-400 manufactured by JSC "Anzheromash" (Fig. 3). The measurements were made by analyzers "Corvet", and the signal was processed on Safe Plant software platform developed by NGO "Diatekh".

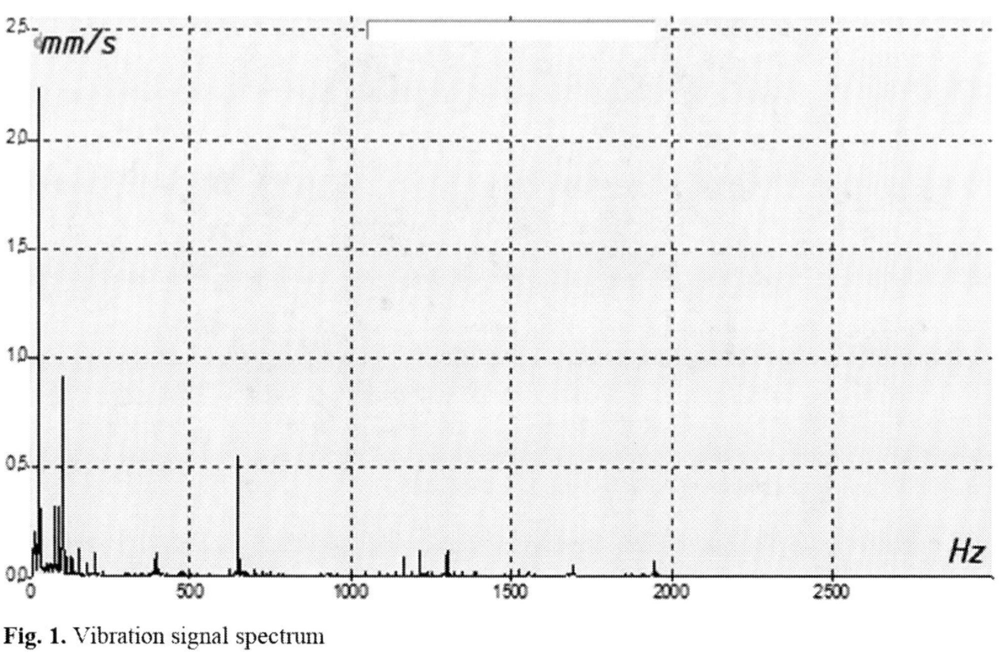

Fig. 1. Vibration signal spectrum

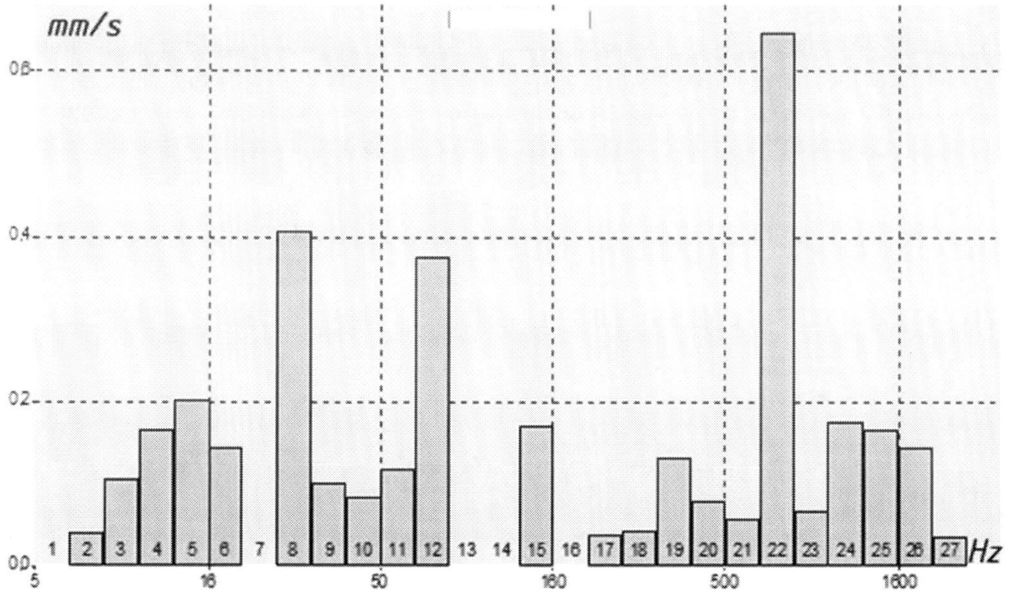

Fig. 2. Vibration signal spectrum mask

In the presented example, the frequency range (2; 3000 Hz) is divided into 27 bands, each of which is normalized by the RMS value of vibration speed *V* determined for both forward and reverse rotation of the output shaft.

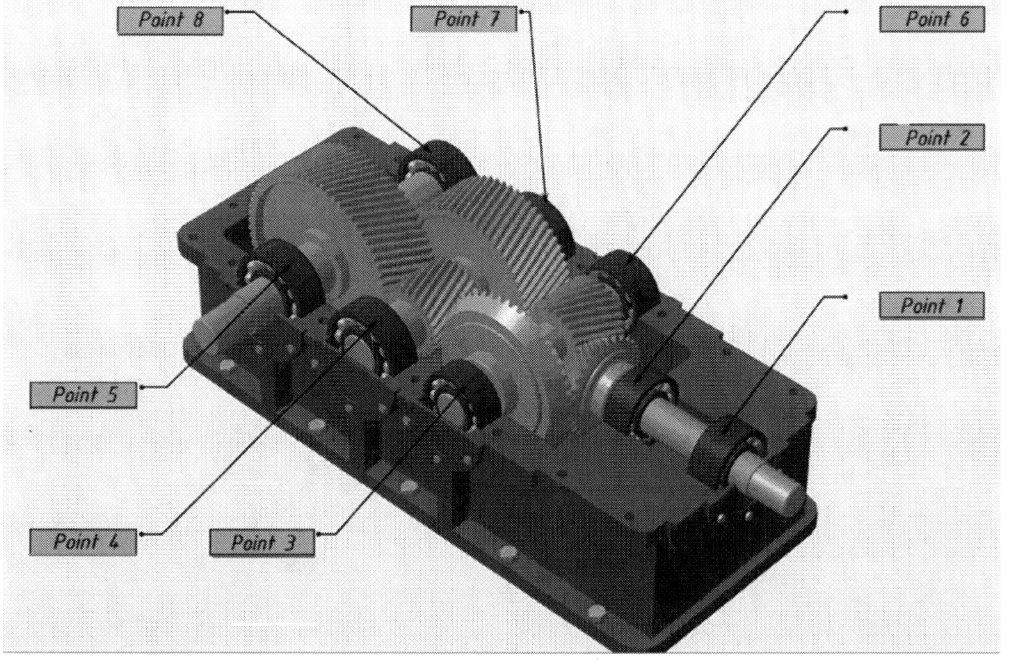

Fig. 3. Vibration measurement points on gearbox RKC-400

The most typical defects inherent to reducers of mining machines occur at frequencies shown in Tables 1 and 2.

Table 1. Frequencies typical for gearing defects in in-line reducers

Frequency	Type of manufacturing defect	Type of assembly defect	Type of wear defect
f_r	Misbalance		
$k\times f_{r1}$ and $k\times f_{r2}$ (k = 1,2, less often 3 and 4), $m\times f_z \pm n\times f_r$ (m, n=1,2…)	Variable gear pitch error	Violation of alignment (misalignment of shafts)	
$k\times f_r$ k = 1, 2...20 and higher		Increased gear backlash	
f_z	Permanent gear pitch error		
$k\times f_z$, $k\times f_r$ growth of noise component $m\times f_m \pm n\times f_r$ (m, n=1,2…)			Abrasive wear
$k\times f_r$, $m\times f_z \pm n\times f_r$, $m\times f_m \pm n\times f_r$ (amplitude fluctuation, n = 0,1,2...)			Teeth chipping
$k\times f_r$, $m\times f_z \pm n\times f_r$, $m\times f_m \pm n\times f_r$ (amplitude fluctuation, n =0,1,2...), growth of noise component			Cracks and (or)broken teeth

Table 2. Defects in the gear part of the planetary reducer and their key diagnostic parameters

Type of defect	Diagnostic signs
Sun pinion shaking	f_o, $nf^*\pm f_o$, $kf_z\pm f^*$
Sun pinion misalignment	$2f_o$, $2nf^*\pm 2f_o$, $kf_z\pm 2f^*$
Sun pinion teeth defect	$knf^*\pm k_1f_o$, $kf_z\pm k_1f^*$
Satellite gear misalignment	$4f_g\pm k_1f_v$, $kf_z\pm 2f_g$
Satellite gear teeth defect	$2kf_g\pm k_1f_v$, $kf_z\pm k_1f_g$
Crown misalignment	$2nf_v$, $kf_z\pm 2nf_v$
Crown teeth defect	knf_v, $kf_z\pm k_1nf_v$
Engagement defect	kf_z
Shaking of final drive carrier	kf_v, $f_o\pm f_v$, $kf_z\pm k_1f_v$
Satellite gear bearing defect	kf_v, $f_o\pm f_v$, $kf_z\pm f_g/2$
Defect of sun pinion bearing	kf_r + growth of RMS MF, impact pulses appear at MF
Bearing lubrication defect	Impact pulses appear at HF, growth of RMS at HF

In industrial conditions (mine Taldinskaya-Zapadnaya of OJSC "SUEK-Kuzbass") an integrated method of the technical state estimation has been tested on the drive units of mine belt conveyor 3LL1600 (conveying length L = 850 m, technical performance Q = 3500 m / h, the belt speed v = 0 -4 m / s) by the parameters of the lubricating oil, vibration and thermal control.

4 Conclusion

The proposed approach to the standardization of mechanical vibration parameters can be used in practice at development of the enterprise standard for normalization of vibration of manufactured products for inclusion in the product data sheet.

The development of a large number of spectrum masks for a wide standard line of mining equipment is one of the prerequisites for release by mining machinery plants of quality products and the transition to new forms of mining machines maintenance and repair.

The results of evaluation of the technical condition of the 3LL1600 conveyor drives using the integrated method based on monitoring of parameters of vibro-acoustic signals, emission-spectral analysis of the composition of working oil and thermal visual control of support units of rolling bearings make it possible to track changes in the technical condition of the reducer elements depending on its load and speed.

The proposed approach will not only increase the accuracy of the assessment of the technical condition of mining equipment gear units but will also help to organize the work on creation of normative and methodological base for building predictive models of changes in the technical condition on the basis of a significant amount of accumulated statistical information on the development of those or other defects of belt conveyor reducers.

References

1. Ansari F., Uhr P. and Fathi M., International Journal of Services, Economics and Management, **6,** 14 (2014)
2. Ierace S., Cavalieri S., Management and Production Engineering Review, **4**, 37 (2013)
3. Shafiei-Monfared S. and Jenab K., IEEE Transactions on Engineering Management, **59**, 293 (2012)
4. Almgren T., Andréasson N., Palmgre M., Patriksson M., Strömberg A.B., Wojciechowsk A. and Önnheim M., *Proceedings of 4th World Conference on Production & Operations Management and the 19th International Annual EurOMA Conference, University of Amsterdam (*Amsterdam, 2012)
5. Franco Jefferds dos Santos Silva, Herbert Ricardo Garcia Viana, André Nasser Aquino Queiroz, Journal of Quality in Maintenance Engineering, 22, 418 (2016)
6. Gajanand Gupta, Rajesh P Mishra, Journal of Quality in Maintenance Engineering, 22, 130 (2016)
7. Dehghanian P. and Aminifar F., IEEE Transactions on Power Delivery, **28**, 761 (2013)
8. Prabhakar D. and Raj V.P., International Journal of Scientific & Technology Research, **2**, 56 (2013)
9. Anil Rana, Journal of Quality in Maintenance Engineering, **22**, 180 (2016)
10. Golmakani H.R. and Morteza P., J of Quality in Maintenance Eng., **20**, 51 (2014)
11. Hussan Saed Al-Chalabi, Jan Lundberg, Majid Al-Gburi, Alireza Ahmadi, Behzad Ghodrati, Journal of Quality in Maintenance Engineering, **21**, 207 (2015)
12. Hartman J.C. and Tan C.H., The Engineering Economist, **59**, 136 (2014)
13. Hussan S. Al-Chalabi, Jan Lundberg , Andi Wijaya, Behzad Ghodrati, Journal of Quality in Maintenance Engineering, **20**, 306 (2014)
14. Gustafson A., Schunnesson H., Galar, D. and Kumar, U., International Journal of Mining, Reclamation and Environment, **27**, 75 (2013)

Determination of Load Performance of Two-Bar Girder Lining

Yuri Glazkov[1], *Anton* Kazantsev[2,3*], *Dmitriy* Nesteruk[3], *Vladimir* Aksenov[2,3], and *Andrey* Efremenkov[2,3]

[1] T.F. Gorbachev Kuzbass State Technical University, 28, Vesennyaya st., Kemerovo 650000, RF
[2] Federal Research Center of Coal and Coal Chemistry of the Siberian Branch of the RAS, Leningradskiy Ave10, 650065 Kemerovo, RF
[3] Yurga Institute of Technology, TPU Affiliate, 26, Leningradskaya st., Yurga 652050, RF

Abstract. The article proposes a base-line design of the system in the form of lattice girders to support peripheral rock masses of workings. The design scheme and its parameters of support and loading conditions are given, the simulation results are shown. Designing a 'reinforcing' support system suitable to be installed into the channel slots formed in the surrounding rock mass and developing a process to construct this support are the prime considerations in tunnels driven using the new mining technology. Helix and longitudinal channel slots excavated by the geokhod in the border zone during tunnel excavation should be used to provide a potential for increasing bearing capacities.

1 Introduction

In our previous works [1-9] we mentioned that our research group has been working on the development of a new class of tunnel boring machines (TBM) called 'geokhod'. During the excavating operations, a geokhod forms a system of channel slots in the surrounding rock mass. Studying the system of excavations creates preconditions for a new approach to the ground support in underground workings and a new design of support systems. This new approach defines a framework for the research: interactions of supports in underground workings and peripheral rock masses as one of the key elements to create a new mining technology. After analyzing the works of foreign scientists [10-21] it is obvious that in recent years different types of lattice girders have been widely used in reinforced shotcrete lining to provide support in shielded TBM driving. Accounting for the available data a design of helix support with girder elements has been designed. The basic idea for creating this support is that it consists of standard bar arch segments to form the tunnel contour perimeter of a full circle (Figure 1). The study subject is a quarter circle segment, and in this case one turn of the lining consists of four identical bar girder elements.

In the process of driving, the lining is made in the way of sequential installation of new sections. The long channel slots are employed for installing longitudinal connecting elements. The purpose of this paper is describing the finite-element model of a typical

* Corresponding author: kazantsev@tpu.ru

lining segment and the received results after analyzing its particular behavior under the influence of a vertical load distributed uniformly (Figure 2).

Modeling was carried using the software package ANSYS Mechanical v14.5.

The BEAM188 finite element with a quadratic function form and two nodes was used. The outer and inner bars (see Figure 1) are divided into segments to indicate the distance between adjacent nodes. A girder bar within one segment or stiffener was modeled by one finite element. The specified mesh refinement was tested by comparing the computational results of the continuous planar annulus regarding the design load as shown in Figure 2 with the results of the analytical solution to the same problem as described in [22]. The comparison showed that with 40 finite elements the discrepancy in the results as to finite element and analytical solutions does not exceed 3 ÷ 5 percent. In further investigations, each bar has been modeled using minimum 40 finite elements. Figure 3 shows an example of the mesh with similar parameters.

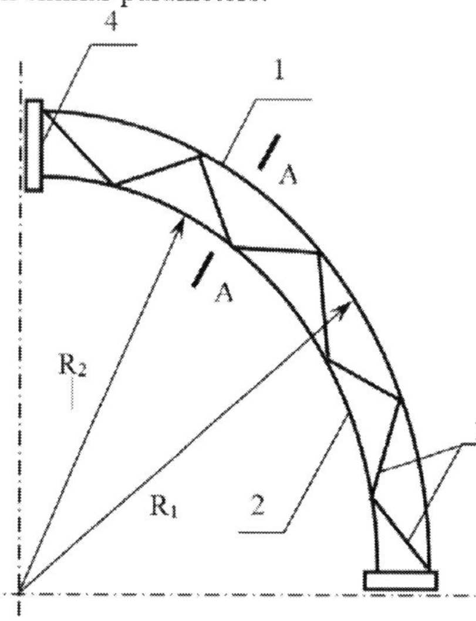

Fig. 1. Schematic design of the lining: 1 –outer bar; 2 –inner bar; 3 – stiffener; 4 –connector

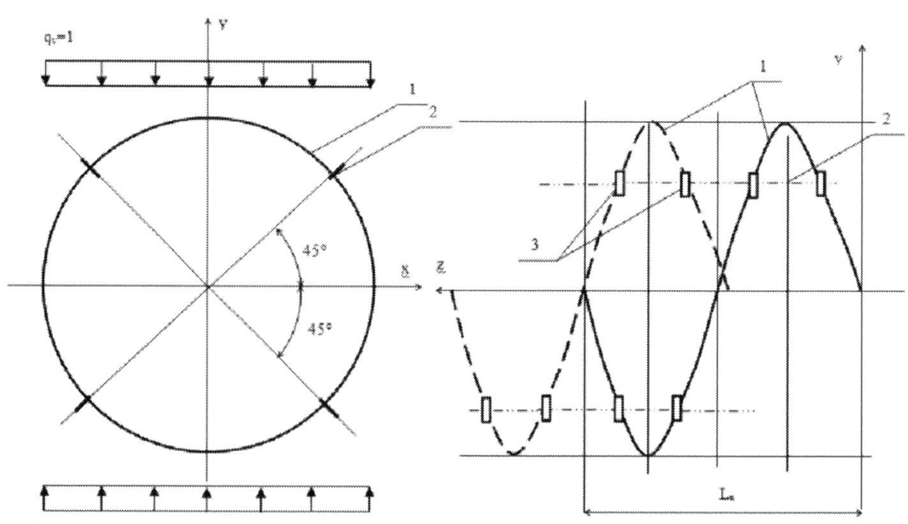

Fig. 2. Design model of the fragmented lining: 1 – support section; 2– longitudinal connecting elements; 3 – nodes of connections of the lining and connecting elements

With the purpose of simplifying the modeling of finite element mesh the nodal connections (p.3 in Figure 2) were not analyzed in detail and taken into account only for setting kinematic links.

Links, limiting displacements, were imposed on the basis of symmetry properties. The design model of lining fragments containing any number of turns has three properties of symmetry.

First, the problem is symmetric in respect to the global coordinate plane x0z. When mesh nodes are uniplanar, motion along the y-axis direction is not possible (UY = 0).

Second, where UX = 0, the symmetry in respect to the plane y0z is similar.

Third, allowing for any lining fragment with a whole number of turns with the start and finish of this fragment lying in the plane of symmetry, oblique symmetry will be performed with the assumptions that the lining is infinite at its fragment ends. For example, for the mesh in Figure 3 rotations about the y and z axes (ROTY = ROTZ = 0) are inhibited at the nodes of end sections in addition to the previously indicated links.

It should be pointed out that it is usually difficult to set correct kinematic boundary conditions for the computation of real-life structures. Regarding this problem, the similar behavior of kinematic boundary conditions with respect to all turns of the lining ensures the correctness of these boundary settings. Below it can be seen all turns behave the same way, both end and middle turns. This fact proves the specified kinematic boundary conditions are accurate.

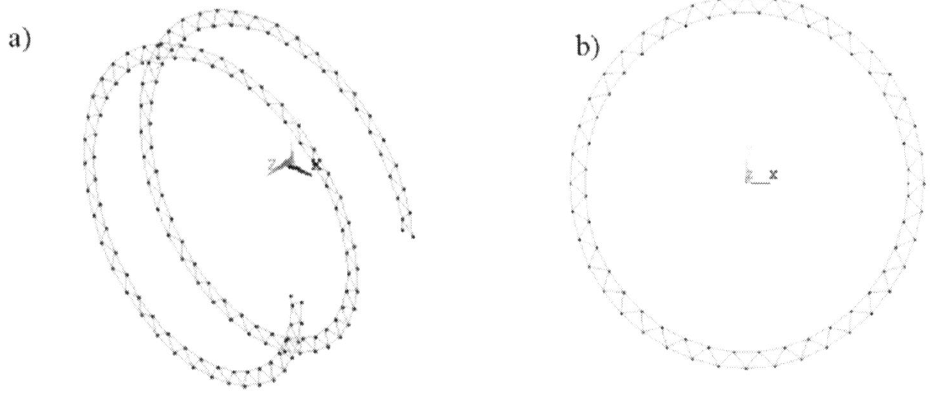

Fig. 3. Schemes of the finite element meshes for the two-turn lining fragment in axonometric (a) and front (b) projections

To unify the multiple computations of lining supports in the context of various parameters, the program-macro FOne.mac has been created using ANSYS Parametric Design Language (APDL). This program has functional capabilities to perform computations and post-process analyses of linings in terms of the design model discussed above.

Below are the results of calculation of a lining variant with parameters given in Table 1 to illustrate the program's capabilities.

When calculating we assumed that a girder element completely fits the helix channel slot and internal fiber of its cross-section had a curvature radius equal to Rк. In this case, the radii were defined as R2 = Rк +d2/2 = 1606 mm for the outer bar and R1 = Rк +d2/2 +H = 1757 mm for the inner bar. Two girder bars with a cross-section diameter d3 of 12 mm were replaced with a round bar having a diameter d3,экв of 15mm what is equivalent to the area of both crossbars.

The rated distributed load was programmatically replaced by its statically equivalent concentrated forces applied to the nodal connections of the outer bar of the lining cross-section. The connecting elements in power scheme calculating were neglected and used as safety factors.

Table 1. Values for a lining fragment

№	Parameter	Designation	Unit	Value
1	Radius of working cross-section	RK	[mm]	1600
2	Diameter of outer bar cross-section	d1	[mm]	30
3	Diameter of inner bar cross-section	d2	[mm]	36
4	Stiffener Diameter	d3	[mm]	12
5	Height of lining cross-section over the axes of bars	H	[mm]	151
6	Length of a turn over the z-axis	LB	[mm]	800
7	Number of segments on a turn		[pc]	40
8	Number of turns in the design model		[pc]	2

In the macro, two options for setting kinematic links in the direction of the global z-axis (longitudinal axis of the working) have been implemented. The first option implies the superposed links on all nodes of lining. This option allows for the potential for fixing the lining in the helix channel slots that results in impossible displacements along the z-axis.

The second option implies the fixing of nodes arranged in sections where the lining and longitudinal connecting elements are connected. This option has been described above. Further in this paper these options of connections will be referred to as the first and second.

Macros make possible to analyze the results of computation in the form of diagrams of internal forces or associated stresses. Thus, each bar generally meets complex resistances to bending with compression and torsion. Within this work only normal stresses applied to cross-sections of the inner and outer bars and stiffeners were discussed. They are defined according to the standard formulas of structural mechanics.

$$\sigma^N = \frac{N}{A};\ \sigma^{M_y} = \frac{M_y}{W_y};\ \sigma^{M_z} = \frac{M_z}{W_z}, \tag{1}$$

where σ^N is the normal stress caused by the axial force; σ^{My} and σ^{Mz} are the maximum normal stresses caused by bending moments; W_y and W_z are the resisting moments of the cross-section caused by bending. When the cross-section is round, we have

$$W_y = W_z = \frac{\pi d^3}{32} \approx 0{,}1d^3 . \tag{2}$$

It needs to note that within this investigation internal stresses and strains are related to the local axes of the finite element. The longitudinal x-axis of BEAM188 element is directed along the nodes at the ends of element. The transverse axis of the element is parallel to the global coordinate plane x0y. In this case, the bending moment My complies with the bending out of cross-section plane, i.e. perpendicular to the plane of acting load. In the cross-section plane (in the plane of acting load) the bending relates to the moment Mz. When analyzing the results of the computation it is necessary to consider that the tensile and compressive stresses are caused by bending at the end points of section. The stress diagram shows stresses with regard to one fiber (section side) that can be positive (tensile) or negative (compressive). Therefore, the calculated total stresses should be determined as the sum of modular values of axial forces and bending moments.

Figure 4 shows the diagrams of all stress components (1) separately for the outer bar. With this, the distribution of stresses due to axial forces σ^N and bending in the plane of load, when the links are superposed, are almost identical. Therefore, these stresses diagrams for the second option are shown here. The main difference in behavior at different variants manifests in the distribution of stresses due to bending out of the plane of load. Figure 5 shows the diagrams where the significant increase in stress components occurs locally in the vicinity of points of the longitudinal connecting elements.

2 Conclusions

The following conclusions emerge from this fact. The behavior of a bar in the helix girder lining significantly differs from that of in-plane lining. In the case of helix girders, there is a tendency for bars to twist relative to each other. Apparently, this fact leads to the appearance of stiffener bending out of the plane of load.

Under the conditions where stiffeners are more flexible then bars, the stresses arisen in the stiffeners are much higher than those in the bars. The stiffeners become dangerous. The influence of twist on bars can be reduced by unfixing of lining in the helix channel slots. This fact is confirmed by comparing the diagrams σ^{My} for the first and second options of securing nodes. The first option of connections reduces σ^{Mz} more than three times as compared with the second option.

The use of spatial girder constructions like 3-bar girders is another way to reduce the effect of the twisting belts.

The bearing capacity of the lining is estimated in respect to the total design stresses in critical sections of girder bars and stiffeners. The results of the design stresses are given in Table 2.

a)

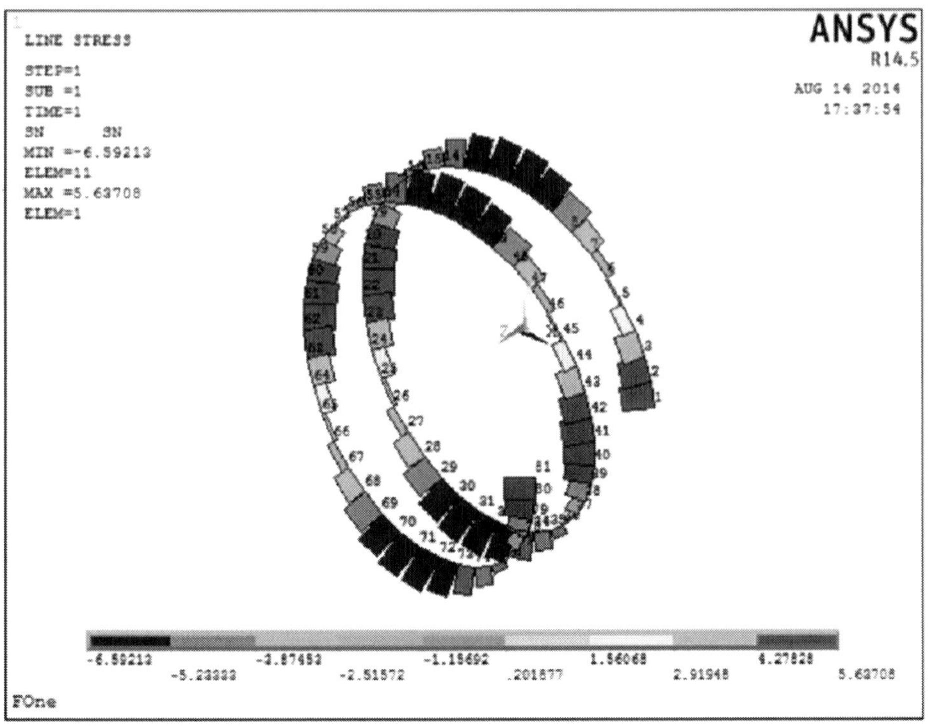

DOI: 10.1051/e3sconf/20171503009

b)

c)

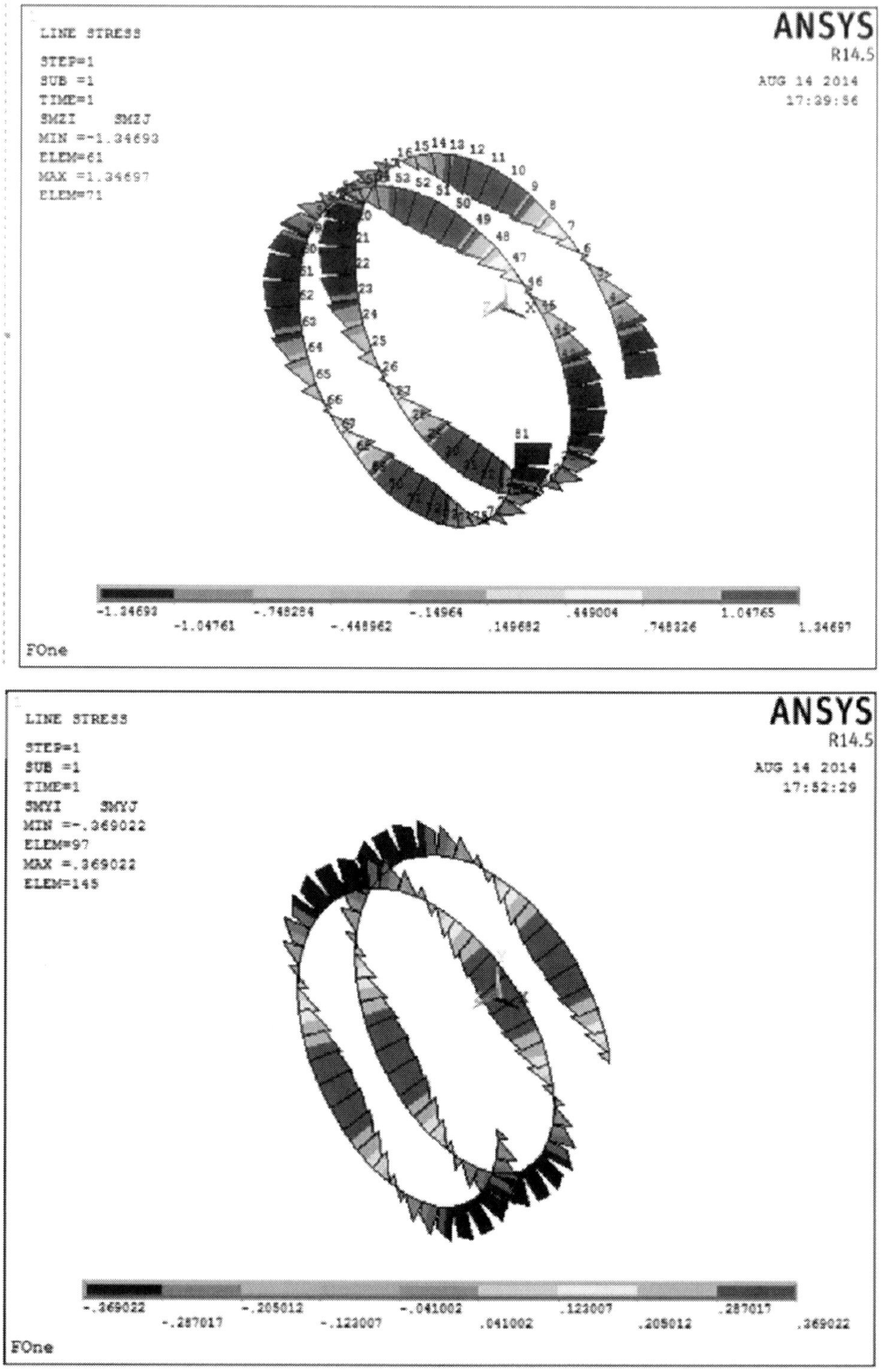

Fig. 4. Stress diagrams for outer bars σ^{N} (a); σ^{M_z} (b) and σ^{M_y} (c)

a)

b)

c)

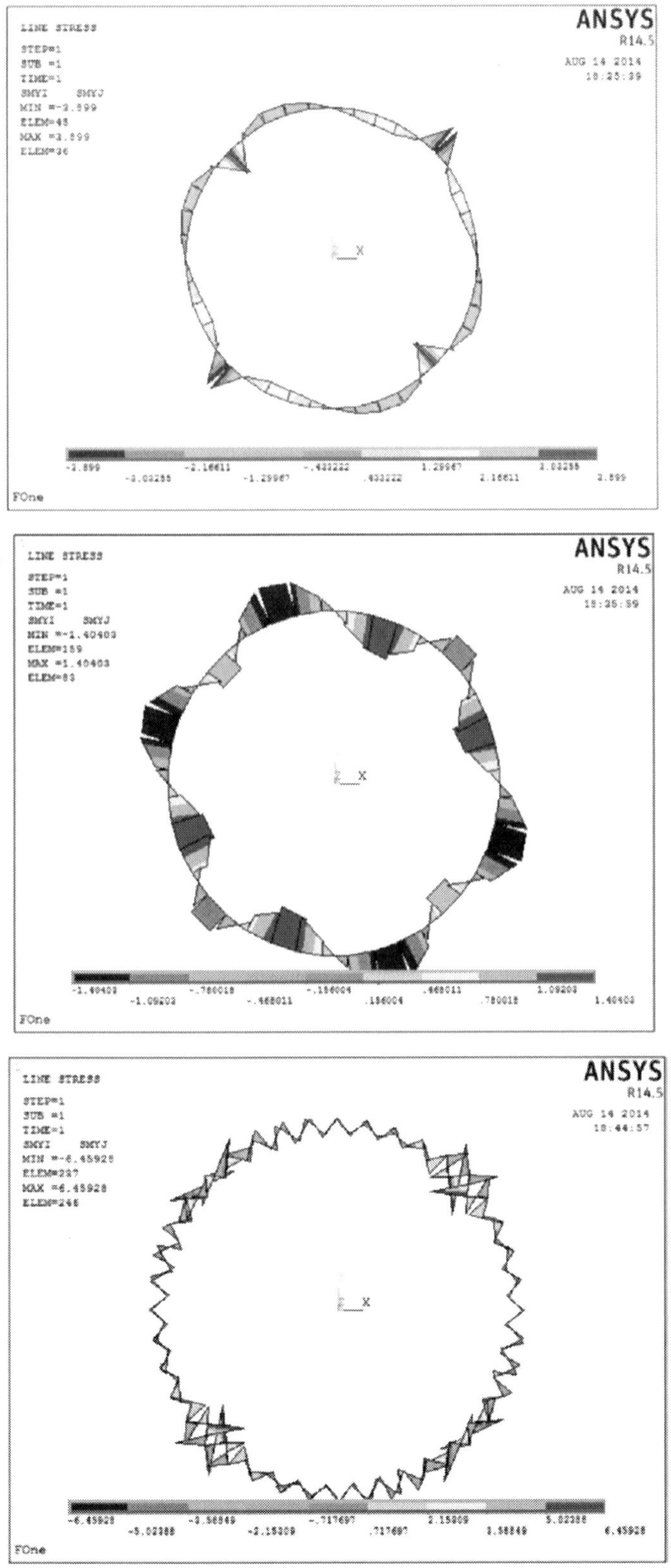

Fig. 5. Stress diagrams σ^{My} for the outer bars (a); for inner bars (b) and stiffeners (c) for the second option of connections

Table 2. Design stresses

Support Element	Stress, MPa	Options of connections	
		First	Second
Outer bar	σ^{N}	6,59	6,59
	σ^{M_z}	1,35	1,35
	σ^{M_y}	0	0
	σ^{Σ}	7,93	7,93
Inner bar	σ^{N}	5,39	5,39
	σ^{M_z}	1,79	1,79
	σ^{M_y}	0	0
	σ^{Σ}	7,18	7,18
Stiffeners	σ^{N}	6,67	7,30
	σ^{M_z}	1,91	2,03
	σ^{M_y}	1,95	6,45
	σ^{Σ}	9,39	14,10

Assuming the design stress under a unit load q = 1 N / mm = 1 kN / m and characteristic strength of the girder material Rn = 240 MPa, the bearing capacity of a single turn of the lining is defined as follows:

$$P_{\lim} = q\frac{R_n}{\sigma^{\Sigma}}2R_1 = 1\frac{240}{14,1}\cdot 2\cdot 1,75 = 60 \text{ kN}. \quad (3)$$

This value correlates well with the bearing capacity of the in-plane annular lining made of SVP17 special section with a clear diameter of 3.00 m [23] that equals 150 kN under tough operating conditions.

References

1. Aksenov V.V., Beglyakov V.Y., Kazantsev A.A., Ganovichev S.I., IOP Conf. Ser.: Mater. Sci. Eng., **127**, 012030 (2016)
2. Aksenov V.V., Beglyakov V.Y., Kazantsev A.A., Saprykin A.S., IOP Conf. Ser.: Mater. Sci. Eng., **127**, 012032 (2016)
3. Aksenov V.V., Beglyakov V.Y., Kazantsev A.A., Doroshenko I.V., IOP Conf. Ser.: Mater. Sci. Eng., **127**, 012031 (2016)
4. Chernukhin R.V., Dronov A.A., Blashchuk M.Y., IOP Conf. Ser.: Mater. Sci. Eng., **91**, 012086 (2015)
5. Chernukhin R.V., Blaschuk M.Y., Chazov P.A., Blumenstein V.Yu., IOP Conf. Ser.: Mater. Sci. Eng., **142**, 012003 (2016)
6. Blaschuk M.Y., Dronov A.A., Ganovichev S.S., IOP Conf. Ser.: Mater. Sci. Eng., **142**, 012128 (2016)
7. Aksenov V.V., Walter A.V., Gordeyev A.A., Kosovets A.V., IOP Conf. Ser.: Mater. Sci. Eng., **91**, 012088 (2015)

8. A.A. Khoreshok, S.A. Zhironkin, M.A. Tyulenev, G.A. Barysheva, V.Yu. Blumenstein, M.C. Hellmer, S.V. Potyagailov, IOP Conf. Ser.: Mater. Sci. Eng., **142**, 012122 (2016)
9. Revuzhenko A.F., Kazantsev A.A., Glazkov Y.F., Dortman A.A., Applied Mechanics and Materials., **682**, 196 (2014)
10. Dimitrios Kolymbas (Springer-Verlag, 2005)
11. Bernhard Maidl, Markus Thewes, Ulrich Maidl, *Handbook of tunnel engineering. Volume I: Structures and Methods* (Wilhelm Ernst & Sohn, Verlag für Architektur und technische Wissenschaften GmbH & Co. KG, 2013)
12. Bernhard Maidl, Markus Thewes, Ulrich Maidl, *Handbook of tunnel engineering. Volume II: Basics and Additional Services for Design and Construction* (Wilhelm Ernst & Sohn, Verlag für Architektur und technische Wissenschaften GmbH & Co. KG, 2014)
13. Austin, S. A., Robins, P. J. and Goodier, C. I. *Construction and Repair with Wet-process Sprayed Concrete and Mortar, The Concrete Society, Technical Report 56* (Draft), Crowthorne, Berkshire (2000)
14. Clayton, C. R. I., Hope, V. S., Heyman, G., van der Berg, J. P. and Bica, A. V. D., Proc. Inst. Civ. Eng. Geotechnical Engineering **143**, 119 (2000)
15. Institution of Civil Engineers. Sprayed Concrete Linings (NATM) for Tunnels in Soft Ground. Thomas Telford, London, ICE design and practice guide (1996a)
16. Institution of Civil Engineers. Specification for Piles and Embedded Retaining Walls. Thomas Telford, London (1996b)
17. International Tunnelling Association, Tunnelling and Underground Space Technology **15**, 303 (2000)
18. John, M. ASCE **III**, 27 (1978)
19. Proctor, R. V. and White, T., *Rock Tunnelling with Steel Supports* (Commercial Shearing and Stamping Co., 1946).
20. Szechy, K. *The Art of Tunnelling* (1967)
21. Terzaghi, K. *Rock defects and loads on tunnel supports in Rock Tunnelling with Steel Supports* (eds Proctor, R. V. and White, T.). Commercial Shearing and Stamping Co., Youngstown, USA, 15–19 (1946)
22. Pisarenko G.S., Yakovlev A.P., Matveev V.V. *Handbook on Strength of Material* (Kiev, Naukova Dumka, 1988)
23. Litvinsky G.G., Gaiko G.I., Kuldirkaev M.I.. *Steel frame supports in mining workings* (Kiev, Tekhnika (Equipment), 1999)

Ways of increasing excavator fleet productivity in Russian coal open pits (Kuzbass case study)

Michael Drygin[1,*], *Nicholas* Kurychkin[1] and *Alexander* Bakanov[1]

[1] T. F. Gorbachev Kuzbass State Technical University, 28 Vesennyaya str., Kemerovo 650000, Russia

Abstract. A steady increase in coal extraction in Russian Federation and worldwide results from objective causes of natural resources availability and development of advanced coal processing technologies. In the context of the long-standing economic crisis, considerable investments in coal industry are not to be expected. Consequently, the reserve for extraction growth lies in increasing productivity of the key equipment applied in Kuzbass coal pits. The production data analysis revealed that the presented excavators working capacity does not depend on the bucket volume and the cycle time. It is hereby specified that working capacity is primarily dependent on the actual operation cycle time. As it is demonstrated, the key reserve for increasing the working capacity consists in elevating efficiency of actual performance time use.

1 Introduction

Significant volumes of coal resources compared to oil and gas resources place the world coal industry in top position in the world fuel and energy sector structure. Similar processes are ongoing in such countries as the USA, China, India, RSA and Australia. The coal industry development is the basis of their economic, technological and environmental policy. The economy orientation to implementing advanced coal processing technologies plays quite a significant (if not the principal) role in the coal industry development. Production of coking coal, sorbents, resins, benzene, methanol, carbon fiber, fertilizers and electric power (on production site) also places the coal industry among the leaders in the Russian economic development.

Kemerovo region is the leading coal region in Russia. 75% of all the Russian coal volumes and 2/3 of all the coking coal volumes are extracted here. The prospective objective is to set up new mines and pits in Kuzbass with overall annual capacity of over 50 mln tonnes of coal by 2030. Therefore, the tendencies and particularities of the Kuzbass coal mining industry fully reflect the tendencies and particularities of the Russian Federation coal industry.

Over 70% of coal in Kuzbass is extracted by open pit mining. This method possesses apparent advantages compared to underground mining and is characterized by higher productivity, low production cost and safe mining operations.

* Corresponding author: mike.drygin@gmail.com

The prospect of consistent increase in open pit coal extraction in the years to come has to be ensured not only by commissioning new facilities but also by increase in labour productivity and consequently by reduction of coal production cost [1]. This result can be achieved in two ways: firstly, by considerable investments in purchase of modern equipment (excavators, bulldozers, mine trucks) and secondly, by reducing idle time periods of the principal mining equipment [2] due to implementation of new advanced methods of the equipment operation and repair organization [3].

The first way is unlikely to be considered as feasible in the context of the long-standing economic crisis and of chronical funds shortage. Fig. 1 shows excavator commissioning / decommissioning trends within the largest Kuzbass coal company UK Kuzbassrazrezugol from 2008 till 2015.

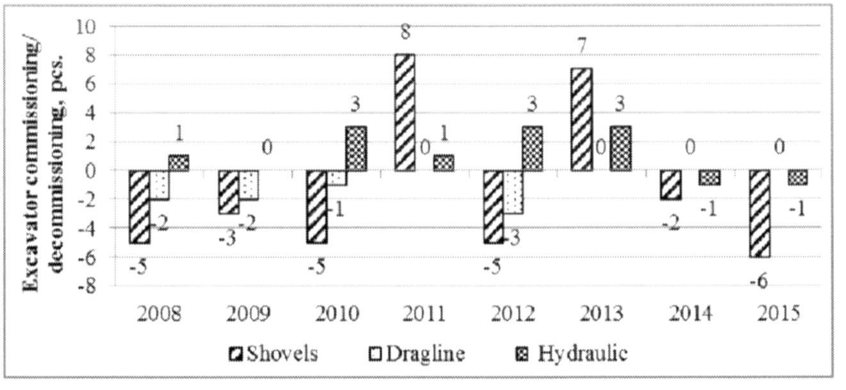

Fig. 1. Excavator commissioning / decommissioning trends in UK Kuzbassrazrezugol

As shown on the diagram, the trends of quality and quantity alterations of the excavator fleet are disappointing. During the 8 year period till 2015, the total of 34 excavators have been decommissioned and only 26 new equipment units have been purchased. With this, decommissioning and putting new equipment in operation is sporadic, with exception of years 2011-2013 when mass commissioning of imported and Russian shovel excavators took place. In recent years the systematic policy of expanding the fleet with single excavator units of high capacity is being implemented. However, despite this policy the mentioned units do not play a significant part in excavation volumes. According to the UK Kuzbassrazrezugol data, only 12% of overburden was moved in 2015 by the new excavators made in Russia with the bucket volume exceeding 10 m^3, whereas 23% was moved by the new imported electric units.

The second way is the most feasible one for the present time. The key parameter which characterizes open pit mining efficiency is the working capacity of principal open pit production equipment, that is of various types of excavators.

2 Results and discussion

The major factors affecting the excavator productivity include:

- difficulty of overburden excavation which depends on the rock type and state and which affects the bucket filling up coefficient [4];
- the excavator technical condition and reliability [4];
- the operator's qualifications [5];
- the face quality estimated by its bench height, truck access to the loading point, illumination [6];

- operations management determined by sufficient fleet size, roadways condition, timely supply of fuel, power, spare parts [7; 3].

Excavator productive capacity includes theoretical (rated), technical and working capacity.

Theoretical productive capacity per hour of continuous operation is calculated by the following formula:

$$P_d = 60gn, \tag{1}$$

where g is geometrical bucket capacity in m^3 and n is number of cycles per minute.

The number of cycles per minute depends on the length of one cycle:

$$n = 60/(t_d + t_s + t_{du} + t_{sb}), \tag{2}$$

where t_d is digging time, 10-20 s; t_s is swinging time, 4-6 s; t_{du} is dumping time, 3-5 s; t_{sb} is swinging back time, 2-3 s [8].

Working productive capacity is the excavator average actual productivity (m^3/h) during operation in specified conditions with account of unavoidable idle time, time losses for shift changeover and equipment inspection, lubrication and rolling stock replacement. The working productive capacity indicator is lower than the technical productive capacity. This indicator shows efficiency of organizing operations of the excavator and all its service equipment:

$$P_w = 60g\,k_f k_o n/k_s, \tag{3}$$

where k_f is the bucket filling up coefficient; k_o is the equipment operating time utilization coefficient which represents the ratio between the actual operations time and the total consumed time; k_s is the ground swell coefficient. Coefficients which make part of formula (3) are assumed equal to: k_f = (0.8 – 1.5) (depending on ground type, humidity and the operating equipment type); k_o = (1.1 – 1.3); k_s = (0.75 – 0.85).

Formulas (1), (2), (3) demonstrate that the excavator productive capacity depends above all on the bucket volume g and on the cycle time [6, 9]. In reality this is revealed in the following way. Fig. 2 presents the year-to-year relationship between the overall bucket volume (m^3) of all the excavators operating in UK Kuzbassrazrezugol (KRU) open pits and their aggregate working productive capacity (mln m^3/year).

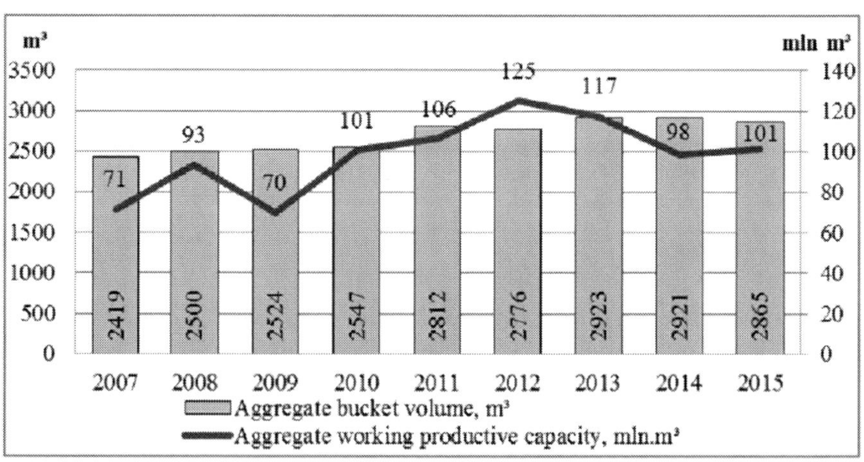

Fig. 2. Aggregate bucket volume and aggregate working productive excavators' capacity in UK KRU

The diagram shows that these indicators have no correlation from year to year. Thus, from 2012 to 2013 the aggregate bucket volume has increased by 5%, whereas the aggregate productive capacity on the contrary has decreased by 6.2%. Comparing years 2014 and 2016, the contrary situation is observed: with decrease in the aggregate volume by 2% the aggregate working productive capacity increased by 3.2%. However, the general tendency of both indicators improvement during 9 years from 2007 to 2015 remains unchanged – with the increase in overall bucket volume by 15.6% the aggregate working productive capacity increased by 29.7%.

Therefore, analysis of the Fig. 2 diagram shows that the excavator working productivity is in minor dependence on its bucket volume [10]. This conclusion is confirmed by analyzing another diagram (Fig. 3).

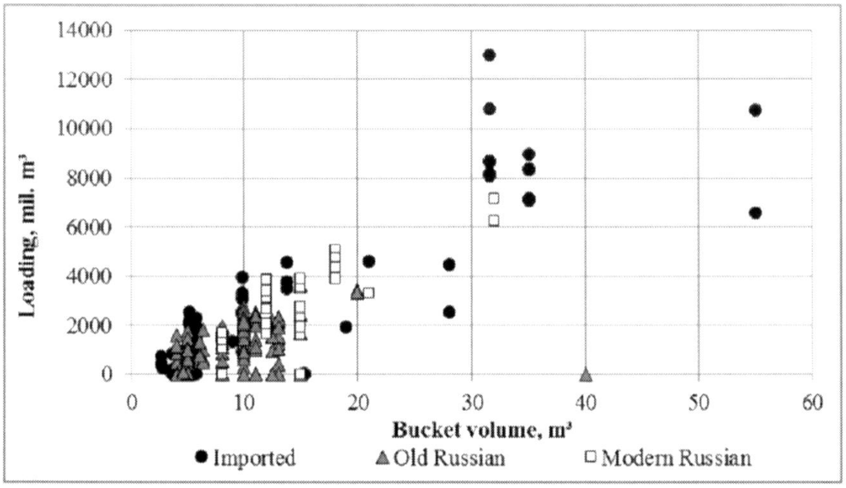

Fig. 3. Relation between the overburden loading volume in 2015 and the bucket volume of excavators in UK Kuzbassrazrezugol

The dot diagram shows that, for instance, in 2015 one of the two imported excavators with bucket volumes of 28 m^3 each has moved twice as much overburden as the second one. Excavators with bucket volumes of 32 m^3 have an annual loading volume variation exceeding 100%. But two imported excavators with bucket volumes of 55 m^3 each have moved less overburden for the year than two excavators with bucket volumes of 32 m^3.

Examining the influence of the cycle time and wear rate of different types of excavators on working productivity in actual conditions; Fig. 4 demonstrates the diagram of specific productive capacity (working capacity per 1 m^3 of bucket volume) for various types of excavators as well as diagrams of equipment wear rate depending on time and performed operation volumes. It is apparent that due to constructive particularities of various excavator types the cycle time of their operation differs [7, 11].

Analysis of the diagrams shows that the specific productive capacity variation for various types of excavators is not significant. The same conclusion can be made for the influence of the excavator wear rate on specific productive capacity. As shown on the diagram, excavators with the wear rate exceeding 100% (EKG-10, EKG-15, ESh-10/70) have the same specific productive capacity as the new excavators with the wear rate of (10-15)% (EKG-18P, EKG-32, P&H-4100XPC, WK-35).

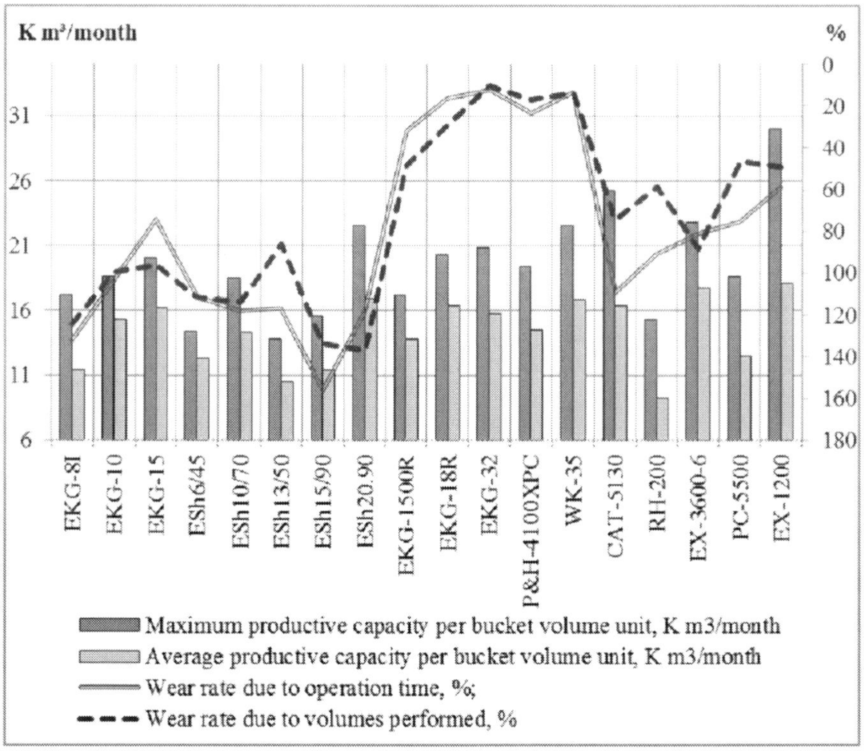

Fig. 4. Specific productive capacity of excavators in UK Kuzbassrazrezugol open pits.

The diagram drawn for dragline ESh 10/70 is illustrative (Fig. 5).The diagram shows that during four years from 2012 to 2015 the working productive capacity of this excavator has steadily been lower than its theoretical productive capacity by 2.5 times on average. However, after the excavator upgrade in the end of 2015, including replacement of the 10 m^3 bucket by a 13 m^3 one and decreasing the boom length to 50 m, the theoretical productive capacity has increased by 75%. On the other hand, the working productive capacity has remained at the same level.

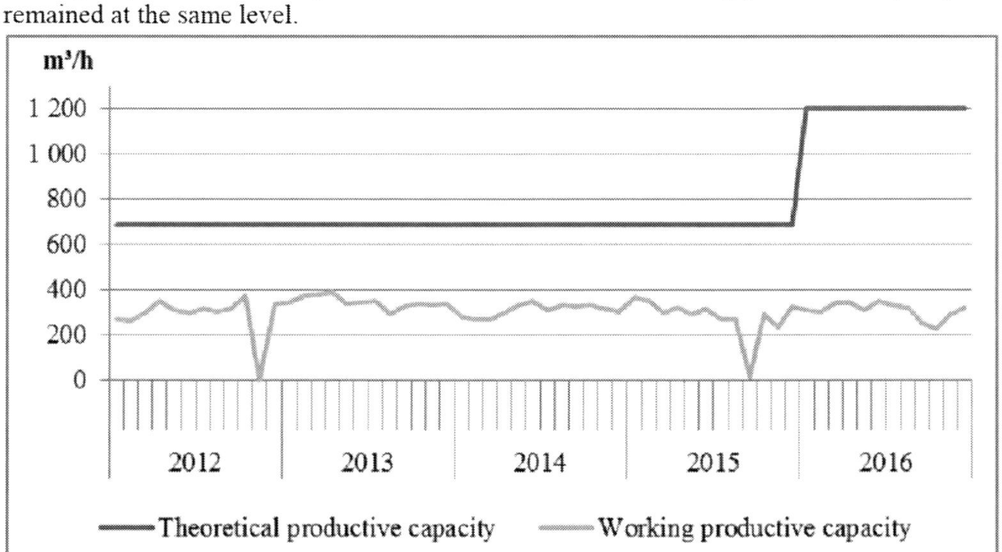

Fig. 5. Theoretical and working productive capacity of ESh-10/70 excavator.

Analyzing the obtained result it can be concluded that after commissioning of new excavators with improved principal operational parameters the overall fleet productive capacity had a lower increase than was expected. Meanwhile, the actual working productive capacity of particular equipment units with similar parameters differs considerably.

Therefore it can be assumed that this result is caused by considerable increase in various scheduled and unplanned idle times [12; 13].

Yet, the below diagram refutes this thesis as well. Fig. 6 demonstrates year-to-year alteration of various idle time percentages of UK Kuzbassrazrezugol excavators for years 2007-2015. The diagram shows that the idle time periods caused by planned repair and technological pauses in operation have the maximum specific percentage in the idle time structure [14,15], amounting to an average of 14% and 10% from overall working time fund respectively.

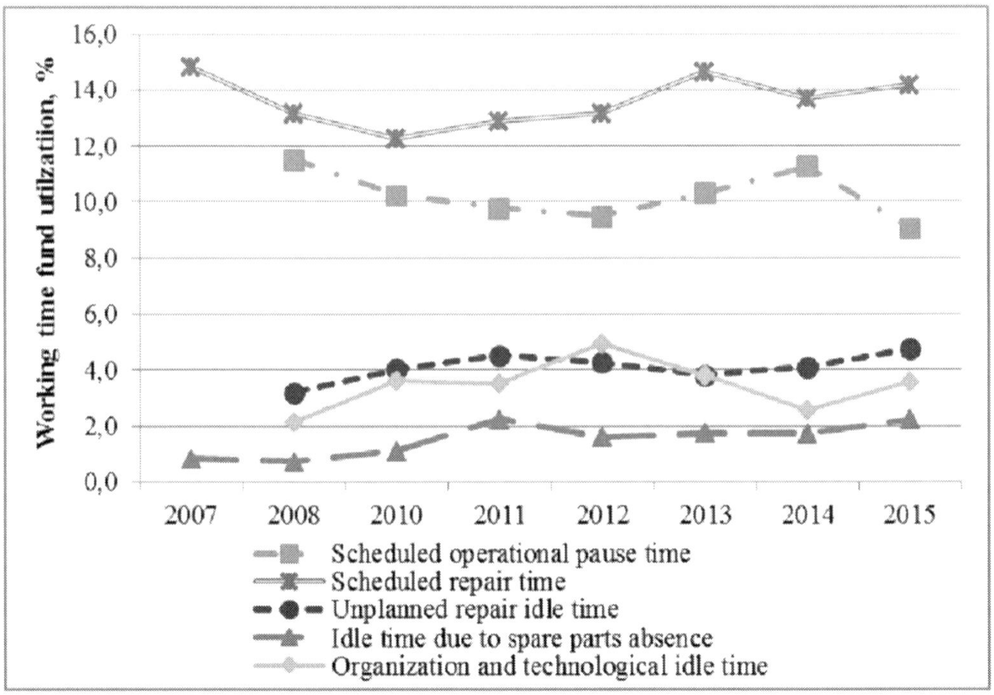

Fig. 6. Trends of working time fund utilization of excavator fleet in UK Kuzbassrazrezugol open pits by years.

However during eight years these indicators have decreased by 5% and 22%. Other types of idle time related to unplanned repair, absence of spare parts and organization and technological events amount to in total to (8-9)% of the working time fund. Their insignificant increase is offset by a decrease in idle time related to scheduled repair and operational pause time. Consequently, no increase in scheduled as well as unplanned idle time is observed.

Therefore, the production data analysis at Kuzbass open pits has revealed that excavator specific working productivity does not depend on the theoretical operation cycle time; moreover, the actual productivity does not depend on equipment wear rate based on operation time and volumes performed.

3 Conclusion

The studies performed made it possible to draw the following conclusions:

1. A steady increase in coal extraction in Russian Federation and worldwide results from objective causes of natural resources availability and development of advanced coal processing technologies.

2. In the context of the long-standing economic crisis and chronical funds deficiency, investments in Russian coal industry are not substantial. Consequently, the tendency of excavator fleet renewal in coal pits is poor, whereas newly commissioned excavators play no significant part in the coal extraction volume increase.

3. One of the principal indicators characterizing the excavator operating efficiency is its specific working productivity which does not depend on the theoretical operational cycle time.

4. The actual working capacity of Kuzbass open pit excavators does not depend on equipment wear rate based on operation time and volumes performed.

5. It was established that the working productive capacity is in primary dependence on the actual operational cycle time. As shown hereby, the principal reserve of the working productive capacity increase lies in elevating efficiency of actual performance time use.

References

1. H. Aykul, E. Yalcin, I.G. Ediz, D.W. Dixon-Hardy and H. Akcakoca, The Journal of The Southern African Institute of Mining and Metallurgy, **107** (2007)
2. V. M. Kalra, T. Thakur, B.S. Pabla, IOSR J. of Mech. and Civil Engineering (2015)
3. P. Rai, Indian Journal of Engineering & Materials Sciences, **11** (2004)
4. C. Kirmanli and S.G. Ercelebi, The Journal of The Southern African Institute of Mining and Metallurgy, **109** (2009)
5. S. Namata, *Performance analysis of heavy earth moving machineries (hemm) in opencast coal mines* (Department of Mining Engineering, Rourkela National Institute of Technology, 2015)
6. H. Naik, The Indian Mining and Engineering Journal, **13** (2015).
7. R. Andreev, *Evaluation of Hydraulic Excavator and Rope Shovel Major Maintenance Costs in Operation* (Department of Civil and Environmental Engineering, University of Alberta, 2015)
8. P. Rai, R. Trivedi and R. Nath. Indian J of Eng & Materials Sciences, **7** (2000)
9. A. Hall, *Characterizing the Operation of a Large Hydraulic Excavator* (The University of Queensland, Brisbane, Australia, 2002)
10. E. Bozorgebrahimi, R.A. Hall and G.H. Blackwell, Mining Technology, **112** (2003)
11. D. Mahambetov, B. Rakishev, G. Samenov, A. Sładkowski, Trans. Problems, **8** (2013)
12. M.E. Michael Arputharaj, Int. J. of Adv. Res. in Eng. and Tech.(IJARET), **6** (2015)
13. A. Lanke, H. Hoseinie, B. Ghodrati, International Scholarly and Scientific Research & Innovation, **8** (2014)
14. S. Elevli, B. Elevli, Acta Montanistica Slovaca, **15** (2010)
15. C. Drebenstedt, R. Singhal, *Proceedings of the 22nd MPES Conference* (Dresden, Germany, 2013)

Strategy of Russian Coal Mining Enterprises' Excavator Park Technical State Correction

Michael Drygin[1,*], *Nicholas* Kurychkin[1] and *Alexander* Bakanov[1]

[1]T. F. Gorbachev Kuzbass State Technical University, 28 Vesennya str. Kemerovo 650000, Russia

Abstract. Coal extraction in Russian Federation increases by 3…4 %. This tendency will remain in the nearest 10…15 years. Opencast mining makes (70...72)% from the total extraction in the country. Though analysis shows that investments in coal mining within the last 25 years were insignificant. In these conditions the priority area for coal production cost reduction and increasing productivity of opencast coal mines is optimization of the main technological equipment (excavator park) maintenance system. The article shows that current system of scheduled preventive repair is not efficient and requires to be changed to the system of repair works due to actual state of the equipment.

1 Introduction

Fuel and energy complex is a vitally important part of the world economy. Consumption rates for energy products increase year to year with the growth of the world gross product [1]. Currently basic energy sources are natural gas, oil, coal, pyroshales and uranium. Oil and gas play the main role but their reserves gradually decrease and will be completely exhausted by the 2050s [2, 3]. At the same time coal deposits are enough for about 200 years even at the maximal extraction rates.

Regardless of the world economy downfall coal is still one of the crucial Russia's energy resources stabilizing the situation in the power industry [4, 5]. In the end of 2015, there were 245 coal mining and processing enterprises in the country; 70 of them are mines, 119 opencast mines, and 56 processing plants. Together there were 373.4 mln tons of coal extracted (Fig. 1), of them 269 mln tons were extracted by opencast method.

Russian Federation is the sixth in the world in coal extraction and the fourth in coal consumption. Kuznetsk coal bearing basin (Kuzbass), is one of the largest in Russia and in the world (Fig. 2) [6].

Thus in 2015, export that was yearly increasing by 5-15% remained practically the same as in 2014 – 151.4 mln tons. Within the last five year steady positive growth of coal export turnover can be observed, and Kuzbass made Russia the third in coal export after Australia and Indonesia [7, 8]. Coal from Kuzbass is exported to 85 countries of the world.

* Corresponding author: mike.drygin@gmail.com

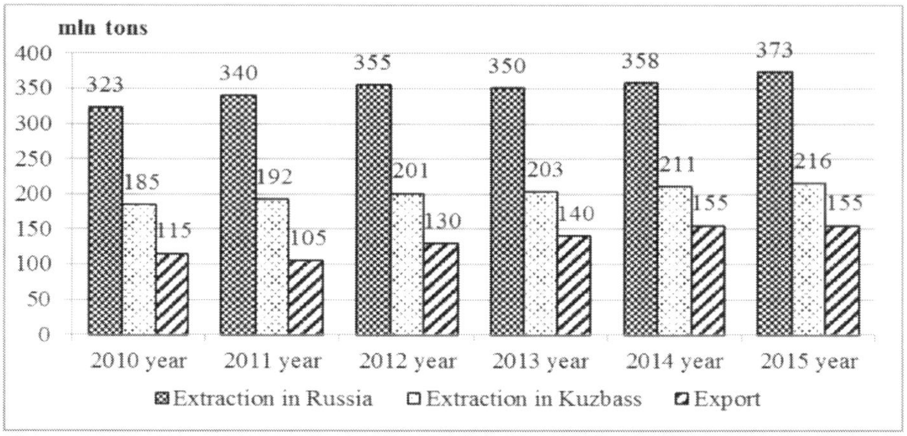

Fig. 1. Coal Extraction Dynamics for Russia and Kuzbass

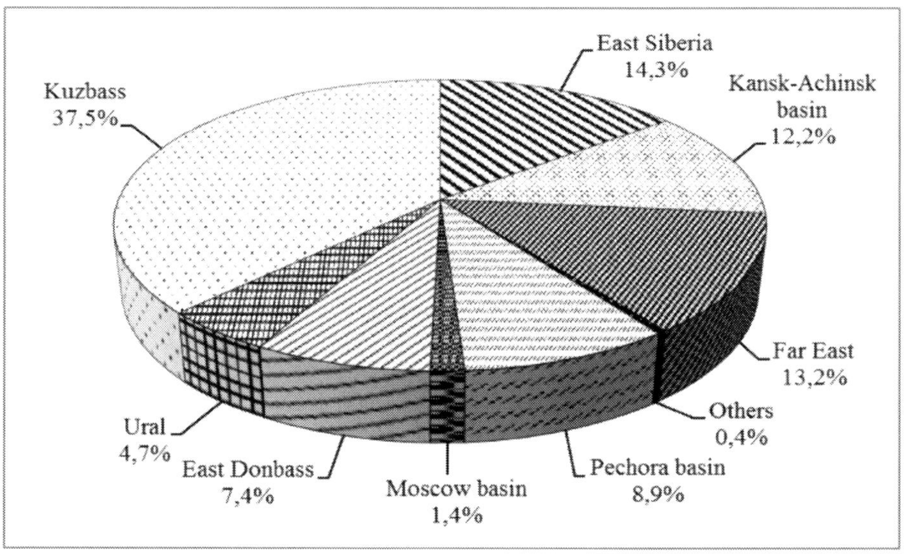

Fig. 2. Extracted Coal Percentage by Regions

The tendency for the increase of coal extraction within the country will remain in the nearest 10-15 years (Table 1) [9, 10]. Though, regarding the crisis in the economy [11, 12], by 2030, coal extraction in Russia will hardly correspond to the forecast 550-650 thousand tons. Extraction of 450-500 thousand tons looks the most real, and around 40% of this coal will be mined in Kuzbass, and 70% of it mined by opencast method [13, 14].

Table 1. Coal Extraction Factors in Russia

Factors	In fact		Forecast by years		
	2010	2015	2020	2025	2030
Coal extraction in Russia, mln tons	360-375	425-445	470-485	515-580	565-650
Kuznetsk basin, mln tons	180-190	215-250	230-250	255-300	275-330
Kemerovo region share in the total extraction in Russia, %	50	50-51	49-51	50-52	49-51

2 Results and Discussion

Opencast method has some obvious advantages in comparison with underground one: higher efficiency, low producing cost and safer mining work. That is the reason why 70% of the coal mined in Russia is extracted by opencast method, and this factor is steadily growing together with total extraction volume. Thus in 2015, it grew by 2% compared with 2014 and by 41% compared with 2000.

Nonetheless investments in the coal industry within the last 25 years were not considerable. Even in the most stable 2010 they remained at the level of 5% of the revenue from coal sales (Fig. 3), though during the post-soviet period practically up to 2000 investments in the coal mining enterprises were minute. At the moment these investments target building of modern technological complexes for coal extraction, processing and enrichment.

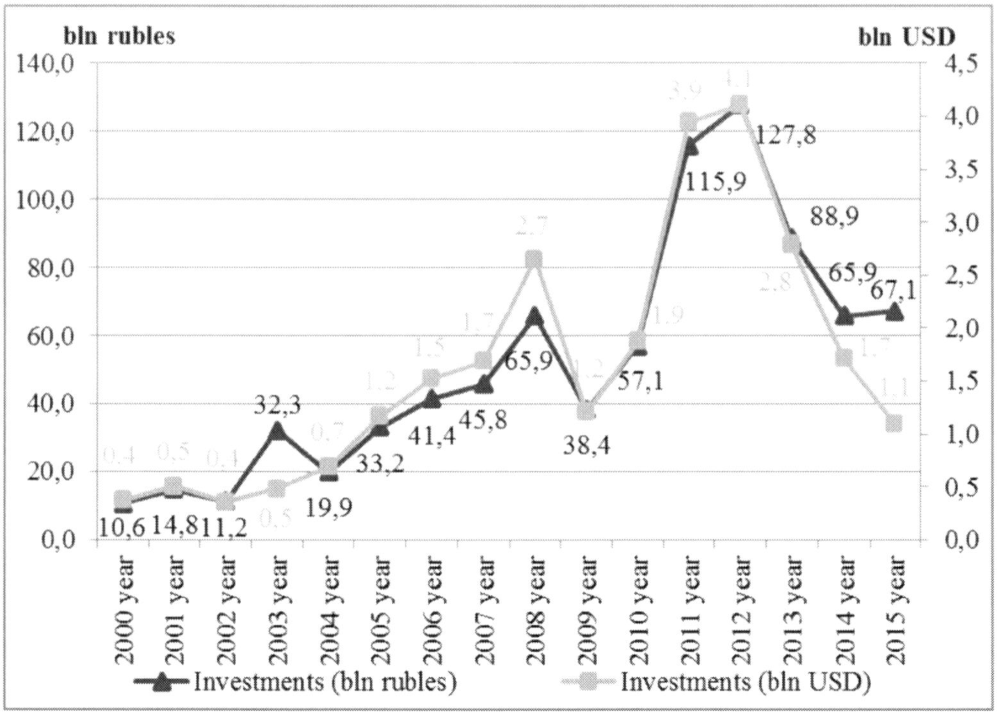

Fig. 3. Investments in Coal Mining Enterprises in Russia

Therefore, while average labour productivity has grown twice within the last decade and reached 200 tons per person monthly (though it is lower than average numbers for the world), the only logical solution for this situation is reducing of coal production cost. Cost reduction is possible only due to growth of the main technological equipment productivity, mainly excavators.

The renewal of the current technological equipment park at opencast coal mines is much slower than it is required. Though Russian and particularly Kuzbass coal mining enterprises are monitored by the supervisory authorities of RosTekhNadzor (Federal Service for Environmental, Technological, and Nuclear Supervision), and 70 % of the equipment is practically every year undergoes the procedure of industrial safety expert review, diagnostic procedures included in the required set of works necessary for equipment service life extension are minimal, formal and fragmentary, and are not enough for its fault-free service.

At mining enterprises basically outmoded equipment and technical devices were replaced but the excavator park was hardly renewed. Due to corrective actions (performed on paper as well) technical devices with expired standard service life are used.

As a rule, meanings that characterize the number of basic production facilities run despite of expired standard service life period vary widely from region to region depending on operating organizations and companies. However, as a part of general tendency, it is possible to highlight single-time renewal of excavator park segment replacing excavators manufactured in the Soviet Union with the equivalent ones manufactured in Russia. This renewal is basically performed by putting into operation shovel dredges manufactured in Russia, USA and China with bucket capacity more than 18-20 m^3.

Analysis of Russian mining enterprises' excavator park shows that the main part of the park includes crawler mounted mining shovels with bucket capacity of 8-15 m^3 manufactured by OMZ IZ-KARTEX (Fig. 4).

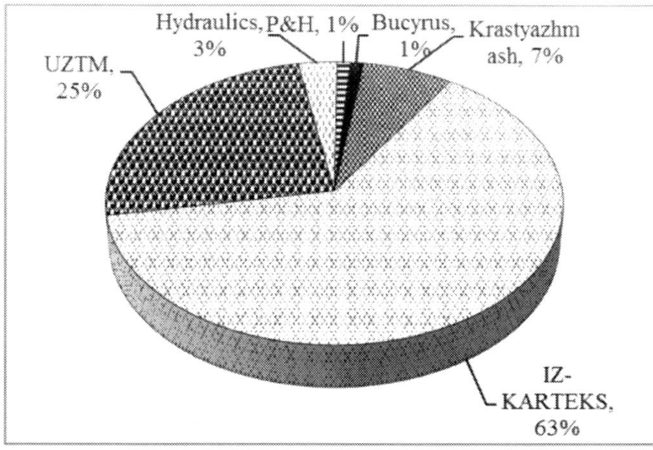

Fig. 4. Structure of Excavator Park of Mining Enterprises

The further analysis is carried out on the basis of the largest coal mining company of Kuzbass as it is the closest to the modern tendencies of Russian coal mining enterprises' excavator park renewal. This is OAO Kuzbassrazrezugol – the second Russian company with the largest coal extraction volume.

According to data from 16 January 2016, opencast mines of OAO UK Kuzbassrazrezugol use 264 technological excavators, of them 169 mining power shovels, 56 walking draglines and 39 hydraulic excavators. 157 excavators of the total number outlasted standard service life period. Within 12 months of 2015, 1 excavator was put into operation and 8 excavators were written-off. Most of the technological excavators are used in Kuzbass. The main current tendencies are removal from operation of the excavators manufactured in the Soviet Union which exceeded their safe usage date, and putting into operation the excavators with bucket capacity from 18 m3 to 40 m^3 and more. However notwithstanding putting the new equipment into operation the average age of the operating equipment is growing and in 2015, it reached 16 years (Fig. 5).

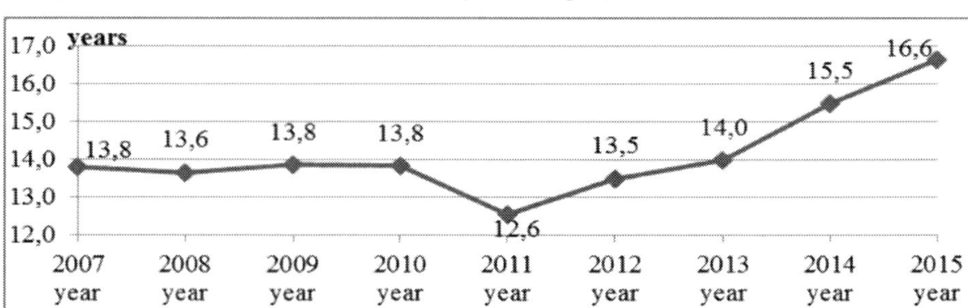

Fig. 5. Average Service Life of Excavators

The tendencies for changing the extent of park wear by service life and the performed score of work practically do not change (Fig. 6). Only in 2011, a lot of new equipment was delivered, and this delayed only for a year reaching the 100% level of excavator park wear extent.

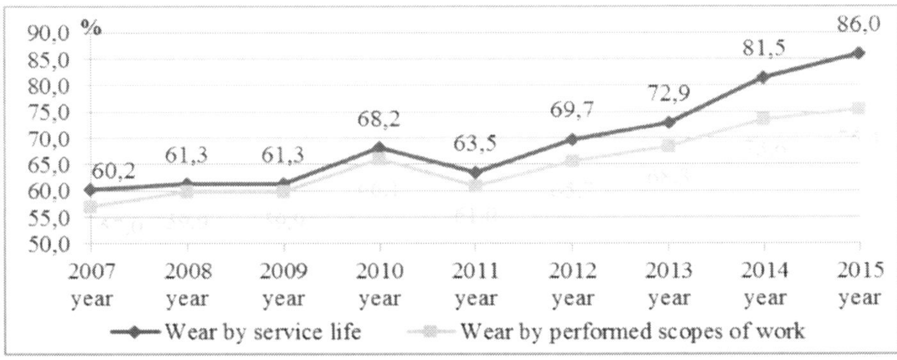

Fig. 6. Dynamic of UK Kuzbassrazrezugol Excavator Park Wear by Service Life and Performed Scope of Work

Thus for 1 January, 2016, the extent of wear for the park of open-pit excavators made 75.4% by the performed scope of work and 86% by the service life.

Average extent of excavator park wear at UK Kuzbassrazrezugol currently exceeded 80%. A great deal of excavators are in operation more than 15-25 years, and service live of more than 45% excavators are in excess of industry norms.

Analyzing the diagram on Fig. 6, it can be said that wear of the excavator park by performed scope of work until 2011, practically mirrors the wear by service life and then begins to deviate significantly, primarily highlights the decrease of total productivity of the park, ad it means that an excavator works out the time but does not work out the scope of work. The further forecasts are discouraging. Without any changes the excavator park of UK Kuzbassrazrezugol will reach the 100% level of wear extent by 2018, and total wear of the park by performed scope of work.

Purchase of new equipment is necessary for park renewal. For instance, yearly purchase with the volume equivalent to 2010, would help to reduce the extent of park wear to 67% and 57% of wear by the service life and by the performed scope of work respectively.

Previously described situation causes the necessity of renewal of Kuzbass mining enterprises' excavator park and search for new organizational and technical solutions of currently used equipment operation. At the moment the following main tendencies can be mentioned. The first is putting into operation of new excavators with large bucket capacity which is required by growing demands for productivity and loading of mining trucks with bearing capacity excising 220 tons.

The second tendency is removal from operation of low-powered hydraulic excavators and excavators manufactured in the Soviet Union usually up to 8 m^3, which had exceeded their safe usage date. Low-powered hydraulic excavators of foreign origin up to 8 m^3 which showed high mobility are usually put in use in contour coal mining [15-18].

Generally, considering the tendencies of the early 2000s based on total replacement of the park of excavators manufactured in Russia with foreign hydraulic ones it can be said that this strategy proved inefficient. The advantages of the foreign equipment during its use during the first year were come to nothing after 4–5 years of use. Thus introduction of foreign equipment except current expenditure demanded significant operating costs, and tough interrepair maintenance requirements caused the necessity of total routine maintenance system revision. Thus in the majority of cases such equipment is under post-warranty maintenance and requires continuous presence of the technical support engineer, and absence of any technical documents and structural complexity of basic mechanisms required plant repair, and, as a rule, this lead to per-unit repair.

One of an excavator's efficient work factors is using of the calendar working time fund, exactly, increase of working time at downtime reduction. It is necessary to understand that total absence of pauses and downtime is impossible, there is a need for proper organization of work for the personnel, including organizational workers, so that the equipment was provided with «routine maintenance», and in case of repair everything necessary for it would be available (Fig. 7).

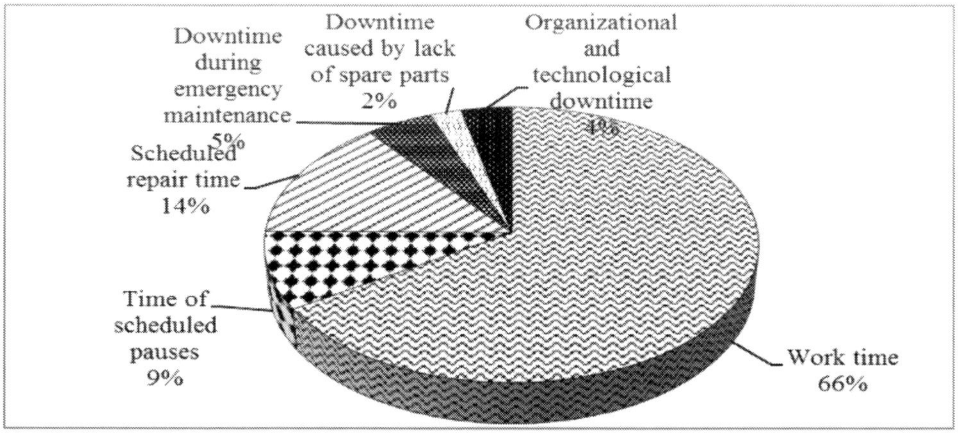

Fig. 7. Using the Calendar Working Time Fund of an Excavator at UK Kuzbassrazrezugol for 2015.

Periodic equipment maintenance presupposes various scope and set time of repair. Incidence rate work and scope of repair are specified by the scheduled preventive repair (SPR) system of the enterprise [19]. In accordance with this system overall annual downtime under scheduled repair make from 50 to 90 days per year and determined by the following factors: extent of equipment wear; qualification of operating and maintenance stuff; quality of delivered spare parts; climatic, mining and geological conditions of the repair works.

Except the SPR system there is system of repair works caused by actual state of equipment [20]. Analysis shows that the system of equipment actual state reacts to the same number of faults that the SPR system but removes troubles much faster (Fig. 8).

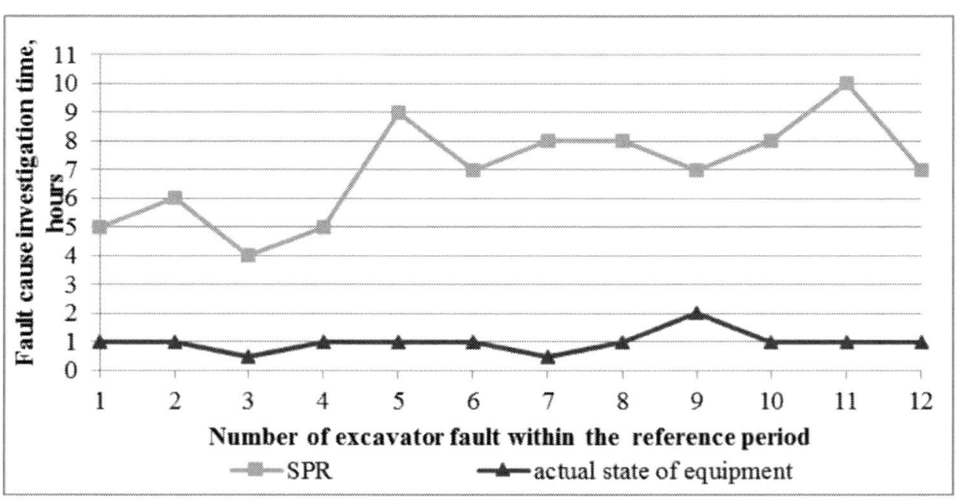

Fig. 8. Comparison of equipment actual state and SPR(organizational and diagnostic)

3 Conclusions

Analysis of the above information allows us to draw the following conclusions.

1. Development of coal mining in Russian Federation comes with extraction volume growth.
2. The tendency for renewal of the excavator park at opencast coal mines is insignificant, and excavators put into operation do not play any major role in coal extraction growth.
3. Annual capacity of excavators drops due to downtime increase.
4. Current system of scheduled preventive repair of excavators is not efficient and requires to be changed to the system of repair works due to actual state of the equipment.

References

1. J. Conti, *The International Energy Outlook 2016* (2016)
2. M. J. Kaiser, Y. Yu, Oil & Gas Fin. J. **9**, 1 (2012)
3. M. J. Kaiser, Y. Yu, Oil & Gas Fin. J. **9,** 2 (2012)
4. H. Arden, A. Tverdov, *Resource and Reserve Valuation Practices in CIS Countries* (2014)
5. Statistic review of Emerging Markets Information Service "Coal Mining Sector Russia" (2013) https://www.emis.com/sites/default/files/Russia/Coal/Mining/Sector/Report.pdf
6. S A Zhironkin, A A Khoreshok, M A Tyulenev, G A Barysheva, M C Hellmer IOP Conference Series: Materials Science and Engineering, **142**, 1 (2016)
7. B. Heath, *Report of Coal Industry Advisory Board International Coal Policy Developments in 2015* (2015)
8. International Energy Agency's technical paper on "Key Coal Trends" (2016) https://www.iea.org/publications/freepublications/publication/KeyCoalTrends.pdf
9. L. Grossman, R. Gebhard, M. Haala, H.-J. Seger, M. Smiarowski, R. Bursch, Coal Age **121**, 52 (2016)
10. Project Survey 2017: Industry Indicators Improve. E&MJ, 2017
11. V. Vorotnikov, Coal Age, **121**, 48 (2016)
12. T. Saywell, The N. Min., **101**, 7 (2015)
13. A. Harris, S. Hall, K. Brown, O. Munnion, *Ditch Coal: the global mining impacts of the UK's addiction to coal* (2016)
14. V.A. Kovalev, A.A. Khoreshok, O.I. Litvin, *The 8th Russian-Chinese Symposium. Coal in the 21st Century: Mining, Processing and Safety*, 287 (2016)
15. M.A. Tyulenev, T.N. Gvozdkova, S.A. Zhironkin, E.A. Garina, Geotech. Geol. Eng., **35:1**, 203 (2017)
16. M.A. Tyulenev, S.A. Zhironkin, E.A. Garina, Int. J. of Mining and Mineral Engineering, **7:4**, 363 (2016)
17. A.B. Efremenkov, A.A. Khoreshok, S.A. Zhironkin, A.V. Myaskov, IOP Conf. Ser.: Earth Environ. Sci. **50**, 012009 (2017)
18. M. Tyulenev, E. Garina, A. Khoreshok, O. Litvin, Y. Litvin, E. Maliukhina, IOP Conf. Ser.: Earth Environ. Sci., **50**, 012035 (2017)
19. Fleet management. Mining Magazine 44 (2013)
20. A. Morshedlou, H. Dehghani, S. H. Hoseinie, JME, **5**, 113 (2014)

The innovative development of machine building as a driver of import substitution

Sergey Zhironkin*[1], *Magerram* Gasanov[1], *Galina* Barysheva[1], *Konstantin* Kolotov[2], and *Olga* Zhironkina[3]

[1] Tomsk Polytechnic University, Department of Economics, 634050, 30 Lenina st., Tomsk, Russia
[2] Tomsk Polytechnic University, Post-Graduate Study, 634050, 30 Lenina st., Tomsk, Russia
[3] Kemerovo institute (branch) of Plekhanov Russian University of Economics, Department of Humanities, 650992, 39 Kuznetsky av., Kemerovo, Russia

Abstract. Within the rent-and-raw material model of Russian economy the state's attempts to stimulate import substitution do not generate innovative type of economic development and rebuilding of its industrial sector, competitiveness on the domestic market. In high entropy of market processes it becomes obvious that without recovery of effective global production chains in Russian economy the risks of economic and geopolitical security of Russia, the final loss of Russian technological identity, the loss of international competitiveness and securing raw material vector of long-term social and economic development are increasing. Therefore, it is machine building, as an industry, being a member of most production chains must become a driver of import substitution. Innovative development of machine building is especially in demand in such industrialized region as Kuzbass. Its major industries – coal, metallurgical, chemical - are in need of machinery products. The article describes the features and tools of machine building innovative development.

1 Introduction

In economy of Kuzbass (old industrial Russian region), despite the high level of urbanization and concentration of basic industries, the problems of imports that are typical for Russian economy as a whole are increasing. They form a complex of negative processes in the region, holding the production of competitive output of raw materials processing, stabilization of employment and budget revenues, and social wellbeing growth [1].

2 Material and Method

The solution of the problems of regional economy requires the development of import substitution, which should be initiated in the regional machine building industry: - the transmission of external shocks in reducing financial stability of basic for Kuzbass

* Corresponding author: zhironkin@tpu.ru

industries enterprises – coal, chemical, metallurgical. As evidence can serve a high proportion of lossmaking medium and large enterprises the proportion of which increases when raw material prices on the world market are decreasing [2]. So, in the region this figure was 32.6% in 2005, 36.1% in 2008, 31.1% in 2011 and 38.1% in 2014. The fact that more than a third of industrial enterprises in the Kemerovo region are unprofitable indicates under-use of the domestic demand potential for smoothing fluctuations of negative foreign economic conditions;

- continuing import growth of production means by the enterprises dominating in the region coal industry along with the recovery of coal production since the beginning of the 2000s;
- non-innovative development of the regional economy, resulting in enhancing its dependence on the global raw material market. Thus, according to the government program of the Kemerovo region "Economic development and innovative economy of Kuzbass for 2014-2016" the share of innovative goods and services in total sales of Kuzbass enterprises decreased from 1.7% to 1.4% (with minimum 1.1% in 2009) from 2006 till 2012. This is in 8-9 times less than in the countries of Eastern Europe and in 12-13 times less than in Germany and the United States;
- raw material export growth, along with an increase of processed products import. In particular, the enterprises of Kuzbass coal industry are exporting to South-East Asia up to 15% of coal mined in the region. At the same time the plastic products import from these countries cover 70% of the regional consumption.

The comparative dynamics of production and export of coal, metal, on the one hand, and machinery import, on the other hand, illustrates the complex problems of neo-industrial import substitution in Kuzbass [3].

The sustainable growth of coal mining in the region (for 109% over the period 1998-2015) is accompanied, on the one hand, with the accelerated replacement of production means for coal mines and quarries. On the other hand, since 2003, as the consequences of ruble devaluation in 1998-1999 are over, faster growth of machinery import (in 4 times in 2002-2012) is observed in Kuzbass. In 2014, a new round of devaluation led to a drop in the regional machine building import (from 1360 to $ 322 million), but in 2015 its volume grew by 8%.

Under these conditions the production volume of machinery in Kuzbass has been declining since 2011 against the unstable investment background. At the same time, the main export-oriented industries of Kemerovo region economy - coal and metallurgical - are increasing their exports (in particular, the export of Kuzbass coal for 1998-2015 had increased tenfold.). This demonstrates the high potential demand for the means of production for the basic industries of Kuzbass economy, which is met today mainly due to the import [4].

In Kemerovo region economy mining and basic industries dominate. In 2014 their share in the gross regional product (GRP) were respectively: the fuel industry - 36.8%, ferrous and nonferrous metallurgy - 29.1%, chemical industry - 7, 0%, machinery - 3,2%. And the peculiarity of Kuzbass machine building is targeting most of its enterprises to the needs of the dominant regional industries, mainly coal and chemical.

3 Results and discussion

Machine building is that branch which neo-industrial transformations of the regional economy depend on [5-7]. Their starting point should be the technological modernization of extracting and basic industries. At the same time, from the beginning of market reforms the entire machine building industry in the Kemerovo region is experiencing significant problems and cannot get the role of a "catalyst" of neo-industrial transformation [8-12].

First, before market reforms machine building industry was a technological leader and one of the main branches of Kuzbass economy, having produced up to 14% of the GRP in 1988-1992. By now there are such major machine building companies as JSC "Anzheromash", LLC "Yurmash", JSC "Sibtenzopribor", JSC "Elektromash-M", LLC "Kemerovohimmash", "Razvitie" operating in the region. They account for 10% of the regional industrial production and 11% of the total value of fixed assets in the region.

Secondly, after almost five-time decrease of the machine building share in GRP for 1998-2015, since 2011 there has been a steady decline in the volume of machinery production - as a result of reduced demand from the regional coal, metallurgical, chemical, power companies as well as unstable investments.

Third, the growth of import-dependence of the key industry of Kemerovo region economy - coal – takes place against the background of the progressive increase in volume of coal mining in the region by 123% over the period 1998-2015 from 94 million tons to 215 million tons.

The share of import in the equipment used in coal mining in Kuzbass reaches 27% for some types of products. At the same time 60% of mining equipment is imported from the "sanction" countries - the US, EU and Australia.

Particularly acute need for import substitution of mining equipment for Kuzbass machine builders creates considerable deterioration of used domestic means of production and the dominance of import in the least deteriorated part of it. Thus, major open pit mining equipment (excavators, drilling machines, pumps, transformer stations) are deteriorated more than 70%, while the most part of the equipment purchased during the last 10 years and having a deterioration rate of less than 30% had been imported.

Problems preventing the development of import substitution in Kuzbass economy also appear to be obstacles on the way of its neo-industrial development, which should begin with the innovative development of the basic industries. These problems are connected with the lack of global competitiveness of the regional companies' products, with the lack of investments into modernization of new equipment output within the region, with undeveloped institutions' integration among research organizations, industrial enterprises, investors in the process of neo-industrial import substitution.

In Kuzbass there is a certain reserve for the formation of institutes necessary for initiating neo-industrial import substitution. In particular, in the region the Coordination center for the development of import substitution and the Regional plan and the list of investment projects in this area were set up.

The issues of import substitution are not reflected in such an important document as the "Strategy of social and economic development of the Kemerovo region up to 2025 (Kuzbass-2015)". Meanwhile, in Section 2.1.3 of the strategy "The risk of technological backwardness of the region" the replacement of the domestic machine building products by the imported ones is clearly indicated: "there is a dangerous situation in which the coal industry begins to depend on foreign suppliers of equipment, spare parts and accessories. According to experts, this is largely due to the practice of so-called "bound" foreign loans, which are burdened by the conditions of purchase exceptionally import equipment."

Also, the Strategy "Kuzbass 2015" lacks the link between the growth of living standards in Kuzbass and modernization and import substitution in the basic industries through the increased productivity and new jobs in machine building, radio electronics and professional equipment, chemical industry.

Therefore, to solve regional problems of neo-industrial import substitution it is necessary to take into account the specifics of the industry structure and the state of productive forces as well as the role of import substitution potential of the industries initiating neo-industrialization of economy. These industries must ensure technological modernization, first of all, extractive and basic industries, which, in turn, will create a basis for import substitution of products in manufacturing and high-tech industries.

With regard to Kemerovo region economy, machine building stands for such industry, in particular its segment producing equipment for mining companies. Concerning this, it is advisable to create neo-industrial import substitution cluster in Kuzbass, oriented to the following tasks solution:

1. Radical reconstruction, technical and technological modernization of industrial regional enterprises, many of which were created in 1960-70ies and have fixed capital depreciation level of about 70%. At the same time for the whole period of market reforms there was no mass replacement of the production means in Kuzbass machine building that enhances the technological gap between the regional producers and foreign competitors, and complicates the start of import substitution.

2. Breaking the vicious circle of investment and production problems: a significant physical deterioration of fixed capital - reducing the international competitiveness of machine building output – decreasing of sales - the lack of investment resources for the modernization of production - further increase in physical deterioration. This requires the formation of the *Single investment program of regional machine building*, reflecting the needs of coal, chemical, iron and steel industries in new equipment, credit and investment opportunities of banks and investment companies, as well as technological possibilities of machine building enterprises.

3. Increasing of machine building regional companies' cooperation, which is now estimated at least 10%, and overcoming their technological disunity. This is particularly important for the production of equipment and production systems with predetermined properties and output with a definite purpose competitive on the regional market. An important role in the development of cooperative relationships within the cluster plays non-profit organization "Association of Kuzbass machine builders" which brings together about thirty machine building enterprises from Kemerovo, Novosibirsk and Tomsk regions.

The formation of neo-industrial import substitution machine building cluster in Kuzbass should include the following activities of the interested parties - coal, machine building enterprises, the regional authorities, research, design and educational organizations.

Firstly, the analysis of technological conditions for machine building enterprises' modernization necessary for import substitution of the equipment for the coal industry, considering the possibilities of inter-firm cooperation should be made. Also it is necessary to analyze investment needs and financial possibilities of the banks and investment companies represented in the region. The result of this analysis should become the basis for the Development strategy of Kuzbass neo-industrial import substitution cluster of machine building.

Secondly, the list of cluster's residents, which includes machine building companies as well as the regional subjects of financial, research and educational spheres, should be compiled. In addition, it is necessary to make partnership agreements with foreign developers of modern machine building technologies, computer software on assignment of property rights, patents and licenses.

Third, it is advisable to make the complex offer of modern machine building products for the needs of coal and other industries of Kuzbass economy capable in the medium term to replace foreign equipment. This, in turn, should be the logical modernization outcome of the regional machine building complex, and involves the implementation of the following actions:

- the organization of the investment consortium, which would include the main consumers of machinery products – enterprises of coal, chemical and machine building industries, making parity investment pool in regional machine building;
- the establishment of research-and-production and engineering firms' network in the industry with participation of the leading international manufacturers of machinery

equipment;

- the establishment of a regional agency to attract and protect investments in import substitution of industrial output, the main focus of which should made on the machine building industry initiating neo-industrialization;

- the structuring of a guarantee investment fund in machinery, the main role in which should be played by the regional authorities.

Fourthly, the list of regional tax exemptions for the residents of import substitution machine building cluster should be compiled.

Currently, there is an urgent need in developing the complex regional target program of neo-industrial import substitution in Kuzbass industry. The program should link the intensification of R&D, investments in modernization of coal, chemical and machine building industries, professional equipment, government support and loan financing guarantees for the enterprises. The main objective of this program should become the adjustment of the target for the import substitution level of technology in these industries and the level of imported by Kuzbass enterprises equipment.

4 Conclusion

The development of neo-industrial import substitution requires setting up the innovative and industrial clusters of a special network type, where companies, research organizations from different clusters would be merged into a single research-and-production complex on the basis of technological platform of import substitution. Innovative and industrial infrastructure should become the unifying elements, as well as investment projects for the production of import substitution output, covering all the residents of the cluster.

References

1. A. A. Khoreshok, S. A. Zhironkin, M. A. Tyulenev, G. A. Barysheva, V. Yu. Blumenstein, M. C. Hellmer, S. V. Potyagailov, IOP Conf. Ser.: Mater. Sci. Eng., **142**, 012122 (2016)
2. G. Barysheva, I. Kashchuk, Procedia Soc. Behav. Sci., **166**, 62 (2015)
3. V.A. Trifonov, V. A. Shabashev, and I. V. Dobrycheva, EpSBS, **VII**, 346 (2016)
4. M. I. Agienko, E. P. Bondareva, G. V. Chistyakova, O. V. Zhironkina, and O. I. Kalinina, IOP Conf. Ser.: Mater. Sci. Eng., 142, 012022 (2016)
5. D. Bell, *The coming of post-industrial society. Venture in social forecasting* (Collman Pub., New York, 1973)
6. C. Clark, *The conditions of economic progress* (Logan Pub., London, 1991)
7. R. Reich, *The work of nations. Preparing ourselves for 21st Century capitalism* (Penguin Pub., New York, 1992)
8. J. Kretschmann, A. B. Efremenkov, and A. A. Khoreshok, IOP Conf. Ser.: Mater. Sci. Eng., **142**, 012124 (2016)
9. V.A. Gogolin, Y.A. Ryzhkov, Soviet Mining Science **17**, 156 (1981)
10. V. Aksenov, A. Khoreshok, V. Beglyakov, App. Mech. and Mat., **379**, 20 (2013)
11. V.A. Gogolin, Y.A. Ryzhkov, Soviet Mining Science, **13**, 17 (1977)
12. V. Kovalev, B. Gerike, A. Khoreshok, P. Gerike, *Mining 2014: Taishan academic forum - Project on mine disaster prevention and control* (Atlantis Press, Paris, 2014)

"Smart Service" as an innovative system of service for mining companies in Kuzbass

Lyudmila Samorodova[1] , *Larisa* Shut'ko[1], *Yulia* Yakunina[1], and *Oleg* Lyubimov[2*]

[1]KuzSTU, Department of economics, 650000 Kemerovo, Russia
[2]KuzSTU, Department of information and automated manufacturing systems, 650000 Kemerovo, Russia

Abstract. The article deals with the issues related to the introduction of "smart service" as an innovative system of equipment service in the business system and "Life-Cycle Management" (LCM) by mining equipment manufacturers. Based on the analysis of the competitive environment an overview description of the competitive environment in the field of equipment sales for surface and underground mining as well as for maintenance and repair of mining machinery for surface mining is presented. The article describes the results of the survey of top managers from company-clients of the company "Joy Global" to identify the demand characteristics of mining companies in using the "Life-Cycle Management" program ranking them in order of priority. To justify the viability of the corporate service package taking into account the individual needs of mining companies within the system "smart service" the contingency of service package and their demand by mining enterprises is used.

1 Introduction

In recent years the global coal industry has been going through a number of problems. Reduced demand for coal leads to lower production volumes which, in turn, affects the companies providing the equipment and technologies for this industry.

Despite the downturn, the mining industry of Kuzbass is still one of the important areas of mutually beneficial cooperation between multinational and national mining companies in the region. It results in the formation of their investment relations and expansion of industrial activity including maintanance service [1-2].

2 Material and method

Multinational corporations presented in the mining industry of Kuzbass are mostly sellers of high quality mining equipment. Direct investments among international companies in the field of coal production and exports in the region are provided by only offshore company "Arcelor Mittal" (see Table 1). Many companies operating in the market of mining equipment and support service, feeling the pressure of the global and regional economies,

* Corresponding author: oleglyub@gmail.com

begin to take steps to differentiate their services, rise to a new management level of maintenance service, which is one of the world trends in modern economy [3-7]. The growth and diversity of services, carefully developed service systems that meet the most challenging requirements of consumers, help them to enter new regional and industrial markets [8-10].

Table 1. Activities of multinational corporations in Kuzbass

Multinational corporations operating in Kuzbass	Area of activity	Area of interest in Kuzbass economy
"Arcelor Mittal" (Cyprus)	Coal mining and processing	Coal mining
"Joy Global Inc." (USA) "Komatsu" (Japan) "Caterpillar" (USA) "Taiyuan Heavy Industry Co." (China) "Liebherr" (Germany) "Ferrit" (Czech Republic)	Manufacturing of industrial and mining equipment	Selling of mining equipment Organization of service centers

The analysis of the competitive environment in the Kuzbass market of sales, maintenance and repair of mining equipment shows that the main competitors are:

– the sales of mining equipment for open pit mining - Caterpillar (USA), Inc. "URALMASHZAVOD" (Russia), JSC "IZ-KARTEKS named after P.G. Korobkov" (Russia), Taiyuan Heavy Industry Co. (China); Liebherr (Germany), "Joy Global Inc."(the USA);

– the sales of mining equipment for underground mining and support service providing - the company "Ferrit" (Czech Republic), which official representative in Kuzbass is JSC "Sibtransservis" [11];

– the support service for surface mining - Belovo Energoremontnaya plant (Russia); "Siberia Service" (Russia); "Pride +" (Russia); Inc. "Anzheromash" (Russia).

The company "Joy Global Inc." is a leader in the mining industry [12-13]. The company's motto is "We are there where coal is mined." "Joy Global Inc." has long and strong relations with Russian coal industry and a significant number of regional offices which were merged into one holding company JSC "Joy Global" in 2013.

The main activity of this company in Kuzbass is the maintenance of mining equipment in surface and underground mining operations. Its long-range strategy is the expansion of services in the field of mining equipment support on the basis of the service center established in Prokopyevsk in 2015 and the development of innovative service technology "smart service". The analysis of international experience shows that the system "smart service" is integrated into the "Joy" business system (JBS), under which the general principles of program "Life-Cycle Management" (LCM) in any region of the world are strictly regulated through direct service network [14].

Expanding the scope of services "Joy Global Inc." offers solutions for the life cycle management of equipment by building partnership relations with the customers to help them to maximize productivity by optimizing the equipment performance throughout its life in the mine. Direct support service, working at enterprises and regional service centers, provide maintenance and repair along with high-quality spare parts meeting the unique needs of each mining enterprise. "Joy Global Inc." teams work in tandem with the staff of mining enterprises and meet the operational and production requirements, focusing on lowering the cost per ton, increasing production, and supporting a zero harm environment throughout the life cycle of equipment [15].

In 2016 the authors surveyed the top-managers of 12 largest mining company-clients of "Joy Global". The purpose of the survey was to determine the nessecaty degree for the enterprises to use the system "Life-Cycle Management" taking into account the innovation made by "Joy Global" - "smart service", i.e. the system of "the most well-organized service."

3 Results and discussion

In the survey of the top managerial groups the current needs of the mining enterprises of Kuzbass were defined and ranked according to their importance. The results are shown in Fig. 1-2.

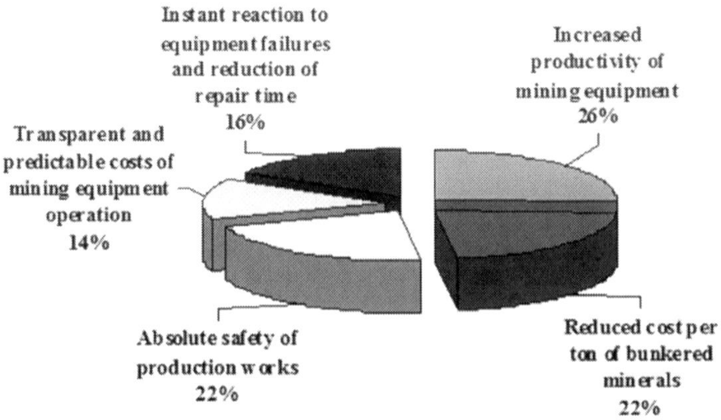

Fig. 1. The importance of the basic needs of the mining enterprises in Kuzbass

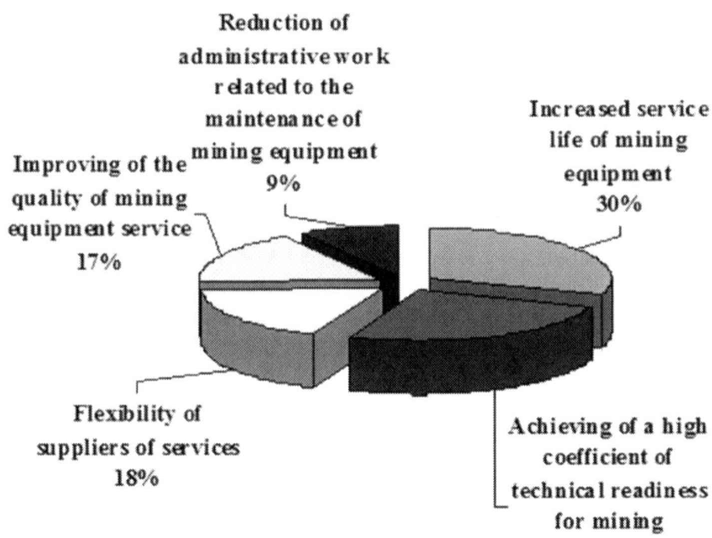

Fig. 2. The importance of the secondary needs of the mining enterprises in Kuzbass

As the basic needs in descending order of importance for enterprises were the following:

– an increase in productivity of mining equipment;
– reducing the cost per ton of mined mineral resource;
– absolute safety of prodution works;
– instant response to the equipment failures and reduction of repair time;
– transparent and predictable operating costs of the equipment.

The secondary needs of the top managers were the following:

- increased service life of mining equipment;
- Achieving of a high coefficient of technical readiness for mining;
- flexiblility of service providers;
- improving the quality of mining equipment service;
- reduction of administrative work related to the maintenance of mining equipment.

The survey showed that the provision of qualitative and in time services, taking into account the diversity of the customers and their needs, requires the formation of a broad set of service offers designed to achieve maximum performance and productivity of mining equipment. As being recommended the following service options to meet the individual reqiements of each regional customer including service package were offered:

- spare parts and consumables;
- assembly and machine rebuilding;
- service program of sharing components;
- maintenance of equipment in operation;
- repair in service center.

Based on the technical base, customers can choose the most effective, in terms of technical and economic indicators, service package for the equipment.

Using the survey results, by the methods of cluster analysis [16] dendogrammies were built. They define contingency of service packages and the optimal sequence of their demand by the mining enterprises.

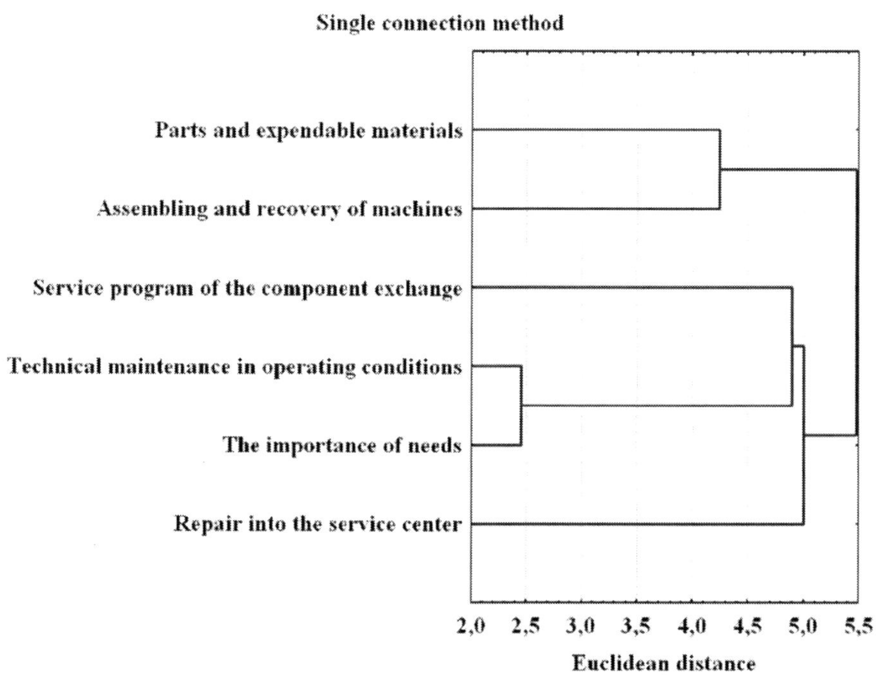

Fig. 3. The contingency service packages defined by the cluster analysis

Having processed the data on the selection of a service package the results showed the following:

1. Decisions about the maintenance of equipment in operation directly depends on the current needs of the mining enterprise.

2. Service package, providing regulated replacement of the parts, is most likely to be extanded to the package of assembly and reduilding of equipment in the service center.

3. In future, these packages can be supplemented by the service program of component exchange and, as a result, direct equipment repair in the service center.

4 Conclusion

Thus, the correlation analysis carried out on the basis of survey showed that today there is a real need for the customers of "Joy Global" in business cooperation and contracts based on the "Life-Cycle Management" program, using the innovative technology "smart service". This technology provides the possibility to control every equipment unit in operation from the surface. It helps to monitor continuously every movement, every change in the operating parameters, allowing to control not only the machinery, mechanisms, eqipment, but human activity, to have a clear idea of what is happening at the moment in each division of underground production.

Investing in the structure of the region, the company "Joy Global Inc" ensures a steady demand for its products. Kuzbass needs in effective, high-tech industries that can provide jobs for qualified personnel, improve their skills. Even today, an American company, based on the service center, organized the production of self-propelled electric cars JOY10SC32-36V intended for "the use in high-aggressive environments." The first copy of the product was introduced in early June in 2016 at the XXIII-th International Exhibition "Russian Coal and Mining". In the long term the service center "Joy Global Inc." will serve the territory not only in Russia, but also in neighboring countries, including Kazakhstan. Stable industrial region is always attractive for investors, including foreign ones. The processes are interrelated, and the understanding of this ensures the sustainable development of the region and benefits for the investors.

In general, to have sustainable competitive position in the regional markets in modern conditions for the manufacturer of mining equipment it is necessary not only to produce high-tech products, but also to ensure its entire life cycle, which results in the development of "smart service" as an innovative service system for mining companies in Kuzbass.

References

1. V. Fomychov, Min. of Min. Depos., **10**, 89 (2016)
2. L. Samorodova, Yu. Yakunina, O. Lyubimov, Adv. in Eng. Res., **92**, 50 (2016)
3. M. Chetveryk, O. Mishina, Min. of Min. Depos., **8**, 519 (2014)
4. M. Stupnik, V. Kalinichenko, V. Kolosov, I. Bah, V. Pozdniakov, Min. of Min. Depos., **8**, 199 (2014)
5. C. Hall, C. Cleveland, *EROI: definition, history and future implications* (ASPO US Conference, Denver, 2005)
6. R. Heinberg, J. Mander, *Searching for a miracle: Net energy limits & the fate of industrial society* (Post Carbon Institute, Santa Rosa, CA, 2009)
7. V. Krasnik, Min. of Min. Depos., **7**, 93 (2013).
8. M.A. Tyulenev, S.A. Zhironkin, E.A. Garina, Int. J. of Mining and Mineral Engineering, **7:4**, 363 (2016)
9. M. Tyulenev, E. Garina, A. Khoreshok, O. Litvin, Y. Litvin, E .Maliukhina, IOP Conf. Ser.: Earth Environ. Sci., **50**, 012035 (2017)
10. S. A. Zhironkin, K. A. Kolotov, A. E. Genin, F. V. Agafonov, S. A. Kovalevsky, IOP Conf. Ser.: Earth Environ. Sci., **50**, 012011 (2017)
11. About company Ferrit, URL: http://www.ferrit.cz/en/about-company/

12. P&H and Joy aligned under Joy Global name, URL: http://www.canadianminingjournal.com/features/p-h-and-joy-aligned-under-joy-global-name/
13. MINExpo 2016 Review, URL: https://www.worldcoal.com/mining-/28122016/minexpo-2016-review-earthmoving/
14. Joy Global Achieves 100% ROI in the First 9 Months, URL: https:// www.meridium.com/clients/joy-global-inc
15. Joy Global Delivers the Smart, Connected Mine, URL: https:// industrial-iot.com/2016/03/industrial-internet-of-things-mining/
16. M. Aldenderfer, R. Blashfild, *Cluster Analysis* (Second Printing, 1985)

Special Modes of AC Drives Operation in the Mining Industry

Lubov Payuk[1,*], *Nataliya* Voronina[1], *Olga* Galtseva[1], *Dini* Zhang[1], and *Alexandr* Rogachev[2]

[1]Tomsk Polytechnic University, 634050, Tomsk, Russia
[2]Francisk Skorina Gomel State University, 246019, Gomel, Belarus

Abstract. This paper presents research of special modes of AC drives operation. The possibility of obtaining of vibrational and vibrational-rotational modes of the AC drives on the basis of multipurpose polyharmonic mode of actuating motor operation is shown. Multipurpose electric motor with improved power characteristics (double-fed machine) has been proposed as the actuating motor of the electroactivator of the coal grizzley. The formation algorithm of polyharmonic operation mode of the double-fed machine is developed. This algorithm allows studying different types of special modes of actuating motor. Research results of dynamic parameters of a universal electric motor of the electroactivator of the coal grizzley are presented. The practical significance of the study of using of double-fed machine in the primary coal processing is shown.

1 Introduction

In modern mining industry a wide variety of high-precision and technological equipment is used. It is based on modern trends using generally accepted drives [1–3]. It is therefore not surprising that in practice the different from "classical" modes of AC drives (ACD) operation are applied. This is mainly vibrational, vibrational-rotational, polyharmonic and other electric drives operating in the special modes [4–6]. They are used in primary coal fragmenting, coal vibro-sorting, vibrational screen separators at the ore beneficiation, lifting mechanisms, etc. Requirements on the mechanical oscillations parameters in the mining industry are relatively high, because its cost depends on the quality of finished materials. The ranges of required frequencies, amplitudes and speed-ups are 3÷150 [Hz], 0.5÷90 [mm], 2÷104 [m/s] for the linear parameters and 15÷100 [Hz] 10÷2.2 [rad], 5÷500 [rad/s] for the angular parameters, respectively [7].

It's worth to mention that the most interesting mode is polyharmonic, because algorithm of its formation (it will be proposed below) takes into account the possibility of obtaining another types of the ACD "special" modes. In the mining industry considered ACD operation mode is used in vibrotransportation and coal vibro-sorting for the primary processing (example, in the grizzlies systems).

According to the Russian state standart GOST 23788-79 the grizzlies are divided on light, medium and heavy. Grizzley is a set of vibrating sieves with the decreasing meshs,

*Corresponding author: lubapa81@mail.ru

which are separated according to the meshs coarseness and the finished product (iron and copper ores, coal, etc.), accordingly. Grizzlies are performed in the following designs: gyratory, inertial, self-balancing and resonance. Motors powers for the grizzlies are from 7 to 20 [kW], which correspond to the line of average powers, oscillation amplitude is 3 [mm] ; the largest size of the finished product is from 100 to 150 [mm], the tilt angle of the sieve is up to 25^0 [7].

Electromagnetic shaker machine of the self-balancing grizzley (Figure 1) transmits up to 3000 oscillations per minute and amplitude of 0.3 [mm] to the sieving surfaces.

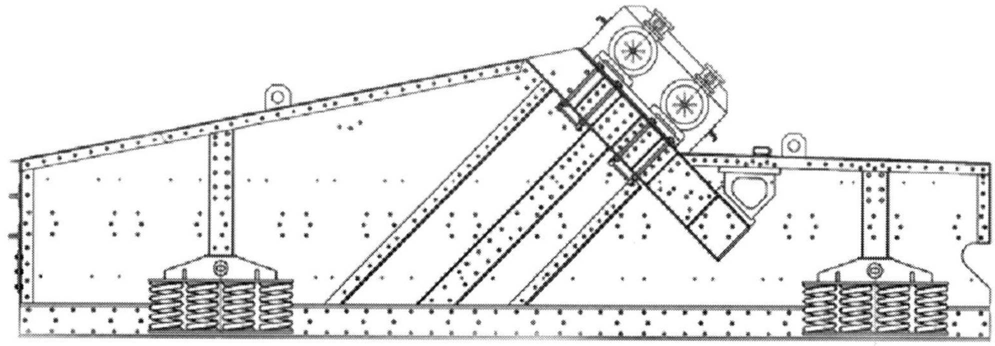

Fig. 1. Grizzley of self-balancing type

Therefore, the basic parameters of grizzley are

- the dimensions of the sieving surfaces,
- frequency and amplitude of oscillations,
- the tilt angle of the grizzley,
- the rotation direction of the shaker machine,
- the sieve motion trajectory.

2 Research method

In this paper a formation method of polyharmonic operation modes of the electromagnetic electroactivator of grizzley is proposed on the basis of the double-fed (DFM) (Figure 2) at different conditions of its implementation; analysis of the load parameters influence on the dynamic indicators of the operating motor was conducted. Well known, DFM using as an operating motor in comparison asynchronous motor [8] and the synchronous [9, 10] is an optimal solution; it is most versatile and has high energy characteristics [11–13]. Formation algorithm of ACD polyharmonic operation mode has been developed; it allows changing the oscillation frequency, phase shift and oscillation amplitude. This method allows to withstand requirements to self-balancing grizzley at the stage of necessary vibrations formation [14–18].

Polyharmonic operation mode (Figure 2) is realized at feeding stator and rotor windings by sinusoidal voltage (currents) with different frequencies, these frequencies are as different on the axes, i.e. $f_1 \neq f_2 \neq f_3 \neq f_4$.

The following statements are true: $\omega_1 = \omega_3 = \omega$, $\omega_2 = \omega`$, $\omega_4 = \omega``$ and $\alpha = \beta = \gamma = \varphi = 0$ at polyharmonic operation mode of DFM.

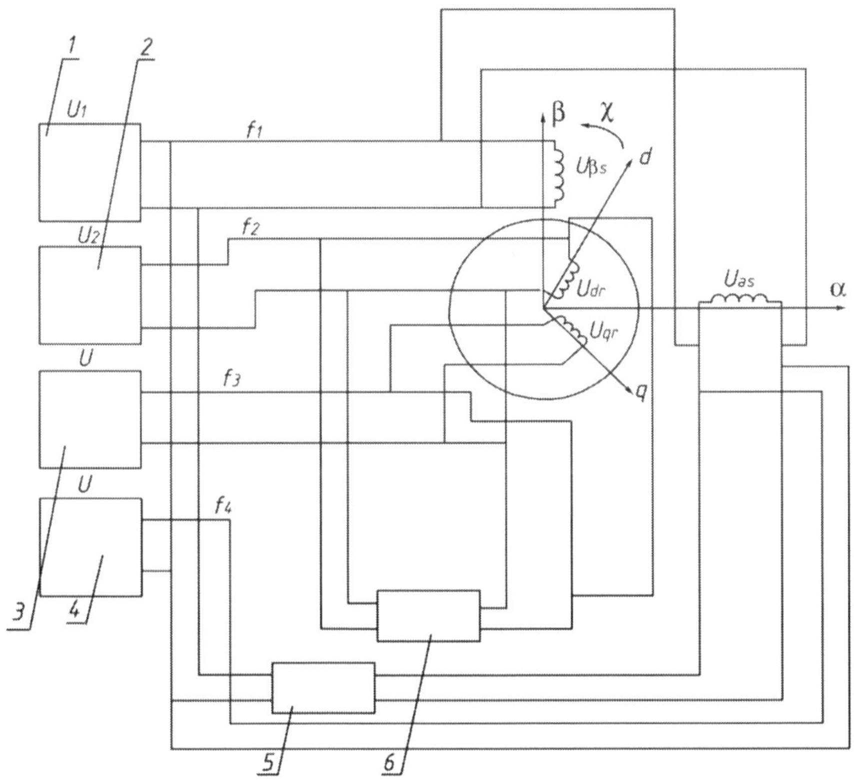

Fig. 2. Scheme of ACD inclusion with polyharmonic operation mode, where 1-4 are supply sources of the stator and rotor windings and 5, 6 are phase-shifting devices in the stator and rotor windings

Equations for the flux linkages of the stator and rotor windings in the transformed coordinate system are:

$$\begin{aligned} \psi_{\alpha s} &= \psi_{ms} \sin(\omega t + \alpha); \\ \psi_{\beta r} &= \psi_{ms} \sin(\omega` t + \beta); \\ \psi_{\alpha r} &= \psi_{mr} \sin(\omega t + \gamma)\cos\chi + \psi_{mr} \sin(\omega`` t + \varphi)\sin\chi; \\ \psi_{\beta r} &= -\psi_{mr} \sin(\omega t + \gamma)\sin\chi + \psi_{mr} \sin(\omega`` t + \varphi)\cos\chi, \end{aligned} \quad (1)$$

where ψ_{ms}, ψ_{mr} are amplitude values of phase flux linkages of stator and rotor windings, t is time, ω, $\omega`$, $\omega``$ are rotation angular frequencies of magnetic fields of stator and rotor along the axes $\alpha$ and $\beta$ and $\omega \neq \omega` \neq \omega``$; χ is motion law of the movable element of control motor.

Equations for squares of real and imaginary parts of generalized flux linkage are:

$$\begin{aligned} \psi_\alpha^2 &= \left(\psi_{ms}\sin(\omega t) + \left(\psi_{mr}\sin(\omega t)\cos\chi + \psi_{mr}\sin(\omega`` t)\sin\chi\right)\right)^2 = \\ &= \left(\psi_{ms}\left(\sin(\omega t) + \mu(\sin(\omega t)\cos\chi + \sin(\omega`` t)\sin\chi)\right)\right)^2, \end{aligned} \quad (2)$$

$$\begin{aligned} \psi_\beta^2 &= \left(\psi_{ms}\sin(\omega` t) - \left(\psi_{mr}\sin(\omega t)\sin\chi + \psi_{mr}\sin(\omega`` t)\cos\chi\right)\right)^2 = \\ &= \left(\psi_{ms}\left(\sin(\omega t) - \mu(\sin(\omega t)\sin\chi + \sin(\omega`` t)\cos\chi)\right)\right)^2, \end{aligned} \quad (3)$$

where $\mu = \psi_{\alpha r}/\psi_{\alpha s}$ is ratio between flux linkages of the stator and rotor along axis α.

Equation for the module of the resultant vector of flux linkage at polyharmonic operation mode is:

$$\begin{aligned}\psi &= \sqrt{(\psi_{ms}(\sin(\omega t)+\mu(\sin(\omega t)\cos\chi+\sin(\omega`t)\sin\chi)))^2+} \\ &\overline{+(\psi_{ms}(\sin(\omega`t)-\mu(\sin(\omega t)\sin\chi+\sin(\omega``t)\cos\chi)))^2} = \\ &= \psi_{ms}\sqrt{(\sin(\omega t)+\mu(\sin(\omega t)\cos\chi+\sin(\omega``t)\sin\chi))^2+} \\ &\overline{+(\sin(\omega`t)-\mu(\sin(\omega t)\sin\chi+\sin(\omega``t)\cos\chi))^2}.\end{aligned} \quad (4)$$

Equation for the motion law of dimensioned vector of flux linkage in the air gap at polyharmonic operation mode is:

$$\begin{aligned}\chi_0 &= \mathrm{arctg}\left(\frac{\psi_{ms}\sin(\omega`t)-\psi_{mr}\sin(\omega t)\sin\chi+\psi_{mr}\sin(\omega``t)\cos\chi}{\psi_{ms}\sin(\omega t)+\psi_{mr}\sin(\omega t)\cos\chi+\psi_{mr}\sin(\omega``t)\sin\chi}\right) = \\ &= \mathrm{arctg}\left(\frac{\sin(\omega`t)-\mu(\sin(\omega t)\sin\chi+\sin(\omega``t)\cos\chi)}{\sin(\omega t)+\mu(\sin(\omega t)\cos\chi+\sin(\omega``t)\sin\chi)}\right).\end{aligned} \quad (5)$$

Hodographs of resultant vector of flux linkage are below (Figure 3) at supplying stator and rotor windings from the power source and the voltage source.

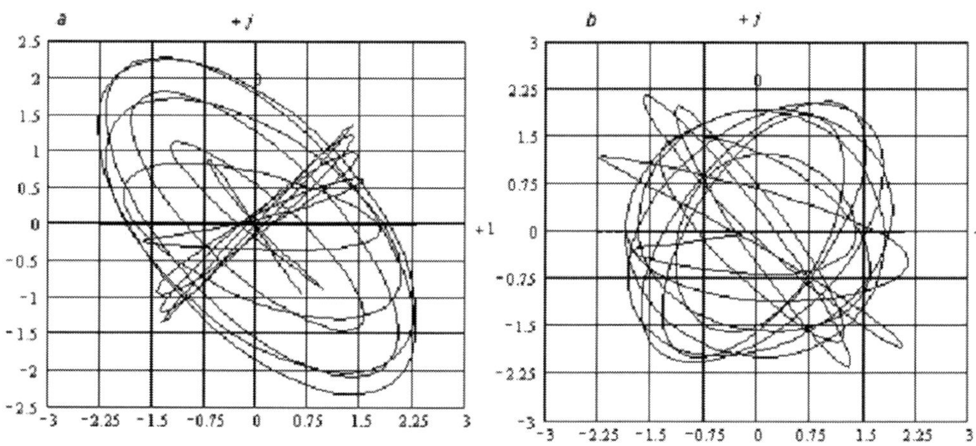

Fig. 3. Vectors hodographs ψ_0 at the feeding of the stator and rotor windings, where a – from the power source and b – from voltage source at the conditions ω_1=1.06, ω_2=1, ω_3=1.16, ω_4=1.1

3 Results and discussion

Equations for voltages on clamps of stator and rotor windings at realization of polyharmonic operation modes and concordant interaction of the electromagnetic fields of DFM in the transformed axes α, β, 0 are:

$$\begin{aligned}U_{\alpha s} &= \nu_1 U_{ms}\sin(\omega_1 t+\alpha); \\ U_{\beta s} &= \nu_2 U_{ms}\sin(\omega_2 t+\beta); \\ U_{\alpha r} &= \nu_3 U_{mr}\sin(\omega_3 t+\gamma)\cos\chi+\nu_3 U_{mr}\sin(\omega_4 t+\varphi)\sin\chi; \\ U_{\beta r} &= -\nu_4 U_{mr}\sin(\omega_3 t+\gamma)\sin\chi+\nu_4 U_{mr}\sin(\omega_4 t+\varphi)\cos\chi.\end{aligned} \quad (6)$$

where U_{ms}, U_{mr} are amplitude values of the phase voltages of stator and rotor windings; ν_1 – ν_4 are control signal coefficients; ω_1 – ω_4 are angular frequencies of rotation of stator and rotor magnetic fields by axes α and β; α, β, γ and φ are initial phase shifts of supply voltages (currents) of stator and rotor windings of operating motor.

The Kron's equations are the basis of the mathematical model of DFM [4]. They describe the energy conversion in an electric machine and are represented by a system of differential equations, consisting of equations of voltages, and the equations of motion of the movable element of motor at general assumptions for ACD, working in special operation modes [4, 5, 14–18]. DFM was implemented on the example of electric machine type 4AK160S8U3 with synchronous rotation frequency n_0=750 [rpm], power P_2=5.5 [kW], nominal stator current I_n=18.35 [A], η=80 [%], $cos\varphi_n$ =0.76 [4].

Algorithms of formation of the vibrational and vibrational-rotational mode of operation to operating motor of ACD, working in " not normal" mode, are presented in Table 1. Also Table 1 shows the ways of polyharmonic operation modes getting (1, 3 and 4 ways) of operating motor for grizzley electroactivator.

Table 1. Algorithm of special modes formation on the basis of polyharmonic

Operation mode of operating motor (DFM)	Control signal coefficients ν_i, [pu]	Angular frequency of rotation of windings magnetic fields ω_i [rad/s]	Phase shift φ, [rad]	Magnetic fields interaction	Application
Vibrational	$\nu_1=\nu_2=1$; $\nu_3=\nu_4=1.225$	$\omega_1=\omega_2=\omega$, $\omega_3=\omega_4=\omega'$	$\alpha=\beta=\gamma=\varphi=0$	Concordant	Ore processing form, soil vibrating
	$\nu_1=\nu_2=1$; $\nu_3=1.225$; $\nu_4=-1.225$	$\omega_1=\omega_2=\omega$, $\omega_3=\omega_4=\omega'$	$\alpha=\beta=\gamma=\varphi=0$	Opposing	
Vibrational-rotational	$\nu_1=\nu_2=1$; $\nu_3=\nu_4=1.225$	$\omega_1=\omega_2=\omega_3=\omega$, $\omega_4=\omega'$	$\alpha=\beta=\gamma=0$, $\varphi=\pi/2$	Concordant	Vibrotrans-portation and coal vibro-sorting
Polyharmonic	$\nu_1=\nu_2=1$; $\nu_3=\nu_4=1.225$ (1 way)	$\omega_1=\omega_3=\omega$, $\omega_2=\omega'$, $\omega_4=\omega''$	$\alpha=\beta=\gamma=\varphi=0$	Concordant	Primary coal fragmenting
	$\nu_1=\nu_2=1$; $\nu_3=\nu_4=1.225$ (2 way)	$\omega_1=\omega_2=\omega$, $\omega_3=\omega'$, $\omega_4=\omega''$	$\alpha=\beta=\gamma=\varphi=0$	Concordant	
	$\nu_1=\nu_2=1$; $\nu_3=\nu_4=1.225$ (3 way)	$\omega_1-\omega_2=\omega_3-\omega_4$ (frequency is not multiple of 5 Hz)	$\alpha=\beta=\gamma=\varphi=0$	Concordant	
	$\nu_1=\nu_2=1$; $\nu_3=\nu_4=1.225$ (4 way)	$\omega_2-\omega_1=\omega_4-\omega_3$ (frequency is multiple of 5 Hz)	$\alpha=\beta=\gamma=\varphi=0$	Concordant	

Figures 4–6 show dependences of transition process time τ_{tp}, amplitude values of angular speed of rotation ω_{sv}, amplitude values of motion law χ_m, on load inertia moment J_n and coefficient of damping moment of load R_g, where dependence 1 is at f_1=f_3=50 [Hz], f_2=52 [Hz], f_4=55 [Hz], dependence 2 is at f_1=50 [Hz], f_2=55 [Hz], f_3= 53 [Hz], f_4=63 [Hz] and dependence 3 is f_1=53 [Hz], f_2=50 [Hz], f_3=58 [Hz], f_4=55 [Hz].

Transient process time at increasing load parameters behaves differently: it increases at increasing load inertia moment J_n and decreases at increasing coefficient of damping moment of load R_g.

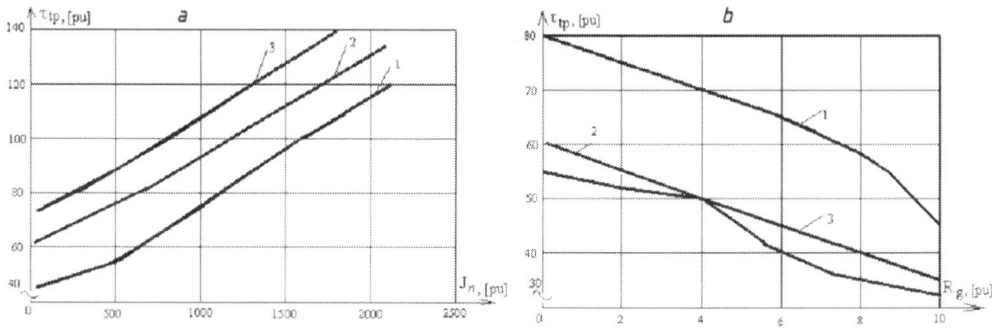

Fig. 4. Dependences of transition process time τ_{tp} on a – load inertia moment J_n and b – coefficient of damping moment of load R_g

Steady-state values of angular speed ω_{SV} are decreased at increasing load parameters, but their character is different: characteristics have a pronounced exponential character at increasing load inertia moment J_n in any method of polyharmonic mode forming. Researched dynamic parameters have more linear character with increasing coefficient of damping moment of load R_g, at that, 4 way of mode formation has the highest nonlinearity (Table 1).

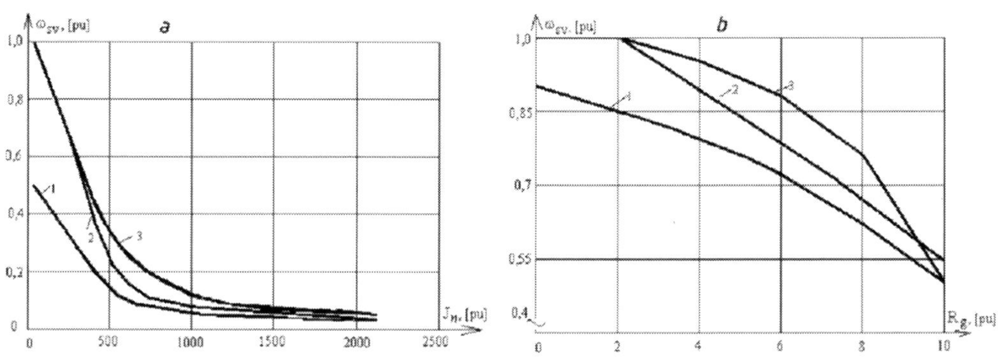

Fig. 5. Dependences of steady-state values of angular speed of rotation ω_{sv} on: a) load inertia moment J_n and b) coefficient of damping moment of load R_g

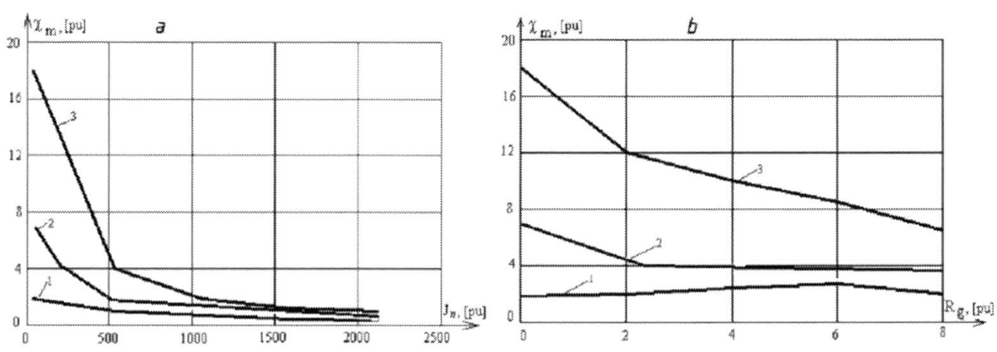

Fig. 6. Dependences of amplitude values of motion law χ_m on a – load inertia moment J_n and b – coefficient of damping moment of load R_g.

Character of amplitude values changes of the motion law of motor shaft χ_m in dependence on the load parameters has exponential nature. Moreover, the load inertia moment J_n has the greatest impact, because χ_m is reduced by 3.5 times; dependence 1 has a resonance peak at the point R_g =6.0 [pu] at increasing of coefficient of damping moment of load (Figure 6b).

4 Conclusions

1. Increase of load inertia moment J_n leads to increasing transition process time τ_{tp} and to decreasing other dynamic indicators with pronounced exponential character.

2. Increase of coefficient of damping moment of load leads to decreasing dynamic indicators of the operating motor: amplitude values for motion law have exponential nature, the rest of the parameters have linear.

3. Proposed algorithms of formation of the special operation modes of the operating motor simply to implement, so they have all prerequisites for use in equipment for mining industry.

References

1. D. Bevz, L. Payuk, N. Voronina, MATEC Web of Conferences **91**, 01037 (2017)
2. A.V. Glazachev, Yu.N. Dementyev, K.N. Negodin, A.D. Umursakova, EPJ Web of Conferences **110**, 01044 (2016)
3. A. Schwery, T. Kunz, BWK–Energie-Fachmagazin **64(7–8)**, 21 (2014)
4. L. Payuk, N. Voronina, S. Korepanov, O. Galtseva, N. Natalinova, MATEC Web of Conferences **79**, 01060 (2016)
5. A.V. Aristov, V.O. Nagorniy, A.M. Gavrilov, Proc. of SIBCON **15287708** (2015)
6. N.M. Natalinova, O.V. Galtseva, E.A. Moldovanova, Proc. of EECEA 7470765, 52 (2016)
7. A.V. Aristov, L.I. Aristova, I.A. Ekkert, Proc. of MEACS **15805203** (2015)
8. E.A. Kolomiyet, A.Y. Chernyshyov, Proc. of SIBCON **15287590** (2015)
9. V.G. Bukreev, S.V. Lyapushkin, M.M. Edgulov, MATEC Web of Conferences **37**, 01012 (2015)
10. I.G. Odnokopylov, V.V. Gneushev, I.A. Larioshina, J. Phys.: Conf. Ser. **671**, 012044 (2016)
11. L.A. Payuk, N.A. Voronina, O.V. Galtseva, J. Phys.: Conf. Ser. **671**, 012044 (2016)
12. S Williamson, A.C. Ferreira, A.K. Wallace, IEE Proceedings: Electric Power Applications **144(2)**, 5604026 (2002)
13. S Williamson, A.C. Ferreira, IEE Proceedings: Electric Power Applications **144(2)**, 5604027 (2002)
14. T. Lugand, A. Schwery, Proc. of ICEM **7732826**, 2191 (2016)
15. L.A. Vaisberg, T.M. Baldaeva, K.S. Ivanov, A.A. Otroshchenko, Mineral processing **1** (2016)
16. G. Wang, X. Tong, Mining Science and Technology (China) **21(3)**, 451 (2011)
17. T. Dyr, P. Wodzinski, Physicochem. Problems of Mineral Processing **35**, 147 (2002)
18. H. Dong, C. Liu, Y. Zhao, Int. J. of Mining Science and Technology **23(1)**, 95 (2013)

Impact of the number of blades of the geokhod cutting body on cutting forces

Vladimir Aksenov[1,3], *Vladimir* Sadovets[2,3*], *Elena* Rezanova[2], *Dmitry* Pashkov[2]

[1] Institute of Coal SB RAS, Kemerovo 650065, Russia, Kemerovo, Leningradsky prospekt, 10
[2] T. F. Gorbachev Kuzbass State Technical University, Kemerovo, 650000, 28 Vesennaya street
[3] Yurga Technological Institute TPU, Yurga 652050, 26 Leningradskaya street

Abstract. The article presents the power dependencies for determining the number of blades of the cutting body of geokhod for destruction of rocks of low hardness. The obtained dependences are based on the developed method of calculation of the blade cutting body of the geokhod. The presented dependences make it possible to choose the rational number of blades of the geokhod cutting body. The intensity of the development of underground space by the countries of the world community, including Russia, is increasing. In this connection, the tasks are to increase the rate of development and to reduce the cost of work.

1 Introduction

One of the areas of solving the problem mentioned above is application of geokhod technology, the basic element of which is the geokhod. At present, a prototype of geokhod is manufactured with the cutting body for destruction of rocks of medium hardness (Figure 1) [1-4].

Geokhods represent a new class of tunneling equipment. A basic feature of the geokhod is the screwing into the contour strata of the body which performs the function of the mobile support. At the same time, the force providing motion and balancing the reaction from the technological operations is formed not by friction forces at the media interface (soil-crawler mechanism) or repulsion from an artificially constructed support element (hydraulic jack-permanent support), but as a result of introducing the propulsion elements in the contour strata and the use of a normal reaction arising from the interaction of the elements of the geokhod with the rock mass [5-8].

When the geokhod is operating, the cutting body moves to the face in strict accordance with the pitch of the helical blade mounted on the outer surface of the head section (Figure 2). In this case, each point of the blade moves along the spiral trajectory.

* Corresponding author: vsadovec@yandex.ru

Fig. 1. Geokhod prototype with drum cutting body (CB)

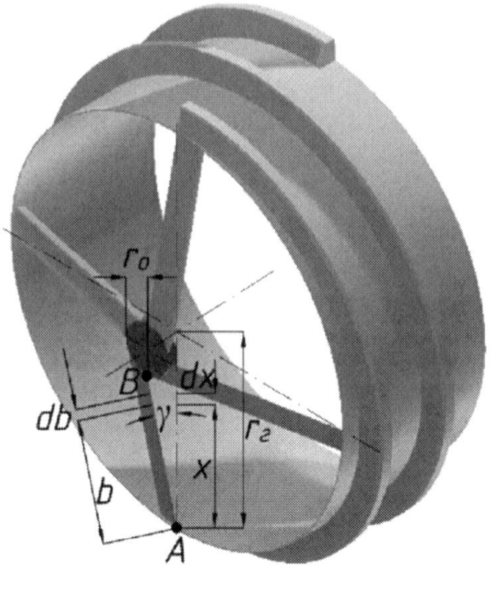

Fig. 2. Blade CB of geokhod

The blade points that are closer to the axis of rotation of the geokhod move to the face at a greater angle than the points on the periphery of the blade (Figure 3). Thus, with the helical motion of the blade cutting body to the face, the points of the blade trace out a helicoid (helical) surface. Consequently, the site of the face after the passage of the blade takes the form of a helical surface. The number of blades determines the number of mating areas with the helicoid surface on the face. Between the two mating helical surfaces, a ledge is constantly formed, which is destroyed by the CB blades as they are fed to the face [9-13].

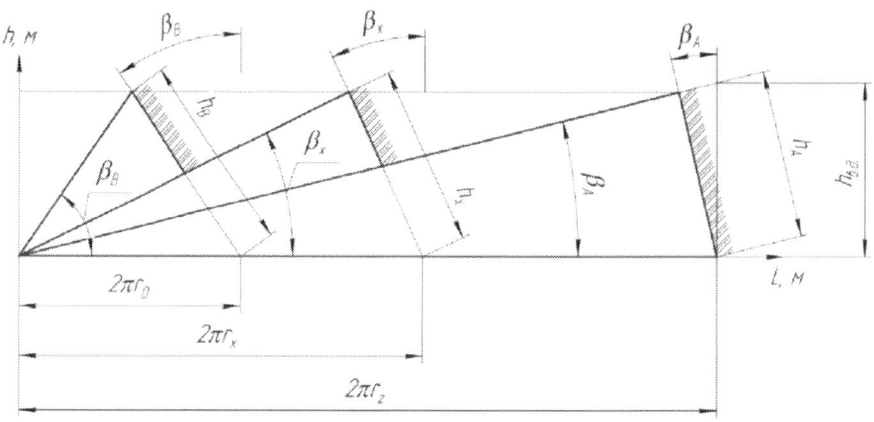

Fig. 3. Direction of movement of the points on CB blades depending on their location on the blade along the radius of geokhod

As follows from the determination of the power parameters of the geokhod blade cutting body presented in [2], cutting forces are affected by the radius of the geokhod, the lifting angle of the blade of the external propeller and the number of blades on the geokhod CB. The impact of the overall dimensions of the geokhod and the lifting angle of the helical blade of the external propeller on its power parameters are determined in work [14].

Based on the above, the task of the presented study is formulated as: determine the impact of cutting forces on the number of blades of the geokhod cutting body.

2 The research techniques

Based on the mining conditions of driving mine workings by the geokhod with the blade CB presented in Table 1 and the expressions used in works [15-17], the full projection of the soil cutting resistance force on the axis of rotation of the geokhod and the plane perpendicular to the axis of rotation were calculated as well as the total moment of resistance to cutting for a different number of sharp blades and the blades having the worn place or dulling.

Table 1. Mining and technical conditions for driving of a mine working by geokhod with a blade CB

Description	Designation	Unit of measure	Value
Radius of geokhod	r_g	m	1.6
Radius of generatrix [13]	r_o	m	0.4
Angle of inclination of CB blades to the axis of geokhod rotation	α	degree	90
Helical blade pitch	H_b	m	0.6
Coefficient taking into account the cutting angle impact [14]	φ		0.59
Specific cutting force in the front of slot at cutting angle 45^0	M_{front}	N/m^2	31000
Soil destruction force in side parts of the slot [14]	M_{side}	N/m^2	9000
Specific cutting force by one of the side edges of blade [14]	$M_{side.spec}$	N/m^2	740
Angle of inclination of the radial blade to the plane perpendicular to the axis of geokhod rotation	γ	degree	0
Parameters characterizing material resistance to elastic-plastic compression [14]	P_O P_u	N/m N/m	11300 61900
Cutting angle	δ	degree	25
Friction angle [15]	φ_{fr}	degree	31.4

Calculation of the power parameters of the blade CB of the geokhod with a radius of 1.6 with a pitch of the helical blade (h_b = 0,6 м) was made in Microsoft Excel.

3 Results

Based on the calculations, graphs were plotted for the change in the projection of the total force of the rock cutting resistance on the axis of rotation of the geokhod (P_O) as well as the projection on the plane perpendicular to the axis of rotation of the geokhod (R_{CB}) and the moment of soil cutting resistance (M_{CB}) for sharp knives with an increase in their number (Figure 4-a, b, c). On the axis of ordinates of the presented dependence, the values of the power parameters of the blade CB (H and Nm) are plotted, and along the axis of abscissae the change in the number of blades.

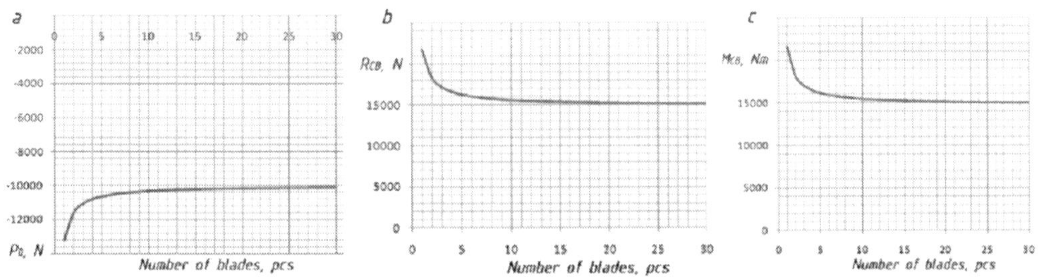

Fig. 4. a) dependence of the projection of the total resistance force of the rock to cutting on the axis of rotation of the geokhod (P_O) for sharp blades on their number on the geokhod CB; b) dependence of the projection on the plane perpendicular to the axis of geokhod rotation (R_{CB}) for sharp blades on their number on the geokhod CB; c) dependence of the moment of soil resistance to cutting (M_{CB}) for sharp blades on their number on the geokhod CB

It can be seen on the graphs that the projection of the total resistance of rock to cutting on the axis of rotation of the geokhod (P_O) increases with the increase of sharp blades, and the projection on the plane perpendicular to the axis of rotation of the geokhod (R_{CB}) and the moment of soil resistance to cutting (M_{CB}) for sharp knives decreases with the increase of the blades quantity.

Figure 5-a, b, c shows graphs of the change in the projection of the total resistance of rock to cutting on the axis of rotation of the geokhod (P_O) as well as the projection on the plane perpendicular to the axis of rotation of the geokhod (R_{CB}) and the moment of soil resistance to cutting (M_{CB}) for dull knives with the increase of their quantity.

On the axis of ordinates, the values of the power parameters of the blade CB (N and Nm) are plotted, and along the axis of abscissae - the change in the number of blades.

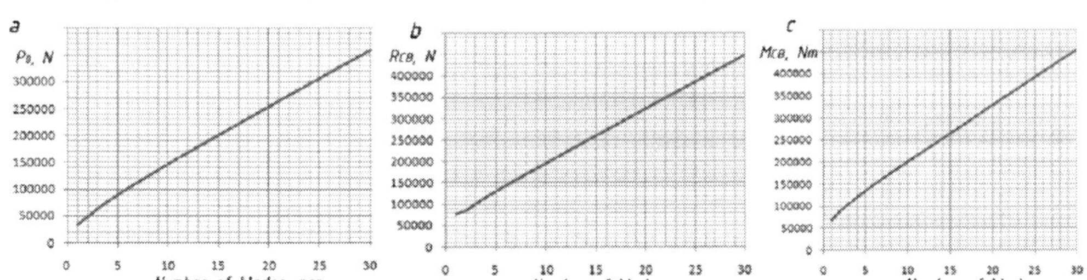

Fig. 5. a) dependence of the projection of the total resistance force of the rock to cutting on the axis of rotation of the geokhod (P_O) for dull blades on their number on the geokhod CB; b) dependence of the projection on the plane perpendicular to the axis of geokhod rotation (R_{CB}) for dull blades on their number on the geokhod CB; c) dependence of the moment of soil resistance to cutting (M_{CB}) for dull blades on their number on the geokhod CB

During operation, the cutting edge of the blade wears out and a blunt area is formed. This area affects the cutting forces negatively. This effect is observed on the graphs shown in Figure 5. Based on the graphs, it is seen that with increase in the number of blunted blades, the cutting forces increase.

4 Conclusion

On the basis of the graphs shown in Figures 4-5, it can be concluded that the rational number of blades geokhod CB is in the range from 2 to 12, as the cutting forces increase with increasing number of blunt blades, and with increasing number of sharp knives, the cutting forces decrease, But after 12 the forces do not practically decrease and further become constant. In this regard, it is not rational to use more than 12 blades due to the fact that the cutting forces are spent more on overcoming the resistance of the blunting area, rather than on cutting itself.

References

1. A.V. Walter, V.V. Aksenov, Applied Mechanics and Materials, **770**, 439 (2015)
2. V.Y. Sadovets, V.Y. Beglyakov, V.V. Aksenov, IOP Conference Series: Materials Science and Engineering, **91**, 012085 (2015)
3. A.A. Kazantsev, O.A. Mikenina, A.F. Revuzhenko, IOP Conference Series: Materials Science and Engineering, **91**, 012090 (2015)
4. A.B. Efremenkov, V.Y.Timofeev, *IEEE* (2012)
5. V.Y. Begljakov, V.Y. Timofeev, M.V. Dokhnenko, Applied Mechanics and Materials, **682**, 282 (2014)
6. R.V.Chernukhin, A.A.Dronov, M.Y.Blashchuk, IOP Conference Series: Materials Science and Engineering, **91**, 012086 (2015)
7. V.Y. Sadovets, V.Y. Beglyakov, A.B. Efremenkov, Applied Mechanics and Materials, **770**, 384 (2015)
8. V.A. Petrova, A.A. Bakanov, A.V. Walter, Applied Mechanics and Materials, **682**, 431 (2014)
9. S.A. Zhironkin, A.A. Khoreshok, M.A. Tyulenev, G.A. Barysheva, M.C. Hellmer, IOP Conf. Ser.: Mater. Sci. Eng., **142**, 012127 (2016).
10. D. Nesteruk, M. Momot, *IEEE* (2012)
11. Yu.V. Lesin, S.Yu. Luk'yanova, M.A. Tyulenev, J. Min. Sci., **46**, 78 (2010)
12. M. Tyulenev, S. Zhironkin, K. Kolotov, E. Garina, Pollution Research, **35**, 221 (2016)
13. A.A. Kazantsev, V.P. Kosykh, A.F. Revuzhenko, IOP Conference Series: Materials Science and Engineering, **91**, 012089 (2015)
14. A.A. Kazantsev, O.A. Mikenina, A.F. Revuzhenko, IOP Conference Series: Materials Science and Engineering, **91**, 012090 (2015)
15. M. Tyulenev, E. Garina, A. Khoreshok, O. Litvin, Y. Litvin, E. Maliukhina, IOP Conf. Ser.: Earth Environ. Sci., **50**, 012035 (2017)
16. M.Y. Blaschuk, A.A. Dronov, S.S. Ganovichev, IOP Conf. Ser. Mater. Sci. Eng., **142**, 012128 (2016)
17. A.V. Koperchuk, A.V. Murin, V.V. Filonov, IOP Conf. Ser. Mater. Sci. Eng., **127**, 012040 (2016)

Comparison of technological efficiency of gravitational devices for preparation of large diluted coal

Vladimir Udovitsky[1,*], *Vladimir* Kandinsky[1] and *Artyom* Begunov[1]

[1]T.F. Gorbachev Kuzbass State Technical University, Department of mineral resources processing, 650000 Kemerovo, 28 Vesennya st., Russian Federation

Abstract. The purpose of the work is to assess the possibility of preparation of poor washability, high-ash coals demonstrated by the example of the Apsat deposit. The following tasks have been accomplished. The analysis of modern equipment produced for washing of large classes of coal: heavy-medium separators and jigging machines. The most preferred were the heavy-media separator SKVP, jigging machines ROMJIG and BATAC. The analysis of mathematical models of processes and technologies of gravitational processing for determining the yield and ash content of the concentrate has been carried out. Algorithms and computer programs have been developed in which two mathematical models of the processes of heavy-media preparation and jigging machines have been used that allow determination of the separation densities at which the maximum yield of the total concentrate of the required ash content is reached. The computational experiment carried out showed that in processing of high ash diluted coals with poor washability, it is expedient to use SKVP separator and a jigging machine ROMJIG.

1 Introduction

Recently, more and more attention has been paid to the processes and technologies of coal preparation. Such technologies belong to the priority direction "Rational Nature Management" and allow a more complete extraction of the combustible mass from run-of-mine coal. Coal plays a significant role in the energy balance of Russia and especially Siberia [1]. In developing deposits, diluted rock mass is formed, which due to high ash content cannot be sold. It is only possible to increase its commodity value, especially for large classes, after its preparation.

2 Materials and Methods

Beneficiation of large classes of energy coals, as a rule, is carried out in one stage with the separation of two products: concentrate and rock. The density of separation depends on the

* Corresponding author: uvi@kuzstu.ru

fractional composition of coal, the quality requirements of the products obtained and is taken equal to 1.65 to 2.05 g/cm^3.

To implement this scheme, heavy-media wheel separators SKVP [2], drum separators WEMCO [3] and jigging machines BATAC and ROMJIG are used to process large coals [4-6].

Since 1975 Teska type heavy-media wheel separators SVKP have found the greatest distribution in Russia in preparation of coals, anthracites and shales with lump size of 13 (25) to 300 mm due to reliability and high efficiency of separation. The designs of separators and technical characteristics are presented in sufficient detail in the reference literature [2].

In 1985 the jigging machine ROMJIG by MBE was successfully tested in difficult conditions during continuous operation at the Emil Mayrisch mine in Germany. The machine was used to separate rocks from anthracite. The size of the processed material was 30 to 400 mm.

In the jigging machine ROMJIG, the unloading of rock and concentrate is carried out by a twin elevator wheel (Fig. 1).

Fig. 1. Jigging machine ROMJIG with a twin elevator wheel

The jigging process in this machine takes place in an water medium. Bed loosening required for preparation is implemented by lifting with the help of hydraulics and the subsequent sharp lowering of the swinging jigging sieve with the material on it. The draw of the rock is performed by a special unloader, which simultaneously restrains the flow of material. To regulate the thickness of the bed on the sieve during its rise in the hydraulic system, the pressure is measured, which is used as an input parameter in the unloader rotation speed control system.

When the concentrate and rock are unloaded by the elevator wheel, there is also a partial dewatering of the material. The fines falling through the holes in the wheel get into the conical part of the machine and are transported through the sluice device for sorting by the size.

At the quarry in Binet (India), the following technology is used to wash coal with an ash content of 41 ÷ 45%. ROM coal is crushed to a particle size of less than 400 mm and is classified on a screen with a sieve hole size of 30 mm. Dry screening (less than 30 mm) is combined further with the product released from the machine through the sluice device and dehydrated on the screen. A large class of 30-400 mm is processed on three jigging machines ROMJIG [5, 6].

The concentrate with a density of less than 1.9 g / cm3 with an ash content of 30-31% is combined with ROM coal to produce a product with 34% ash content and shipped to a power plant.

For underground operations, a ROMJIG machine with a small constructional height has been designed. In this version, the machine is equipped with a scraper conveyor, unloading the rock, on the canvas of which the preliminary water removal takes place. The lighter material above the rock falls into the tray and is unloaded from it by a large-diameter screw conveyor. The material, which has passed under the jigging sieve, is removed by means of a screw device and lifted by a bucket elevator with simultaneous dehydration.

The BATAC jigging machine produced by MBE for preparation of coals with a particle size of 13-150 mm (Fig. 2) is successfully used at the Ziminka preparation plant in Prokopyevsk, Russia.

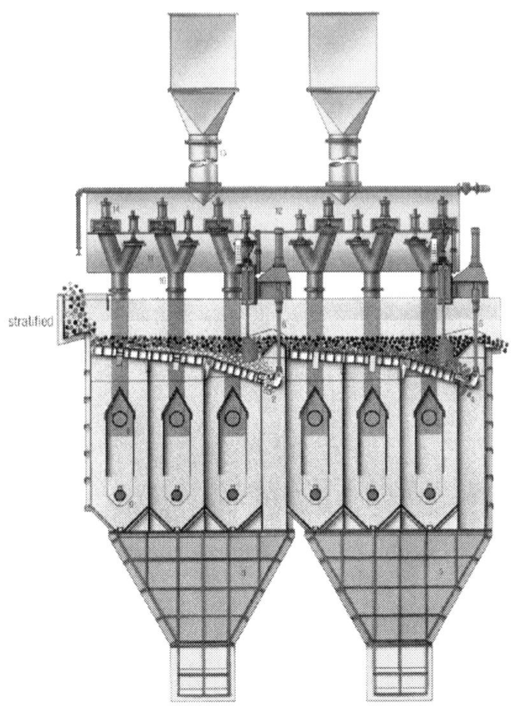

Fig. 2. The BATAC jigging machine

In its design the machine is based on air pulsation. Vibrational movement of water is set by means of air supplied to the air chambers located under the jigging sieve. Compressed air is periodically supplied to the chambers using electronically controlled pulsators, which expands to raise the water level, creating an upward flow relative to the sieve. The material on the sieve (bed) rises with the water and separates, depending on the density and size.

Small and light particles are in the upper layers of the bed, heavy and large - in the lower layers.

Further, the air supply by the pulsators is stopped and a time of pause occurs, during which the separation of the bed ends.

Then, the rotary dampers of the pulsators (depending on the design of the machine) release air from the air chambers into the atmosphere, creating a descending flow relative to the sieve. There is a condensation and movement of the material by transport water along the inclined sieve to the discharge pouch. Small heavy particles are discharged through the holes of the jigging sieve, large ones through the pocket. The lighter particles, together with the water, move to the next jigging compartment, in which the stratification process is similar.

In addition, sub-sieve water is supplied to the bottom of each jigging compartment, which enhances the ascending and weakens the downward flow of water.

The sub-grid arrangement of the air chambers allows achieving equal rates of ascending flow over the entire area of the jigging sieve, which contributes to an increase in the efficiency of separation of mineral particles in the machine.

2.1 Initial data

As raw material, coal of 25-100 mm in size was sampled at the Apsat deposit [7], the fractional composition of which was determined by the SibNII-Coal Preparation Institute (Table 1).

Table 1. Fractional composition of coal with lump size of 25-100 mm

Density of fractions, g/cm^3	Yield, %	Ash content, %	Floaters		Sinkers	
			Yield, %	Ash content, %	Yield, %	Ash content, %
< 1.3	8.3	7.4	8.3	7.4	100.0	27.4
1.3–1.4	45.4	13.1	53.7	12.2	91.7	29.2
1.4–1.5	20.8	20.9	74.5	14.6	46.3	44.9
1.5–1.6	6.3	27.3	80.8	15.6	25.5	64.5
1.6–1.8	3.5	38.8	84.3	16.6	19.2	76.7
>1.8	15.7	85.1	100.0	27.4	15.7	85.1
Total	100.0	27.4	–	–	–	–
Washability very difficult, $T = 36.3$ %						

2.2 Mathematical model

The use of a probabilistic method for the distribution of fractions of various densities into products of preparation for calculating the yields (1) and ash contents (2) of the concentrate in gravitational apparatuses first proposed by the authors of the publications [8-10] is mentioned in the work [11]:

$$\gamma = \sum_{i=1}^{n} \gamma_i \cdot \Phi(x_i)\,, \tag{1}$$

$$A^d = \frac{\sum_{i=1}^{n} \gamma_i \cdot A_i^d \cdot \Phi(x_i)}{\sum_{i=1}^{n} \gamma_i \cdot \Phi(x_i)}\,, \tag{2}$$

where n – number of fractions; $\Phi(x_i)$ – the value of the probability integral; γ_i, A^d_i – yield and ash content of i-th fraction; x_i – deviation of average density of fractions ρ_i from the intended separation density ρ.

The value x_i for the jigging concentrate is calculated by the formula (3), for the heavy-media separator SKVP - by the formula (4):

$$x_i = \frac{0.675}{\ln\left(I^2 + \sqrt{I^2+1}\right)} \cdot \ln\frac{\rho - 1}{\rho_i - 1}; \quad (3)$$

$$x_i = 0.675\frac{\rho - \rho_i}{E_p}. \quad (4)$$

3 Results and Discussion

At computer calculation [12, 13] of processes of gravitational coal washing, in technological schemes dilution of washing products in heavy-media separators SKVP is determined by the formula (5):

$$E_{pm} = 0.01\rho + 0.02. \quad (5)$$

The values of the error of separation (imperfection) I in the formula (3) for the jigging machines ROMJIG® and BATAC® are taken according to the data of MBE, the machine manufacturer [6], and are equal, respectively: 0.11 and 0.09.

The mathematical model and computer implementation of the probabilistic method of fractional distribution in calculating the technological scheme of coal enrichment in three machine classes in order to achieve the maximum yield of the total concentrate of the required ash content is shown in the works [14, 15].

The publication [16] shows the realization of the well-known theorem of Reinhardt on the achievement of a maximum yield of the total concentrate of a given ash content in the enrichment of several coals.

In mathematical models of processes of gravitational enrichment the formulas published in [14-16] were used. Table 2 shows the results of computer simulation of the processes of gravitational enrichment of coals of the Apsat deposit, the fractional composition of which is presented in Table. 1. The values of the yields and ash contents of the concentrate for heavy-media enrichment and jigging machines were calculated at separation densities varying from 1.40 to 1.80 g/cm^3.

Table 2. Values of concentrate yield and ash content calculated for a separator and a jig

Separation density, g/cm^3	SKVP		BATAC		ROMJIG	
	Yield, %	Ash content, %	Yield, %	Ash content, %	Yield, %	Ash content, %
1.40	49.74	12.67	50.48	12.78	–	–
1.45	63.10	13.67	63.16	13.73	–	–
1.50	71.97	14.53	71.37	14.53	70,85	14.52
1.60	79.75	15.50	79.29	15.50	79.02	15.49
1.70	82.53	16.12	82.32	16.09	82.20	16.08
1.80	84.18	16.56	83.76	16.46	83.68	16.46

DOI: 10.1051/e3sconf/20171503016

4 Conclusions

Computational experiments established:

1. The calculated value of the concentrate yield of the SKVP separator is higher by 0.2-0.3% of the concentrate yield of the jigging machines at a separation density of 1.7 g/ cm^3.

2. For the mechanized separation of the rock in order to reduce the ash content of steam coals of large classes, it is advisable to use SKVP or ROMJIG, since BATAC can wash coal with a particle size of less than 150 mm.

3. It should be noted that the design features of the jigging machine ROMJIG allow to achieve: a minimum separation density of 1.5 g/cm^3; completeness of separation of the rock to 90% at the separation density of 1.85 g/cm^3; specific consumption of electricity – 0.2 kWh/t; specific water consumption – 0.03 m^3/t.

4. Particularly it is necessary to emphasize that ROMJIG has elevator unloading of the rock, which is very important when enriching the diluted rock mass when the yield of the rock is very large.

References

1. Information on http://www.gks.ru/
2. J.W. Leonard, B.C. Hardinge, *Coal Preparation* (SME, Littleton, CO, 1991)
3. G.J. Sanders, D. Ziaja, J. Kottmann, Coal Prep. **22** (2002)
4. M. Wesp, H. Thelen, H. Müller, G. Wüster, Aufbereit, Tech. **30** (1989)
5. J. Winckler, G. Mustafi, in: *Mining Challenges of the 21st Century* (APH Publishing Corp., New Delhi, 2000)
6. K. Tromp, *Neue Wege fur die Beuerteilung der Aufbereitung von Steinkohlen* (Gluckauf, 1937)
7. A. Terra, *Revue de lindustrue minerale* (1939)
8. A. Terra. *Sens de la courbe de partage anamorphosee et de lecart probable dans le controle des lavoirs* (R. I. M., 1955)
9. B.S. Gottfried, Int. J. Miner. Process. **5** (1978)
10. L.H.E. Weyher, Mining Congress Journal, **7** (1971)
11. T. Gouri Charan, *Coal beneficiation* (CRC Press, Boca Raton, FL, 2015)
12. V. Udovitsky, *XX International Mineral Processing Congress* (Aachen, Germany, 1997)
13. V. Udovitsky, *Regional APCOM'97 Symposium: The Moscow State Mining University* (Publishing Center Moscow, Russia, 1997)
14. I. Korolev, V. Udovitsky, *XVIII International Coal Preparation Congress* (Saint-Petersburg, Russia, 2016)

The intelligent mechatronic system for open pit mining to increase the operation life of equipment

Irina Semykina[1,*], *Valeriy* Zavyalov[1,2] , and *Veniamin* Kashirskikh[1]

[1]T.F. Gorbachev Kuzbass State Technical University, Department of Electric Drive and Automation, 650000, 28 Vesennyaya Street, Kemerovo, Russia
[2]National Research Tomsk Polytechnic University, Department of Electric Drive and Automation, 634050, 30 Lenin Avenue, Tomsk, Russia

Abstract. This article proposes the intelligent mechatronic system that is based on the controlled induction motor drives. This system is effective to increase the technical operation life of mining machines for open pit mining. It allows control of the dynamic state of the mining machine mechanical components, thus the dynamic loads in the transmission are reduced and the accumulated intensity of fatigue wear becomes smaller. In the beginning, the article describes the concept of such system. The central parts demonstrate the control algorithms of the spring forces in the transmission and observe the estimation of immeasurable mechanical values needed for the mechatronic system operation. In the end of the article, the authors confirm the effectiveness of the proposed solutions experimentally using the data from the acting electric rope shovels.

1 Introduction

In the open pit mining technology, the mining machines take a central place, and among them the mining excavators are widely distributed. [1]. Despite of the worldwide annual unit production of electric excavators is reducing year by year and the worldwide annual unit production of hydraulic excavators is increasing [2], in Russia not the hydraulic but the electric excavators are most common [3] because of the features of the climate and the production organization. Obviously, the reliability of electric excavators directly affects the mining efficiency, and therefore reliability is the focus of the equipment producers for excavators [4].

By design, any electric excavator is a robotic system with electric drives that perform the operation of movement, rotation and digging [5]. The key feature of this robotic system is the load on the effector that can stochastically vary over a large range until the full locking of the effector when it contacts with solids during the operation of digging. In addition, there is the elastic compliance in the transmission. Together with the variable load the elastic compliance contribute to continuous transient processes in the mechanical

* Corresponding author: siyu.eav@kuzstu.ru

elements in transmission [6-14]. As the consequence and one of the main causes of the excavator transmission breakdowns there is accumulation of fatigue damage in them.

The service life of the mining machines for open pit mining can be increased without any changes in the design of mechanisms by using intelligent tools [15]. Such tools may be the introduction of additional measurements or using the controlled induction motor drives, which, for example, control the spring forces in the transmission. The authors believe the latter is most effective, because it is a comprehensive tool. In this case, electric drive can provide a given performance of the excavator by maintaining the mechanical load on the transmission elements in the permitted range, and at the same time it can diminish the accumulation intensity of fatigue damage by control of the dynamic loads, so controlled electric drives can really reduce the likelihood of breakdowns. Noteworthy is that the repair of mechanical transmission breakdowns is the most time consuming [11].

The development of such systems is especially important for Russia, where the average equipment wear rate of excavator fleet is very high. For example, in the largest open-pit coalmines in Kuzbass the wear exceeds 90 % of service life and exceeds 65 % on operating time.

2 Structure development

Among the excavator operations the most difficult is the operation of digging, in which two main drives move the bucket through the rock. In the simplest case, the operator controls the movement manually and there is no coordinate control of main drives. Thus, the characteristics of the motors and the properties of the destructible rock both determine the torque and the speed in the transmission elements. There is a strong possibility of motor overload and there are always the significant dynamic loads in transmission.

To avoid this effect, some researchers suggest the more complex control systems where the two main drives of the excavator are working interconnected. For example, the author of [10] proposes the interconnected control system of the drag drive and the hoist drive of the dragline to minimize the energy consumption. Unfortunately, [10] does not regard the improving of the excavator dynamic as a problem.

Another direction of the control systems improving is the individual optimization of one of the main drives in terms of operation life criteria without their collaboration. The most common optimization methods are the identification of the relationship between the parameters of the drives, which provides the best conditions of oscillation damping in transmission. For the transmissions with the elastic compliance there are also the optimization methods based on the subordinate control system with the introduction of additional feedbacks on the spring forces and the speed of the effector. One more optimization method is using of polynomial equations to reduce the rate of response of the control loops. All these methods are united by passive reduction of the dynamic loads.

The most promising control systems of the excavator electric drives are the system with the active limitation of the elastic oscillations where depending on the transmission state the motor forms the electromagnetic torque to suppression of the elastic oscillations. For example, the authors of [12] develop the principles of the control systems for the active limitation of the spring forces oscillations.

So, generalizing the best practices of the control systems of the excavators, the criteria of the intelligent mechatronic systems have been formulated. The system should:

1. Use the perfect type of electric drive.

2. Control the spring forces. That means the system will be able to limit the load on the transmission elements caused by whether the resistance forces of rock destruction or the acceleration and the deceleration and it will be able to reduce the operator influence on the formation of the dynamic loads in the transmission.

3. Provide the required performance. This means the system must be able to control the bucket speed.

The implementation of these criteria is the basis of the concept of intelligent mechatronic systems, which is shown at Fig. 1 for rope shovel. The diagram denotes: superscript * means the reference value; ω_h – the speed of the dipper handle about the saddle block; v_c – the crowd speed; subscripts $_c$ and $_h$ are related to the crowd drive or the hoist drive values, respectively; F – the spring force; T – the motor torque. In general, the proposed system is quite complex. So, authors have held the decomposition on its mathematical description and for each item the local control tasks are solved.

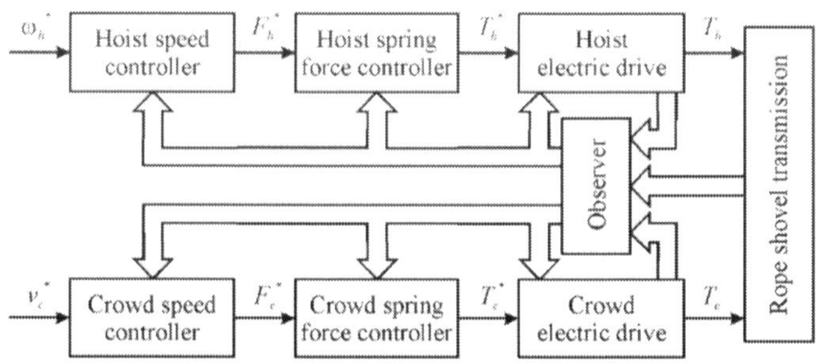

Fig. 1. Block diagram of the proposed intelligent mechatronic systems.

3 Mathematical model

From the reliability point of view, the most important for the considered robotic system is the mathematical description of the excavator transmission participated in the operation of digging, because it is a nonlinear system with internal interconnection and variable parameters. To obtain this mathematical description the authors propose to use the model of an interconnected multimass system with elastic coupling. Thus, the mathematical description is based on the transmission kinematics by using the Lagrange equations.

Let us look at the mathematical description of the rope shovel as the multimass system, obtained for the transmission kinematics that is shown in Fig. 2.

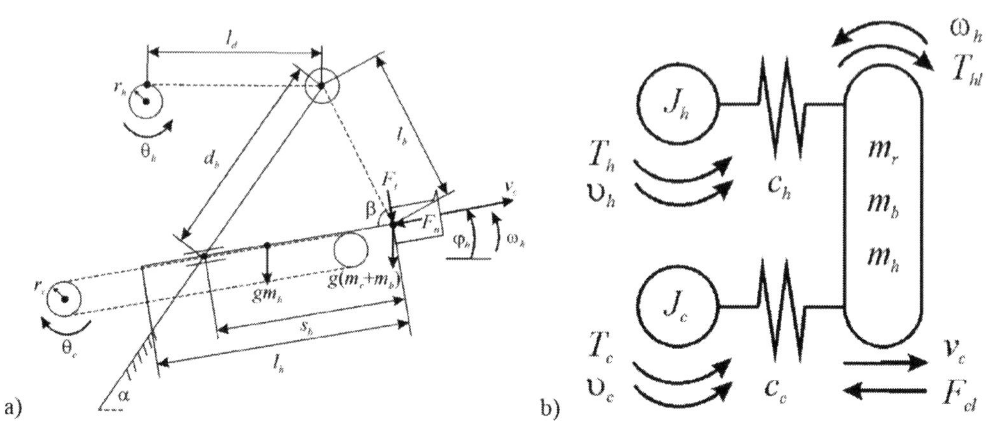

Fig. 2. Rope shovel schemes: a) the transmission kinematics, b) the model of the multimass system.

The schemes denote: α – the boom angle to the horizon; β – the hoist cable angle to the dipper handle; l_h – the dipper handle length; d_b – the boom length from the saddle block to the head block; l_d – the hoist cable length from the hoist drum to the boom point sheaves; l_b – the hoist cable length from the boom point sheaves to the bucket; s_h – the dipper handle stroke; φ_h – the dipper handle angle to the horizon; F_t, F_n – the tangential and the normal resistance forces of rock destruction; m_r, m_b, m_h – the mass of the rock, the bucket and the dipper handle; g – gravitational acceleration; F_{lc} – the crowd drive load force; T_{lh} – the hoist drive load torque; subscripts $_c$ and $_h$ relate to the value of the crowd drive or the hoist drive respectively; r – the drum radius; θ – the drum angle position; υ – the drum speed; J – the total specific inertia including the motor, the torque-limiting clutch, the brake, the gear box and the drum; c – the rope specific rigidity rate. A feature of the model is that m_r increases during digging.

The mathematical description is the following system of differential equations:

$$\begin{cases} \frac{d\theta_c}{dt} = \upsilon_c; \frac{d\theta_h}{dt} = \upsilon_h; \frac{ds_h}{dt} = v_c; \frac{d\varphi_h}{dt} = \omega_h; \\ \frac{d\upsilon_c}{dt} = \frac{T_c - r_c F_c}{J_c}; \frac{d\upsilon_h}{dt} = \frac{T_h - r_h F_h}{J_h}; \\ \frac{dv_c}{dt} = \frac{\begin{pmatrix} F_c + m_h\omega_h^2(s_h - l_h/2) + (m_b + m_r)\omega_h^2 s_h - \\ -F_{cl} - F_h \cos\beta - v_c(dm_r/dt) \end{pmatrix}}{m_h + m_b + m_r} = \frac{F_c - F_{\Sigma cl}}{m_\Sigma}; \\ \frac{d\omega_h}{dt} = \frac{\begin{pmatrix} F_h s_h \sin\beta - T_{hl} - 2m_h\omega_h v_c(s_h - l_h/2) - \\ -2(m_b + m_r)\omega_h v_c s_h - \omega_h s_h^2(dm_r/dt) \end{pmatrix}}{m_h(s_h - l_h/2)^2 + m_h l_h^2/12 + (m_b + m_r)s_h^2} = \frac{F_h s_h \sin\beta - T_{\Sigma hl}}{J_\Sigma}, \end{cases} \tag{1}$$

where the spring forces are described by:

$$\begin{aligned} &F_c = c_c(r_c\theta_c - (s_h - s_{hi})); \\ &\begin{cases} F_h = c_h(r_h\theta_h - (l_{bi} - l_b)), \ \mathit{if} \ l_b > l_{bi} - r_h\theta_h; \\ F_h = 0, \ \mathit{if} \ l_b \le l_{bi} - r_h\theta_h, \end{cases} \end{aligned} \tag{2}$$

where l_{bi} – the initial hoist cable length; s_{hi} – the initial dipper handle stroke, and the variable parameters are computed as:

$$\begin{aligned} l_b &= \sqrt{s_h^2 + d_b^2 - 2s_h d_b \cos(\alpha - \varphi_h)}; \\ c_h &= \frac{ES}{l_b + l_d}, \end{aligned} \tag{3}$$

where E – the tensile modulus; S – the total cross section of the hoist cable.

The external loads in (1) are also variable parameters, where the forces F_{cl} and T_{hl} are random and gravity force is deterministic:

$$\begin{aligned} F_{cl} &= F_n + g(m_h + m_b + m_r)\sin\varphi_h; \\ M_{hl} &= F_t s_h + g m_h \cos\varphi_h (s_h - l_h/2) + g(m_b + m_r)\cos\varphi_h s_h. \end{aligned} \tag{4}$$

Despite the description, (1)-(4) applies only to electric rope shovels, the same ideology of obtaining the mathematical description is also valid for the draglines.

4 Control algorithms

In the synthesis of the control algorithms, the authors believe that the modern electric drive like the induction motor drive with the direct torque control is able to provide such performance of the motor as to equate the electromechanical part of the mechatronic system to the inertia-free source of electromagnetic torque. Then, the internal is a control loop of the spring forces, and the outer is a control loop of the bucket speed. Each control loop is equipped with a controller. As shown in Fig. 1, each spring force controller should calculate the required motor torque. The analytical design method of aggregated regulators allows us to synthesize the desired control algorithms for spring force controllers. This method allows providing the aperiodic transients with a given rate of response for interconnected nonlinear systems with a complex structure.

Let us consider in more detail the synthesis of the control algorithm of the crowd spring force. The following invariant manifold provides the implementation of the criteria of the spring force response:

$$\Psi_{1c} = F_c - F_c^*(t) = 0, \tag{5}$$

and the following functional equation ensures aperiodic transients:

$$T_{1c}\dot{\Psi}_{1c} + \Psi_{1c} = 0, \tag{6}$$

where T_{1c} – the time constant, which determines the rate of approach to the invariant manifold Ψ_{1c}.

Then, solving (6) relative Ψ_{1c} and using (1), the following equation can be obtained:

$$T_{1c}c_c(r_c\upsilon_c - v_c) - T_{1c}\frac{dF_c^*}{dt} + F_c - F_c^* = 0. \tag{7}$$

To link the equation (7) and the reference crowd motor torque T_c^*, it is necessary to use (7) to form a new invariant manifold in such a variable as directly dependent on the T_c according to (1):

$$\Psi_{2c} = \upsilon_c - \frac{1}{r_c}\left(\frac{1}{T_{1c}c_c}\left(F_c^* - F_c + T_{1c}\frac{dF_c^*}{dt}\right) + v_c\right) = 0, \tag{8}$$

and it is necessary to solve the new functional equation in analogy to (6):

$$T_{2c}\dot{\Psi}_{2c} + \Psi_{2c} = 0, \tag{9}$$

where T_{2c} – the time constant, which determines the rate of approach to the invariant manifold Ψ_{2c}.

By completing these calculations, the required control algorithm can be synthesized:

$$\begin{aligned} T_c^* = {} & \frac{J_c}{r_c}\frac{dv_c}{dt} + \left(r_c - \frac{J_c}{T_{1c}T_{2c}c_c r_c}\right)F_c - \frac{J_c}{r_c}\left(\frac{1}{T_{1c}} + \frac{1}{T_{2c}}\right)(r_c\upsilon_c - v_c) + \\ & + \frac{J_c}{c_c r_c}\left(\frac{d^2F_c^*}{dt^2} + \left(\frac{1}{T_{1c}} + \frac{1}{T_{2c}}\right)\frac{dF_c^*}{dt} + \frac{F_c^*}{T_{1c}T_{2c}}\right). \end{aligned} \tag{10}$$

Similarly, the control algorithm of the hoist spring force can be synthesized:

$$T_h^* = -\frac{J_h}{r_h}\left(\frac{d^2 l_b}{dt^2} + \left(\frac{1}{T_{1h}} + \frac{1}{T_{2h}}\right)\frac{dl_b}{dt}\right) + \left(r_h - \frac{J_h}{T_{1h}T_{2h}c_h r_h}\right)F_h - J_h \upsilon_h\left(\frac{1}{T_{1c}} + \frac{1}{T_{2c}}\right) + \\ + \frac{J_h}{c_h r_h}\left(\frac{d^2 F_h^*}{dt^2} + \left(\frac{1}{T_{1h}} + \frac{1}{T_{2h}}\right)\frac{dF_h^*}{dt} + \frac{1}{T_{1h}T_{2h}}F_h^*\right). \qquad (11)$$

In accordance with Fig. 1, the spring force controllers provide the operation of the external speed control loop, which in turn determines the performance of the excavator. To provide the required performance of the rope shovels, it is necessary to control the speed of the dipper handle about the saddle block and the crowd speed by the hoist drive and the crowd drive with a reasonable accuracy. At the same time, the control system should generally provide the reduction of the dynamic loads. To solve both of the tasks the fuzzy logic controller or the adaptive controller can be applied.

These speed controllers allow the deviation of the instantaneous values of the bucket speed and the handle speed on the reference values within the limits defined by the random component of the load (it is usually no more than ±5%). However, the controllers and support the average rope shovel performance and minimize the dynamic component of the spring forces transients in the transmission, primarily in the ropes. These properties are fully consistent with the main idea of the proposed intelligent mechatronic system.

5 Observer

The above-mentioned controllers according to (1) require information about mechanical values, but some of them are immeasurable because of rope shovel construction. To estimate the immeasurable values the authors propose the observer.

The estimation basis on the drum angle positions θ_c and θ_h that are measurable and the motor electromagnetic torques T_c and T_h that are given by frequency converters [17] or obtaining from the special diagnostic computer system [35]. The estimated variables are the spring force F_c и F_h, the crowd speed v_c, the speed of the dipper handle about the saddle block ω_h and the hoist cable length from the boom point sheaves to the bucket l_b. Besides, the observer estimates the total crowd drive load force $F_{\Sigma lc}$ and the total hoist drive load torque $T_{\Sigma lh}$. The estimated values are calculated from (1) by means of the inverse model method. The structure of the observer is shown in Fig. 3.

The practical realization of this observer is called the numerical differencing. This operation can lead to increasing of high frequency noises from the measuring system. To avoid potential estimation errors and provide high quality of controlling to relay on estimated values, the authors suggest using the numerical filter-differentiator with finite pulse response [14] according to the methodology [15].

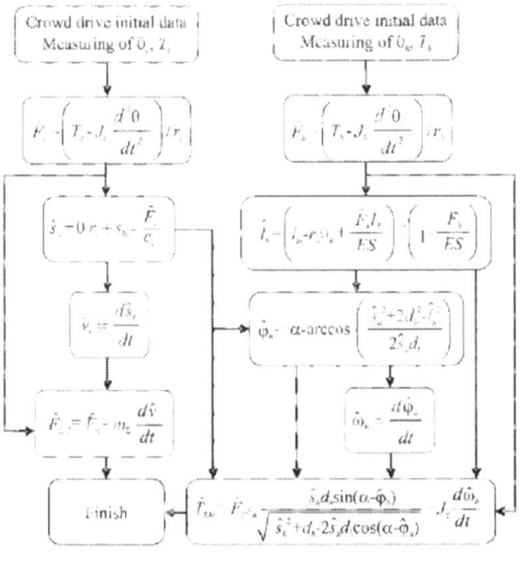

Fig. 3. The observer structure

6 Experiments

To prove the effectiveness of proposed development the authors have carried out two kinds of experiments. The first kind is experiments on location which have been carried out on the acting electric rope shovels. It confirms the adequacy of rope shovel model (1). The second kind is computational experiments. They help to investigate the effectiveness of the intelligent mechatronic system.

6.1 Experiments on location

The object of experiments on location was the rope shovel EKG-10 with mobile measuring system. This system provides the synchronous measuring of the drive currents and the angle positions of the crowd and hoist drums, the saddle block and the head block. Mounting of sensors was made with magnets and data exchange was organized with radio channel. The sensors location points are shown in Fig. 4. The processing of data was carried out with both the microcontroller and the laptop by means of a special program.

The rope shovel where the experiment was carried out was equipped with DC motor that's why to obtain the drive torques T_c and T_h it was necessary to measure the drive currents. The current sensors LT 1000-SI were used to measure current. The current data were transmitted to the measurement station where the data were analog to digital converted and then sent to the microcontroller of basic station. The angle position was got by the sensors based on two accelerometers. The sensors also included microcontroller ATmega-48 which was used for sending angle position data to the basic station.

DOI: 10.1051/e3sconf/20171503017

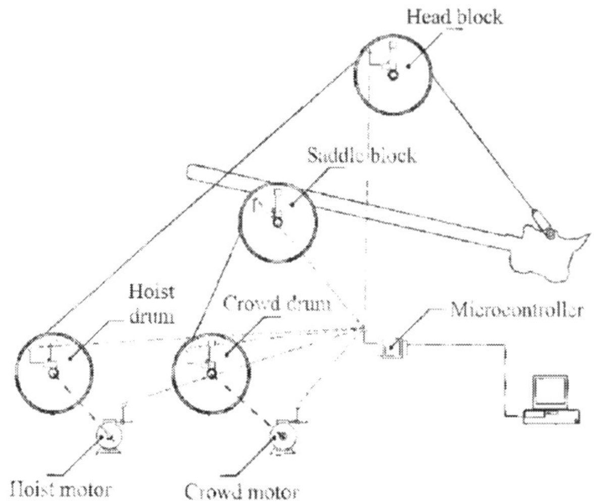

Fig. 4. The sensors location points

All above-mentioned data were read with laptop that realized the model (1) and at the same time calculated the speed of EKG-10 revolving elements from their angle positions as well as the spring forces. The comparison of these values from the model and from the direct calculation has confirmed the adequacy of (1). The maximum deviation was in the crowd speed at the start of digging and amounted to 25%. The average deviation was 6.4% for the crowd speed, 3.4% for the speed of the dipper handle about the saddle block, 6.1% for the crowd spring force, and 8.8% for the hoist spring force.

6.2 Computational experiments

The result of the experiments on location allowed moving to the computational experiments. The object of these experiments was the same rope shovel EKG-10. The experiments were carried out with computer simulation of (1) by means of the self-made program. The parameters of EKG-10 model are presented in Table 1. The parameters of control algorithms (10) and (11) were selected based on the drive overload capacity of EKG-10. Thus the time constants T_{1c}, T_{2c}, T_{1h} и T_{2h} were set on values from 0.005*s* to 0.05*s*. The observer contained the filter-differentiator with finite pulse response that had the cut-off frequency 4*Hz* and the window to weigh with 500 counting breadth.

Initially the authors carried out the investigation of the spring force control quality with, the first, independent and, the second, corresponding operation of each drive. The independent operation result has shown in the maximum reference spring force the transient time is 0.27*s* for the crowd drive and 0.23*s* for the hoist drive. On the whole range of reference spring force the control error did not exceed 2% for the crowd drive and 3.5% for the hoist drive. The average observer estimation errors for all the values was 1.5..3% and the maximum error didn't exceed 10%. This observer accuracy has been sufficient condition for the controllers operation. The corresponding operation result has shown the spring force control quality almost did not depend on the total load of the both drive equally. It has confirmed the drive interconnections don't influence the spring force controllers operation mode and the accuracy of spring forces control.

 DOI: 10.1051/e3sconf/20171503017

Table 1. The parameters of EKG-10 mechanical subsystem

Parameter	Hoist drive	Crowd drive
Transmission coefficient	24	41
Drum radii, m	0,64	0,45
Rope specific rigidity rate, N/m	—*	10,5.105
Total specific inertia, kg·m2	10,3.103	17,5.103
Nominal bucket speed, m/s	1,1	0,81
Maximum rope force, kN	1274	588
Boom angle to the horizon, rad	0,79	
Boom length from the saddle block to the head block, m	8,32	
Dipper handle length, m	11,37	
Dipper handle mass, kg	15.103	
Bucket mass, kg	17,2.103	
Rope specific rigidity rate C_h is calculated in accordance with (2), where: E=1,7.10^{10}Pa; S=1,8·10^{-3}m^2; hoist cable length from the hoist drum to the boom point sheaves l_d=23,9m.		

Finally, the circle of digging was simulated. In the simulation the trajectory of bucket moving was determined by the experiment on location. The simulation result has borne out the high quality of the spring force controllers operation. Comparison of the simulated spring force of each drive and the experimental one (Fig. 5) has noted the simulated values changed only because of operation mode technological requirements and they didn't carry additional oscillations.

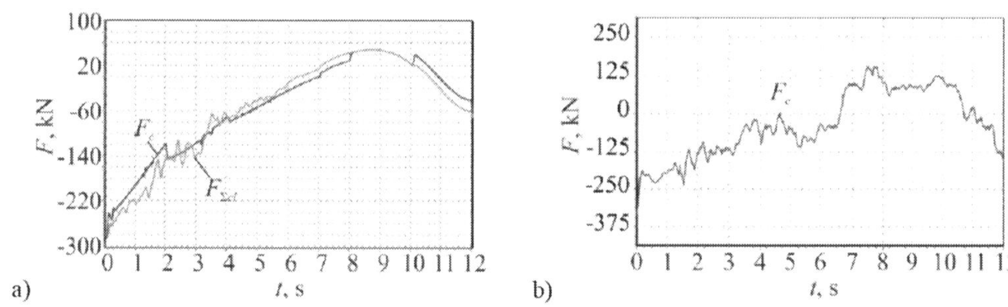

Fig. 5. Crowd rope spring force: a) simulation, b) experiment

Using the comparative criteria of dynamic loads decreasing that is based on amplitude spectrum analysis, the authors have found out that the dynamic load of crowd rope decreased by 19% and for the hoist rope the decreasing was 31%.

Thus, the proposed intelligent mechatronic system for rope shovels ensures the high quality of controlling the dynamic state of the mechanical elements that avert drive overloads and high dynamic loads in the rope. It means this system performs all of prescribed criteria.

7 Conclusions

Notwithstanding rope shovels are the complicated robotic system, the intelligent mechatronic system based on mechanical subsystem nonlinear model decreases the dynamic loads in the transmission. The same ideology can also be successfully implemented for the Draglines. Realization of above-mentioned system requires the specific control algorithms, the estimation of the mechanical values and the new kind of measurement system. However, all these novelties can be embodied with electric drive equipment that is applied in the novel excavators.

The authors thank the staff of open pit mine "Mezhdurechye" and open pit mine "Vakhrushevsky" for the given opportunity to carry out the experiments on location.

References

1. K. Uno, K. Imaie, K. Maekawa, G. Smith, A. Suyama, Hitachi Review, **62**, 99 (2013)
2. Graham Lumley, Matt McKee, *Mining for efficiency* (in PwC Australia, 2014)
3. *Reliably Safely Sustainably Electrical Solutions For Mining* (GE Energy, 2012)
4. S.E. Baybakov, Questions of electronics, **4:2**, 112 (2012)
5. A.A. Artemyev, V.S. Potapenko, S.L. Ivanov, E.A. Kremcheev et al., Mining Equipment and Electromechanics, **9**, 31 (2007)
6. *CRCmining annual report* (2004-2005)
7. Aditya Tirta Pratama, Armand Omar Moeis, Farizal., *Proceeding of International Seminar on Industrial Engineering and Management,* 7 (2011)
8. *Enhancing efficiency in mining Safe and reliable control of motor-driven applications* (ABB, 2014)
9. V.M. Zavyalov, I.Yu. Semykina, *Proceedings of symposium "Advances in geotechnical and structural engineering"*, Qingdao, China, 445 (2008)
10. S.V. Doronin, Proceedings of the universities. Mining Journal, **5**, 86 (2007)
11. D.A. Kaminskaya, Proceedings of the universities. Mining Journal, **5**, 154 (1975)
12. V.I. Kluchev, L.V. Zhiltsov, Yu.T. Kalashnikov, Electricity, **7**, 28 (1981)
13. A.A. Bronov, V.P. Kochetkov, A.A. Kolesnikov, R.G. Reyfman, Proceedings of the universities. Mining Journal, **7**, 106 (2007)
14. V.M. Zavyalov, I.Yu. Semykina, International transaction and application "Electric technic, electromechanics and electric technology", 69 (2007)
15. V. Kashirskikh, A. Gargayev, V. Zavyalov, I. Semykina, *Proceedings of the 8th Russian-Chinese Symposium «Coal in the 21st Century: Mining, Processing, Safety»,* 274 (2016)

The principles of municipal industrial clusters' establishment on the territory of advancing social-and-economic development of mono-town

Olga Ivanova[1*], *Gennady* Antonov[2], *Sergey* Bereznev[3]

[1] Institute of Economy and Management, Kemerovo State University, 650000 Kemerovo, 6 Krasnaya St., Russia
[2] Institute of personnel training and retraining, 650000 Kemerovo, 13A Michurina St., Russia
[3] T.F. Gorbachev Kuzbass State Technical University, 650000 Kemerovo, 28 Vesennyaya St., Russia

Abstract. Overcoming social-and-economic problems of single-industry towns by diversification of their economy is possible by creating territories of advancing social and economic development (TASED). In contrast to the researchers claiming that the formation of a cluster with participation of enterprises of more than one municipal territory strengthens mono-profile of town, the authors justify synergies effect from using advantages of TASED and-cluster approach in diversification of single-industry towns' economy. Regional (municipal) industrial cluster, in contrast to the previously known concepts, the authors consider to be formed with an active role of the regional and municipal authorities for implementing projects of mono-town economy diversification and investment projects that meet the requirements of TASED. As a result of problems and risks systematization concerning clusters' functioning in Russian and foreign practice the basic principles of building clusters on TASED in mono-towns were defined. The research results present a contribution to the cluster theory, the regional economy in terms of methodology development of building regional (municipal) industrial clusters within TASED for economic diversification of mono-town economy. The practical value of the study lies in the possibility of using the results of work in the activities of regional and municipal authorities in the formation of cluster as residents of TASED in single-industry towns.

1 Introduction

Reducing the number of single-industry towns included in the list approved by the state, from 335 in 2009 to 319 in 2015, is not accompanied by a decrease in the number of accumulated problems. In addition, 148 single-industry towns are areas in which there are risks of worsening economic and social situation. The way out of these difficult situations is the establishment of TASED in single-industry towns. TASED along with clusters, special

* Corresponding author: ivanova@mail.ru

economic zones are considered to be the mechanisms to stimulate industrial and innovative development. In economic literature there is an opinion that the clustering of enterprises with the participation of more than one municipal area intensifies mono-profile of the town. But the practice shows the positive experience of regional clusters, for example, being developed with a state support automobile industrial cluster in the Kaluga region. So it makes sense to say that on TASED in mono-towns it is possible to identify and develop the union of enterprises with the potential to be transformed into regional clusters. Up to now there is no experience of establishing clusters in the framework of programs for creation TASED in mono-towns, there is no concept of cluster applying to the forms of business organization for the development of TASED in single-industry towns. The liberal model does not associate cluster with a special economic zone. However, it is possible to assume that the provision of favorable conditions for business development, for example, TASED, will become a factor stimulating the birth and development of clusters as residents of such territories. This determines the relevance and practical significance of this study.

2 Objects and methods

The object of our study is cluster projects, the experience in clustering in industry. The research of domestic and foreign scientists is mainly directed to the development of methods for clusters identification. But they do not evaluate the potential of businesses, in particular, single-industry towns, which can become a member of the cluster's core; the methods of clusters' establishment are equated only with the methods of their identification.

The development of principles of methodical approach to the formation of clusters on TASED in single-industry towns was based on the following theoretical concepts and research: the possibility of more efficient intra-control on the basis of inside information, the regulation and coordination of activities of combined entities as a result of the organization of intra-corporate financial market (the concept based on the in-house capital market); the effectiveness of establishing suppliers, manufacturers, consumers networks and technological cooperation networks (D. Ernst); the advantages of industrial networks formation, commodity chains, supply chains and value chains (E. Yourdon, M. Christopher, T.J. Gallin and M. Hendon, M. Garrett and P. Dyusyuzha).

3 Results and discussion

Cluster is a concentrated on a specific territory set of enterprises, interrelated, but belonging to different industries. M. Porter [1] pays attention to the interaction between the cluster participants (the main company-manufacturers, supporting (attendant) industries, universities, research organizations and government structures).

R. Brough defines a cluster as "cross-sectoral concentration of firms that creates jobs, exports goods and services, has common basic economic needs and brings together public sector of economic development, universities, colleges, educational community, funds and all other stakeholders" [2].

Established on TASED of mono-town regional (municipal) industrial cluster is proposed to consider as: the tool of industry structuring and its network organization; the union of organizations formed to implement economic diversification and investment projects for mono-town that meet the requirements of TASED; networking of enterprises and organizations connected by the production relations on the territory of mono-town; the kind of territorial cluster, the regional industrial cluster, industrial district. We consider it necessary to highlight its main features: local economic structure, where the interaction of social-and-cultural and production-technical factors takes place, ensuring their competitiveness and dynamism; the combination of competition and cooperation, market and organization; the form of industrial organization characterized by the concentration of economic and entrepreneurial activity and limited by the municipal district, urban district; participants are mono-town's enterprises specializing in the production of competitive products, suppliers of raw materials, services, the cluster's infrastructure creating the final product and value added; the internal competition that distinguishes cluster from integrated entities.

To the most successful and dynamic clusters formed by means of creation the special economic zones, technological parks and business incubators can be referred the following clusters in mono-towns: motor industry cluster in Tolyatti and Naberezhnye Chelny, "Titanium Valley" in Verkhny Ufaley in the Sverdlovsk region, chemical cluster in Nizhny Tagil, timber cluster with the latest technology in Sokol city in the Vologda region and others.

The most important factors in the effective clusters' functioning are: management quality [3]; mechanisms and forms of organization for the accumulation and dissemination of knowledge and accumulation of social capital [4]; the presence of at least 30-50 specialized companies in a cluster to implement the diffusion potential of innovations.

To describe the community of technologically interrelated sectors in France term "die" [5] is widely used. It is a form of interaction with the features of innovative clusters. The term «clusters of innovation» has gained wide popularity among the public and private sector leaders in the United States after the establishment of the cluster which was called «Clusters of Innovation» [6].

The methodology and implementation of cluster policy in Russia as a whole correspond to the conceptual framework similar to the European programs, particularly French and German. Since 2012 the RF Ministry of Economic Development has conducted competitive selection of projects for the development of clusters in the regions of Russia. The competition was attended by about 100 cluster initiatives, 25 of them were selected for the pilot support. During 2013-2014 for the development of clusters on co-financing conditions from regional budgets 3.8 bln rubles were provided from the federal budget. For example, an innovative regional cluster in the field of information and communication technologies in the Novosibirsk region has received 269 million rubles, and the cluster of information technologies in St. Petersburg - 1.3 mln rubles. The average cluster size for subsidies amounted to about 100 million rubles. The analysis of study results of Russian pilot innovation clusters conducted in 2015 by the method of European initiatives of improving clusters indicates that the clusters are located mainly in Russian regions with a high level of innovative development (from 21 examined clusters 13 (62%) are located in the regions - "strong innovators", 5 (24%) - in the "medium-strong innovators", 1 cluster is in the region - "medium in-innovator" and 2 (9.5%) - in the "medium-weak innovators"); in new sectors (information technologies, biopharmaceuticals and new technologies) there are 11 clusters, whereas 12 clusters can be referred to the traditional high-tech industry the foundation for which was laid during the Soviet era.

The limits for the development of clusters in Russia became not only a lack of knowledge and inability to use the world experience in the local environment, focus on obtaining quick results, difficulties in finding investments and worn out fixed assets, personnel problem, but also weak strategy elaboration in choosing the development priorities.

Often, as the key points of growth all sectors of the territory are selected which leads to the dissipation of forces, the lack of resources for the implementation of all projects. The problem of weak innovation of the formed clusters is significant.

The world experience shows that the formation of clusters exclusively by the initiative of the state is one of the main risks - neglect of the business development trends, as well as its economic interests. Artificially built cluster exists only as long as it has government support. Due to the artificiality of some clusters considered by the regional authorities as an instrument of public support in Russian clusters there is little horizontal communication, co-operation among the participants is not developed. Russian risks of cluster approach coincide with the risks described by the foreign experts. In particular, the T. Man-Wen and R. Voyer [7] highlight the most significant risks which foreign governments face in establishing and promoting the development of clusters: changing macroeconomic conditions, weak innovation character of clusters due to the lack of production mechanisms and adaptation of new knowledge, insufficient development of internal and external scientific and technological relations which weakens the synergetic effect of clustering, the use of inefficient management techniques.

The program of establishment and development TASED in mono-town requires the list of economic activities that can be internal points of growth. Obviously, in mono-town there can be formed clusters around one or a group of companies. Using cluster approach to the development of the program TASED in mono-towns one should consider not only the international experience of cluster management "European Cluster Excellence Initiative", the results of research of government management practices in industrial clusters, the processes of formation of business clusters and high-tech clusters [8-15], but identified during the national experience study of clusters' functioning their sustainable development factors.

Backgrounds for the regional (municipal) industrial clusters within TASED of mono-towns are: the lack of enterprises' own possibilities for the effective organization of production activities on the innovation basis; the need for diversification of the economy; demand for the modernization of production facilities; import substitution and the possibility of using favorable conditions for development of private entrepreneurship.

The scenario approach to the establishment of municipal clusters within TASED pre-supposes priority of "top-down" movement with the initial development of the cluster's strategy and its support within TASED. For example, M. Wickham [14] considers the role of the state in the creation of clusters as the most important. In addition, it was noted [15], that more often the initiators of cluster's establishment are governments (32%), business (27%) and together business and government (35%), financing for clusters' initiatives by the state (54%), business (18%), and together business and government (25%).

The basic principles of regional (municipal) industrial cluster's establishment as a part of TASED in mono-towns are defined:

1) Selecting cluster to assess the concentration of enterprises conducting profile, related and supporting activities, as their high value promotes innovations (priority for support of mono-towns on TASED is given to clusters operating in sectors that have economic indicators considerably exceeding the average in the country);

2) To use networking strategy in order to benefit from coordination network structures, adapting to changes, rapid response to changing market conditions, specialization, cost reduction;

3) To provide attractive conditions of urban environment for high-skilled personnel and innovative entrepreneurship in order to avoid the dominance of city-forming enterprises and to diversify the city's economy;

4) To set up specialized management companies implementing cluster management functions;

5) To use outsourcing, i.e. cluster formation through cooperation, involving a lot of independent companies and competition between them around the core businesses, including, for example, the city-forming enterprises, large factories;

6) To develop the internal competitive environment.

4 Conclusion

It was proposed to consider regional (municipal) industrial cluster on TASED of mono-town as a tool for industry and its network organization structuring, uniting the organizations to implement the economic diversification of mono-town and investment projects meeting the requirements of TASED. The features distinguishing regional (municipal) industrial cluster from the other types are systematized. It is proved that scenario approach to the establishment of municipal cluster within TASED is necessary and it should be based on priority "top down" movement with the initial development of the cluster's strategy and its support within TASED. The basic establishing principles of municipal industrial cluster as a part of TASED in single-industry town, which differ from existing by a combination of: the priority evaluation criteria when selecting the cluster being able to develop within TASED of mono-town; networking stage by stage; the use of TASED opportunities to create attractive conditions for urban environment; coordination of participants' strategies; outsourcing; the development of internal competitive environment as a form of competition among entities and the element of market mechanism introduced in the functioning and development of the cluster's system.

References

1. M.E. Porter, Economic development quarterly, **14(1)**, 15, (2000)
2. R. Breault, *Photonics Tech Briefs,* (2000)
3. European Cluster Excellence Initiative. The quality label for cluster organizations – criteria, processes, framework of implementation, **8,** (2015)
4. S.A. Rosenfeld, *Regional Technology Strategies* (Carrboro, North Carolina, USA, 2002)
5. J. Toledano, Revue d'Economie Industrielle, **6(4)**, 149 (1978)
6. M. Porter, *Council on Competitiveness, Monitor Company, Clusters of Innovation: National Report* (2001).
7. T. Munn-Venn, R. Voyer, The Conference Board of Canada Report, **32**, (2004)
8. B. Derudder, F. Witlox (editors) *Commodity Chains and World Cities* (Wiley-Blackwell, 2010)
9. R. Le Heron, Agri-Food Commodity Chains and Globalising Networks (The Dynamics of Economic Space), **258,** (2016)
10. J. Michael, *Innovation clusters and interregional competition* (Springer Berlin Heidelberg, 2003)
11. M. Perry, *Business Clusters: An International Perspective (Routledge Studies in Business Organizations and Networks)* (Routledge, 2005)
12. P. Braunerhjelm, M.P. Feldman (editors) *Cluster Genesis: Technology-Based Industrial Development* (Oxford University Press, 2006)
13. J. M. Valdaliso, J. R. Wilson (editors) *Strategies for Shaping Territorial Competitiveness (Routledge Studies in Global Competition)* (Routledge, 2015)
14. M. Wickham, *CRIC Cluster conference. Beyond Cluster – Current Practices & Future Strategies* (Ballarat 2005)
15. The Cluster Initiative Greenbook: New Findings on the Process of Cluster-Based Economic Development. http://www.cluster-research.org/greenbook.htm

The innovative strategy of social and economic development of mining region

Sergey Bereznev[1*], *Olga* Zonova[2], and *Elmira* Lubkova[2]

[1] T. F. Gorbachev Kuzbass State Technical University, Production Management Department, 650000 Kemerovo, 28 Vesennya st., Russian Federation
[2] T. F. Gorbachev Kuzbass State Technical University, Finance and credit Department, 650000 Kemerovo, 28 Vesennya st., Russian Federation

Abstract. Depressed state of Russian economy in recent years, the unstable functioning of most regional social and economic systems in the conditions of increased uncertainty of the global economy makes strategizing of the development based on a new model of economic growth a highly important issue. It assumes the tasks of modernization and restructuring of the economy, as well as improving the quality of management. It is known that the solution of these conceptual target problems was suggested by the concept 2020, but the global crisis of 2007-2009 and other geopolitical factors in the post-crisis period pushed their implementation back. The results of 2016 show an adaptation of Russian economy to the difficult internal and external challenges. In this article, the authors analyze the strategizing process by the example of the "Strategy of social and economic development of the Kemerovo Region up to 2025", as well as analysis comparatively the strategies for the development of the national economy, the Siberian Federal District and the region. Based on the study results the authors formulated and substantiated conclusions and recommendations to improve strategizing process and the effectiveness of strategic management of the development of social and economic systems.

1 Introduction

The main legal acts regulating the procedure of the formation and content of documents for economic development strategizing of the national economy and regional social and economic systems of the Russian Federation subjects include the following: the federal laws "On state forecasting and programs of social and economic development of Russia» № 115-FZ on 20.07.1995 (rev. on 09.07.1999) and "On the strategic planning in the Russian Federation» № 172-FZ of 28.06.2014, as well as the Concept of long-term social and economic development of the Russian Federation for the period till 2020, technical standard "The requirements for social and economic development of the Russian Federation subject" and the regional laws.

The federal law №115-FZ and the corresponding legal acts provide for the following stages for strategizing: 1) the forecasting of social and economic development of the

* Corresponding author: bereznev45@mail.ru

administrative and territorial unit in the long and medium term; 2) concept formulation of the social and economic development in the long term based on the previous forecast; 3) designing the program of social and economic development in the medium and long term (in order to ensure the continuity of economic and social policy of the state). Later, the federal law №172-FZ made some amendments concerning the preparation order and content of the documents, according to which the regional law of the Kemerovo region "On strategic planning" №103-OZ of 23.12.2016 was adopted.

2 Methods

The development methodology for the strategy of social and economic development is based on the combination of two approaches: 1) economic-and-mathematical approach [1], involving the collection, processing and analysis of objective and subjective indicators (statistical data and data obtained from interviews) characterizing the social and economic development of the administrative-territorial unit; 2) scenario approach [2], involving an analysis of possible deviations from the targets of the administrative-territorial unit development and defining the activities aimed at implementing the most favorable for the regional development scenario and leveling the risk events which may impede the realization of a particular scenario.

In foreign and domestic theory and practice the opinion about the content of the main components of the strategizing system is considered to be well established: 1) the business model of the region (the subject of the Russian Federation), which reproduces the dynamics of social and economic processes; 2) the system of strategic management forming and working out the administrative decisions on the model of the region; 3) economic agents in the model as an interactive set of economic activities (in accordance with the division on the national Classification of Economic Activities); 4) households and the agent "invisible hand of the market", responsible for the balance of supply and demand in the markets of goods and factors of production [3, 4].

3 Results and Discussion

The analysis of currently implemented “Strategy of social and economic development of the Kemerovo Region up to 2025” approved by the Kemerovo region law № 74-OZ on 11.07.2008 (amended on 02.11.2012) revealed the following: 1) strategy of social and economic development of the Kemerovo region is basically consistent with the objectives of the “Concept of social and economic development of the Russian Federation until 2020” and the “Energy strategy of Russia until 2020”, as well as with the priorities of the “Strategy of long-term social and economic development of Siberia until 2020”; 2) policymakers made SWOT-analysis and used scenario approach to justify the selection of the target (active) scenario of strategic development; 3) the purpose of the strategic development till 2020 was the following: "the increase of the regional competitiveness and population wellbeing growth on this basis". However, it should be noted that the strategic goal, first of all, is not linked with the mission of social and economic development of the region; 4) the program block of the strategy provided a number of tasks the solution of which implied the achievement of the strategic development target.

As it follows from the mentioned above the developers complied with the basic formal characteristics and rules for the development of regional development strategy, including the coordination of regional target with Siberia development priorities.

However, the choice of the so-called active scenario, which, in fact, implies the change of the technological layer, coincided with the period of the global financial and economic crisis of 2007-2009. During this period, even without taking into account the challenges and risks generated by the global crisis, the list in the SWOT-analysis of existing restrictions for the development of regional social and economic system in depth, nature and scale of problems is actually institutional conglomeration of factors (capital, labor, management and etc.), which has the highest inertia. It is not only the problems and peculiarities of the natural resource region, but also the problem of Russian national economy as a whole, which complicates the solution of strategic tasks in the context of subsidiary relationship between the RF subject and the federal authorities in tax revenue distribution. It is known that now there is a sharp imbalance of tax revenues in favor of the federal budget. So from 1 ruble which generates a large investment project, 85 kopecks go to the federal budget, only 14 kopecks to the regional one, and, accordingly, 1 kopeck to the municipality. It is obvious that such situation and chronic imbalance in the regional budget does not increase investors' confidence and reduces the possibility of regional authorities in cooperation with the business within public-private partnership.

Another aspect in our opinion deserves deeper assessment and consideration in the implementation of tasks to achieve the goal of "active" scenario. It is an issue of resources and, in particular, about own financial possibilities of regional businesses. As the experience of previous years showed funding of short and medium term investment programs in the real sector of the Kemerovo region was provided by business on 80% from its own profit. In the pre-crisis period production assets indicators and profitability of basic industries actually reached their limits by market conditions. During these years, there was a favorable situation on foreign markets for the export group of products (coal, metal, fertilizer) the highest saving rate of about 30% of the GRP was achieved [5]. During the development and adoption of the strategy the world economy, especially the markets of energy products, showed clear signals of reducing conditions and therefore there was no possibility to rely on the former exporting model by the developers. Moreover, the enlarged assessment of funding necessary to solve the strategic tasks with a planning horizon up to 2025 made by the authors showed that the rate of accumulation had to be increased up to 50-60%, which is unrealistic in a post-crisis economy and long-term external challenges.

Unfortunately, the forecast quality, the assessment of the resource base and the attainability of chosen scenario, its target, is considered to be one of the "narrow" places in the national practice of strategizing social and economic development. We believe that these shortcomings are fully applicable to the development strategy of the Kemerovo region. The limits of this article does not allow talking about these and other aspects of the strategizing process, however, we should note one more fundamental observation which follows from the analysis: "wellbeing growth" of the population in the region is, of course, a good and dominant purpose of the whole philosophy of the social and economic development. This is an internationally recognized value which is directly connected with other criteria and indicators: quality of life, life expectancy, population reproduction, etc. [6, 7]. The social and economic sphere of the Kemerovo region has its own specific characteristics: the dangerous production, multi-shift mode of employed in basic industries and as a consequence of these and other conditions - low average life expectancy (68 years which is below the national average level on almost 3 years). Kemerovo region, one of the few industrial regions, which does not provide a simple reproduction of the population and shows the trend of steady decrease in total population (2006-2016 on 88,314 people) and a very high occupational diseases. These are real internal challenges, which do not only find a proper solution in the scenario of coal mining expansion, but also form a vicious circle. According to the Institute of coal and coal-chemistry of RAS, increased coal production by 10 million ton per year will increase by 1% the incidence rate in the Kemerovo region, which entails the growth of budget expenditures on health care and the implementation of environmental protection measures. That is why we consider that one of the main requirements for strategy development should become not formal but thorough and in-depth implementation of the correlation of the interests of authorities, economic agents and society.

In 2012, the strategy of social and economic development of the Kemerovo region, taking into account internal and external challenges was adjusted, and its goal was not changed. What are the reasons for making adjustments in the strategy? Let us refer to the workbook of the "Strategy of social and economic development of the Kemerovo region", posted on the Kemerovo region Administration website. It proves the necessity of adjustments: risks faced by the implementation of the strategy and the emergence of new economic growth drivers.

Special attention deserves the strategic statement "the development of coal industry". Thus, the increase of coal industry role in the national fuel and power complex and Russian economy as a whole is projected. The growth of domestic demand and export potential of the Asia-Pacific region markets stipulate the necessities of increasing the competitiveness of industry. An interesting fact is that the economic development strategies, adopted at the federal level, do not set a goal to increase coal exports, which cannot be said about the strategic documents of the Kemerovo region. According to the mentioned above, we believe that the regional "development" must still have parametric environmental restrictions in accordance with the principles of the Concept of Sustainable Development of the United Nations organization [8].

4 Conclusion

The analysis of strategizing of social and economic development of the region revealed a number of "narrow" places, leveling which will lead the development of strategic documents to a new level. The authors are aware that it becomes difficult to predict the future [15]. Perhaps the way out of this situation would be timely and systematic monitoring of the strategy implementation. It is surprising that after understanding (in 2011) that the strategic target of social and economic development of the Kemerovo region is unreachable, neither target nor scenario was changed. In our view, the most appropriate at the present time would be the implementation of the transition scenario, aimed at the effective interference of raw material and innovative model of development of the regional economy.

References

1. L.V. Kantorovich, Manag. Sci., **6**, 4 (1960)
2. K. Riahi, A. Grübler, N. Nakicenovic, Technol Forecase Soc, **74**, 7 (2007)
3. V.K. Mathur, Econ. Dev. Q., **13**, 3 (1999)
4. A. Lagendijk, J. Cornford, Geoforum, **31**, 2 (2000)
5. S. Bereznev, M. Kumaneeva, M. Makin, *8th Russian-Chinese Symposium. Coal in the 21st Century: Mining, Processing and Safety* (Atlantis Press, 2016)
6. O. Zonova, E. Nekhoda, E. Slesarenko, *8th Russian-Chinese Symposium. Coal in the 21st Century: Mining, Processing and Safety* (Atlantis Press, 2016)
7. E. Vershkova, T. Filippova, E. Nekhoda, SHS Web of Conferences, **28** (2016)
8. J. Friedmann, M. Douglass, *Agropolitan development: towards a new strategy for regional planning in Asia* (United Nations Centre for Regional Development, Nagoya, 1976)

The analysis of strategies for the mining regions' development in Russia as a condition of effective management of economy

Natalya Zaruba[1,*], *Natalya* Egorova[1] and *Peter* Kosinskij[1]

[1]T.F. Gorbachev Kuzbass State Technical University, Department of State and Municipal Management, 650000 Kemerovo, 28 Vesennya st., Russian Federation

Abstract. The conceptual issues of a new approach in the implementation of strategic management development of the coal-mining region as conditions of effective government regulation of economy at the macro level are considered in the article. The purpose of the study is to justify the use of marketing techniques in the strategic management of the region, clustering on the basis of the territorial concentration and combination of all available resources, the integration of regional economic networks. A comparative analysis of the main strategic directions of development of the coal-mining regions from the point of view of the leading economic development strategies is carried out. The main result is that the estimation of value of synergy effects occurring when the resources of combining sectors and industries in the region are united has been made. The results of the study can be recommended for usage in the development of strategies for sustainable development of «mono-territory».

1 Introduction

The relevance of analyzing the strategies of mining regions as a condition of effective government regulation of Russian economy is justified, first of all, by the problems of the economy itself.

The current state of Russian economy requires new solutions in the theory and practice of management, including the development of innovative strategies for regional development, as the promising areas of competitive production of definite goods (and services), jobs and high living standard for the population.

Modern economic problems cause the need for taking measures adequate for market conditions in the regional strategic management. They should become conditions for effective state regulation of the economy as a whole. Such measures we believe will help to ensure effective functioning and future development of the most complex structure-forming areas.

The centers where such transformations take place are the regions where the real management tools of economic systems are concentrated. A comprehensive innovative approach to the management of such territory implies not only the system of goals and

* Corresponding author: zaruba@kuzstu.ru

objectives of the authorities in management of all sectors of the economy, but also mechanisms for their implementation, which requires, in turn, the need for analysis of the strategies of such territories.

Rational regional policy management directed to the development and implementation of such mechanisms must be based on multidimensionality and importance of its role in the development of regional economy. At the same time the priority of regional development strategy should be given to that in which the region acts as a single economic system, integrated territorial product.

We believe that such strategy should be based on an innovative approach to management which, in our opinion, can be considered marketing approach to clustering technology, as it allows, first of all, taking into account the specifics of territories, their resource, social and economic and other features, and also it is based on the possibility of getting a synergistic effect in management.

The scientific novelty of the work lies in the theoretical and practical substantiation of diversification of regional strategic management by clustering and the synergies from the combination of all its resources.

2 Results and Discussion

The need for the formation of a new management strategy of regional development comes from the ever-changing economic situation in and around the region.

It should be emphasized that the highlighted problems today are quite popular in the scientific literature. Many experts are studying the issues of strategic management of regions, but, despite this, a number of problems remain under discussion [1-5]. Particularly relevant and unsolved, in our opinion, is the problem of mono-territory management.

In addition, the existing approaches to the definition of "strategic management" do not reflect, in our opinion, the individuality of territories. For example, O.S. Wihansky defines the regional strategic management as a certain philosophy or ideology of business and management, not pointing to a specific decision making and regional specific. [6]

It is important to note that today there is a variety of regional development strategies, which are aimed at ensuring social and economic indicators growth without defining it as a complex product able to represent its individual specifics on the market. In connection with this we emphasize that the solution of the problem of the territory's sustainable growth is possible only through an integrated approach to the strategic management of the regional economies, thus it becomes objectively necessary to develop a concept territory's promotion, which would structurally reflect and address all industries and sectors of economy.

In addition, the development of the clustering concept on the basis of the territorial concentration and combining all available resources we should take into account the differentiation of territories. We should distinguish them by not only the availability of resources, mineral resources, specialization, but estimate the reserves for the innovative development of regional economy. Particularly vulnerable in this sense are the regions in which heavy industries have been and remain the basic ones, without reducing the role of these industries and their importance in the development of the regions, we consider it necessary to "modernize" management in terms of its effectiveness growth.

To these regions, in our view, should be referred the mining regions of Russia. It should be noted that in the Russian Federation various types of coal (brown, anthracite and hard coal) are mined. As for coal reserves Russia is one of the leading economies in the world. Two thirds of the total volume of coal is hard coal. Russia is on the fifth place in coal production and it is important to note that a substantial volume of coal (the half of Russian coal production) is concentrated in Kuznetsk Basin. Almost 95% of reserves are concentrated in the east of the country, more than 60% in the Siberian Federal District (SFD).

In 2015 84.5% of total coal amount was produced in Siberian Federal District, 9.4% - in Far Eastern Federal District; 4% - in North-West Federal District; 1.3% - in Southern Federal District; 0.5% - in Urals Federal District; 0.2% - in Volga Federal District; 0.1% - in Central Federal District. As we can see, the "lion" share of mined coal is accounted for SFD.

All this testifies the positive dynamics of the development of coal industry, which is the factor determining the growth of the Kemerovo region's economy as a coal-mining region. However, despite the positive trend of coal industry development in Russia as a whole the demand for coal will rise to 2020 as analytics predict. Then this kind of fuel consumption will gradually decline, which is associated with increasing consumption of natural gas in future.

These possible changes in market conditions are not in favor for coal mining industry and today coal-mining region authorities, in particular, the Kemerovo region, should pay attention to possible risks and threats of "mono-dependent" economy.

Defining innovative steps in the regional management, we believe it is necessary to conduct a comparative analysis of strategies for social and economic development of two the most leading mining and coal reserves territories: Kemerovo and Sakhalin regions.

Analyzing the strategies of Russian coal mining regions, we consider being necessary to examine the management advantages and disadvantages in order to identify the so-called "bottlenecks" and provide their subsequent elimination, reduction to the optimum denominator.

It should be noted that based on the objectives and goals of the studied development strategy, Kemerovo region should be referred to the export-oriented regions of Russia. The basic sector in the structure of regional economy is the extraction and primary processing of natural resource - coal.

At the same time, it is important to emphasize that the strategic development of Kemerovo region, as well as other coal-producing regions, in the first place will depend on the dynamics of the social and economic situation in the country, on the global market and regional product (coal) competitiveness.

This means that in the near future, the inertial development of the region only by strengthening the competitiveness of the coal industry is most likely to be. Realizing the importance of keeping the coal industry as the base for the regional economy, we believe that in the long run it is especially important to take into account the risks that may arise in industrial mono-specialization, such as the dependence of the entire regional economy on the situation on a separate market segment (coal), especially globally.

Moreover, such limitation of the regional economy within only heavy industries will lead to disproportionate structural shift in general and increasing "mono-dependence" of economy.

We believe that especially "vulnerable" in this sense is the Kemerovo region, which today has an industrial structure with a vivid "mono-dependence" from the coal mining industry, which already makes one think about the "rehabilitation" of the region through an integrated solution of the problem removing its primary acuteness.

One of the limiting factors in the Kemerovo region development is, in our opinion, a low diversification and "innovation" degree in strategic management of the regional economy, which, as it was noted, has a clear resource orientation (coal mining and low processing of natural resources). This was confirmed by the analysis of the Kemerovo region development strategy, where, in contrast to Sakhalin, primarily fundamental is the coal industry development.

We believe that diversification approach to "mono-region" management becomes particularly important in the recession of Russian economy. Current problems determine the need for the innovations development and implementation in management which should contribute to the complex development of the region.

It must be emphasized that diversification in management of narrow-specialized (coal mining) regions should be directed not just to the economy profile-changing, but rather to changing the economic structure. In this sense, we consider it necessary to transfer the center of gravity in the economic structure from the traditional for the region metallurgical, coal, petrochemical industries, without diminishing the importance of the latters, to new and emerging trends, such as services. Such an approach would solve the problem of promoting the region as an integrated product that includes both basic and supporting industries.

It is important to emphasize that the Kemerovo region has not only coal reserves, which put the ground for its industrial development (coal, chemical, metallurgical industry), but also significant natural and recreational resources, which, despite some structural changes in the regional economy are not fully used.

Differentiation of regions by various social and industrial-and-economic aspects, mono-structure and low volatility of their development processes make it necessary to use individual forms, methods and strategic management approaches to identify, use and diversify the competitive advantages of every particular region [7] .

We believe that in order to solve the problem of region's promotion, to improve its investment climate and innovative management it is necessary to use regional marketing technology. It is advisable to direct the vector of economic development of the Kemerovo region in the direction of complex use of all available resources, taking into account their specificities and existing potential.

In this case, the marketing strategy of the Kemerovo region development should be based on the use of marketing tools, taking into account a possible "targeting" of the region and to become the basis for its promotion concept. The importance of marketing approach in the concept of complex promotion of the region, its sustainable development is dictated, in the first place, by increasing competition, both on the domestic and global markets.

In a number of works scholars have pointed to the need for using marketing and its tools in an innovative strategic development of the region together with the diversification policy in management [2, 8-12].

It is especially important to take into account that it is strengthening of the competitive advantages which in modern conditions becomes the main source of the regional strategy. In this sense, we should take into account the so-called cluster approach in management (marketing technology), which in our view is still not used in the strategic management of the region. The concept of cluster management of regional economy allows identifying the factors of competitive advantages of the region, contributes to the development of priority sectors defining the investment attractiveness of the region [13].

The cluster approach should be reasoned and, above all, by the fact that in this administration a regional innovation system is being formed. It is based on public - private partnerships, enhancing regional economy innovativeness and its investment attractiveness.

The cluster in our understanding will represent a region as a complex product, which management should be based on the regional concentration and combination of all available resources, integrated subjects of the regional economic network.

Proposed regional agglomeration will provide a synergistic effect, which determines its future competitiveness. At the same time, we believe that the clustering of the region will show the synergistic effect from a mutual combination of all the regional resources, industries and sectors of economy. The synergy can be calculated by appealing to nonlinear theory, in accordance with which a total combined effect is not equal to a single element impact [14, 15].

Using regression analysis method, we construct a model that takes into account not only the direct contribution in obtaining the impact effect, but also the combined effect of the elements of resource provision for management processes, in our case by the Kemerovo region.

The simulation results show that at the impact on regional social and economic system of x_i, x_j... parameters, the greatest positive synergistic effect can be observed when combining social policy, national economy, related to both the total income in the region and the profit from the organizations. The presented combination determines, in our view, management effect of clustering on the sustainable development of the region.

3 Conclusion

Based on the study the authors came to the following conclusion:

- The comparative analysis of strategies for the development of coal-mining regions of Russia showed that the basis for the sustainable development of the studied regions remains the coal industry, the strategy weakly represent other sectors of regional economy.
- The logic of scientific research in the field of strategic management of regional development allowed determining the need for management modernization;
- The problems identified in the strategic management of "mono-territories" (the Kemerovo region) indicate the inertia of its further development as a result of "mono-dependence" from the coal-mining industry;
- The cluster approach in coal-mining region management makes it possible to define it (region) as a complex product the management of which should be based on the ground of territorial concentration and combination of all available resources, integrated subjects of the regional economic network;
- The use of marketing technologies (clustering) in strategic management of coal-mining region serves as a condition for the government regulation of economy as a whole;
- The calculation of synergetic effects of using clustering in management has proven the need for using marketing technologies (clustering) in the strategic management of coal-mining region.

References

1. I. Ansoff, *Strategic management* (1989)
2. N.A. Zaruba, N.N. Egorova, *Coal in the 21st Century: Mining, Processing and Safety,* 39 (2016)
3. V.M. Tumin, A.G. Koryakov, E.P. Nikiforova, World App Sci Journal **25,** 6 (2013)
4. J.R Galbraith, Human Resource Management, **22,** 63 (1983)
5. G. Johnson. *Strategic Change and the Management Process* (Black-well, 1987)
6. S.M. Puffer, D. McCarthy, O.S. Vikhansky, A. Naumov, Org. Dynamics. **34,** 3 (2007)
7. A. Granberg, I. Zaitseva, Problems of Economic Transition, **45,** 8 (2002)
8. V.M. Tumin, A.V. Polyarus, S.N. Lomachenko, A.S. Troshin, S.Y. Doroshenko, Biosciences Biotechnology Research Asia. **12(1)**, 445 (2015)
9. N.P. Rebrova, B.P. Volovikov, A.I. Kovalev, O.V. Frik, European Journal of Economics and Management Sciences, **2**, 10 (2016)
10. P. Kothler, Business Horizons, **79** (2000)
11. D. Day, *Strategic marketing* (2002)
12. J. Hulbert, Norman, E. Toy, Journal of Marketing, **12** (2011)
13. J. Medina, M. Duffy, Journal of Product and Brand Management. **7(3)**, 223 (1998)
14. H. Sislakova, Acta Univ. Ags. Sil-vicult. Mendelianae Brunensis, **48(2)**, 129 (2000)
15. A. Thompson, *Strategic management: Awareness and change* (1990)

Increasing the efficiency of coal mining based on the concept of Shewhart-Deming variability management

Vadim Mikhalchenko[1,*], *Yuryi* Rubanik[2]

[1] T.F. Gorbachev Kuzbass State Technical University, Institute of Economy and Management, Department of Economics, 28, Vesennaya str., Kemerovo, Russia
[2] New Management Technologies Centre, Moscow, Russia

Abstract. The article reveals the key moments and technology of the administrative decision-making on the basis of concepts of the theory of variability that is an integral part of the "Deep Knowledge System" by E. Deming. This system is the theoretical basis of modern management, i.e. quality management. Quality management shifts the focus of managers from machines to people and the quality of all processes implemented in the business system. The main idea of quality management is continuous improvement of processes, reduction of variability and search for solutions aimed at maximizing customer satisfaction. In order to proceed to practical steps to solve the problems of business efficiency, it is first of all necessary to achieve an appropriate level of theoretical understanding of the origins of the formation of this effectiveness. Understanding the fundamentals of the theory of variability is a necessary condition for making a rational management decision. Control cards are an effective tool for developing the right management solution, they are designed for continuous monitoring and diagnostics of the state of business processes. The examples of control charts considered in the article in a generalized form characterize the "resulting" variability in coal mining enterprises, and the results of their analysis indicate to managers the direction of searching for ways to increase the efficiency of coal mining.

1 Introduction

In accordance with the provisions of the modern management theory – quality-based management [1-8], the success of the organization is determined to 95% by system factors and, thus, the latter are a potential reserve to improve its effectiveness. In this case, all or the vast majority of system factors, like the purpose of the system, are under control and can only be changed at the highest level of management of the organization, i.e. by top management. Recognition of this circumstance caused a radical revision of the role of the leaders of the organization. It consists in creating conditions within the organization that maximize the full use of its resources including human resources.

* Corresponding author: v.mikhalchenko@mail.ru

The well-known management specialist Myron Tribus described the situation as follows: "People work in the system. The task of the manager is to work on the system, improving it with their help. ... Only top-level managers are authorized to make changes to the system, and if you personally do not do this as a manager, then the matter will not be done. Your production as a whole will be feverish. As a manager, you can not absolve yourself of the responsibility for the quality of management of production processes for which you are required to respond. And if you can shift this responsibility to someone, then do we need you? "[9].

From the above, it follows that to increase the efficiency of business it is important to establish not the one who is to blame, but what needs to be done to improve the system and achieve better results.

Solving of this problem requires constant monitoring of the state of the system, monitoring of the variability (change) of parameters of its processes, analysis of collected statistics, development of improvement programs and their implementation.

2 Methods

Until recently, the so-called "engineering" approach is widely used in management practice, when the assessment of the state of the system is carried out in terms of compliance of process parameters with established requirements, norms and tolerances. With this approach, the results of processes and the state of systems are evaluated as "satisfactory" or "unsatisfactory", "bad" or "good".

The limitation of the traditional approach is that it does not provide guidance and necessary information on optimal ways to improve the process, how the manager should act.

The solution of this problem in the early 20s of the twentieth century was proposed by W. Shewhart [10, 11], developing the concept of variability management, which was subsequently developed and supplemented by E. Deming [12].

The concept of the Shewhart-Deming variability is based on the principle of separating all the causes of variability into two main types:

- "common" causes of variability - are potentially inherent in the process or system, manifest constantly, hour after hour, day after day and affect everyone who participates in a particular process;
- "special" causes of variability - are not a permanent part of the system, they can only detect themselves at certain periods of time, do not appear on all participants in the system and are the result of certain factors.

For example, the productivity and quality of labor of all workers in the shop are influenced by factors that are common to all: indoor lighting, temperature, ventilation quality, adopted processing technology, the adopted system of staff selection and training, labor remuneration, etc. But there are factors that are specific to individual workers. Some of them may feel ill due to illness. Someone has problems in the family, and all his attention is focused on these problems, the other has gone wrong with the machine, and as a result, the quality of his work has significantly decreased. This example demonstrates the importance of understanding the type of variability that one has to deal with in a particular case before any management decision is made. A process or system whose results are influenced only by "common" causes of variability, W. Shewhart proposed to call "stable" or "manageable". Such systems are in a state of "statistical control". This does not mean that there is no variability at the output of the process or that the degree of this variability is small and is within the tolerances. No! Stability of the process means only one thing: the variability of its outputs is statistically predictable, that is, the probability of their being found in a certain range of values can be indicated.

The process, the results of which are determined by the joint action of "common" and "special" causes of variability, is called unstable. Instability does not mean that this process produces results with a high degree of dispersion. It is called unstable because the magnitude of this spread and, consequently, the probability of its various values appearing varies unpredictably in time.

The approach to the analysis of manifestations of variability, in terms of the type of causative factors, is fundamentally different from the traditional one, based on the "engineering" approach. As shown by W. Shewhart, methods of variability management should differ significantly for stable and unstable processes.

In particular, W. Shewhart justified and formulated the following rules of variability management:

1. It is ineffective to attempt to identify and eliminate the causes of individual deviations, when variations arise in a stable process.
2. It is effective to identify and eliminate the cause of individual large deviations in an unstable process.

W. Shewhart proposed an effective tool for classifying the type of causal factors of variability and determining the state of the process "stable" or "unstable" - the so-called "control card" (Fig. 1).

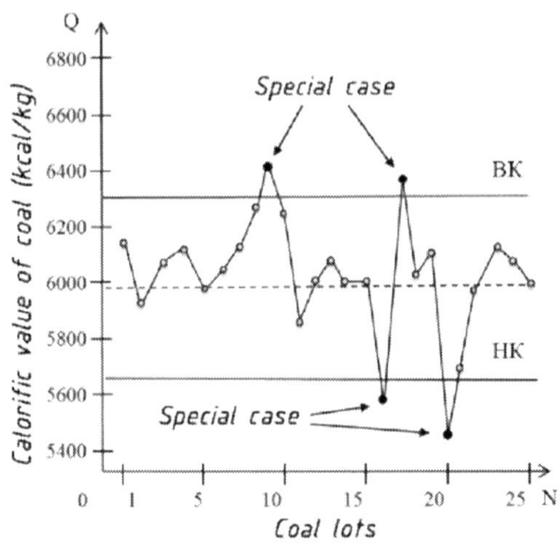

Fig. 1. Control card for coal calorific value in lots for export

The control card can be constructed on the basis of data on the change in time of the controlled parameters or characteristics in the process of interest to us. It can be quality parameters (calorific value, moisture, ash content of mined coal), production indicators (volume of production, processing or shipment for individual calendar intervals), statistics of injuries, absenteeism, etc. These statistical characteristics are plotted on a control card.

The center line on the control card is the average value of the statistical characteristic for the parameter under study. The upper (BK) and lower control boundaries (HK) are calculated according to certain relationships based on the same data. In the simplest case, the upper and lower control boundaries are defined by the relations:

$$BK = \bar{x} + 3\sigma_x; \; HK = \bar{x} - 3\sigma_x,$$

where $\bar{x} = \frac{\sum_i^N x_i}{N}$ – mean value of the studied variable in the sample under analysis { x1, x2..., xn};

$\sigma_x = \sqrt{\frac{1}{N-1}\sum_{i}^{N}(\bar{x}-x_i)^2}$ – RMS deviation estimated for this sample.

The process is considered stable if the values of the variable are within the control limits, in spite of the fact that some of the points, being within these boundaries, do not satisfy the requirements for this parameter. The control boundaries fix the range of possible fluctuations in the values of the variable being studied. Points that fall outside the control boundaries indicate the appearance of special causes of variability, the instability of the process.

The general algorithm of actions of the manager for the implementation of the concept of continuous improvement of processes and systems is given below.

At the first step of the process monitoring, based on the analysis of the control card data, the type of the system is set.

If the system is "unstable", it is necessary to identify and eliminate the causal factors that predetermine the occurrence of "special" cases and, thus, lead the system in a "stable" state.

If the process (system) is "stable", the decision maker does not respond to individual deviations of the process (system) parameters, but performs continuous monitoring and evaluation of the effectiveness of this process (system).

In case of unsatisfactory performance of the "stable" system, it is necessary to identify the most significant "common" causal factors of variability and eliminate their negative impact on the process (system). In the future, if the parameters of the process satisfy the effectiveness requirements, the decision-maker monitors the indicators until the moment of its completion without interfering in the process of its development and does not react to individual deviations of its parameters.

The main responsibility for identifying the specific reasons for the variability of the unstable process rests with the workers who directly implement this process, although they may need help from managers and engineering staff. Responsibility for revealing factors of common causes of the variability of a stable process lies with the engineering staff and the leaders.

At the same time, the responsibility for introducing changes into the system lies almost entirely with the leaders of the organization.

Heads of organizations are responsible for creating a system that should initiate and implement the process of change. The process of change presupposes training of all employees in the organization, from top managers to workers, on the basics of the concept of variability management and methods of control and management.

It is necessary to emphasize that the enormous potential opportunities that open before organizations that consistently realize the knowledge of the basics of the concept of variability can be realized in full only if mastering and practical use of this knowledge begins with the top management. The control cards of key performance indicators of the enterprise built by the managers of these enterprises present the greatest value and importance.

To date, the techniques of statistical analysis and, in the first place, the toolkit of control cards at all levels of management and in various spheres of business activity have been well studied and a large number of publications on this issue have been presented in considerable detail (13-16).

As in other industries, the concept of variability can be used in the coal mining industry. Moreover, in comparison with other industries, in coal mining there is a certain specificity consisting in the fact that the effectiveness of its processes is largely influenced by factors that do not depend on human will, namely, the variability of natural conditions. Laid over the technical, technological and organizational variability generated by the imperfection of human activity, this "wave", increasing and causing significant losses, spreads to related industries, reaches final consumers, ultimately determining the level and quality of people's lives. In this regard, mastering the fundamentals of the Shewhart-Deming variability concept by the managers of coal mining enterprises and the implementation of this knowledge in practical activities acquires high relevance and becomes a critically important and necessary condition for ensuring competitive advantages and survival of coal-mining enterprises in the current economic conditions.

3 Results and discussions

Let's consider some examples of application of the concept of variability in real management practice at the coal-mining enterprise. At the same time, we will focus not so much on the technique of constructing control cards, but on those conclusions that follow from understanding the concept of variability for engineering and management practice.
From the point of view of the end user, the performance of the coal mining enterprise can be assessed by a set of characteristics that determine the consumer value of its product and include the following three main groups.

The first group is the product quality indicators that characterize its functional purpose (calorific value, chemical composition, ash content, etc.).

The second group of indicators characterizes the terms of delivery, including the most important of them - the prompt delivery, its reliability (accuracy of compliance with the schedule).

The third group of indicators is economic. The most significant of all possible economic indicators for the end consumer is the price of products at the point of consumption.

Figures 1, 2, 3, and 4 show the data characterizing the typical for coal mining operations level of variability of the indicated groups of users.

In particular, Fig. 1 illustrates the variability in the level of caloric value of Russian coal received from Kuzbass, a large coal-mining region of Russia, by one of the foreign consumers. As it follows from Fig. 1, from the standpoint of the end user, the process of formation of the functional properties of coal is unstable.

Deviations from the average level of quality cannot be explained by the presence of only system-wide variations, since special reasons for variations also occur. At the same time, fluctuations are possible both in the direction of reduction and in excess of the expected "natural" spread of the calorific value of coal, which indicates the absence of a system in the formation of the quality of coal supplied.

In this situation, in accordance with the recommendations discussed above, the optimal strategy involves conducting an immediate analysis and eliminating the reasons for undesirable fluctuations in the quality of the supplied coal in cases where it exceeds the control limits identified on the control card. The consistent detection and prevention of further manifestation of the specific causes of variability will allow stabilizing the situation - to form a system for ensuring the quality of the supplied coal.

Fig. 2 presents data characterizing the delay in the implementation of the agreed schedule of coal supply to consumers.

This indicator directly characterizes both the efficiency and reliability of the coal supply process. As follows from the figure, the delivery process can be considered as stable: all points on the control card fit within the control limits (the НКГ has negative values and is assumed to be 0). This means that the observed variability is completely determined by the existing system of organization, management and delivery. However, the statement of the conditions of stability does not mean that the consumer is satisfied with the existing state of affairs. As the control card shows, the average delay is 1-2 months relative to the agreed delivery schedule. At the same time, delays in the beginning of the delivery, reaching 3-4 months, are observed steadily and predictably. Moreover, we can expect a delay of 4-5 months within the framework of the current system. In this case, such a large deviation cannot be regarded as a special case - it fits perfectly into the control boundaries. It can be said that the existing system of organization and management of the supply process is "ideally adapted", that it is stable to give the observed backlog from the delivery schedule.

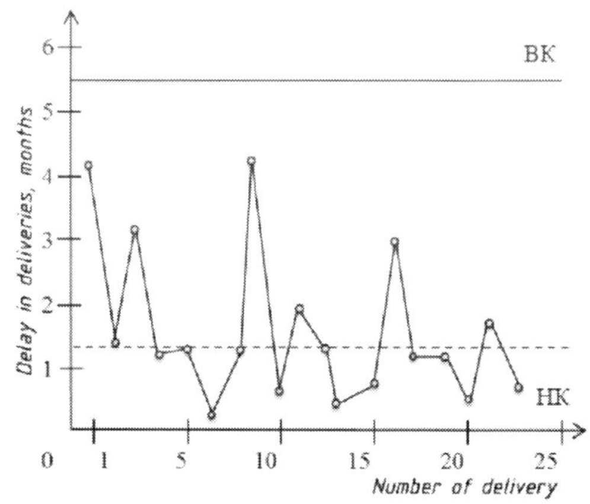

Fig. 2. Control card for the indicator of deviation from the start of coal deliveries for export as per the agreed schedule

The theory of variability determines the optimal actions in this situation: it is pointless to find out the reasons and a person to be blamed for large delays in the delivery under individual contracts. In Figures 3 and 4 there are control cards for the indicators of technical and economic efficiency of a group of coal-mining open-pit mines of Kuzbass. As it follows from the analysis of these control cards, despite the significant differences in the technical and economic indicators that characterize various enterprises, they all belong to the same system. In this case, this means that from the point of view of the applied design methods, the technology of mining, the methods of management and organization of labor, all these enterprises are the same. Differences in technical and economic efficiency are due to some random coincidence, which is characteristic for this moment, as well as a complex interconnection of factors common to all enterprises.

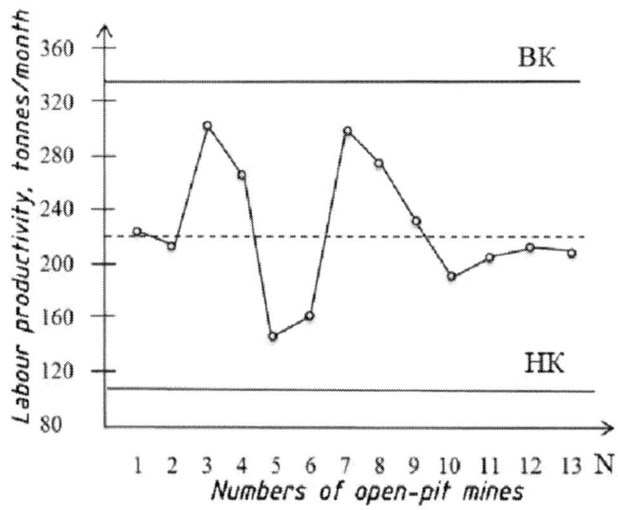

Fig. 3. Control card for workers' labour productivity indicators of a group of coal mining enterprises of Kuzbass

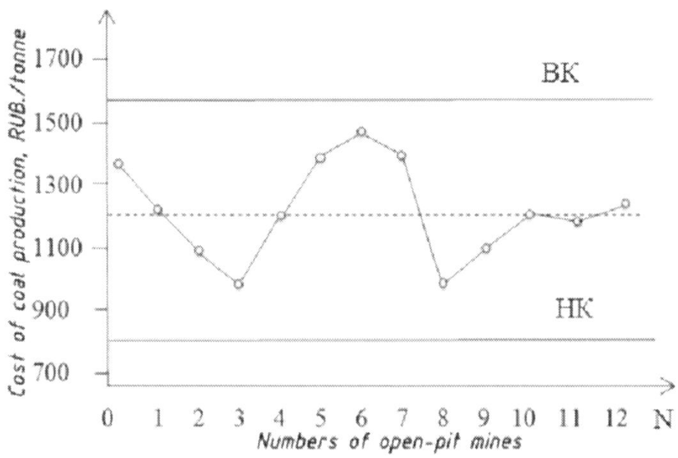

Fig. 4. Control card of indicators for cost of 1 tonne of coal production in a group of coal mining enterprises of Kuzbass

This means that in different time periods the magnitude of the spread of the characteristics considered between individual enterprises is likely to remain, but the position of different enterprises may change, that is, both the leaders and outsiders can change places. As in the previously considered examples, this state of affairs does not at all mean it's satisfactory. On the contrary, the comparison of the obtained characteristics with the requirements of the market shows that some of the enterprises considered do not satisfy the conditions of competitiveness. From the point of view of the traditional logic of management, one should expect an inadvertent reaction from the leadership to find out the causes and who to be blamed for "low efficiency" of these enterprises, to carry out measures for their recovery or closure. It is possible that in a situation of "extinguishing a fire" in the event of a threat to the survival of the whole company, such actions may be justified. However, from the perspective of the theory of variability, this approach will be equivalent to using the traditional "rejection" method. This "management trick", creating the illusion of going out of the situation, often at the expense of significant additional costs, leaves aside the consideration of the causal factors of the appearance of the problem.

Namely: the existing coal mining system, which includes the methodology of design, management and organization of labor, the technology of mining will inevitably lead to the appearance of this average level and dispersion and labor productivity, and the cost of 1 ton of coal at individual enterprises. The solution of the problem, therefore, is not in the search for the allegedly "special reasons" of the poor performance of individual enterprises, but in a qualitative change in the system common to all, which determines their technical and economic efficiency.

4 Conclusion

A well-known consultant in the field of modern management and one of the authors of the new generation of management science - quality management E. Deming, very briefly and clearly formulated the main idea of business success as follows: "If I had to express my message to management in only a few words, I would say: "The whole point is in decreasing the variations".

Developing control cards, W. Shewhart set the task to improve the quality of products by reducing the variability in technological operations. E. Deming developed the idea of W. Shewhart, placing it in the basis of the philosophy of quality management. He proceeded from the fact that the principles of management in social and business systems were subject to the same objective laws of variability as processes in technical systems. E. Deming convincingly showed that those who didn't understand the fundamentals of the theory of variability would not be able to work out the right management decision and, thus, could not effectively manage his business.

Using the Shewhart-Deming concept of variability management, managers focus their attention on the "causal factors" that limit the effectiveness of the business system, and, improving processes, achieve a decrease in the variability of parameters and characteristics of the latter. They achieve better use of working time (less breakdowns, adjustments, downtime), better use of materials (less rejects), lower costs for re-work and control. This allowed simultaneously to increase the degree of uniformity (quality) of products, and to reduce its cost. On this basis, the competitiveness of producers is being created and the chain reaction of a qualitatively different, much more effective economic strategy is triggered - the strategy of continuous improvement of processes and systems.

The Russian industry and the economy as a whole have yet to understand and master these principles. Obviously, the way of their development will not be easy, but it must be walked for survival in an open market economy, for cooperation and rivalry with those for whom these principles have become the norm of life. Knowledge and adherence to these principles is what will ultimately separate the ascent to success from the trajectory of fall for any organization and including the operations in the coal industry.

References

1. V. Feigenbaum, *Total Quality Control* (McGraw Hill Inc. 1983)
2. J. M. Juran, Juran on Leadership for Quality, New York City : The Free Press (1983)
3. W. E. Deming, Out of the Crisis, Cambridge, Mass.: MIT Center for Advanced Engineering Study (1986)
4. T. Ohno Toyota Production System, Productivity Press (1988)
5. John S. Oakland, *Total Quality Management. The Route to Improving Perforanans. Second Edition* (University of Bradford: Management Centre 1993)
6. J.P. Womack and D.T. Jones *Lean Thinking. Banish Waste and Create Wealth in Your Corporation* (New York at al:, Free Press, 2003)
7. V. Mikhalchenko, Y. Rubanik, IOP Conf. Ser.: Earth and Environ. Sci., **45**, 1 (2016)
8. V. Mikhalchenko, Y. Rubanik, N. Osokina, A. Mikhalchenko, *Coal-16. Advances in Engineering Research* (Atlantis Press, 2016)
9. M. Tribus, *The Germ theory of management* (SPC Press, 1992)
10. W.A. Shewhart, *Economic Control of Quality of Manufactured Product* (Milwaukee, WI: ASQ Quality Press, 1931, reprint 1980)
11. W.A. Shewhart, *Statistical Method from the Viewpoint of Quality Control* (N.Y., Dover Publications, Inc., 1939, reprint 1986)
12. Henry R Neave, *The Denting Dimension* (SPC Press, Knoxville, Tennessee 1990)
13. R. Futami, Rep. of Statistical Application Research, JUSE, **33**, 2 (1986)
14. Y. Nayatani, Rep. of Statistical Application Research, JUSE, **33**, 1 (1986)
15. D. Wheeler, *Understanding variations. The Key to Managing Chaos* (SPC Press, 1991)
16. D. Wheeler, *Advanced Topics in Statistical Process Control* (SPC Press 1995)

Evaluation of consumer satisfaction with the quality of training of young professionals by the universities for enterprises of coal-mining complex

Vadim *Mikhalchenko*[1*], Irina *Seredkina*[1]

[1] T.F. Gorbachev Kuzbass State Technical University, Institute of Economics and Management, Department of Economics, Vesennaya str., 28, Kemerovo, Russia

Abstract. The article presents the study results of evaluation of customer satisfaction (a big coal mining company) with the quality of training of young specialists with higher education. Research is based on the model of Kano, one of the most effective tools for assessing the quality of products within the framework of modern management model - Total Quality Management. Based on the results, "quality profile" of training of young specialists with higher education is built and the factors that determine its level and characteristics are identified. The findings show universities the need to include such transformations and changes into the system and the processes of creation and positioning of their "product". It will give a higher added value to the "product" and, therefore, provide the manufacturer and the consumer with the higher competitive advantage in the current economic conditions.

1 Introduction

The modern world economy is showing an ever-increasing pace of change and innovation. Experience and knowledge of previous generations are depreciating too quickly. Almost no company can believe into its secured future any longer, or ensure the sale of their products. Tom Peters, one of the authors of the research in the field of modern methods of management that is widely known in the west [1], mentioned in his later works: "... The pace of change has become too fast to make at least one company feel safe" [2].

Thus, the continuous change in enterprises' environment has become one of the fundamental characteristics of a modern economy. According to the well-known theorist in the field of corporate strategy Kenichi Ohmae, "... the only reliable way to ensure competitiveness in an environment with a high degree of uncertainty and variability is to make this variability be a way of life" [3]. In the context of rapid changes and tough competition only highly adaptive systems can remain viable, i.e. systems that can quickly **learn** and **change**. In constructive terms, this means that the business system should possess a degree of flexibility equivalent to a changing environment at all stages of

* Corresponding author: V.Mikhalchenko@mail.ru

production and management levels, which provides it with the ability to feel itself "like a fish in barrel" with regard to this kind of uncertainty. From the point of view of the implementation of this concept in practice, the closest approach is based on the creation of so-called flexible manufacturing systems (agile manufacturing) [4]. The term "flexible manufacturing" characterizes the organizations, which have business systems that include processes, tools (machines) and the staff training to respond quickly to changing customer needs and trends in the market, and at the same time provide the product manufacturing or services without significantly increasing costs or reducing quality.

Many researchers consider flexible production system as a further step in the development of production systems after the lean-production [5] and Toyota production system (TPS) [6].

The implementation of this approach requires the reengineering of traditional production systems. In particular, in the mining industry, as studies have shown [7, 8] the implementation of this approach provides enterprises, which operate in conditions of high uncertainty and volatility parameters of the economic environment, with a significant increase in efficiency.

The conclusion about the necessity of continuous change, predefined high rate of technological progress is the most significant in the situation with a rapid change in social, political and economic conditions. Russian companies are going through such a period, and the problem of rapid changes, adaptation to new environmental conditions for them is relevant as ever.

The foregoing allows us to see the problem of finding effective management principles from a new angle. It is useless to copy the style of behaviour and management of successful companies. The ability to **learn** and **change** is becoming mandatory for organizations seeking to achieve or maintain success. At the same time, it is not just about learning and change. The issue of life is to achieve the necessary pace, the speed of these processes. The main means of survival and the achievement of perfection for organization is **the maximum high speed of its training.**

In the circumstances outlined, skill development plays an important role in ensuring the viability of the business. As a reaction to it, the market of educational services is continuously expanding. In recent years, for example, only in the Kemerovo region 29 educational institutions of higher education were preparing specialists in more than 200 areas and specialties.

In an increasingly competitive environment the client satisfaction with the quality of trained specialists becomes the determining factor in gaining competitive advantage on the market of education for universities. Thus, as it was noted by guru of modern management E. Deming: "... it is absolutely not enough to have a consumer who is just satisfied ... success in business comes from regular customers who are loyal to exactly your product or service and bring you their friends" [9].

2 Methods

The research of the labour market satisfaction with the quality of training of specialists with higher education was carried out based on Professor N. Kano's model [10]. The method is aimed at identifying the needs and expectations of consumers, their structuring and drafting of "quality profile" of young professionals.

The Kano model considers different types of emotional reaction of consumers to the individual characteristics of the product [11]:

Type 1. "Attractive features" (desired, advancing quality)

Attractive features are a group of parameters of quality which give the consumer unexpected value of the offered products, the existence of which he could only dream of, not even suggesting the possibility of their implementation. Attractive characteristics (if they are in the product) cause feeling of satisfaction and delight. However, if there are no such characteristics, the users do not experience dissatisfaction. Attractive characteristics are unexpected, they satisfy previously unmet needs. Their presence in the product provides favourable conditions for a breakthrough to the market and further improvement of the product.

Type 2. "Dimensional characteristics" (required quality)

These characteristics cause satisfaction (if there are such) or dissatisfaction (if there are no such). Usually they are measured directly by the consumer, and primarily affect the value of the product according to his point of view. This is a set of quality indicators, which has both technical and functional characteristics of the product.

Type 3. "Required characteristics" (basic, main quality)

According to consumers, these characteristics belong to the group of those qualities that must be in the product, and therefore, expecting them in the product, he does not consider it necessary to talk about them with manufacturer beforehand. Their absence may result in a negative reaction of the consumer. Manufacturer has a risk to lose its image and subsequent business if he does not pay adequate attention to the basic indicators of the quality of the product.

Type 4. "Unimportant characteristics"

Availability of unimportant characteristics causes a mixed reaction of users, but in general, they do not care whether these characteristics of the product exist or not. The return on investment in these characteristics is low.

Type 5. "Undesirable characteristics"

The presence of a part of undesirable characteristics in the product nullifies the positive impact of attractive and desired characteristics.

Understanding the requirements of the consumer by the manufacturer, properly identified before the start of product design and production, ensures the success of this product in the market and reduces the time of its creation.

The OJSC "SUEK-Kuzbass", one of the largest industrial complexes of the Kemerovo region and which brings together mining and processing of coal, acts as a customer during research, which was conducted in the period of 2015-2016.

Within the framework of sociological research on consumer satisfaction with the quality of training specialists with higher education, respondents were asked only 4 questions:

1) What kind of qualities of young professionals who got education in universities and got a job in your organization (structural division) **do** you like?
2) What kind of qualities of young professionals who got education in universities and got a job in your organization (structural division) **don't** you like?
3) What is **not necessary** (not required) for specialists in the program which they take in the universities?
4) What kind of skills are **required** at work, however, universities do not include them into the program to prepare specialists?

The responses were grouped according to their "kinship" in the framework of the method used "affinity diagram" [12], one of the tools of the total quality management [13]. Pareto analysis was used to identify main problems [14].

3 Results and discussion

According to the results of the study, the respondents mentioned the following as the characteristics of the training of specialists for the enterprises of coal-mining complex causing *satisfaction* (required quality), which can be roughly grouped into the following groups:

1) professionally significant qualities of the person (interpersonal skills, commitment, desire to achieve the highest results, responsibility, curiosity, initiative, activity, sense of duty, discipline, diligence, ability to engage in the work quickly, and so on);

2) developed personal qualities (kindness, outlook, optimism, perseverance, hard-working, flexibility, reliability, ability and willingness to learn, adaptability, etc.);

3) availability of professional knowledge (knowledge of modern technologies, the ability to work with search tools and information processing, understanding of the profession);

4) uncommon thinking (a fresh perspective on task solution, production, innovative thinking).

35% of respondents noted the absence of a close connection between the content of educational programs and the requirements of the customer, excessive theorization of education, stereotyped training. These characteristics in the preparation of specialists with higher education are *indifferent* in accordance with the Kano model for employers. However, the presence of such characteristics reduces the level of customer satisfaction with the quality of products offered to them.

Among the *unacceptable qualities* for specialists with higher education the respondents named:

1) lack of formed professionally significant qualities of the person, such as: "not able to take independent decisions", "lack of discipline", "lack of zeal to work", "frivolous attitude to work", "lack of initiative", "lack of subordination", etc.;

2) presence of a number of personality qualities that impede the normal implementation of tasks, decision-making, conflict-free communication in the team: infirmity, hesitation, excessive ambition, "claims in the official position do not meet existing knowledge";

3) lack of specialized knowledge in the chosen specialty;

4) lack of practical training.

As *attractive* characteristics of specialist with higher education, which give unexpected value to the employer and could cause admiration, the respondents indicated:

1) developed individual professionally significant personal qualities (responsibility, discipline, "the seriousness with respect to given task", team work, desire of self-development, etc.);

2) general professional knowledge ("human resource management", "computer skills", "the ability to search for information");

3) specialized knowledge of chosen profession;

4) practical skills of the specialist;

5) consumer demands are matching with specialization.

These findings allowed to form a "quality profile" of training specialists with higher education for the enterprises of coal-mining complex (see Figure 1).

Quality parameters, which are required for the specialists with higher education according to employers' answers (the "A" position), are based on the answers to the questions: "What kind of qualities of young professionals who got education in universities and got a job in your organization (structural division) do you like,?" and "What kind of skills are required at work, however, universities do not include them into the program to prepare specialists?"

In turn, the quality parameters, underdevelopment of which cause negative reaction for the consumer (the "B" position) were obtained in the framework of the answers to the questions: "What kind of qualities of young professionals who got education in universities and got a job in your organization (structural division) don't you like?" and "What is not necessary (not required) for specialists in the program which they take in the universities?"

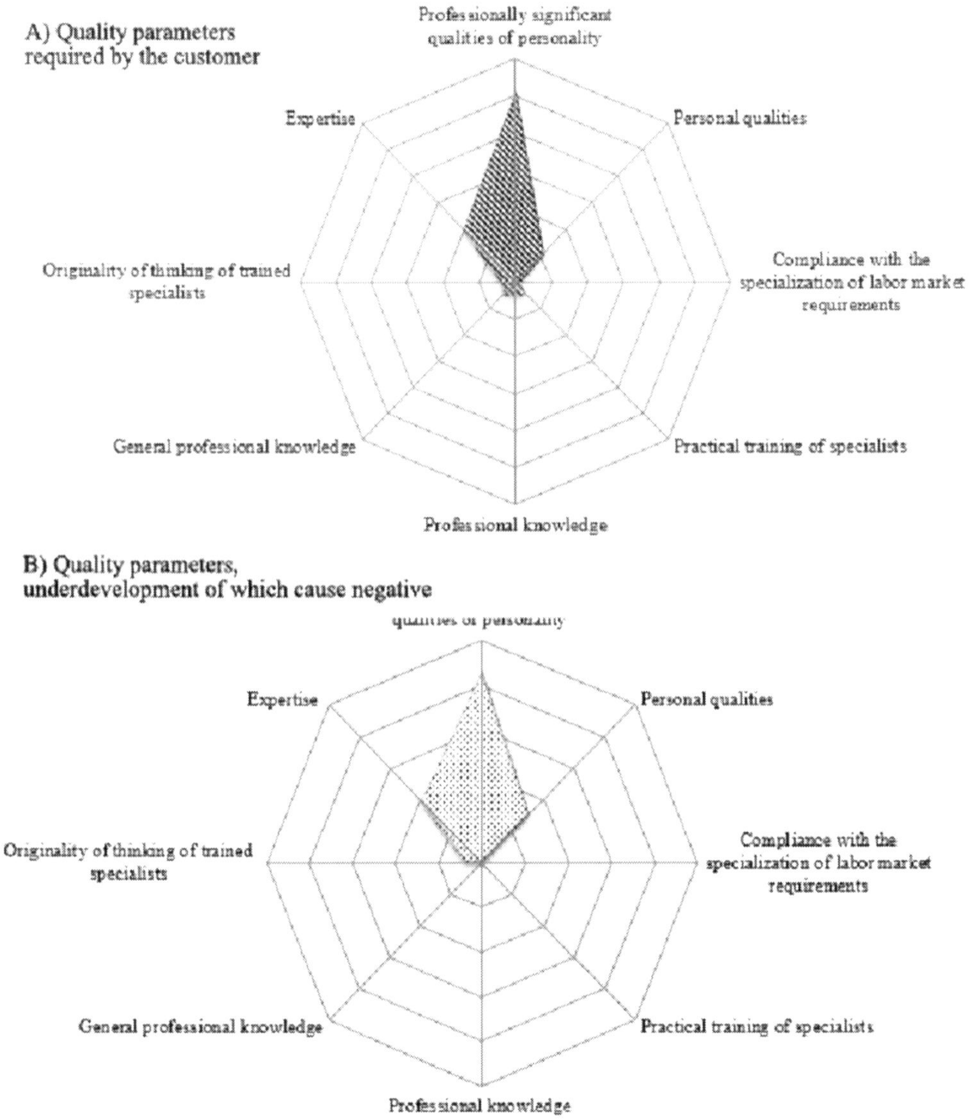

Fig. 1. "Quality profile" of training specialists with higher education for the enterprises of coal-mining complex

Comparing the formed quality profiles of training of specialists with higher education, it should be noted that for consumers, when they hire specialists with higher education, extremely important are:

1) development of professionally significant qualities of the person. If the specialists with higher education do not have such qualities, it causes extreme dissatisfaction of employers, at a sufficiently high level of development of such qualities - the admiration and affection;

2) availability of practical skills. If the specialists with higher education do not have such skills, employers consider a serious disadvantage of higher education, at a sufficiently high level of development of such skills – make them excited;

3) availability of expertise. The high level of development of such skills of specialists with higher education is able to arouse the admiration of the employer;

4) developed personal skills.

These qualities are included in the *basic* category.

4 Conclusion

Results of the study show that the employer is not satisfied with the quality of training specialists with higher education. He does not get on a regular basis the "product", which would be fully consistent with his ideas about the high level of training in all aspects of the quality required for the implementation of the organization's strategic goals. In support of this, it may be noted, for example, that some specialists can have the same qualities which are formed at a high level (professionally significant qualities of personality, personal qualities, professional knowledge, practical skills), which causes the affection and admiration of the employer, and other specialists do not have such qualities and skills and it causes criticism by employers.

In the existing system of relations, the employer is not burdened with obligations, not interested and does not consider it necessary to inform universities about what they should do to form desired qualities of specialists with higher education in the process of their education.

Analysis of "quality profile" of training specialists gives an opportunity to the universities to "hear the voice of the customer", define strategic positions and organize work on the further quality function deployment [15] of training of specialists with higher education, demanded by the labour market.

However, to achieve the desired result it is necessary to build a system of cooperation of universities and consumers of their "products" so that it is possible not just only manage the process of training and improve its quality, but also provide high-speed training of "consumer". It is critically important and necessary to ensure competitive advantage and survival of the coal-mining companies in the current economic conditions.

References

1. T. Peters, R. Waterman *In search of Excellence* (Wings Books, 1980)
2. T.Peters, *Triving On Chaos* (Wings Books, 1995)
3. K. Ohmae, *The Mind of The Strategist: The Art of Japanese Business* (McGraw-Hill, Inc., 1982)
4. P.M. Noaker, Manufacturing Engineering, **113**, 40 (1994)
5. J.P. Womack and D.T. Jones, *Lean Thinking. Banish Waste and Create Wealth in Your Corporation* (Free Press, 2003)
6. T. Ohno *Toyota Production System* (Productivity Press, 1988)
7. V. Mikhalchenko, Y. Rubanik, IOP Conference Series: Earth and Environmental Science, **45,** 1 (2016)
8. V. Mikhalchenko, Y. Rubanik, N. Osokina, A. Mikhalchenko, Advances in Engineering Research, **16** (2016)
9. W. E. Deming, *Out of the Crisis* (MIT Center for Advanced Engineering Study, 1986)
10. N. Kano, *Guide to TQM in Service Industries* (Asian Productivity Org, 1996)
11. Jan Moorman, *Leveraging the Kano Model for Optimal Results,* available from https://uxmag.com/articles/leveraging-the-kano-model-for-optimal-results
12. Y. Nayatani, Rep. of Statististical Application Research, JUSE, **33**, 1 (1986)
13. John S. Oakland, *Total Quality Management. The Route to Improving Perforanans. Second Edition* (University of Bradford: Management Centre, 1993)
14. H. Gitlow, S. Gitlow, A. Oppenheim, and R. Oppenheim, *Tools and Methods for the Improvement of Quality* (Homewood, III.: Irwin, 1989)
15. James L. Brossert, *Quality Function Deployment* (ASQ Quality Press, 1990)

Innovative development of the economy as the most important factor in ensuring the financial security of the mining region

Natalia Kudrevatykh[1,*], *Tatiana* Snegireva[1] and *Anastasiya* Tselischeva[2]

[1]Branch of T. F. Gorbachev Kuzbass State Technical University in Prokopievsk, 653039, Kemerovo region, 19a Nogradskaya str., Prokopievsk, Russia
[2]JSC SUEK-Kuzbass, Mine Management Taldinskoye-Zapadnoye, Russia, 652700, Kemerovo region, Prokopievsky district, village Bolshaya Talda

Abstract. The financial security of the region, being one of the main elements, has a significant impact on the economic security of the Kemerovo region, the largest mining region in the country. The task of the study is to assess the real situation in the financial security of the region and justify the need to develop measures to strengthen it. The article assesses the financial security of the Kemerovo region according to its main criteria: revenues and expenditures of the consolidated budgets of the constituent entities of the Russian Federation per capita; the number of operating credit organizations and their branches per 1000 people of the population; the amount of deposits of legal entities and individuals in credit institutions and bank branches; average amount of individuals' deposits on ruble accounts with the Savings Bank of the Russian Federation; the balanced financial result of organizations' activities, etc. In the course of the work, threats to the financial security of the region were identified, which can be solved by conducting innovations. The results of the study can be used to develop a strategy for the development of the region.

1 Introduction

The problem of economic security worried the minds of many philosophers, economists and politicians, among them Democritus, Aristotle, Plato. The basis of modern views on the security of society in the XX century was laid by the works of V. Pareto and the outstanding Russian scientist A. Bogdanov.

A considerable number of papers have been devoted to economic security in general and to its certain elements [1-14].

The following types of regional security are distinguished as part of the economic security of the region: industrial, socio-economic, financial, environmental, energy, transport, information, food, etc.

* Corresponding author: knv.fk@yandex.ru

In spite of the fact that financial security is the basis of the social and economic security of the region, this aspect is not sufficiently studied at present, and a number of issues remain sharply controversial.

Finance plays an important role in the effective functioning of any socio-economic system; therefore the analysis of the financial sphere should be part of the assessment of the overall level of economic security of the territory.

The purpose of the work is to assess the financial security of the mining region and justify the need to develop measures to strengthen it.

To solve the set goal, the following objectives were solved:

1. The assessment of the financial security of the region by the main criteria is given.
2. The main threats to the financial security of the Kemerovo Region have been identified;
3. The need to increase the level of regional innovative potential for the purpose of ensuring financial security as the basis for economic security of the region has been justified.

2 Materials and methods

During the research, such scientific methods as: statistical observation, analysis and synthesis, a systematic approach were used.

3 Results and discussions

Considering the Kemerovo region from the standpoint of financial security, it should be noted that this region has been and remains single-industry. The basis of financial security is coal, which makes it dependent on the situation on the foreign market. Today, the Kemerovo region accounts for 58.3% of all coal production in Russia, 72.7% of all coking coal production, and 100% for a whole group of highly valuable coking coal grades. Over the years, the coal industry has allowed the region to maintain a high level of socio-economic well-being. The realities of today dictate the need for the development of new types of industries. The coal industry, despite its leading position, does not allow the region to maintain a high level of financial security. This is confirmed by the results of the evaluation of statistical indicators for a number of recent years.

To assess the financial security of the Kemerovo region, a comprehensive approach based on the application of the system of indicators and comparison with other territories of the Siberian Federal District (SFD) was used. In this approach, it is necessary to analyze financial security through the following assessment: the state of the regional budget (the ratio of revenues and expenditures of the consolidated budgets of the constituent entities of the Russian Federation per capita); the scale of the banking sector in the region and the well-being of credit institutions; effectiveness of the activities of organizations and enterprises operating in the territory of the entities; financial well-being of the population (by comparing data on the amounts of bank deposits and arrears on loans). The indicators characterizing the financial situation in the region are divided into two groups in accordance with their positive or negative impact on the state of the territory's economy.

The first group of the indicators analyzed in this block is presented in Table 1.

In the table: 1 – Revenues of consolidated budgets of subjects of the Russian Federation in per capita terms, thousand rubles; 2 – Expenditures of consolidated budgets of subjects of the Russian Federation in per capita terms, thousand rubles; 3 – Number of operating credit organizations per 1000 people of the population; 4 – Number of operating branches of credit organizations per 1000 people of the population; 5 – The amount of deposits of legal entities and individuals in credit institutions and bank branches, million rubles; 6 – The average amount of individuals' deposits on ruble accounts in the RF Savings Bank, rubles; 7 – Balanced financial result of organizations, million rubles

Table 1. Indicators characterizing the level of development of the financial sector in the regions of the SFD in 2015 (the first group)

Region	1	2	3	4	5	6	7
Republic of Altai	79,1	82,2	9,3	28,0	7072	4254	-41
Republic of Buryatia	53,9	55,1	1,0	8,2	38509	18135	30571
Republic of Tyva	67,2	71,1	3,2	9,5	5875	3186	-4949
Republic of Khakassia	52,8	62,2	3,7	3,7	21843	10939	8597
Altai territory	40,1	41,2	2,9	5,5	117944	64590	32534
Trans-Baikal territory	49,0	55,1	0,0	4,6	46994	29800	-291
Krasnoyarsky krai	73,9	81,2	1,7	10,5	208090	99141	318276
Irkutsk region	53,5	57,6	2,9	7,9	183329	90356	184288
Kemerovo region	48,6	51,9	2,2	4,0	181614	88513	-15324
Novosibirsk region	47,9	52,7	2,2	21,1	301973	92103	47290
Omsk region	39,9	42,6	2,5	10,1	133121	53204	27967
Tomsk region	57,1	59,7	1,9	15,8	89665	31672	88459

Analyzing the data presented in Table 1 it should be noted that in terms of per capita income and budget expenditures the Kemerovo region shows values that are slightly below the average. However, if we correlate data on these indicators for all subjects of the Siberian Federal District, we can note the absence of an explicit direct link between the level of incomes and expenditures of the budget per capita and the overall socio-economic well-being of the territory. Thus, some of the highest values for the level of per capita incomes and expenditures of budgets were recorded for such regions of the Siberian Federal District as the Altai and Tyva republics. When analyzing the level and quality of life of the population of these republics, they show the worst values for the group of analyzed subjects.

According to the results of the assessment of other components of economic security, these republics often appeared in the second half of the list and, on the contrary, many subjects of the Siberian Federal District, assigned to the leaders, are on average or even lower in terms of per capita incomes and expenditures of budgets. Among the latter, there is also the Kemerovo region. This situation, apparently, is explained by the different degree of effectiveness of the formation and expenditure of financial resources of the analyzed territories.

Starting with the analysis of indicators characterizing the state of the banking sector of the regional economy, first of all, we note that according to the number of operating credit organizations and their branches, reduced to the population, the Kemerovo region is below the average recorded by the district values. And although the situation is somewhat distorted by the value of the Republic of Altai (which is more determined by the population size, rather than by the number of operating credit institutions) by the first of the mentioned indicators, one can note that even the exclusion of the named subject from the analysis does not significantly change the situation.

However, drawing attention to the following indicator of the analyzed group (the amount of deposits of legal entities and individuals in credit institutions and bank branches), it is evident that for this indicator the Kemerovo region exceeds the average by the county, behind Novosibirsk, Irkutsk region and Krasnoyarsk region.

This ratio of indicators in general is typical for Kuzbass and reflects its specifics, since it is associated with a larger scale of organizations operating in the region than the average for the district.

According to the average size of the deposit of individuals on ruble accounts with Sberbank of Russia, Kuzbass exceeds the average steady-state value by almost 82%. On the one hand, this confirms the previous conclusion, on the other hand, it allows us to speak about a sufficiently high level of financial well-being of the population of the region in comparison with other territories of the Siberian Federal District.

According to the last indicator of the first group - the balanced financial result of organizations - the Kemerovo region is below the average. Analyzing this indicator in absolute terms, we note that the balanced financial result of the organizations of Kuzbass was negative, which affected the location of our region among other regions. A negative financial result was recorded in a number of other regions, including the Republic of Altai, the Republic of Tyva and the Trans-Baikal Territory. However, the largest negative balance was observed on the basis of the performance of organizations of the Kemerovo region.

Indicators of the second group, characterizing the degree of severity of negative trends in the financial sector of the region are presented in Table 2.

The data of Table 2 allow us to note that the severity of negative trends in the financial sector of the Kemerovo region is at a fairly high level, not one indicator falls below the average recorded by the Siberian Federal District values. According to the last two indicators of this group of indicators, the significance of the Kemerovo region can be conditionally recognized as overestimated.

Table 2. Indicators characterizing the level of development of the financial sector in the regions of the SFD in 2015 (the second group)

Region	Debt on loans in rubles provided by credit institutions to legal entities (including IE), million rubles	Debts on loans in rubles provided by credit institutions to individuals, million rubles	Specific weight of unprofitable organizations, %	Overdue accounts payable of organizations, million rubles	Overdue accounts receivables of organizations, million rubles
Republic of Altai	12750	13956	38,8	277	210
Republic of Buryatia	62513	91342	33,9	7952	3823
Republic of Tyva	3634	21419	48,9	4704	1710
Republic of Khakassia	15577	35004	38,4	4998	3486
Altai territory	130307	155099	29,3	4834	3683
Trans-Baikal territory	20864	90368	41,5	8325	5853
Krasnoyarsky krai	382927	290587	29,3	30317	91137
Irkutsk region	179628	236203	27,9	18869	15031
Kemerovo region	231636	209274	36,7	127675	81597
Novosibirsk region	248911	248186	28	7281	4630
Omsk region	137361	143160	20,9	5797	3897
Tomsk region	75200	91053	28,6	16794	10082

This is due to the existence of a regularity connecting the size and economic power of the region with the amounts of debts of legal entities and individuals to credit institutions. In addition, the first of the analyzed indicators is calculated based on the location of the borrower, which also somewhat distorts the picture, characterizing the financial well-being of business entities rather than the banking sector of the region. However, on the other hand, the voiced dependence cannot be considered absolute and not always a larger number of loans entails the appearance of large amounts of debt. And the business and banking sector of the region are so strongly connected with each other that, in our opinion, the indicator in question is still quite informative from the point of view of analyzing the financial sphere of the territory. In this regard, the current situation with the amounts of arrears on loans should be recognized as potentially dangerous from the point of view of preserving the economic security of the Kemerovo region.

The spread of regions of the Siberian Federal District in terms of the proportion of loss-making organizations is low - there are no obvious leaders and outsiders, and Kuzbass is located approximately in the middle of the list of analyzed territories. However, considering that about a third of the enterprises of the Kemerovo Region are unprofitable, such a situation can hardly be considered successful, although this corresponds to the values of other regions taken for consideration.

Turning to the last indicators of this group, characterizing the amounts of overdue receivables and payables, we note that they are also not translated into relative values (as well as the first two indicators of this group). Therefore, for a more reliable analysis, we give the values of the shares of accounts receivable and payable in the total amount of debt (Table 3).

Table 3. Indicators of payables and receivables on subjects of the Siberian Federal District in 2015, %

Region	Share of overdue accounts payable as a percentage of total debt	Share of overdue accounts receivable as a percentage of total debt
Republic of Altai	8,4	5,7
Republic of Buryatia	19,1	6,7
Republic of Tyva	52,6	26,6
Republic of Khakassia	13,2	6,9
Altai territory	4,5	3,8
Trans-Baikal territory	17,1	11,5
Krasnoyarsky krai	6,5	17,8
Irkutsk region	8,5	4,3
Kemerovo region	26,0	16,6
Novosibirsk region	1,9	1,8
Omsk region	3,8	3,2
Tomsk region	9,0	6,1

Analyzing the data in aggregate, it can be concluded that the high place corresponding to the Kemerovo region in terms of overdue accounts receivable and accounts payable is due not only to the more intensive implementation of entrepreneurial activities. Thus, Kuzbass shows one of the worst values across the entire Siberian Federal District (26.0%) (the lowest is 1.9% (Novosibirsk Oblast), and the average is 14.2%), according to the share of overdue payables. According to the share of overdue accounts receivable, the situation cannot be called good either - 16.6% with the least - 1.8% (Novosibirsk region), the average - 9.3%. Separately, it should be noted that in regions that constitute a serious competition to the Kemerovo region based on the analysis of other blocks of economic security, there are markedly more favorable values of the indicators in question, which make their competitive status stronger, weakening the position of Kuzbass.

4 Conclusion

Summarizing the above, it can be noted that the problem of financial security for the Kemerovo region remains extremely topical. The most serious threats are:

1) low budget revenues and expenditures per capita;

2) a small number of operating credit institutions and their branches per 1000 people;

3) a negative net financial result of the regional organizations;

4) a high amount of debt on loans in rubles provided by credit institutions to legal entities and individuals;

5) more than 36% of regional organizations are unprofitable;

6) high amount of overdue accounts receivable and accounts payable.

The foregoing allows us to conclude that financial security, to date, requires considerable attention and the development of a set of measures aimed at stabilizing it, which will allow it to become one of the main factors for the sustainable development of the Kemerovo region, and this will positively affect the economic security as a whole. This can be achieved by raising the level of regional innovation capacity.

References

1. Bussire M., Imbs J., Kollmann R., Rancire R., American Economic Journal: Macroeconomics, **5**, 75 (2013)
2. Capone R., El Bilali H., Debs P. et al., Journal of Food Security, **2**, 13 (2014)
3. Capone R., El Bilali H., Debs P. et al., Journal of Food Security, **2**, 1 (2014)
4. Echterling F., Eierle B., Ketterer S., Int. Review of Financial Analysis, **42**, 235 (2015)
5. Glover S.M., Prawitt D.F., Messier W.F., *Auditing & Assurance Services: A systematic approach* (McGraw-Hill, 2014)
6. Galbraith, J.K. *Economics and the Public Purpose* (Boston, Houghton Mifflin Company, 1973)
7. Karen, D.E., Journal of Economic Perspectives, **23**, 49 (2009)
8. Kevin M.M., Topel R.H., American Economic Review, **103**, 508 (2013)
9. Kormishkina L.A., Kormishkin E.D., Semenova N.N., Koloskov D.A. Mediterranean Journal of Social Sciences, **6**, 163 (2015)
10. Kudrevatykh N., Sheveleva O. *Coal in the 21st Century: Mining, Processing and Safety,* 120 (2016)
11. Perroux, F.L., *L'Economie du XX siècle* (Presses Universitaires Grenoble, 1961)
12. Pottier, P., Revue Economique, **14**, 63 (1961)
13. *Sustainable Economic Development. Resources, Environment and Institutions* (Massachusetts: Academic Press, 2015)
14. Yu.A. Fridman, G.N. Rechko, A.G. Pimonov, Regional Res. of Russia, **2**, 206 (2012)

Theoretical Foundations of the New Industrialization of the Mining Region under Globalization

Ludmila Kusurgasheva[1], *Olga* Nedospasova[2], and *Evgeny* Zhernov[1,*]

[1] Department of Economics, Institute of Economics and Management, T.F. Gorbachev Kuzbass State Technical University, 28 Vesennyaya Street, 650000 Kemerovo, Russia
[2] Department of Economics, Institute of Social and Humanitarian Technologies, Tomsk Polytechnic University, 30 Lenina Avenue, 634050 Tomsk, Russia

Abstract. This study is intended to establish theoretical foundations of the new industrialization of the mining region under the globalization. The urgency of the problem is due to a significant de-industrialization of many branches of the Russian economy as a result of the neoliberal macroeconomic paradigm in the years of radical market reforms. As a result, a number of Russian regions has formed the structure of industrial production, which is characteristic mainly for the countries of the raw materials periphery. The authors see the new industrialization (neo-industrialization) as a strategic tool for solving the problem. The trend that has formed in the Russian regional policy to transfer the responsibility for the economy modernization to the regional level has attached the special importance to regional studies. Exhaustion of predominantly raw-materials export model of economic growth has put Kuzbass, the mining region in Western Siberia, in the most difficult situation due to its tight dependence on the state of the world market of energy resources. This article presents a theoretical justification of the need, feasibility and advisability of neo-industrialization for the mining region in the conditions of globalization.

1 Introduction

The purpose of the study is to establish theoretical foundations of the new industrialization of the mining region as the key element of the Russian economy modernization in the conditions of globalization. **The scientific novelty** of the study is in the development of the concept of neo-industrialization of the Russian economy in relation to the resource-extracting regions and specific features of the present stage of their economic development determined by global trends. **The scientific significance** is in identification and systematization of the theoretical (conceptual) foundations of neo-industrialization of the mining region as illustrated by the Kemerovo region - an important mining region in Western Siberia.

* Corresponding author: zhee.eti@kuzstu.ru

The practical significance of the study is related to the possibilities to use the proposed concept for justification, formulation and adoption of solutions on strategic issues of socio-economic development of the resource-extracting subjects of the Russian Federation.

This goal is achieved by means of identification and scientific substantiation of the necessity, opportunity and feasibility of new industrialization of the resource extracting region, by the comprehensive analysis of the international experience in changing the socio-economic role of this type of region in modern conditions.

The relationship of this study with the previously published works. The entire array of specialized literature, which seems important for the research topic, can be divided into three blocks. The first block consists of classical studies of foreign authors of the twentieth century on general problems of the regional economy and state regulation of socio-economic development of regions including resource extracting ones. The second block includes the study of neo-industrialization as the strategic direction of economic development at present. In the third block, the issues related to the globalization of the economy are studied. To give the theoretical justification for the implementation of the mining region neo-industrialization, let us make a brief analytical review of classical theories and opinions of regionalist economists. This will reveal the basic theoretical principles for the development of the concept of the coal industry neo-industrialization as the impulse of neo-industrialization in the mining region. The appeal to the classical writings (theories) is justified, as they have become the basis of social and economic transformation of nowadays highly developed countries. These studies include the writings of I. Thünen, W. Launhardt, A. Weber, W. Christaller, A. Loesch, W. Alonso, W. Izard et al. Let us highlight the issues disclosed by them that are important for achieving the goal of our work.

Placing a single enterprise in relation to the basic factors of production in order to minimize costs has been reviewed in the work of Weber "Pure theory of industrial location" (1909), which is the development of the Standort theories of production placement by I. Thünen and Wilhelm Launhardt. Quite rightly for that time, A. Weber considered the costs of labor, raw materials, fuel resources, and costs of their transportation to each other as the traditional factors affecting the placement. A. Loesch outlined all the known theories of the location of production in the book "The spatial organization of the economy" (1940), he extended the theory of central places of W. Christaller and developed the theory of the organization of economic space. His provisions can be used to justify the "corridor of opportunities" of the spatial development of the Kemerovo region with an area of 4% of the territory of Western Siberia.

W. Izard tried to create an integrated model of territorial projections of socio-economic life of society in major works "Location and Space Economy" (1956) and "Methods of Regional Analysis" (1960). There, the economic landscape of the territory determines the combination of consumer demand, its geographical distribution and the availability of these or other economic resources. W. Izard was a supporter of the technical and economic approach, in which special emphasis was placed on the raw materials and the transport component of the cost of production and its exposure to the scale of production. This provision remains relevant in relation to the resource-extracting Kemerovo region (the region coincides with the greater part of the territory of Kuzbass - the Kuznetsk coal basin - one of the largest coal deposits in the world; now the name of "Kuzbass" is the second title of the Kemerovo region) with its median geographic location in the country.

The new theory focuses not so much on the traditional factors of placement (transport, material and labor costs), but on the intangible factors - innovations. "The new combinations", or innovations, in the theory of economic dynamics of J. Schumpeter ("Theorie der wirtschaftlichen Entwicklung", 1911) include the development of new sources of raw materials. In the capitalist reality, according to the influential Austrian-

American economist, the predominant importance lies in the competition based on the discovery of not only new technologies, but also new sources of raw materials. This competition ensures a significant cost reduction and (or) quality improvement. Therefore, in this study, we consider the innovative component as a factor in the distribution of productive forces crucial for neo industrialization of the raw material coal industry as the key sector of the Kemerovo region industry.

The theory of growth poles by F. Perroux ("Economic Space: Theory and Applications", 1950 [1]) has received wide recognition in the studies of regionalistics, thus allowing us to draw a conclusion about the importance of the sectoral structure of the regional economy and, above all, about the recognition of the leading role of industries creating new goods and services. The centres that host the operations of the leading industries become the poles of attraction of production factors, as they provide their most effective utilization. That is what leads to the concentration of enterprises and to the formation of economic growth poles, which enable to develop natural resources comprehensively, creating a technological chain of production together with modern infrastructure facilities.

For the Kemerovo region, it is also relevant to give a detailed presentation of the poles of economic growth by J.R. Lasuen ("On Growth Poles", 1969 [2]): the growth pole can be represented by an integrated group of regional enterprises connected not only with the leading sector, but also with the export of the region. Export accounts for almost 95% of the Kuzbass foreign trade, and two-thirds of the export are the products of the fuel and energy complex, in which coal, in its turn, accounts for almost 70%.

In regards to the Kemerovo region, to justify the need for dynamic development of the economy which is characteristic of neo-industrialization, for practical purposes experts offer to rely mainly on the theories of local growth and endogenous development. The latter are included as component parts in the concept of the regional self-development. The theory of innovations diffusion deserves no less attention, in their opinion. The model endogenous development is based on the maximum use of local resources: manpower; capital accumulated at the local level; entrepreneurial skills; specific knowledge of production processes. The availability of the network of production interactions between different sectors and within the production sectors (clustering), and the susceptibility of the regional economy to innovation make the basis of development. In this model, great attention is paid to various aspects of the development of the domestic regional market and to the solving of specific problems of the territory. However, that does not mean that the economy is of closed nature, although all activities are aimed primarily at creating conditions for the steady expansion of the use of local resources and, above all, labor. The approach means renouncing of the absolute character of the overall economic growth as the basis for development. In this context, the main task of the regional policy is to create conditions for the formation of local growth potential. Thus, the theory of local growth and the concept of endogenous development are organic elements of the concept of self-development in the region, according to which self-development mechanisms are built into the policies of the federal and regional authorities. This provision is consistent with the policy of neo-industrialization.

The second block of the literature includes the study of neo-industrialization as a strategic direction of economic development in modern conditions. The neo-industrialization concept in this context is actively developed by the following authors. A.P. Sussan and R.D. Kassira [3], in the first place, have made an assumption of leveling of the economy and social development as a result of re-industrialization, which is important for smoothing of the socio-economic inequalities of the Russian regions. Secondly, they have disclosed the changes in the essence of competition and the role of technologies. S.A. Aryeetey-Attoh, P.S. Lindquist, W.A. Muraco, and N. Reid [4] considered the problems of re-industrialization and emissions reduction in relation to each other, which is particularly important for environmentally problematic Kuzbass. A. Etzioni [5] drew attention to the social and psychological aspects of re-industrialization. R. Rothwell and W. Zegveld [6] outlined a wide range of problems in industrial development. Many authors explain the so-called "Asian miracle", the rapid growth of the economies of such countries as Korea and Taiwan, by the possibilities of transferring the human capital from developed to developing countries. This phenomenon is described in detail and determined as one of the locomotive growth factors in the work of A.B. Shifa [7]. In the case of Russia, and especially Siberia, a reverse negative trend is observed.

In the third block of the literature the problems associated with globalization of the economy were studied: M. Allais [8], P. Hawken, A. Lovins, H. Lovins [9], T. Piketty [10], et al. The impact of globalization on Kuzbass is due to the current dependence of the region on the state and trends of the global economy development. Thus, the complex current situation in the region's economy is largely associated with a reduction in coal prices at the world markets in the conditions of the global crisis.

The theoretical analysis has shown that in the current literature the state and development trends of a resource extracting region and the new industrialization of the economy are often considered separately, outside the context of globalization processes. Consequently, the problem of theoretical justification of new industrialization of the extracting region of Western Siberia in the conditions of globalization is relatively new in the economic science.

2 Results and Discussion

Western Siberia in general and the Kemerovo region, in particular, become particularly important for the neo-industrial turn of the Russian industry for the following reasons.

1. Kuzbass is an old industrial region with the established industrial structure and traditions, which is at the same time capable to start a new stage of industrial development. Natural climatic, orographic, landscape, cultural, and historical features of the Kemerovo region allow us to consider it following the accepted methodology as the region with the high recreational potential.

2. At present, the coal resource base of the country is concentrated on the territory of Kuzbass: this basin mines about 56% of bituminous coal and up to 80% of coking coal; earnings from coal sales are of great importance for the economy of Russia. Therefore, Kuzbass is prepared for the initial phase of the new industrial turn, provided technical and technological re-equipment of its industrial capacity and increase in production capacity of the processing sector. In the future, they can be used for the development of industrial clusters and technological centers in other regions of the country.

3. The special emphasis on new industrial development in Kuzbass is associated with the solution to stop the ongoing migration outflow from its territory.

The increasing share of knowledge-intensive and high-tech industries, the imbalance in the structure of training of highly qualified personnel and their involvement in the regional economy, low labor productivity in the coal industry in general and the high risks to the ecological safety of the region remain to be the problematic issues. That is why in Kuzbass accumulation of the indicators of the neo-industrial economy goes slower than in similar regions of the leading European countries, such as the Ruhr region in Germany. Its positive experience [11-13] demonstrates the need to intensify work in the field of import substitution and to stimulate high-tech products exports, especially in the framework of the clusters being formed. In this case, as the latest research of the experience of the regions of Europe shows [14], it is required to consider the dependence of the performance of innovative activity in the real sector of the economy on the quality of government control.

The important goal of the neo-industrialization is to restore the Kuzbass economy as a reproductive integrity. This implies a gradual and consistent rejection of the practice of the organization of manufacture of individual products and support independent businesses, even related to the global technological networks. This is due to the fact that such companies tend to perform auxiliary and service functions with a minimal contribution to the creation and appropriation of the added value. This situation is natural, since global TNCs have been controlling for a long time all the main technological niches. Therefore, it is advisable to mining regions to participate in the creation of interconnected and integral production and processing chains within the Russian economic circuit. And only after the production is reorganized in this way, they can enter the world market. The more that Russia has the best compared with other countries variant of the formation of multi-link production and processing chains with the multiplying result in the form of increasing value-added capacity. Such a structural approach in the modern economic science is becoming more common and it is considered as one of the useful theoretical bases of the neo-industrialization policy [15].

Mining and deep processing of raw materials should be implemented on the basis of the new technology, which can be further used in the production of intermediate goods with high quality characteristics. This, for example, relates to chemical products developed in Kuzbass. In terms of economy modernization based on innovations, coal becomes primarily raw material for production of expensive chemical products. They are the necessary material for the deployment of a multi-processing industries. The latter, in turn, produce the necessary equipment for extractive industries and production of materials, etc.

In modern conditions, the models of regional development should be based on knowledge-intensive products, technologies, services, and current knowledge, which are not only strategically important resources, but also the equity. Therefore, regional growth poles should be based on educational and research centers, where all the innovative, scientific research, information and advisory, educational, and cultural activities are concentrated aimed at priority development of the region. This role in Kuzbass in the future should be performed by the innovative research and production center for development and implementation of new technologies in coal utilization - Kemerovo UgleNaukoGrad - a specialized division of the Siberian Branch of the Russian Academy of Sciences. It will deal with the development of fundamental and applied coal science in the country. Such centers in the peripheral regions are sources of knowledge for innovative firms [16]. In contemporary practice, the idea of growth poles is also realized in the creation of free and (or) special economic zones, territories of advanced socio-economic development, and industrial parks. It should be noted that the Kuzbass academic campus will be developed in close proximity to Technopark in the field of high technologies of the Kemerovo region which is the modern site for implementation of academic institutions studies. Creation of scientific institutions in the vicinity of the research object will contribute to the intellectual development of the region in the direction of neo industrialization, as well as to the transformation of the socio-cultural and educational spheres. Around the scientific center there should be highly specialized enterprises with small-scale and customized individualized production located using modern labor-saving technologies. These small and medium-sized firms are not only a guide for the formation of the modern industrial structure of the region, they also increase the region's innovation attracting new large research laboratories [17]. The experience gained in the implementation of this project can be used in other resource-extracting regions. This is how the authors see a possible way of revival of high-tech, economically and socially integrated system of regional economy.

The established theoretical foundations have allowed us to identify the following main areas for the implementation of the neo-industrialization of the mining region: 1) progressive changes in the structure of the regional industry; development of regional cluster structures; 2) growth of innovation and investment activities of resource extracting companies; 3) the greening of resource extracting production. The identified theoretical foundations of the neo-industrialization present the scientific basis of the specially developed conceptual approach to the justification of the necessity, possibility, and principal advisability of the new industrialization of resource extracting regions of the Russian Federation in the context of globalization. This is the main result of the work.

3 Conclusion

This work has defined the basic theoretical principles of transition to neo-industrial type of development of the regional economy, which, in our opinion, can be taken as the basis for justifying the neo-industrialization of a mining region. It is proved that the new industrialization of the leading coal mining complex in Western Siberia can and must become an impetus to accelerate the innovation development of the entire economic system of Russia. However, this is only the first step towards the structural modernization of the Russian economy. It is the government that must become the driving force behind the innovative type of modernization of the national economy in its main sectors and strategically important regions by conducting an active balanced industrial and regional socio-economic policy of the neo-industrialization.

References

1. F. Perroux, The Quarterly Journal of Economics, **64**, 89 (1950)
2. J.R. Lasuen, Urban Studies, **6**, 137 (1969)
3. A.P. Sussan, R.D. Kassira, *The Re-industrialization Strategy in the Global Economy* (2009)
4. S.A. Aryeetey-Attoh, P.S. Lindquist, W.A. Muraco, N. Reid, *Northwestern Ohio: re-industrialization and emission reduction* (Cambridge University Press, 2003)
5. A. Etzioni, Review of Policy Research, **2**, 677 (1983)
6. R. Rothwell, W. Zegveld, *Reindustrialization and Technology* (1985)
7. A.B. Shifa, *Essays on growth, political economy and development* (2013)
8. M. Allais, *La mondialisation: la destruction des emplois et de la croissance. L'évidence empirique* (1999)
9. P. Hawken, A. Lovins, H. Lovins, *Natural Capitalism: Creating the Next Industrial Revolution* (1999)
10. T. Piketty, *Le Capital au XXIe siècle* (2013)
11. D.K. Carter (ed.), *Remaking Post-Industrial Cities: Lessons from North America and Europe* (2016)
12. Information Technik Nordrhein-Westfalen, https://www.it.nrw.de/index.html (2017)
13. Struktur und Entwicklung der sozialversicherungspflichtig Beschäftigten von 1980 bis 2004 (2006)
14. Rodriguez-Pose, M. Di Cataldo, Journal of Economic Geography, **15**, 673 (2015)
15. J. Lin, *New Structural Economics, A Framework for Rethinking Development and Policy* (2012)
16. H. Pinto, M. Fernandez-Esquinas, E. Uyarra, Regional Studies, **49**, 1873 (2015)
17. A. Agrawal, I. Cockburn, A. Galasso, A. Oettl, J. of Urban Economics, **81**, 149 (2014)

Problems and Prospects of Sustainable Development of Mining Regions

Irina Levitskaya[1*], *Natalia* Pastukhova[1], and *Olga* Dubrovskaya[1]

[1]T.F. Gorbachev Kuzbass State Technical University, Mezhdurechensk branch, Department of Economics and Management, 652870 Mezhdurechensk Kemerovo region, 36 Stroiteley st., Russia

Abstract. At times of global financial crisis starting from 2008, nearly all single-industry towns had a different set of problems the main of which was a sharp decline in investment activity, the temporary suspension of investment projects, and from time to time the refusal to implement them. Russia's economy is represented at a significant level by city-forming organizations in single-industry cities. Core enterprises of single-industry cities provide a large part of the raw materials industries, particularly fuel and steel industry. Regional economic development strategy as a system of measures aimed at the realization of long-term objectives of social and economic development of the state, taking into account the contribution of sustainable regions in addressing these challenges is determined by the actual prerequisites and restrictions on their development. Anti-recessionary measures aimed to improve the situation will have to reduce spending on social programs and personnel, to discharge employees, so they caused drastic consequences.

1 Introduction

The objective to develop the social and economic potential of the regions is to develop the leading economy sectors in advance. Particularized employment evolution determines the location of the region in the structure of the world economy and contributes to the maximum realization of regional competitive advantages. In this regard, the leading role is played by regional and branch development that takes into accounts the specifics and peculiarities of each specific federal entity.

The recent years are characterized by the increasing coal significance in the world and in Russia. Confirming we refer to the World Coal Institute experts' views that for the next 25 years coal is sure to be the driving force of the world economy. The demand for carbon will increase at least for 50% [1].

Total probable coal reserves in Russia are around 4 billion tons, i.e. 10% of the world, general balance reserves are estimated at 200 billion tons. The raw material base of the coal industry is considered to be proven reserves in categories A + B + C1 operating, under construction coal mines as well as thoroughly explored land reserve for the new coal-

* Corresponding author: levitskaya@mail.ru

mining enterprises construction. According to these standards the coal industry raw material resource accounts 106 billion tons coal reserves [2].

Basic volume (80%) of balance reserves falls on Western and Eastern Siberia areas. Predicted coal production volumes in the country, as well as other energy resources, will vary depending on one or another alternative of socio-economic Russian development, but all embodiments guarantee high rate coal consumption growth compared with other fossil fuel types. In favorable conditions (optimistic and favorable development options) coal production in Russia may increase to 410-445 million tons by 2020. In a smaller extent favorable or in unfavorable external and internal circumstances and environment sequence (moderate and precarious development options) coal production in the country will be less than 310-375 million tons in 2020 [3].

Till 2020, to provide production growth according to a moderate development option, 130 million tons fresh capacities, of which in the Kuznetsk Basin 55 million tons, in the Kansk-Achinsk 40 million tons will be required. In optimistic case the need in new capacities introduction will be 200 million tons, of which in the Kuznetsk Basin – 75 million tons, in the Kansk-Achinsk – 70 million tons, in Far East deposits – up to20 million tons. Prediction entry capacity in the Far East will depend on the hard coking coal Elga deposit absorptive capacity in the Republic of Sakha (Yakutia), as well as the new gas finds development pace in this area.

Future coal-mining development in Russia is defined by the Russian Energy Policy for the period until 2030 [1], which preserves constant purpose and main principles of the state's energy strategy. The Policy implementation provides three phases: 1st – till 2013-2015, the second up to 2020-2022 and of the third – to 2030. In this case 1st stage is connected with the crisis phenomena in economy and energy overcoming, the 2nd – with a general increase in the economy and energy effectiveness, and the third one – with a highly efficient traditional energy resources use [4].

2 Materials and Methods

Speaking about the strategic development indicators in the coal industry [5], it is necessary to allocate the cost-effectiveness of coal mining. It should be mentioned that high targets achievements to increase the production per person employed in the sector and the pace of growth per face output provided by the Policy [6] were possible on the passed in the last decade the coal industry development stage, over which in accessible mining regions with developed infrastructure a number of deposits with favorable mining and geological conditions were put into operations which, with the growth of coal prices on the world market significantly increase the investment attractiveness of the coal. However considering the high proportion (variously estimated 50-70%) of low-tech reserves, i.e. unfavorable conditions for the development stocks, also if we take into account the fact that most attractive explored deposits supplied by transport and infrastructure are currently exploited, one can predict that maintaining high growth rate of these specified parameters will require in the future significant long-term investments directed to develop new districts of Kuzbass, Eastern Siberia and the Kansk-Achinsk basin, it is not possible only with the attracting their own financial reserves of coal companies. Besides the economic feasibility of such investments will be largely determined by price for the coal which, in turn, depends significantly on the other power resources cost [7].

The necessary conditions for the Russian coal industry development are favorable coal market conditions, providing the required level of investment attractiveness in terms of increased costs related both to the reserves development in difficult geological conditions and to significant cost for transportation the share of which in some cases is notably higher than coal production costs. It should be noted that the low investment attractiveness of the coal industry projects during the crisis causes the owners to give up most development programs production and reduce the cost of production due to the suspension of work on the preparation new areas and horizons inventory, this fact will be the main after crisis limiting factor in the coal mining development [8].

Also in connection with the investment attractiveness decline in coal production, in a number of cities and regions social problems related to the suspension of work and closure of mines and quarries become aggravated. The most acute this problem is for the mining towns and settlements, where the majority of enterprises are engaged in the coal extraction, processing and coal mining industries serving.

In the current market situation, to compete successfully at the international trade of coal mining products, to save (increase) the level of coal export, as well as the viability of some Russian coal mining companies it is to improve the technique and technology of coal production, to reduce production costs and finally increase the technical and economic performance, while ensuring a high level of safety, which in turn depends on the investment attractiveness of the coal mining projects. In this regard, priorities for the coal industry development are: to ensure involvement in testing of high-tech stocks and large-scale modernization of production, which will significantly improve the efficiency of coal mining and bring Russian coal industry to the level of the leading coal-producing countries.

At present, the following main problems in the coal industry can be identified:

- the loss-making coal industry;
- the high level of injuries;
- ecological problems.

In spite of all existing problems, the coal industry in the world today has kept the role of one of the most important basic industries. The value of coal as one of the main types of energy at the turn of the third millennium is caused by the action of the following leading market factors:

- power engineering remains the priority sector of the economy;
- stable and extensive resource base;
- the possibility for coal exporters operate on different markets;
- low cost of coal compared to the cost direct substitutes, coal price stability. Stability and predictability of coal prices provides consumers the convenience of planning costs.

Looking forward to 2020, international experts do not expect a significant reduction in the role of coal as an essential energy.

Moreover, due to the very likely reduce in oil and petrochemicals consumption and reconsideration to the nuclear energy use development of in many countries there may be a slight increase in its share in the energy consumption mix.

At the current level of coal consumption proven reserves will be enough for about 200 years, compared with about 50 years for gas and 30 years for oil [9].

3 Results and Discussion

In recent years, the independence of the regions is increasing; they are becoming responsible for economic development results of the region. Socio-economic status of regions is defined as the objective (macroeconomic conditions, the position of the region in the social division of labor, geographical location), as well as subjective factors, and first of all - the methods of regional management.

Planning for social and economic development of Russian regions is a complex and ongoing process that defines the guidelines for the adoption of any, including tactical, current solutions. The presence of long-term economic development plan allows you to make decisions on the basis of sound and elaborated [10].

Russia's economy is represented at a significant level by city-forming organizations in single-industry cities. The contribution of single-industry towns to the country's GDP is estimated at 20-40%. City-forming organizations provide in physical term 64% oil, 83% gas, 53% coal, 50% of steel products, including 66% of the steel and coke production, 65% of iron which is an important part of the nonferrous metallurgy production. As well as 90% of nickel and 100% alumina, 71% of passenger cars, 84% of the potash fertilizers production and so on. This is truly a unique Russian geographical and economic phenomenon. And unique in turn causes of this phenomenon are natural – the large size of the country.

Core enterprises of single-industry cities provide a large part of the raw materials industries, particularly fuel and steel industry. The products of these industries are dominated in Russia's exports in recent decades, indicating a solid economy problem and the fact that this problem is to be fixed. It is possible to implement by developing innovations, high-tech processing of raw materials, carrying out modernization and diversification. So single-industry cities should be focused on. The possibility of raw materials deep processing as well as production modernization depends precisely on their resources (investment, personnel and infrastructure) and of course local authorities and management of core enterprises motivation plays an important role. The situation on the job market of single-industry cities has also an impact on the socio-political stability in the country. Due to the market tightness core enterprises crisis cause large-scale and long-term local unemployment. Unemployment in single-industry towns has effects not only on these settlements (reduction of the taxable base, subsidized budget, the increase in crime, lumpenization proportion of the population, and others.), but also threatens the social and political stability in the whole country. In crisis periods single-industry towns become extremely dangerous centers of protest storms, of dissatisfaction with local and federal authorities, of the fall in confidence to the social and political institutions of the country. These attitudes have all chances to be caught up, warmed and applied by radical political forces to exacerbate social conflicts [11].

Single-industry towns with high scientific and technical capacity and the city main research and production complexes have a significant impact on the academic and practical science development in Russia and on the innovation system formation in the state. Research and Production Complexes of Science Towns manufacture goods competitive on world markets. It is not accidental that the highest position of Russia on the high-technology world market appears in the areas Science Towns skills – aerospace manufacturing, bioengineering, chemistry, energy production, the nuclear complex, armament. [12] The share of high technology products in the total volume of production in Science Towns is, as a rule, 50%. While the share of innovative goods and services in Russia on average are less than 10%. In Russia there are 75 areas with the highest concentration of intellectual, scientific and technological capabilities, of which 13 settlements are closed cities and the official status of Science Towns have 12 settlements.

At present the most effective management tools of the socio-economic development for the region is the strategic planning and forecasting. Strategic planning can be successfully applied not only in the development of comprehensive programs of social and economic development of the regions, but also in the implementation of anti-crisis measures in the regions and cities in the management of large-scale infrastructure projects and other areas of development in the regions. Creating a favorable environment for economic partners is today one of the major factors of socio-economic development of the region as a whole.

For Russia, the development of single-industry towns is particularly important [13]. Single-industry cities account 14% of all cities in the country and are the home to about 35 million people, i.e. a quarter of the Russian population. It should be noted that a large proportion of single-industry settlements do not have the characteristics of the deflationary, and the problem of their upgrading (repurposing and diversification) is directly related with a direct link to innovation economics. However, in the recessionary single-industry towns there is a high level of social problems exacerbation. Thus, the development and modernization of single-industry towns are closely coordinated with the optimization of the settlement system and the distribution of the productive power of the country, socio-economic districts and regions development [14].

4 Conclusions

Regional economic development strategy as a system of measures aimed at the realization of long-term objectives of social and economic development of the state, taking into account the contribution of sustainable regions in addressing these challenges is determined by the actual prerequisites and restrictions on their development.

Industrial organizations in the majority of single-industry towns have two significant features. Firstly, they are city-forming, and the city has not only been created at the enterprise, but for a long time was developed by enterprises and funded by enterprises. Secondly, almost all the town-forming enterprises are worn to the extreme and do not have any upgraded technology. During the long period of the single-industry towns, with periodically increased internal and external problems of their functioning, effective ways of work done by the authorities to the imbalance monospecialized settlements, adequate to their complexity and severity have not been established. At times of global financial crisis starting from 2008, nearly all single-industry towns had a different set of problems the main of which was a sharp decline in investment activity, the temporary suspension of investment projects, and from time to time the refusal to implement them. Anti-recessionary measures aimed to improve the situation were to reduce spending on social programs and personnel, to discharge employees, so they caused drastic consequences for the people [15].

However, there is range of negative factors. There is a population decline in single-industry towns with a population since 2010 (by 661 thousand people, or 4.4%), while the number of economically active population dropped by 392 thousand people (or 6.4%). The lag of revenues growth from the average figure is characterized by the fact that as of 01.01.2015 in the whole country the average salary increased by 13.2% (from 18.6 to 21 thousand rubles) and in single-industry towns only 9.3% (from 16.1 to 17.6 thousand rubles) [16]. The decline in business activity in the 2015-2016 is expressed in the number of small and medium business employees fell from 1 million 381 thousand to 1 million 080 thousand people (which amounts 21.8%). Disinvestment in fixed assets reached 31% (from 715 billion rubles in 2009 to 493 billion rubles in 2015). The Ministry of Regional Development of the Russian Federation offers several options for the identification problems in single-industry towns, the common for them is the dominant methodological principle «from the company to the city», providing the following alternatives [17]:

- there are no problems in single-industry towns (or they are insignificant), if the city-forming enterprise operates successfully;
- single-industry towns' problems arise in relation with the occurrence of problems in the city-forming enterprise;
- to solve the problems of single-industry towns, you need to solve the problems of city-forming enterprise.

In accordance with the approach the Russia Ministry of Regional Development, carrying out state policy in the sphere of socio-economic development of the regions of the Russian and municipal formations, has developed several options for strategic decision making in single-industry towns. First, the solution of a problem with the owners of core enterprises to implement their perfect set of social functions on the content of single-industry towns (decision single-industry towns problem is exacerbated due to the fact that businesses bear the heavy burden of social responsibility to ensure vital functions criterion in the towns, the need to maintain cost-inexpedient levels employment, does not meet the conditions of production and marketing). Secondly, the assistance in the expansion of markets for the city-forming enterprise (implicitly supposed that the more successful in the market is the city-forming enterprise, the better it will be to solve social problems a single-industry town, without incurring problems). Third, the enterprise re-forming (the transition to a competitive product on the assumption that the city-forming enterprise to qualitatively different work profile will remove a single-industry town caused problems). Fourth, the elimination of the main enterprise and relocation of the population of a single-industry town [18].

It is obvious that unresolved problems of urban governance specifics of single-industry type tows remain in the development and negotiation and the proposed policy options remains, as single-industry town management strategy is at the most sophisticated level of socio-economic system and does not coincide with the control strategy of the main enterprise as the subject of market municipal system that brings profit.

So, planning for social and economic development of Russian regions is a complex and ongoing process that defines the guidelines for the adoption of any, including tactical, current solutions. The presence of long-term economic development plan allows you to make decisions on the basis of sound and elaborated.

References

1. L. S. Chernova. Studies on Russian Economic Development, **18**, 600 (2007)
2. Yu. V. Saveliev. Regional Research of Russia, **3**, 211 (2013)
3. A. R. Belousov. Studies on Russian Economic Development, **17**, 1 (2006)
4. Problems of Economic Transition: Economics of Eastern Russian Regions, **58**, 604 (2016)
5. V.A. Malakhov, T.G. Dubynina, Studies on Russian Economic Development, **27**, 429 (2016)
6. G. V. Kalabin. Journal of Mining Science, **51**, 416 (2015)
7. V.A. Kryukov, G.P. Litvintseva, M.V. Khairullina. Problems of Economic Transition: Economics of Eastern Russian Regions, **58**, 598 (2016)
8. A.G. Aganbegyan. Studies on Russian Economic Development, **25**, 319 (2014)
9. S. S. Artobolevskiy, Transactions of the Institutions of Min and Metal, **112**, 27 (2003)
10. S. A. Surkov. Studies on Russian Economic Development, **19**, 80 (2008)
11. I.A. Pogosov, E.A. Sokolovskaya, Studies on Russian Econ Dev, **27**, 669 (2016)
12. T.F. Remington, I. Soboleva, A. Sobolev, Europe-Asia Studies, **65**, 1855 (2013)
13. V. N. Leksin, B. N. Porfiryev, Studies on Russian Econ Development, 27, 621 (2016)
14. A. Piliasov. Problems of Economic Transition, **46**, 32 (2003)
15. D. Moskovskaya, I.V. Soboleva. Studies on Russian Econ Development, **27**, 68 (2016)
16. N. A. Kazakova, A. I. Bolvachev, A. L. Gendon, G. F. Golubeva, Studies on Russian Economic Development, **27**, 638 (2016)
17. B. Zamaraev, A. Kiiutsevskaia , A. Nazarova, E. Sukhanov. Problems of Economic Transition, **57**, 13 (2014)
18. V.A. Malakhov, K.V. Nesytykh, Studies on Russia Econ Development, 27, 528 (2016)

Innovations as borders of stages of coal industry historical development

Elena Sigareva[1], *Sergey* Popov[1], *Sergey* Baturin[1,*], *Nadezhda* Sidorova[1], *Margarita* Borisova[1]

[1]Kemerovo institute (branch) of Plekhanov Russian University of Economics, Department of Humanities, 39 Kuznetsky av., Kemerovo, Russia

Abstract. The article deals with the historical development of mineral deposits exploitation which has very old history. Several thousands of years hid from us where and when people first began to extract minerals from the interior of the earth. With the development of human society the extraction of minerals has been increasing steadily. Nowadays the growth of extraction is accompanied by the deterioration of mining and geological conditions of mining operations, reducing mineralization in the deposit. As a result of intensive development mineral reserves located near the soil surface have changed significantly. Modern open-casts, open pits, and mines are highly productive and automated production complexes, including very extensive and diverse mining, various industrial buildings and facilities, the system of energy-mechanical equipment, automation and control with the use of microcomputer equipment. The paper describes a number of stages which modern mining in its development had passed and at which innovative breakthrough radically changed the place of the mineral sector in the society, and made the work of miners safer and more productive. For the systemization of the historical development of coal industry it is important to identify those technological breakthroughs that radically changed the technological and technical conditions of mining.

1 Introduction

Solid minerals are mined from the interior of the earth in two main ways - opencast and underground. Moreover, the level of open-pit mining in the world is about 80% [1]. The innovative development of mining is largely due to the open pit mining extension [2, 3]. It is explained by low, compared with the underground method, cost price of 1 ton of extracted raw material. The main technological conditions of innovative development of mining industry are favorable mining and technological conditions: relatively shallow bedding, large capacity of mineral resources, and the ability to use highly productive stripping, mining and transport equipment, powerful machineries (excavators, drilling machines, etc.), dump trucks [4-5].

* Corresponding author: language1994@mail.ru

2 Material and Methods

The origin of mining science appeared in ancient times. Even Pliny the Elder left the work in mining, where he gave some facts related to the extraction of ore and building materials. In the same way another ancient encyclopedist Vitruvius described the origin of mining. In his book "On Architecture" he wrote about the terrible evaporation of noxious gases (carbon dioxide) from the bottom of mine draw-well and how the miners burned them out with the help of torch. He also wrote about the discovery of underground rivers.

The first innovations that determined the very appearance of mining science became mechanisms which facilitated the work of people. The history of mechanics begins with the works of Greek and Egyptian scientists. In remained up to the present day papyrus books there are the results of the simplest tasks of statics, which put the ground for the theory of equilibrium. Some guesses and provisions made by the outstanding philosopher Aristotle (384-322 BC.) did not lose their significance today. That was Aristotle who introduced the concept of "mechanics" that defined the area of human knowledge within which we study the motion of material bodies being in nature or appeared as a result of human activity.

Among the famous scholars of ancientry a special place is occupied by Arhimed (287-212 BC) who used the first mathematical method of solving many theoretical and practical problems. He developed the foundations of equilibrium doctrine, the principles of statics of solid bodies, hydrostatic, and made a large number of inventions.

As a result of the birth of ancient mechanics the first mines received lifting machines, grinders, and the first type of cutting and chipping tools.

Further innovative development of mining operations is marked by the creation of a number of simple lifting, ventilation, dewatering machines. Their thorough description is given in the works of Georg Agricola (1494 - 1555), the mining science founder.

In times of Agricola mining development gained a new impetus in connection with the first attempts to systematize the production of raw material technology. In particular, the principles of opening the deposits, depending on the shape and bedding of ore body - groves, vertical and inclined bores were defined. The bores usually had a rectangular cross section with dimensions of 3 per 4.5 m. Their depth sometimes reached up to 145 m. Inclined working were widely spread, as well as horizontal one, conducted on the mineral vein.

The first mention of the systematic coal mining in Europe refers to 1100-1200. The destruction of forests in densely populated areas was a background for the development of coal deposits. The earliest systems of coal development were chambered, applied on the flat seams of average power.

In XVII-XVIII centuries rails (initially in the form of wooden beams with a metal plate, and then from the angle profile of iron) were used in transporting.

In connection with the development of coal industry ventilation of mine working was improved. Ventilation with air heating was widely used.

Since the XVIII century the Industrial Revolution gave rise to the innovative activities: steam engines appeared; coal was widely used as an energy source; steel became the main structural material; the railways were being built. The XX century was characterized by more significant qualitative changes in industry, as a result of which coal went to the first place among energy products.

The first steam turbines and internal combustion engines, electrical machines appeared. Motor drive made it possible to solve the problem of mechanized delivery of minerals. Underground equipment was provided with electric motors. Their use brought great changes in the organization of drainage and ventilation.

Started at that period scientific and technological revolution continues today, changing the face of mining enterprises.

The coal industry is one of the branches of modern fuel and energy sector; it has been existing for not one century. However, the study of its history is relatively recent. The main flow of works falls on the second half of the 40ies of the XX century. It was the World War II which revealed a particularly significance of the fuel and energy complex, not only for the development but also for the existing of nations, their security and independence. During this period scientists gave a general description of periods in coal industry development, its current situation and future prospects. Later, in 1970-80ies the scholars started to consider the process of institutional evolution of industry, the social-and-economic conditions of miners. Current studies include the possibility of coal use in various industries, the historic "competition" of oil and coal in modernizing economy of industrialized countries, technological development, and others. It is obvious that during the existence of the industry it has been going through various stages. Labor conditions, the cost structure, the profitability of coal production and its use were changing. Modern needs in technical-technological development of domestic production require paying special attention to the historical experience of innovative development of the coal industry.

3 Results and Discussion

As a result of the retrospective analysis of the mining industry genesis we identified the following stages in its development.

The first stage is associated with the formation of the industrial demand for coal, which promoted the innovative development of industry. In fact, until the early 18th century coal mining had "pre-industrial development". It was associated with the use of it as a fuel in ancient China and Greece, as well as throughout the Middle Ages. For example, in Europe in 1095 for the first time the fact of using coal for heating in France was documented. In 1537, in Zwickau (Saxony) the first association for coal production was founded. At the end of the pre-industrial period a new direction in using the product was formed. In 1589 J. Peterson received the coke from coal, used later in the melting, and in 1709 A. Derby for the first time smelted iron using coke produced from mineral coal instead of char coal.

The second stage (XVIII c.) or an early industrial development stage is characterized by the appearance of the first mechanical devices to facilitate the manual labor. However, their level made low productivity growth. At that time, the coal industry got modern features during the destruction of medieval crafts, changing craftsman-miners into the hired workers and emergence of mining and metallurgical manufactures. In 1740 the first industrial development of coal began in the United States, and in 1761 the patent for a mechanical device to undercut the coal seam was granted. In addition, the second stage is characterized by the first changes in the technology of coal mining process; the basic operations of the coal enrichment appeared.

The third stage marked the initial technological development of coal industry. The industrial breakthrough of the late XVIII - early XIX centuries served as an incentive for increasing mineral extraction as an energy basis in industrial production of the developed countries: England, Germany, France. The demand for machines caused the development of metallurgical industry. The demand for iron ore and coal for coke production and use of fossil fuel increased. The Industrial Revolution played an important role in expanding the scope of coal usage: during the first half of the XIX century coal began to be used as fuel for steam boilers for water and rail transport.

In the XIX century, in relation to the development of mining, innovations were developed primarily to improve productivity in coal industry. In 1835 in the United States the main machine patent was issued for the open development method - single-bucket steam excavator on rail travel, and in 1852 in England disk cutting machine for coal mines was patented. In 1860 French engineer M. Cuvre invented the multi-bucket excavator, and in 1864 in the UK the first chain cutting machine was produced; in 1876 in Germany, a hydraulic rotary drilling machine was invented. In 1880 coal was for the first time used to generate electricity, and in 1884 the electric drilling machine was invented.

At the same time in the United States the first in the world dragline for open pit mining was made. In 1902-1906 in the coal mines of England drag bar and extensible belt conveyers began to appear, and in Germany - shaking conveyors to transport mined coal. However, the machinery in the coal industry appears not only to increase productivity, but also to improve the safety of work. In 1815 H. Davy invented the miner's lamp, which led to the reduction of the explosion risk in mines. Due to the use of fine-granular metal grid, which was placed on an oil lamp the work in mine became more safety.

The fourth stage in the development of coal industry is connected with further development of equipment and production technology. But in this period the attention is given not so such to the equipment (wide-web winning by combined machine with individual metal single prop support, 1945-1965) as to the technology. In the industry to reduce the costs the drilling-and-blasting technique with mechanical coal cutting started to be used widely (1930-1940). At the fourth stage of development of coal industry there was the slowdown in productivity growth and improving the production technologies. In 1950ies the technological development of coal industry continued due to the introduction of narrow-web winning by combined machine and skimmers with individual metal single prop support (1955-1970), and the use of complex-mechanized winning by combined machines and skimmers with hydroficated metal single prop support (1961-1980). This stage is characterized by the active displacement of coal from the fuel and energy complex of the industrialized countries by hydrocarbon fuel - cheaper and more multifunctional type of energy.

The fifth stage, which continues up to now, is associated with re-orientation of innovative activity of scientists from extraction stage to other stages of production process (enrichment, environmental safety, the improvement of coal quality characteristics). The priority areas of coal industry at this stage and up to now is to improve the quality and other consumer properties of coal products with new coal enrichment technologies and its deep processing. Technologies of complex processing for getting domestic termo-brickets from brown coal and coal-rock sludge and screenings are promising. Coal-wastes would be burned in a mini steam-electric plant with technology of burning sludge in the circulating boiling layer, placed near mine, open pit, or ore-processing plant. This will reduce emissions of sulfur dioxide, nitrogen oxides, aromatic hydrocarbons and benzopyrene.

The International Energy Agency made three forecasts of coal industry for the next few decades:

1. The world coal consumption will continue to grow and by 2035 will increase by 70%.
2. The demand for coal will fall sharply by 2020 because of the strict environmental requirements of the majority of developed countries.
3. The peak consumption of coal will be in 2020, and further the demand for this resource will steadily decline.

The latest forecast is the most likely to be. Environmental problems connected with coal mining and falling of gas prices for the majority of countries are sufficient arguments in favor of reducing the consumption of this resource [6, 7]. The projected growth in consumption in India and some other Asian countries cannot compensate the decline in demand for coal in Europe, China, Russia and the United States.

Compared with oil and gas, coal is cheap raw material, so most mining companies operate with low margins. The only way out for the companies in this industry is the development of innovative technologies for deep coal processing, which allow significantly increase the profit.

Tough environmental standards in some cases may be not obstacles, but incentives for the development of coal mining [8, 9]. The Kyoto Protocol established quotas on emissions of methane - gas that is present in mines. Utilization of methane has become profitable for many enterprises which decrease emissions in the mines, and then sell their quotas to other companies. Profit from these projects makes tens millions of dollars [10].

Another promising technology is coal gasification, which is actively developed in China. It is planned that by 2020 the country will produce 50 billion tons of synthesis-gas, which is equivalent to 25 billion tons of natural gas, taking into account the difference in calorific power [11]. Moreover, according to official figures the cost price of million caloricity units of synthesis-gas is almost twice lower than the cost of one million caloricity units of natural fuel imported into the country.

Coal can even replace oil: currently in South Africa several plants that produce motor fuel from coal are operating. These technologies have been used for a long time. South Africa started to develop the production of synthetical fuel as early as the 1980s, when Western sanctions prevented petroleum imports.

4 Conclusions

Summing up it should be mentioned that the strategy of development the coal industry cannot ignore the identification of stages in its innovative development. First of all, they concern not the equipment but technologies of coal mining, enrichment, ecological safety, improving coal quality characteristics. These factors should be the ground of competitiveness management in modern market conditions of the coal enterprise.

References

1. M. A. Tyulenev, T. N. Gvozdkova, S. A. Zhironkin, E.A. Garina, Geotech. Geol. Eng., **35:1**, 203 (2017)
2. M. Tyulenev, S. Zhironkin, K. Kolotov, E. Garina, Poll. Res., **35:2**, 221 (2016)
3. M.A. Tyulenev, S.A. Zhironkin, E.A. Garina, Int. J. of Mining and Mineral Engineering, **7:4**, 363-370 (2016)
4. T. Gvozdkova, M. Tyulenev, S. Zhironkin, V. Trifonov, Yu. Osipov, IOP Conf. Ser.: Earth Environ. Sci., **50**, 012010 (2017)
5. M. Tyulenev, E. Garina, A. Khoreshok, O. Litvin, Y. Litvin, E .Maliukhina, IOP Conf. Ser.: Earth Environ. Sci., **50**, 012035 (2017)
6. M. A. Tyulenev, Y. V. Lesin, *Mining 2014: Taishan academic forum - Project on mine disaster prevention and control,* 441 (Atlantis Press, Paris, 2014)
7. Yu. V. Lesin, Luk'yanova S. Yu., M. A.Tyulenev, J. Min. Sc., **46:1**, 78 (2010)
8. A. A. Khoreshok, S. A. Zhironkin, M. A. Tyulenev, G. A. Barysheva, V. Yu. Blumenstein, M. C. Hellmer, S. V. Potyagailov, IOP Conf. Ser.: Mater. Sci. Eng., **142**, 012122 (2016)
9. S. A. Zhironkin, A. A. Khoreshok, M. A. Tyulenev, G. A. Barysheva, M. C. Hellmer, IOP Conf. Ser.: Mater. Sci. Eng., **142**, 012127 (2016)
10. S. A. Zhironkin, K. A. Kolotov, A. E. Genin, F. V. Agafonov, S. A. Kovalevsky, IOP Conf. Ser.: Earth Environ. Sci., **50**, 012011 (2017)
11. V. Trifonov, O. Loyko, D. Nesteruk, S. Zhironkin, E. Strekovtsova, AIP Conf. Proceed., **1800**, 050009 (2017)

Using innovative interactive technologies for forming linguistic competence in global mining education

Galina Chistyakova[1,*], *Evgeniya* Bondareva[1], *Kseniya* Demidenko[1], *Ekaterina* Podgornaya[1], and *Oksana* Kadnikova[1]

[1] Plekhanov Russian University of Economics, Kemerovo Branch, Department of Humanities, 650992 Kemerovo, 39 Kuznetsky Av., Russia

Abstract. Globalization of mining education imposes new requirements for mining engineer competence. Nowadays linguistic competence is one of the most demanded. It guarantees technical university graduates the possibility of global employment, on the one hand, and the chance of getting cutting edge education in leading training centers of the world, on the other hand. Distance education is actively developing all over the world and is widely used in technical colleges and universities, as well. Interactive method that involves active engagement of students appears to be of the greatest interest due to introduction of modern information and communication technologies for distance learning. The paper presents step-by-step implementation of several interactive technologies (jigsaw, case study, brainstorming, and role-play) that can be used in distance education in the process of teaching subjects in foreign languages with the help of information and communication technologies. In response to the changes in the conditions of educational process, the implementation of the methods has been transformed to a combination of traditional (in-class) and distance (online) learning.

1 Introduction

It is generally accepted that distance learning is an integral part of modern education alongside the traditional in-class studying. A number of authors in their research papers consider innovative structures and components of distance learning in education [1]. Mining engineers training is no exception with the world leading higher technical schools and universities offering distance-learning services. The main language of educational programs is English, which requires students' knowledge of not only general, but also professional vocabulary.

The tendency of gradual conversion of global mining education to the distance learning format is quite understandable if one looks at last decades' intensive development of technical means. Functioning of XXI Century University, in our opinion, is not possible without new resources, including modern information and communication technologies that

* Corresponding author: galvik06@mail.ru

enable to increase the number of participants in educational process. The development of distance learning technologies ensures the possibility of providing educational services without being tied to a specific geographical location (Open and Distance Learning).

2 Material and method

Even though distance learning technologies in the sphere of mining engineers training are innovative, they have proved to affect the process positively. Thus, positive relation between students' usage of information technologies for educational purposes and their involvement in learning and interaction with lecturers has been acknowledged [2, 3, 4]. Another study has shown that the use of Twitter for educational purposes has a positive effect on students' learning interest and the average semester grades [5]. Some authors analyze the effectiveness of distance learning basing on students' own evaluation of the educational process [6, 7] and evaluation of students by their teachers [8]. A group of scientists have compared traditional (off-line), on-line and mixed forms of education and have proved that educational results in groups with mixed education methods are the highest [9].

In the context of implementing distance learning technologies into the training of mining engineers, the process of educational design (specifically targeted planning of training courses) should be adjusted towards finding some learning forms and methods that can encourage students to self-study of foreign languages alongside with university training.

The target mentioned can be reached by applying e-learning and distance learning that can be successfully used for forming linguistic competence of future mining engineers. E-learning can be applied in a traditional form (in-class face-to-face learning), when students use electronic devices and resources of the university or their own mobile devices as educational tools. Distance learning allows students to acquire knowledge by accessing university virtual educational environment with the help of any computer connected to the Internet.

Educational design of the courses with elements of distance education requires a certain level of professional creativity from the educators teaching courses in foreign languages.

Among the innovative methods of foreign language teaching, the interactive approach appears to be the most promising in terms of distance technologies. It is based on active interaction of students. Teachers act as consultants and coordinators of students' educational and research activities. They step aside while students' work intensifies and dominates the process.

Some research works are dedicated to the importance of the interactive approach in learning using materials and information about real situation in economy in general and in mining industry in particular [10, 11]. Researchers pay attention to the theoretical description of the method analyze specific industrial technologies [12]; or consider the possibility of using interactive methods when teaching particular subjects concerning mining industry [13]. Evidently, it is of interest to contemplate the usage of the interactive approach in distance education.

New teaching methods, based on using modern information and communication technologies (webinars, interactive multimedia, online learning, etc.), allow developing distant forms that do not involve teacher's direct physical participation in the process. It is impossible to overestimate modern information and communication technologies: they provide access to the sources of information previously unused; significantly increase efficiency of students' independent work; provide completely new opportunities for creativity, identification and demonstration of capabilities, acquiring and consolidating different skills, both of a teacher and a student; allow implementing essentially new forms and methods of training (local and global information networks, conference calls, e-mail, e-libraries, forums, chats, and others). It is worth mentioning that distance learning technologies can noticeably reduce the level of panic and stress experienced by students when taking tests and exams in traditional forms, increase students' initiative and motivation, as well as their psychological comfort.

Using information and communication technologies in educational process enables to address the following didactic tasks more effectively: forming and improving reading, writing, speaking, and listening skills, as well as extending students' active and passive vocabulary. The authors particularly emphasize that students are provided with an extra opportunity of facing social and cultural environment of foreign languages they study (speech etiquette, verbal behavior, culture issues, traditions of foreign countries, etc.), as well as increasing academic motivation through the use of authentic materials [14]. Undoubtedly, that contributes to forming skills useful for solving professional problems in a foreign language as a means of direct as well as indirect communication.

Both in traditional and distance educational systems, fulfillment of educational goals in disciplines taught in foreign languages depends on the quality of learning materials, facilities of the educational environment, professional skills as well as information and communication competence of teachers. An important role is also attributed to the correct choice of tasks, which demand using information and communication technologies, such as working with e-learning materials, using Internet resources in foreign languages, performing online tests and exercises, etc.

3 Results and discussion

When summarizing authors' long-term practice in using innovative methods for teaching foreign languages, it can be stated that it is the interactive approach that demonstrates maximum efficiency in developing linguistic competence in a foreign language through the use of distance technologies.

In terms of distance learning, among interactive teaching techniques the main interest is aroused in technologies and methods to be implemented within two stages: preparation and presentation. It is possible to use distance learning technologies at the preparation stage, when students communicate, for example, by means of Skype-based video conferencing. This technology is used in synchronous distance education which offers an environment closer to the traditional in-class environment, and allows students to establish visual and voice communication with teachers [15]. During video conferences students discuss their tasks, choose the leader of the working group, distribute authorities, produce and edit texts, etc. A set of actions performed at the preparation stage depends on the technology used. Teacher's participation in such video conferences is optional.

This article focuses on step-by-step description of implementing traditional interactive learning technologies applied with the help of innovative information and communication technologies in the course of foreign language teaching. The technologies under consideration are: jigsaw, case study, brainstorming, role-play.

Jigsaw technology was developed by Dr. Elliot Aronson in 1978. Using this technology assumes that students are divided into teams of 4 to 6 people to work with some training material (for example, a text on the section of the subject), which is divided into fragments (logical or semantic units). It is recommended to perform the division into the teams and explain the task during in-class activity.

During the preparation (distant) stage the following set of tasks is performed:

– each team member examines a part of the material (offline independent out-of-class activity);

– "experts meeting", when students from different teams who explored the same part of the material meet and exchange information acting as experts in this section (online team out-of-class activity);

– experts return to their teams and share discovered information with other team members. Thus, each expert acts as a "tooth of a saw" (online out-of-class activity).

Each participant of the educational process (each expert) is responsible for the extent to which all members of the team acquire new information. The work of each team member (expert and / or student) is reflected in their final evaluation.

The presentation (traditional) stage, evaluation of the acquired knowledge, is held by the teacher during the in-class work (face-to-face learning). Testing is carried out individually and / or as a team. The choice of a particular task type depends on the level of the linguistic skills of the academic group [16], the type of the material under consideration, time available, etc.

«Case study» technology was first applied in the mid-twentieth century at Harvard Business School. The principles of the technology are as follows: students receive a training case; after considering the task and the problem it poses, which, as a rule, has no unique solution, students are supposed to propose their own decision, based on the acquired knowledge and skills. Applying case-technology during language classes leads to improvement of students' speaking skills as a result of discussing a problem under consideration.

The preparation stage of training based on «Case Study» technology consists of two stages: organization and implementation. It is advisable to carry out the organizational work as a traditional offline classroom team activity. At this stage, the teacher explains the task and introduces materials to be examined. The implementation stage (students' detailed acquaintance with the information on the case, task execution, and decision-making) is held in online form (online independent out-of-class work). At the same time, students have access to additional resources that might be required during the preparation.

The final (traditional) stage, presentation of the case solution, is carried out in class (offline team classroom activity). The teacher assesses the level of language skills developed and provides the team with feedback on their performance. Each member's personal contribution to the teamwork is defined by the students themselves.

«Brainstorming» technology is one of the most popular technologies used in foreign language training. It is effective for stimulating students' creativity. Participants are encouraged to propose a great variety of suggestions, including most improbable ones. The discussion of the problem usually lasts from 1 to 5 minutes; afterwards, the most feasible solution of the problem is chosen.

It is generally considered that this technology can be used offline only, since it is necessary to write the ideas down on the board or a piece of paper for the purpose of selecting the best. However, modern information and communication technologies provide a way of not only seeing and hearing one another during the communication process, but also actively interacting with a variety of online graphic editors, available to all the participants simultaneously.

The preparation stage may be carried out in two ways:

1. Offline in-class team activity: the teacher divides the students into teams, appoints the leaders of each team (members of the teams who are responsible for organizing the conference, recording ideas, managing the discussion), and provides them with the tasks for consideration.

2. Online out-of-class teamwork: the teacher organizes the conference, appoints the leaders, and provides the task. This activity is followed by an online discussion in which students choose the best ideas and prepare to defend their point of view in front of other

teams. The online discussion, during which students select the best ideas and prepare for speaking up for their point of view, concludes the preparation stage.

The presentation (traditional) stage is held as an offline classroom team activity during which the results are presented by the teams. As an additional activity, one idea can be chosen and developed into an independent project (Project technology).

«Role-play» technology is another effective way of forming linguistic competence which makes it possible to introduce foreign language training to the model of mining engineers' future work. During this activity students learn how to deal with problems that they might encounter in their professional activity.

At the preparation stage (offline team in-class activity), the teacher forms teams, explains the tasks for the role-play, and determines the degree of students' creativity freedom. Discussing key positions (distributing roles, determining the course of the game, etc.) and writing an open-ended script takes place in a distant form (online team out-of-class activity). The teacher reviews the script via e-mail.

The presentation (traditional) stage is carried out in the classroom (offline team in-class activity) in the form of presenting the game for the teacher and other academic groups. At the stage of discussing the results of the activity (offline team in-class work) students have an additional opportunity to communicate in a foreign language.

Questioning students reveals that they have been using information and communication technologies for homework preparation without extra motivation from their teachers. This fact indicates that it is university teachers, not students, who should reconsider their position regarding the use of distance learning technologies in teaching foreign languages. Success of students' remote communication depends largely on teachers' motivation and competence in implementing information and communication technologies in educational process.

4 Recommendations

Developing forms and methods of distance learning as a part of designing and implementing professional training programs still raises a number of issues. At present, strengthening the position of distance learning with the help of information and communication technologies in the process of teaching foreign languages requires the following:

- conducting regular surveys of students on the issues of training with elements of distance learning technologies;
- working out guidelines on using distance learning technologies in teaching;
- holding seminars for university teachers devoted to distance learning technologies in teaching;
- including in educational content tasks that require using information and communication technologies;
- developing a mechanism of monitoring students' out-of-class activities with the use of distance learning technologies.

5 Conclusions

Like any living organism, educational process can develop and improve only within the terms and the conditions the society lives in. Modern society cannot be imagined without information and communication technologies extensively used in education, including distance learning. The next step in developing distance education is transforming methods used in traditional education into new forms of teaching that can be implemented with the use of information and communication technologies. Interactive methods should be particularly valued among them, as they are positively perceived by students and demonstrate high effectiveness in forming linguistic competences of graduates.

References

1. M. Leontev, MATEC Web of Conferences **7307005 7**, 07005 (2016)
2. T.F.N. Laird, G.D. Kuh, Research in Higher Education, **46(2)**, 211 (2005)
3. T. Krasnova, A. Popova, SHS Web of Conferences **28**, 01059 (2016)
4. S. Alumu, P. Thiagarajan, Indian Journal of Science and Technology, **9(6)** (2016)
5. R. Junco, G. Heiberger, E. Loken, J. of Computer Assisted Learning, **27(2)**, 119 (2011)
6. I. Rudaleva, I. Kabasheva, E. Kovaleva, SHS Web of Conferences **26**, 01128 (2016)
7. M. Kuimova, A. Kiyanitsyna, A. Truntyagin, SHS Web of Conferences **28**, 01129 (2016)
8. A.Y. Awadh, Al-Qahtani, E. Higgins, Journal of Computer Assisted Learning, **29(3)**, 220 (2013)
9. L.-T. Lee, J. Hung, Human-centric Computing and Information Sciences, **5**, 15 (2015)
10. M.A. Tyulenev, Y.V. Lesin, *Mining 2014: Taishan academic forum - Project on mine disaster prevention and control*, 441 (Atlantis Press, Paris, 2014)
11. Yu.V. Lesin, S.Yu. Luk'yanova, M.A. Tyulenev, J. Min. Sci., **46**, 78 (2010)
12. S. Zhironkin, M. Gasanov, K. Kolotov, EpSBS, **XIX**, 761 (2017)
13. M.A. Tyulenev, T.N. Gvozdkova, S.A. Zhironkin, E.A. Garina, Geotech. Geol. Eng., **35**, 203 (2017)
14. N.V. Ivleva, E.V. Fibikh, IOP Conf. Series: Materials Science and Engineering **70**, 012012 (2015)
15. S. Ghazal, Z. Samsudin, H. Aldowah, Indian Journal of Science and Technology, **8(30)** (2015)
16. A.A. Khoreshok, S.A. Zhironkin, M.A. Tyulenev, G.A. Barysheva, V.Y. Blumenstein, M.C. Hellmer, S.V. Potyagailov, IOP Conf. Ser.: Mater. Sci. Eng., **142**, 012122 (2016)

Technological convergence and innovative development of natural resource economy

Felix Agafonov[1,*], *Anton* Genin[2], *Oksana* Kalinina[3], *Olga* Brel[4], and *Olga* Zhironkina[5]

[1]T.F Gorbachev Kuzbass State Technical University, Post-Graduate Study, 650000 Kemerovo, 28 Vesennya st, Russia
[2]Branch of JSC "SO EES" Siberian ODU 650000, Kemerovo, 29 Kuzbasskaya st., Russia
[3]T.F Gorbachev Kuzbass State Technical University, Department of History, Philosophy and Social Sciences, 65000 Kemerovo, 28 Vesennya st, Russia
[4] Kemerovo State University, Department of Geology and Geography, 650000 6 Krasnaya st., Kemerovo, Russian Federation
[5]Plekhanov Russian University of Economics, Kemerovo Branch, Department of Humanities, 650992 Kemerovo, 39 Kuznetsky Av., Russia

Abstract. An integral component of the innovative development of natural resource economy is the integration of technological solutions and investments in industrial production, the expansion of national manufacturing complex in the world market with the following change in the export structure. Implementation delay of convergent technologies may lead to the loss of Russian technological identity, strengthening raw vector of long-term social and economic development. In this regard, the relevance of the study of problems and obstacles to the implementation of convergent technologies in the economy of countries which specialize in production and export of mineral resources is increasing. To start the innovative development in Russian economy it is important to restore the dominance of modern industrial complex in it. This will lead to the decrease of pre-industrial economy and will contribute to the formation of post-industrial technological layer. Therefore, convergent and technological transformation of natural resource economy is imposed with the strategic task of improving profitability, investment attractiveness and international competitiveness of industrial production.

1 Introduction

In the course of market reforms some basic principles of innovative development of Russian economy had been violated. They are: the priority of modernization of production basis and transformation of economic relations towards neo-industrialization, the integrated use of program-target and market forms of regulation of the business sector by the government in accordance with the depth of structural problems.

As a result, Russian economy has rental-resource model of development. It allows, on the one hand, achieving short-term budget macroeconomic goals, but does not contribute to the economic development through the sequence change of technological layers. As a re-

* Corresponding author: kemshi-ise@mail.ru

sult, in mineral resource economy investments with short-term effects connected with conjunctions of commodity and financial markets dominate over the long-term investments leading to the change of technological layers.

2 Material and method

The driving forces of innovative development of economy - the development of economic relations in terms of investments, innovations, the labor market, in cooperation between the state and business in Russia are restrained by a number of pro-problems [1, 2, 3]:

a) The uncertainty of the "rules of the game" for the subjects of high-tech-industries, innovative sphere, advanced processing of raw materials in economy, in which the major part of government and private sector revenues is based on the extraction and redistribution of mineral resource rent. Under these circumstances of institutional "vacuum" and high entropy of the commodity market, uncertainty and risks of investment the technological level of production, new demand for the labor force, the labor itself are repeatedly increasing;

b) Significant underfunding by the state the fundamental researches - the first stage of innovative process, ensuring the development of current and start of the future technological layers. For example, in 2010-2013 public financing of fundamental and applied science in Russia amounted to 0.2-0.3% of GDP, in the USA – 2.7%, in France – 3.5% [4];

c) The lack of the 6th technological layer experimental "production facilities", allowing to carry out applied research and experimental production. Existing in Russia technological platforms, funded mainly by the state, are concentrated in the sectors of the 5th technological layer, and the business does not have an appropriate effect on their development. In Russia R&D is funded by the state on 75-80%, while in the USA and the EU countries on 20-25% where the rest part falls on the business [4];

d) Generated in Russian technological platforms innovations do not reach mass production, and the large enterprises in clusters of innovative development continue to produce traditional products, including raw materials. The share of innovative output produced by coal and metallurgical enterprises of Kuzbass cluster does not exceed 1% of their output, by machine-building enterprises of Altai cluster - 3%, petrochemical enterprises of Tatarstan – 4%. Among Russian small firms only 1% compared with 9% in the USA and 8% in France is engaged in innovative activity, while the volume of venture capital investment in these countries in 2012 amounted 0.73, 29.5 and 1.2 billion dollars respectively [5];

e) Existing Russian innovative tax advantages do not attract investors in the most advanced technologies of the 6th technological layer and scientific-production special economic zones contribute very little to their implementation in mass production [6-8]. Also Russian tax system does not support innovation development of environmental saving technologies [9-10].

f) There is no state support for the production and promotion of technologically advanced, innovative products in the domestic and foreign markets (subsidies, lobbying, and government investment in the companies of the 6th technological layer). Meanwhile, today, the majority of Russian innovative clusters have public-private partnerships allowing coordinate the actions of the state and business in bringing research results to innovation [11-12]. However, by themselves they cannot remove a significant obstacle in modernization process of Russian economy - the gap between innovation and mass production to overcome which the massive system changes in reproduction, industry, market-competition, institutional structures are required.

3 Results and discussion

To solve these problems it is necessary to identify a number of areas for implementation the principle of active participation of the state in the structural transformation of natural resource Russian economy:

1. Bringing the system of formal and informal institutions in compliance with the requirements of neo-industrial structural changes, interaction harmonization of institutions of the state and the market in order to eliminate barriers and protect the interests of high-tech companies in the domestic and foreign markets.

2. The transition from the fragmented support of public research organizations, public companies in defense sector to the broad arrangement of preferential conditions for private investments in innovation, particularly in the field of breakthrough technologies.

3. The involvement of financial capital movement in the modernization process of industrial production, the reorientation of investments from the financial market into the innovative sphere, the transformation of property relations in the process of vertical integration of raw material, manufacturing industry and the scientific sector, the use of emission mechanism for the financing of the most competitive technologically advanced industries.

4. Changing the role of state ownership in economy and its gradual withdrawal from raw material and financial sectors and the reorientation to high-tech industries.

5. Promotion of innovative initiatives on the labor market, changing relations associated with staff hiring and wages in the direction of the full motivation of in-house innovation process, employees' participation in the creation, investment and promotion of innovations.

6. Reducing the investment and innovative risks of technological industry modernization placing high demands for production structure of fixed capital, the level of its depreciation.

7. The encouragement of effective demand for domestic goods and services of manufacturing and high-tech industries and the development of their import substitution.

The basis of structural transformation of natural resource Russian economy includes a system and multi-subject nature of its structural reforms, the combination of normative, indicative regulation and market self-regulation, the presence of neo-industrial attractor, goal setting and strategic programming taking into account the depth of the structural problems that impede neo-industrialization of economy, import substitution development and formation of the target social group. The advanced subjects of national innovation system, first launched in the European Union, are technological platforms that integrate business, science and the government. The European Commission understands technological platforms as the directions of cooperation among universities, innovative companies and large manufacturers. It is in these areas public funds for fundamental and applied research are allowed. The main feature of the technological platforms of the EU is the participation of not only scientific organizations and industrial companies, but also banks, investment companies, European concerns. The state participation in technological platform is not limited by direct financing, but includes government guarantees, government investments, leasing and credit subsidies. The initiators of the European technological platforms creation are the so-called sectoral associations of industrial corporations - High Level Groups.

Since the beginning of the 2000s the EU has created more than 40 technological platforms (including 26 - in sectors of the 6th technological layer, such as «Future Manufacturing Technologies», «Embedded Computing Systems», and «European Nano-electronics Initiative Advisory Council» working for more than 10 years) [13]. National Support Groups of technological platforms are being founded in the European Union countries. They bring together leading scientists and officials and fulfill the following tasks:

- technological breakthrough in high-tech industries;
- the development of new processing chains in the traditional raw material sectors;

- the development of new sectors of economy by overcoming the fragmentation of science and lack of structuring of business interests.

In Russian economy the development of technological platforms began only in 2011 on the initiative of the Russian Government. Government Commission on High Technologies and Innovations approved about 30 platforms [14], the founders and coordinators of which were industrial companies, the government research institutes and state corporation "Rosnano", "Rostechnologii" [15-16]. It is obvious that Russian technological platforms do not follow the main condition of their participation in economy neo-industrialization - providing technological breakthrough (the latest technologies of the 6th layer) required for outgoing to the path of innovative development. On the contrary, they involve basic and extractive industries such as power industry, transport, mining, metallurgy, chemistry, construction [17]. And in industries that can widely use high technologies (biochemistry, photonics, manufacturing materials with special nano-properties and supercomputers), only a few platforms were formed. Financing the innovative enterprises in these industries, which are mostly state-owned, is directly made by the state, without motivating business to participate in this process. Therefore, we consider further development of cooperation links between producers and processors of raw materials on the one hand and high-tech firms on the other on the basis of techno-platforms to be necessary. The development of high-tech service industries specific for neo-industrial economy (the 5th technological layer) requires massive public and private investments in the creation of new and the development of existing techno-platforms (especially bio-technological, information and communication).

But as in Russian economy massive private and corporate investments in high-tech sector are almost absent, there are underdeveloped investment ties, scientific and industrial cooperation capable to alter the territorial structure of the productive forces. The potential use of innovative clusters in Russian economy neo-industrialization is not enough.

The issues of structural convergence (a special type of structural changes in the course of which new forms of network and cluster structures can be formed, and in future - new industries) were considered by M.A. Gasanov. It must be based on digitization of most production processes of material and non-material values. It means changing of technical, administrative, financial processes into a single digital format clear for modern computers, eliminating the differences between various types of information.

Today, convergent technologies are the subject for study of biologists, physicists, chemists, experts in the field of information technologies and micro-robots. We define the development of convergent technologies as an important technological condition of innovative development of Russian economy in the process of its structural changes.

The content of technologies' convergence consists in the interpenetration and combination of different innovations, united by a common digital form of designing, printing and diffusion. Also, technological convergence means the development multi-platform innovations in existing clusters, developed in different technological platforms, but able to give a united product. We have identified the following areas of convergent technologies development:

- Technologies at the interface of biochemistry and energy (bioenergy), allowing in the nearest future, move to the non-waste production of environmentally clean fuel replacing non-renewable hydrocarbon resources. The development prospects of these technologies we see in the cross-platform interactions of such techno-platforms as "Medical and biotechnologies", "Power industry", "Extraction of natural resources and oil and gas processing";
- Technologies that connect the manufacturing of processor equipment and software (distributed computing). They will be able to create an integrated telecommunication networks, significantly advance the diagnostic medicine, and accelerate the creation of new equipment by computational design and multi-factor computer simulation. Their cross-

platform character should be revealed in the interaction of such techno-platforms as "Information and Communication Technologies", "Electronics and Machine Building";

- Laser informatics technologies that can create data depositories, surpassing modern ones in thousands of times. This will speed up the digitization of technological basis of industry, initiate robotics of such traditional for Russia industries as coal, oil and gas, chemical, heat power industry. Cross-platform relations here can be implemented in such platforms as "Photonics", "Information and Communication Technologies".

4 Conclusions

Implementation of convergent and technological conditions of innovative development of Russian economy is restricted by a vicious circle of problems in which the business does not have an attractive innovation investment technological basis, and the development of convergent technologies is constrained by the lack of strategic interaction of the state and business. To solve the problems of innovative development of Russian economy and to move from the natural resources export to the production of high-tech products it is necessary to increase investments in fundamental research, the development of connections between R&D and production, the participation of banks and investment companies in the innovation financing, the staffing of the innovative process.

References

1. A. Zhavoronok, M. Gasanov, S. Zhironkin, SHS Web of Conferences, **28**, 01144 (2016)
2. M. Gasanov, S. Zhironkin, M. Hellmer, EpSBS, **XIX**, 745 (2017)
3. S. Zhironkin, M. Gasanov, O. Zhironkina, EpSBS, **XIX**, 753 (2017)
4. M. Gasanov, S. Zhironkin, EpSBS, **VII**, 117 (2016)
5. A.A. Baryshev, G.A. Barysheva, EpSBS, **VII**, 352 (2016)
6. S. Zhironkin, M. Gasanov, O. Zhironkina, EpSBS, **VII**, 124 (2016)
7. S. Zhironkin, V. Guzyr, EpSBS, **VII**, 160 (2016)
8. V. Trifonov, O. Loyko, D. Nesteruk, S. Zhironkin, E. Strekovtsova, AIP Conf. Proceed., **1800**, 050009 (2017)
9. M.A. Tyulenev, Y.V. Lesin, *Taishan academic forum – Project on mine disaster prevention and control*, 441 (2014)
10. Yu.V. Lesin, S.Yu. Luk'yanova, M.A. Tyulenev, J. Min. Sci., **46:1**, 78 (2010)
11. S. Zhironkin, E. Gasanov, O. Zhironkina, E. Taran, SHS Web of Conferences, **28**, 01145 (2016)
12. S. Zhironkin, M. Gasanov, K. Kolotov, EpSBS, **XIX**, 761 (2017)
13. S.A. Zhironkin, K.A. Kolotov, A.E. Genin, F.V. Agafonov, S.A. Kovalevsky, IOP Conf. Ser.: Earth Environ. Sci., **50**, 012011 (2017)
14. M. A. Tyulenev, S. A. Zhironkin, E.A. Garina, Int. J. of Mining and Mineral Engineering, **7:4**, 363 (2016)
15. M.A. Tyulenev, T.N. Gvozdkova, S.A. Zhironkin, E.A. Garina, Geotech. Geol. Eng., **35**, 203 (2017)

16. M. Tyulenev, E. Garina, A. Khoreshok, O. Litvin, Y. Litvin, E. Maliukhina, IOP Conf. Ser.: Earth Environ. Sci., **50**, 012035 (2017)
17. M.Y. Blaschuk, A.A. Dronov, S.S. Ganovichev, IOP Conf. Ser. Mater. Sci. Eng., **142**, 012128 (2016)

Structural problems of mining region innovative development (Kuzbass, Western Siberia)

Elena Dotsenko[1,*] *Natalya* Ezdina[1]

[1]Russian University of Economics, Department of Political Economy and History of Economic Science, 117997, Moscow, 36 Stremyanny lane, Russia

Abstract. At present, the issues of overcoming the negative structural shift in Russian economy, accelerating economic growth, reducing technological and social- and-economic gap between Russia and the developed countries are strategically important. Modern structure of Russian economy which had been developed as a part of market model does not generate the innovative type of development. In this situation, mining regions, the structure of which was formed in the early 20th century and is represented by the extractive industries, are undergoing the greatest problems of innovative development. Therefore, despite high urbanization and industrial development level the economy of Kuzbass (Western Siberia, Russia) is characterized by significant structural problems. They are associated with primary extractive nature of the regional industry, in which coal and iron ore, metallurgical and chemical industries dominate. They highly depend on Russian and global market of raw materials, widely use low-productive technologies, and they are characterized by high level of fixed capital depreciation and insufficient pace of innovative infrastructure development. The solution of the structural problems of economic development of mining region is connected with technological modernization of extractive industries and use of modern materials for the production of high-tech products.

1 Introduction

The structural problems of mining region are mainly connected with the "old industrial" origin of its economy, when from the beginning of the 20th century the mining (mostly coal) and metallurgy industries had been extensively developing, and later - the power industry and connected with these branches machinery. All these was accompanied by a high level of investment, output, gross income and tax revenue in two main industries of the regional economy - the coal and metallurgy ones.

2 Material and method

The analysis methodology of structural problems of the region is based on the concept of a multi-sector economy, outlined by D. Bell [1], C. Clark [2] and Robert Reich [3], the theory

* Corresponding author: ktyf110372@rambler.ru

of technological layers by S. Glaziev [4], the ideas of neo-industrial accelerated development by M. Gasanov and S. Zhironkin [5-9]. The structural problems of Kemerovo region economy are considered in the papers of K. Kolotov, A. Zhavoronok and others [10-11].

Now the most important structural problems of economy in "old industrial" areas are ecological issues [12-13]. They are connected to national economy structural problems through institutional issues [14-15] among them can be found the following:

1. De-formalization of rules in the interaction between government and business, its transition to "black" (corruption) and "gray" (implementation by the authorities of public functions by evasion of law) zones.

2. A significant role of the administrative resource, in which practically no one institution can acquire legitimacy without the governor's approval. Accordingly, all attempts to influence the reproductive, industrial, competitive structure of the regional economy will inevitably come across with the necessity of agreement with the highest regional authorities, each of "white" and "gray" and "black" interaction with business.

3. Incomplete institutional reforms in major industries of Kuzbass economy. This concerns in the first place the structure of the property. Despite the fact that by the end of 2008 in the form of ownership 41960 (81.8%) enterprises in the region were private, 4,700 (9.2%) were municipal, 1681 (3.3%) were owned by the state, the majority of coal-metallurgical Kuzbass assets were distributed among several of the largest companies in the country. Such groups as "Evraz Holding", "Mechel", "Gazprom", "Severstal" are the main investors in the region. This concentration of ownership in the hands of an extremely limited group of businesses prevents the development of new market institutions required for structural transformation of Kuzbass economy and makes business dependent on informal agreements with the authorities.

4. Unformed state and business institutions necessary to initiate structural reforms (major venture funds and innovative banks, techno-innovation zones, guarantee system for investments in innovation, etc.). Established in 2007 Kuzbass Technopark is focused on the extraction of raw materials, whereas its participation in the development of deep raw material processing depends entirely on the "subjective will" of corporations - the owners of coal and metallurgy enterprises.

3 Results and discussion

The structural problems of Kemerovo region industry can be presented as the following:

A) Physical depreciation and obsolescence of fixed capital. Despite the fact that one-third of the equity capital of Western Siberia is centered in Kemerovo region, its average depreciation for regional economy is 43%, in industry - up to 55%, with an annual increase on 1-2%.

The maximum depreciation is typical for fixed capital of regional key sectors - power industry (in 2006 - 59%, in 2013 - 53% for power equipment), coal industry (mainly mining: in 2008 - 48%), machinery (in 2000 - 70%, in 2008 - 67%, in 2013 - 62%). During the reforms the average depreciation of fixed capital at the enterprises of Kuzbass main sectors have increased on the average of 11%, far exceeding the critical level of 40%.

Mainly, depreciation growth of the production means in Kuzbass is associated with an investment problems of regional machinery, which fixed capital modernization is going very slowly, with the annual disposal ratio of 1.5% and the implementing ratio - no more than 10% (with maximum 12, 7% in 2006). That is, keeping such rate of investment in Kuzbass machinery the full update of its production means will require more than 60 years.

B) Lack of investments in renewal of fixed capital in traditional for Kuzbass primary industries and insufficient capital flow in processing sector. Over the entire period of reforms the problem of predominance of enterprises' own funds as a source of investment had not been solved in the region.

On the average for the period of 1998-2010 own investments in the fixed capital were unstable and varied from 75% in 1998-2002 to 69% in 2004-2010. This indicates that the reproduction of fixed capital at the own funds expense exhausted itself. This is comparable with the investments of outside investors (in 2002-2012 - 20% on the average), which are also not able to meet the needs of economic subjects in a speedy renewal of fixed capital.

To a large extent, this was due to the fact that the main investors in the extractive and processing industries in the region, as well as in the services and infrastructure, are the largest coal-mining and metallurgical companies, which in 2007-2008 accounted for more than 70% of investments. Foreign investments, the maximum of which was in 2007 (US $ 579 million), were mainly trade and other loans (92.2% of total revenue).

B) Over-concentration of investments in coal industry. Industry investments in equity capital in Kemerovo region also have "pro-mineral", or rather "pro-coal" character. On the average in 2006-2013 the structure of investments in Kemerovo region economy was the following: extraction of mineral resources (primarily coal and iron ore) - 53%, processing production - 16% (including metallurgy - 5%, chemical industry - 4%, machinery - less than 1%), power industry - 6%, transport and communication - 11%, construction - 1%, R&D and education - less than 1%, agriculture - 3%, real estate, trade, housing and utilities infrastructure, and other socially important services - 11 %. Achieving the "top" of coal mining output led to the fact that, since 2004, there had been a slowdown in investment in the region – on 45% in 2005, on 1% in 2008 and on 34% in 2009 (including small enterprises and the informal sector of economy according to preliminary data).

Mainly, slowing inflow investments in regional economy is closely connected with the concentration of capital in coal and metallurgy industries, in which in 2006-2013 only 22% of investments were directed for the purchase of high-quality new and more efficient equipment, and 6% for the full technologic upgrade, while 45% for major repairs and 26% for the replacement of the means of production by their analogues. In such circumstances the possibility of decreasing the fixed capital depreciation in the leading industries and the development of R&D, machinery, deep processing of minerals investments are rather weak.

During reforms the sectored structure of regional economy also comes across the large-scale problem of mineral resource orientation and the lack of factors to overcome it in the nearest future. On the contrary, the prospects for consumption growth of mineral resources by global and Russian economies make favorable for Kuzbass marketing conjuncture of deepening "metal-coal" profile (largely due to the projected by 2020 deficit of natural gas on the domestic market by 120 billion m^3). When investments in advanced processing of material resources depend on the agreement between the governor and the main owners the dependence of the regional budget from unstable situation in the global commodity markets is increasing. In particular this applies to iron and steel industry which, using coal, heavily depends on the world prices and demand for its products. Coal mining itself, first of all, is connected with the danger of man-made accidents, and secondly, is strongly tied to the iron and steel industry. This reduces the stability and forecasting the sources of revenue for the regional budget and deprives the regional authorities of confidence in the possibility of financing the structural transformations in future. Thus, three main Kuzbass industries (coal, metallurgy and chemical industry), generating more than 70% of GRP, provide only 54% of taxes. The main "generator" of Kemerovo region GRP is the industry (over 85%) and in its structure - fuel, metallurgy, energy industry (in total - more than 70%). At the same time, these industries generate about 50% of the federal and regional tax revenues.

Consequently, the issue of critical depreciation of fixed capital in basic sectors of the regional economy should be added by the detraction of their contribution to budget revenues, inadequate to their high share in GRP and investment.

The main feature of the innovative and technological structure of Kemerovo region economy is the prevalence of 19th - early 20th centuries technological layers: the 2nd (coal mining, steel-making), the 3rd (road and urban construction, engineering industry), and the 4th (power and chemical industry). These industries account for more than 75% of Kuzbass GRP (up to 90% of its production part). Along with that the 5fth and 6th layers are hardly represented. On the contrary, in neighbor regions there is quite another picture. So, in Novosibirsk region the 2nd layer is less than 5% of GRP, the 3rd and 4th are about 55%, the 5th is 25%, the 6th is about 2%. In Tomsk region, in spite of a significant part of the mineral resource industries, the sectors of the 5th layer produce about 20% of GRP.

Outdated technological layers, prevailing in Kuzbass economy and corresponding with the end of the 19th – the beginning of the 20th century, give rise to a number of social-and-economic problems, such as the prevalence of low-skilled labor, unequal in terms of added value regional exchanges, low labor productivity in terms of value, the presence of mono-town-satellites of major industrial enterprises with a difficult social situation and others.

The development of enterprises of modern 5th and the newest 6th layers in Kuzbass is inevitably connected with the development of R&D. But in this aspect region lags behind other regions of Western Siberia. So, innovative activity reached its peak in 2006, when, according to the Department of Education and Science of Kemerovo region Administration, 275 applications for inventions were filled in the region (the 4th place), giving the way to Novosibirsk (625), Krasnoyarsk (424) and Tomsk (406) regions. Diversification of the regional economy, described in the Investment passport of Kemerovo region, in one way or another, is connected with mineral resource sectors and do not provide technological increase higher than the 4th layer (the development of power complexes on the basis of coal enterprises, underground coal gasification, methane production from coal seams, service and technological support of the mining industry and specialized engineering, new small production of special grades of steel, as well as metal-processing complexes). The main investors in the framework of these areas are again the largest commodity groups - the main owners in Kuzbass.

Consequently, within the framework of existing sectoral, institutional and technological structure of the regional economy the transition to new technological layers cannot be possible, together with abandoning single-industry mineral resource cities, monopole dominance of large holdings, the use of outdated and physically worn-out equipment. It is important to note that the structural problems of Kuzbass economy significantly worsened in today's crisis. In late 2008 - early 2009 the situation in Kuzbass economy was going to turn out from crisis into disastrous. So, of 30 million tons of unsold by 2009 coal only 12 million was finally sold. 12 coal mining and 3 metallurgical enterprises (producing 11% and 33% of coal and metal in the region respectively) were on the verge of bankruptcy. By mid-2009 58% of industrial and 39% of civilian capital construction had been frozen; the number of unemployed had increased by 4%. As a result, production and investment imbalance of regional economy toward primary industries in 2008-2009 became even more: investments in coal processing in 2008-2009 amounted only 18% of coal industry in the region, in machine-building - 11% of total investment in the regional metallurgical complex. Regionally, the proportion of coal processing and coal chemistry in export is 6% of exported mined coal, and the volume of exported machinery equipment is in 9 times less than exported outside the region metallurgy industry product. As a result, the processing enterprises in 2008-2012 received only 17.5% of total investment for regional industry and deterioration level of their fixed capital exceeded 55%.

4 Conclusions

Thus, the structural problems of Kemerovo region economy are typical for the industrial area and identical national ones. They include permanent depreciation of fixed capital, and inadequate sources of investments for its updating, over-concentration of investments in coal industry, de-formalization of rules in the field of interaction between the authorities and business, unfinished institutional reforms in major sectors of regional economy and the important role of administrative resource, monopolies in the key sectors (power, metallurgy, chemical industry), the dominance of the 2nd and obsolesce the 3rd and 4th technological layers.

Neo-industrialization of regional economy means overcoming its de-industrial structural problems in innovative development of coal, metallurgical and chemical industries, the emergence of new forms of investment relations, changing the role of the state. The technological basis of neo-industrial structural reforms of Kemerovo region economy is a convergence of innovations in the key sectors of regional economy. This means the establishment of new branches of advanced processing of raw materials produced in Kuzbass – bio-coal fuel, composite polymers and semiconductors, ultra-light alloys, the reconstruction of coal machinery and automation, the development of heat power industry recuperation.

Investment component of neo-industrialization of regional economy includes the development of public-private investment partnerships in innovation, guaranteeing investments in advanced raw materials processing on the basis of convergent technologies, issuing the subordinated loans for the establishment of regional guarantee and insurance funds for industry innovation investments.

References

1. D. Bell, *The coming of post-industrial society. Venture in social forecasting* (Collman Pub., New York, 1973)
2. C. Clark, *The conditions of economic progress* (Logan Pub., London, 1991)
3. R. Reich, *The work of nations. Preparing ourselves for 21st Century capitalism* (Penguin Pub., New York, 1992)
4. S. Glaziev, Russian Economic Journal, **2**, 3 (2012)
5. M. Gasanov, S. Zhironkin, Procedia - Social and Behavioral Sciences, **166**, 97 (2015)
6. M. Gasanov, S. Zhironkin, EpSBS, **VII**, 117-123 (2016)
7. S. Zhironkin, M. Gasanov, O. Zhironkina, EpSBS, **VII**, 124-131 (2016)
8. S. Zhironkin, V. Guzyr, EpSBS, **VII**, 160-166 (2016)
9. S. Zhironkin, E. Gasanov, O. Zhironkina, E. Taran, SHS Web of Conferences, **28**, 01145 (2016)
10. A. Zhavoronok, M. Gasanov, S. Zhironkin, SHS Web of Conferences, **28**, 01144 (2016)
11. M. Gasanov, S. Zhironkin, M. Hellmer, EpSBS, **XIX**, 745 (2017)
12. M. A. Tyulenev, Y. V. Lesin, *Taishan academic forum – Project on mine disaster prevention and control,* 441 (Atlantis Press, 2014)
13. Yu.V. Lesin, S.Yu. Luk'yanova, M.A. Tyulenev, J. Min. Sci., **46**, 78 (2010)
14. S. Zhironkin, M. Gasanov, O. Zhironkina, EpSBS, **XIX**, 753 (2017)
15. S. Zhironkin, M. Gasanov, K. Kolotov, EpSBS, **XIX**, 761-770 (2017)

Neo-industrialization of Kuzbass economy in innovative development of coal industry and machinery

Anna Balabanova[1], *Vladimir* Balabanov[2], *Elena* Dotsenko[3,*] and *Natalya* Ezdina[3]

[1]Russian Academy of Entrepreneurship, Rector, 109544, Moscow, 15, Malaya Andronievskaya st., Russia
[2]Russian Academy of Entrepreneurship, President, 109544, Moscow, 15, Malaya Andronievskaya st., Russia
[3]Plekhanov Russian University of Economics, Department of Political Economy and History of Economic Science, 117997, Moscow, 36 Stremyanny lane, Russia

Abstract. As part of rental-resource model and access restrictions to capital and new technologies, the attempts to encourage import substitution made by the government do not generate innovative type of economic development and reconstruction of economy's manufacturing industry competitive on the domestic market. In conditions of searching for a new model of Russian economy development the analysis of a number of problems such as industry degradation, reducing the competitiveness of Russian producers, the growth of structural imbalances and overcoming the negative macroeconomic trends becomes strategically important. This raises the value of neo-industrialization as the integrator of resource security, scientific and industrial potential of Russian economy and high capacity of the domestic market in the innovative development system. A special role neo-industrialization plays for the economic development of old industrial regions, which include Kuzbass (Western Siberia, Russia). Neo-industrial development of Kuzbass economy should be initiated by the technological modernization of mining machinery and innovative development of coal industry.

1 Introduction

Modern structure of Russian economy, being developed within the framework of market model, does not generate an innovative type of development. In conditions of high entropy of market processes it becomes evident that without neo-industrial transformation, the recovery of manufacturing industry on a new convergence-technological base, raising the share of high-tech industries the instability of macroeconomic system is increasing, the risks for economic and geopolitical security are growing up [1]. Also the growth of technological gap between Russia and developed countries leads to the accumulation of environmental problems in resource extracting regions like Kuzbass (Western Siberia, Russia) [2, 3].

* Corresponding author: ktyf110372@rambler.ru

Establishing innovative paradigm of old industrial regions increases the requirements to the institutional structure of economy, raises the priority of cooperation between the state institutions, science and business, investment innovations. The leading role is played by the analysis of clustering process of innovative activity and structuring the competitive advantages of Russian economy, which are not connected with mineral resource export.

An integral economic component of neo-industrial development of mining areas is the integration of technological ideas and investments in industrial production, the expansion of national manufacturing complex in the world market with the following change of the export structure. Lagging of processing import-substitution threatens the loss of Russian technological identity, strengthening raw vector of long-term social-and-economic development. In this regard, the relevance of neo-industrial transformations of Russian economy in the process of changing its structure is increasing.

2 Material and method

Today it is important to restore the dominance of industrial complex in Russian economy, which will decrease the share of pre-industrial economy and will contribute to the formation of post-industrial layer. This, characteristic for the latter, estrangement from the material production should be the consequence of the high level of technological development of industry. Therefore, neo-industrial structural transformation of Russian economy is put on with the strategic task of improving profitability, investment attractiveness and international competitiveness of industrial production.

Neo-industrialization of economy is a systemic state policy of structural reforms in three main areas [4-6].

Firstly, the creation of institutional, financial, social, technological, market conditions for innovative development of the basic branches of Russian economy - coal, oil and gas, chemical, metallurgical industries, machinery, as well as restoring the production of the 5th technological layer (electronics, aircraft and ship building) on a new technological basis.

Secondly, the profound transformation of technological and sectoral economic structure - the development of the industries of the 6th technological layer based on convergent technologies. Their resource base and, at the same time, consumers of "breakthrough innovation" should become modernized basic and manufacturing industries.

Third, the profound changes of reproduction, institutional, market-competitive, social structure, which, on the one hand, should create the necessary conditions for neo-industrialization, on the other hand, become its result.

3 Results and discussion

The purpose of neo-industrialization is to overcome the de-industrialization of Russian economy (reduction of technological level of industry and its share in GDP, loss of its infrastructure, fixed assets and intellectual basis) and to come out at the same technological trajectory with the leading countries.

One of the key questions that have to be answered is the role of fuel and energy complex in the economy. In particular, whether the fuel and energy complex to be its donor or driver? Here we should have a look at the model of other countries. In the USA, the fuel sector plays a minor role in terms of fiscal revenue, exports and GDP, but in the total investments the sector accounts for about 15%, i.e. it significantly affects the development of related industries. In Saudi Arabia, by contrast, the share of oil industry in GDP exceeds 50%, while revenues from the fuel and energy complex are 90%. In fact, the rest of economy merely serves the oil industry. Now let us look at Russia and Norway, the indicators of which are sufficiently close: in Russia the share of the fuel and energy complex in GDP is 21% (in Norway – 24%), in investments – 24% (in Norway – 29%), in the consolidated budget – over 30 % [7]. In both cases the oil and gas industry is important both for the economy as a whole and for the budget [8].

Which model of industrial development should Russia aim - Saudi Arabian, Norwegian or North American, or maybe just not to change anything? To do this it is important to understand the differences between two fundamentally different models of fuel and energy complexes. The sector plays the role of donor, when it has a high level of tax burden, when the budget is highly dependent on oil and gas revenues, as well as, in some cases, when consumers are subsidized with low energy prices [9]. Fuel and energy complex becomes a driver when energy production and consumption are based on new technologies.

In this system Russian (and China) takes an intermediate position between Saudi Arabia and North America. In the North American model (the USA and Canada) the high level of GDP per capita is not due to the oil and gas sector, but due to the economy diversification. In Saudi Arabia, on the other hand, fairly high standard of living is achieved mainly due to the production of hydrocarbons [10-11]. The similar situation is in Norway, where oil and gas sector is, however, the donor and the driver of economy. But Russia, even with the full involvement of resources, cannot reach the level of the Norwegian production of carbohydrate per capita: 142 million people live in Russia compared with 5 million in Norway. Thus, if the goal is to increase the level of life, fuel and energy complex should help neo-industrialization of economy.

To what extent Russian fuel and energy complex is the donor, and to what extent is the driver of neo-industrial development? Of approximately 20 trillion rubles of its annual revenues about 8 trillion is spent on taxes and fees and, according to our estimates, of about 4 trillion rubles on subsidizing the domestic market due to regulated low gas prices and customs duties on petroleum products (causing domestic prices to be lower than the world). Thus, the oil and gas sector donation can be relatively estimated at 60% of its revenue [12].

The remaining 40% is the potential to transform the sector into economic driver. The part of its expenditures including operating costs and capital investments have already been playing this role: they are directly converted into the income of related industries and population, stimulating their growth and consumer demand. However, incompletely - capital investment can be carried out abroad; the services can be provided by foreign companies. According to our estimates, the localization level in Russian fuel and energy complex is about 80% (the share of equipment and services of foreign companies is about 20%) [13]. This can be considered as evaluation of driver's effect from the fuel and energy complex activity [14]. Regarding to Kuzbass economy (Kemerovo region), in which coal industry dominates, neo-industrialization driver is an innovative development of machine-building. A special role is played by its segment producing equipment for the mining companies [15]. In this regard, we propose the development of Kuzbass neo-industrial cluster directed for the solution of the following tasks:

1. Radical reconstruction, technical and technological modernization of the regional machine-building, much of which was created in 1960-70-ies and has an average level of fixed capital depreciation about 70%. At the same time for the whole period of market reforms the mass replacement of the production means in Kuzbass machine-building haven't occurred that enhances the technological gap between regional producers and foreign competitors, and complicates the start of import substitution.

2. Breaking the vicious circle of investment and production problems: significant physical deterioration of fixed capital - reducing the international competitiveness of machinery production – decreasing of sales - the lack of investment resources for the

modernization of production – further increase of physical depreciation. This requires the formation of a single investment regional machinery program that reflects the needs for the new equipment of coal, chemical, metallurgy industries in the region, credit and investment opportunities of banks and investment companies, as well as technological possibilities of machinery enterprises.

3. The increase of cooperation of the regional machinery enterprises, which is now estimated less than 10%, and overcoming their technological disconnection. This is particularly important for the production of equipment and production systems with preliminarily defined options and addressed products competitive in the regional market. An important precondition for the development of cooperative relationships within cluster is the presence of non-profit organization "Association of Kuzbass Machinery Engineers" in the region which unites about thirty machinery enterprises of Kemerovo, Novosibirsk and Tomsk regions. The formation of neo-industrial machine-building cluster in Kuzbass should include the following actions of the interested parties – coal and machine-building enterprises; the regional authorities; research, design and educational organizations.

Firstly, the analysis of technological conditions of modernization of the machine-building enterprises necessary for the replacement of importing coal industry equipment taking into account the inter-firm cooperation. Also it is necessary to analyze investment needs and financial capabilities of banks and investment companies represented in the region. The results of this analysis should be the basis of the Development strategy of Kuzbass neo-industrial import substitution cluster of machine-building.

Secondly, making the list of cluster's residents which should include machine-building companies as well as entities of financial, research and educational spheres of the region. In addition, it is necessary to sign partnership agreements with foreign developers of modern machine-building technologies, computer software for providing intellectual property rights, patents and licenses.

Third, the formation of complex supply of modern machine-building products for the coal industry needs, as well as other industries of Kuzbass economy, capable in the middle term to replace the foreign equipment. This, in turn, should become the logical outcome of the regional machine-building complex modernization, and it involves the implementation of the following actions:

- the organization of the investment consortium, which should include the main consumers of machine-building products – enterprises of coal, chemical and machine-building industries, forming on a parity the investment pool in regional machine-building;
- the development of research-and-production and engineering firms' network with the participation of the leading international manufacturers of machinery;
- the establishment of the regional Agency to attract and protect the investments in import substitution of industrial products, the main focus of which should be placed on the machine-building industry, initiating neo-industrialization;
- the establishment of the guarantee fund of investments in machine-building, the main role in which should be played by the regional authorities.
- making the list of regional tax exemptions provided for the residents of import substitution machine-building cluster.

Currently, there is an urgent need for the development of complex regional target Program of neo-industrial import substitution in Kuzbass industry. The program should link the intensification of R&D, investments in modernization of coal, the chemical and machine-building industries, professional equipment, the state support and guarantee of loans for enterprises. The main objective of this program should be improvement of the target for import substitution technological level of these industries up to the level of imported by Kuzbass enterprises equipment.

4 Conclusions

Neo-industrialization of economy in mining region requires the development of relations between science, production and financial sector. To promote their development it is necessary to provide vertical integration of firms and research organizations, industrial enterprises, banks and financial companies into neo-industrial holdings. They should integrate the entire production chains of goods from mineral resource extraction, the output of materials and components up to promotion of finished products in the domestic market. Neo-industrial holdings can be based on production companies, as well as to be established by the transformation of the public-private partnerships into the business groups uniting industrial enterprises around commercial bank, concentrating in its hands the cash flows of enterprises and providing them with long-term loans, funding of leasing and bonds.

Along with the changes in the sectoral structure, neo-industrialization implies the formation of innovative-industrial clusters of network type, combining enterprises into a single research and production complex. In the innovative cluster the transfer of innovation into production must be implemented by centers of inter-cluster interaction and technological cooperation, as well as cross-platform contract centers. At the regional level there is an evident need for the formation of neo-industrial mining machinery cluster in Kuzbass. For its formation it is necessary to create an institutional structure of machine-building import substitution, including the investment consortium of machinery products consumers, an international network of industry innovation, research-and-production and engineering firms, regional Agency to attract and protect investments in import substitution of industrial products, guarantee investment fund in machine-building, complex regional tax exemption for the residents of import substitution cluster.

References

1. M. Gasanov, S. Zhironkin, Procedia - Social and Behavioral Sciences, **166**, 97 (2015)
2. Yu.V. Lesin, S.Yu. Luk'yanova, M.A. Tyulenev, J. Min. Sci., **46:1**, 78 (2010)
3. M.A. Tyulenev, Y.V. Lesin, *Mining 2014: Taishan academic forum – Project on mine disaster prevention and control,* 441 (Atlantis Press, 2014)
4. S. Zhironkin, V. Guzyr, EpSBS, **VII**, 160 (2016)
5. M. Gasanov, S. Zhironkin, M. Hellmer, EpSBS, **XIX**, 745 (2017)
6. S. Zhironkin, M. Gasanov, O. Zhironkina, EpSBS, **VII**, 124 (2016)
7. M. Gasanov, S. Zhironkin, EpSBS, **VII**, 117 (2016)
8. S. Zhironkin, M. Gasanov, O. Zhironkina, EpSBS, **XIX**, 753 (2017)
9. F. Janszen, *The age of Innovation* (London, Prentice Hall, 2000)
10. S. Zhironkin, M. Gasanov, O. Zhironkina, E. Taran, SHS Web of Conferences, **28**, 01145 (2016)
11. S. Zhironkin, M. Gasanov, K. Kolotov, EpSBS, **XIX**, 761 (2017)
12. A. Zhavoronok, M. Gasanov, S. Zhironkin, SHS Web of Conferences, **28**, 01144 (2016)
13. A.S. Yakovlev, Natural resource bulletin, **2** (2014)
14. E.S. Demidenko, E.A. Dergacheva, *Technogenic development of society and biosphere transformation* (Moscow, Krasand, 2010)
15. M.Y. Blashchuk, A.A. Kazantsev, R.V. Chernukhin, App. Mech. Mat, **682**, 418 (2014)

The problems of correlation the life quality and interpersonal dialogue in legal practice of mining regions

Vladimir Zolotukhin[1*], *Evgenia* Stepantsova[1], *Marina* Kozyreva[1], *Anastasia* Tarasenko[1], and *Alexandr* Stepantsov[2]

[1]T.F. Gorbachev Kuzbass State Technical University, Department of History, Philosophy and Social Sciences, 650000 Kemerovo, 28 Vesennya st., Russian Federation
[2]T.F. Gorbachev Kuzbass State Technical University, Master Course Study, 650000 Kemerovo, 28 Vesennya st., Russian Federation

Abstract. The paper presents an analysis of the impact of life quality at the level of interpersonal dialogue in legal practice of mining regions as the ratio of different subjects there characterizes social-and-cultural environment in general. The attention is given to the fact that the development of social-and-economic potential and independence of the subjects in mining regions in the Russian Federation and other countries makes important the social and cultural factor influencing the development of social stability. The attention is given to the fact that law enforcement on nature and environmental issues depends on the social-and-cultural environment of a particular region and from mentality, legal and environmental awareness, etc. of a particular subject. It was determined that the slowdown in economic growth and an increase of the interests in environmental pollution problems have led not only to the reassessment of the role of coal in China's economic policy, but also touched the problems of social policy, including issues of life quality and law enforcement.

1 Introduction

Today, the research of coal industry influence on the environment mainly deals with the problems of environmental pollution, inefficient use of land and air pollution. But social and cultural aspect of the activities of coal mining enterprises is not much studied, whereas in terms of social and economic independence of subjects of the Russian Federation the social-and-cultural factor becomes of great importance as it affects the development of social stability. In this case it is advisable to analyze the system of formation of social-and-economic and social-and-cultural potential of mining regions, as a result of the interaction of these phenomena. Ultimately, these phenomena analysis will allow forming an idea about the quality of life in the region, to identify positive and negative aspects of the influence on it.

* Corresponding author: zvm6@mail.ru

2 Materials and methods

To study the quality of life the works of J. Galbraith, O. Toffler and other scholars were used. To reveal the features of interpersonal dialogue in the law enforcement practice the research made by N. Bobbio, C. Ford, F. Boas and others were analyzed. The scientific literature review helped to cover the most important studies of life quality and its relationship with the specific social-and-cultural environment, including the coal-mining region, as well as the dependence on the dominant type of intercultural dialogue and the law enforcement practice. In the description of current trends and principles of functioning of the various aspects of life quality and interpersonal dialogue in law enforcement the following methods were used: the rational-logical analysis, structural-functional approach, conceptual analysis, historical and logical analysis.

3 Results and discussion

The most important factor in the implementation of social-and-cultural and economic integration is a civil society. In this respect it is required adopting the policy of improving life quality not only at the federal level but also at the level of the Russian Federation subjects, at the local level, as well as the impact on individuals. This mechanism requires correlations between "sectoral and territorial production relations taking into account regional and national interests" [1]. As a rule, the Russian Federation subject's specialization is defined by the resource factor that determines, ultimately, its social-and-cultural development. Consequently, all areas of regional policy, aimed at improving the level and quality of citizens' life, should be formed on the basis of the real social-and-economic indicators and factors of development and strategic objectives of regional development.

Considering the subject of the Russian Federation as a member of civil society it is necessary to define the role of the individual in this system. In conditions of democratization of social institutions the individual ceases to be a subject influenced by the State, he (she) is increasingly becoming an independent and active subject of social activity, taking part in the development of civil society. However, a comprehensive regional policy must respond to the needs and interests of society, focusing both on its objective and subjective needs [2, 3].

One of the social-and-cultural problems of modern society is the problem of moral development. Actualization of needs and interests of all social-and-economic groups and their subsequent implementation in democracy is possible with high level of law and legal awareness of citizens. The state provides citizens with the statutory level of natural freedom, which gives the latter the possibility to implement social-and-cultural and creative potential. Thus, in our opinion, social-and-cultural environment is formed. In the context of social-and-cultural aspect of the study of life quality social-and-cultural environment is the level of creative abilities of individuals, including the fundamental values, the importance of certain skills and knowledge necessary for the certain activities and performance of individual roles. Social-and-cultural environment directly generates subjective aspect of life quality.

In the study of social-and-cultural environment a special value is acquired by industrial literacy, or industrial culture, the carrier of which is the "engineer-technocrat" - trained and educated employee, meeting the requirements of the technical changes. Within the framework of the theory of technocracy Galbraith notes that a technocrat is an elite manager, having not only specialized knowledge of the natural sciences, but also humanitarian ones. Engineer-technocrat fully meets the requirements of industrial and post-industrial society [4-6].

Considering the "life quality", it is worth mentioning the impact of the standard of living vector on the studied area. Today, the economic aspect of the citizens' life and social well-being determines the impact on the cultural perception, the perception of the environment and "green economy" [7-8]. But, in turn, the social-and-economic status of an individual strengthens the cultural experience, or, on the contrary, exacerbates it [9]. In our view, the legal framework creates, delivers and controls the subjective aspect of life quality. In general, it is necessary to enumerate the documents directly regulating mining activities. These acts include: the RF Constitution, the Law of the Russian Federation dated February 21, 1992 № 2395-1 "On Subsurface Resources", the Federal Law dated January 10, 2002 № 7-FL "On Environmental Protection", Tax Code (Chapter 26), the Criminal Code of the Russian Federation (Article 255), the Land Code of the Russian Federation (Article 88).

In addition, a number of regulation documents specifies the requirements for the health and safety protection for the mining industry workers. In particular, Labor legislation, ordinance of the USSR State Labour Committee dated February 5, 1991 № 23 "On approval of the List of industries, jobs, professions and positions, working in which provides additional holiday for underground, hazardous and difficult working conditions for industrial personnel employed in mining enterprises, associations and mine-building, mine-sinking organizations of metallurgical industry". It provides that summing of the total duration of the additional holiday for harmful and difficult working conditions may not exceed the duration specified in this document. The order of Ministry of Health and Social Development of Russia dated July 6, 2005, № 442 "On approval of the Rate Standards for free providing with certified special clothing, footwear and other personal protective equipment employees engaged in work with harmful and (or) dangerous working conditions, as well as work carried out in special temperature conditions or connected with pollution in the steel industry organizations" (including scale issue of special clothing to employed in mining industries depending on the staff category).

These normative documents, in one way or another, affect not only the life quality of the individual worker in mining industry, but also the qualitative development of a particular region. In terms of regional policy that is accompanied by the activities aimed at the restoration and constant monitoring of the environment, for example, the Round table: "Environmental problems of the Kemerovo Region" (Novokuznetsk, April 6, 2013), "Some indicators of the environment and health of the Kemerovo region population", "Guidelines for the restoration and use of degraded land and the placement of industrial wastes". These documents stated that "54% of population in the Kemerovo region live in urban areas with "high" and "very high" level of air pollution".

The impact of the regional specialization on the life quality of citizens on example of the Kemerovo region is the following. In 2015, the share of coal mined in the Kuznetsk Basin was 57.8% of coal produced in Russia. The share of coal mined in the Kuznetsk Basin in the total volume in Western Siberia in 2015 was 97.7%. In 2015 in the Kuznetsk Basin 215.6 million tons was mined. At the same time the growth rate by 2015 was 187.3% [10].

As on January 1, 2016 there are 95 coal mining companies (47 mines and 48 open pits) and 50 ore-processing plants and facilities, which employ about 99 thousand people in the Kemerovo Region. The proportion of open coal mining is about 30%.

Despite the large amount of coal production in the Kemerovo region, there are a number of legal, social-and-cultural and environmental problems. The main problem is an uncontrolled issue of license for coal mining to inefficient subsoil users in the region by the Russian Ministry of Natural Resources. The main requirement for coal mining is its quality and safety of production.

The Governor of the Kemerovo region has repeatedly appealed to the Russian Ministry of Natural Resources with the suggestion about termination of the right for using subsurface resources by coal enterprises which do not fulfill the conditions of subsoil use. Failure to comply with the license requirements (violation of the initial terms of coal production, the reduction of coal production rates (without the agreement with the regional authority), the requirements of technical projects, failure of reporting to Rosnedra – all this speaks about low culture, including the production and legal one [11]. In this regard it is well to recall the words of Alvin Toffler, who believed that the growing lawlessness, crime, cruelty and arbitrariness witness that the law served humanity well in the past, now turns into an obstacle, and becomes inconsistent [12]. As a result of this problem, there is a threat of disruption of the heating supply for the region's population and social facilities in the area.

Since the mining industry is the main activity in Kuzbass, therefore this sphere provides jobs for a substantial part of its population. The main requirement for coal mining is its quality and safety of production. The Kemerovo region is interested in building new highly productive jobs, corresponding to new production technologies. In connection with this the governor criticizes the delay in putting into operation the new fields, insisting on the withdrawal of retired capacities.

The social-and-cultural aspect is also not less important. From this point of view, the land and its minerals are considered to be valuable for the future generations. Their using practice forms an industrial and technological culture of behavior of different subjects (public authorities, local governments, legal entities, citizens, etc.) operating under the law, but belonging to different social-and-cultural groups. In the end, on the one hand, "cultural diversity endangers the model of society welfare" [13], and on the other it creates the culture as practice [14]. The practice of this legislation depends on the social-and-cultural environment of a particular region (mining regions of Russia or China) and on mentality, legal awareness, environmental awareness, etc. of a particular subject. [15] From this perspective, the issues of state environmental and sanitary control and monitoring, as well as the efficient use and protection of natural resources faced by the various subjects should be solved on the basis of interpersonal and intercultural dialogue and for further reproduction of the population.

The problem of authorities' interaction at the federal and regional levels, increasing the coal production with the permission of the federal government requires a constructive analysis and dialogue between the two levels of government. In addition, it is necessary to organize the interaction "individual-authority" within the professional activity effectiveness. It is not only the interaction of administrative staff and authority, but also social interaction of personnel directly performing its working functions.

Studying the environmental component of coal mining, we should refer to the practice of "energy revolution" in China. Slowing economic growth and an increase of interest to the environmental pollution problems have led not only to reassessment of the role of coal in China's economic policies, but also touched the problems of social policy, including issues of life quality of the population, law enforcement in the area of environmental legislation. The share of coal in energy consumption is expected to decline gradually, as coal-fired power plants will lose their prime importance and give way to nuclear energy.

Culture in the context of the technocracy theory takes the form of not only natural-science knowledge, but also competent and comprehensive policy of implementation of this knowledge in combination with legal regulations. Moreover, N. Bobbio noted that the negative result of law "off-differentiation" from spirituality leads to the suppressing the individual way for adoption of cultural values, which is the ground for "turning the perspective of civil society development" [16]. Technological culture is a range of industrial and intellectual achievements of people. Industrial society needs require the production to be at higher level of development [17-18], therefore, better and modern knowledge. The development of production, as well as many other areas of human activity, directly depends on the level of science development and scientific achievements.

We also believe that there is a direct impact of culture on production, which is likely to increase the level of production culture (regardless the emergence of new scientific results and achievements - they may not be at some considering stage). But culture cannot crucially affect the increase of the production level, the emergence of new technologies, methods, etc. This prerogative is given to science and scientific knowledge.

4 Conclusion

Thus, the life quality level in the mining region depends not only on social-and- economic stability, but also on the social-and-cultural environment, formed by the interaction of various subjects in the framework of existing legislation. One of the conditions for maintaining social stability is interpersonal, intercultural dialogue aimed at reconciliation of interests and the realization of needs of public authorities, local governments, employers, both public and private sector workers, as well as representatives of human rights organizations, expressing the interests of civil society.

References

1. V. Trifonov, O. Loyko, D. Nesteruk, S. Zhironkin, E. Strekovtsova, AIP Conf. Proceed., **1800**, 050009 (2017)
2. S. Zhironkin, M. Gasanov, O. Zhironkina, E. Taran, SHS Web of Conferences, **28**, 01145 (2016)
3. A. Zhavoronok, M. Gasanov, S. Zhironkin, SHS Web of Conferences, **28**, 01144 (2016)
4. S. Zhironkin, M. Gasanov, O. Zhironkina, EpSBS, **VII**, 124-131 (2016)
5. S. Zhironkin, V. Guzyr, EpSBS, **VII**, 160-166 (2016)
6. M. Gasanov, S. Zhironkin, EpSBS, **VII**, 117-123 (2016)
7. S. Zhironkin, M. Gasanov, K. Kolotov, EpSBS, **XIX**, 761-770 (2017)
8. M. Gasanov, S. Zhironkin, M. Hellmer, EpSBS, **XIX**, 745-752 (2017)
9. S.A. Zhironkin, A.A. Khoreshok, M.A. Tyulenev, G.A. Barysheva, M.C. Hellmer, IOP Conf. Ser.: Mater. Sci. Eng., **142**, 012127 (2016)
10. M. Tyulenev, S. Zhironkin, K. Kolotov, E. Garina, Poll. Res., **35:2**, 221-227 (2016)
11. S. Zhironkin, M. Gasanov, O. Zhironkina, EpSBS, **XIX**, 753-760 (2017)
12. A. Toffler, Ecounter, **41(1)**, 13 (1973)
13. W. van Oorshot, Int. J. Soc. Welfare, **16**, 129 (2007)
14. John Gal, Idit Weiss-Gal, Health & Social Care in the Community, **19(2)**, 158 (2011)
15. V.M. Zolotukhin, V.A. Gogolin, M.Yu. Yazevich, M.I. Baumgarten, A.V. Dyagileva, IOP Conf. Ser.: Earth and Envir. Sci., **50**, 012027 (2017)
16. N. Bobbio, *Teoria dell'ordinamento giuridico* (Torino, 1960)
17. A.B. Efremenkov, A.A. Khoreshok, S.A. Zhironkin, A.V. Myaskov, IOP Conf. Ser.: Earth Environ. Sci. **50**, 012009 (2017)
18. M. Tyulenev, E. Garina, A. Khoreshok, O. Litvin, Y. Litvin, E. Maliukhina, IOP Conf. Ser.: Earth Environ. Sci., **50**, 012035 (2017)

Demographic and migration policy in the mining region and its impact on the ecological consciousness of the population

Vladimir Zolotukhin[1*], *Anton* Bel'kov[1], *Evgenia* Stepantsova[1], *Marina* Kozyreva[1], and *Anastasia* Tarasenko[1]

[1]T.F. Gorbachev Kuzbass State Technical University, Department of History, Philosophy and Social Sciences, 650000 Kemerovo, 28 Vesennya st., Russian Federation

Abstract. The paper presents an analysis of demographic and migration policies at the regional level and the mechanism of its influence on ecological consciousness, social-and-cultural environment and mentality. The attention is given to the necessity of increasing life expectancy by reducing both premature and preventable mortality and prolongation of a healthy (active) life by reducing disease and injury, improving the life quality, etc. The process of migrates' adaptation in the social-and-cultural environment is connected with the peculiarities of the regional mentality, which leads to the need of reaching an agreement, and, in the first place, is important for the sustainable social-and-economic development of the region. In this regard, attention is drawn to the mental foundations of law enforcement, in particular, environmental legislation, legislation on mineral resources and their use. The article proves the thesis that the agreement is possible through dialogue, even within the "forced consensus", to improve the integration process and, at the same time, to reduce differentiation.

1 Introduction

Current social transformations in modern society of the Russian Federation have vivid negative characteristics. They include: unfavorable demographic and infrastructural conditions, legitimation of illegitimate behavior, marginalization and depopulation that are typical for a number of regions, including mining. This aspect has an impact on the social-and-economic, demographic, social-and-cultural development of the region, including environmental consciousness. The problems of state management, connected with the management and organization of the economic, social-and-cultural life are also relevant. It is noteworthy that various strategies of behavior (economic, social, cultural, etc.) in Russia is often irrational, which leads to the fact that the benefit is sometimes overlooked due to the influence of subjective perception. Particular significance here has the mentality formed not only in everyday life but also "forced" by institutional authorities. Formation of mentality as a permanent component of social organization has the following task: to give

* Corresponding author: zvm6@mail.ru

law enforcement subjects the ability to live in society, to keep it by their life activity as a whole on the basis of reaching an agreement. The main aim here is the effective control and manipulation of the people's consciousness.

2 Materials and methods

To study the demographic and migration policy and its impact on environmental consciousness the works by R. Berger, T. Luckmann and other thinkers were analyzed. To identify the mechanism of achieving the consent as the Russian "civilizational archetype" features of interpersonal dialogue in the law enforcement practice the research of T.V. Dronov and others were studied. It made possible to cover the most important mechanisms of demographic and migration policy and its relationship to the specific social-and-cultural environment, including the mining region. In the description of current trends and principles of functioning of the various aspects of life quality and interpersonal dialogue in law enforcement the following methods were used: the rational-logical analysis, structural-functional approach, conceptual analysis, historical and logical comparison and synthesis, method of hypotheses and their subsequent verification as well as correlation analysis methods.

3 Results and discussion

The formation and development of demographic processes in Kuzbass are largely stipulated by the natural resource and industrial profile of the region. After the fall of the Soviet Union Kuzbass became the largest coal basin in the country. As on January 1, 2016 there were 95 coal mining companies and 50 ore-processing plants and facilities, which employ about 99 thousand people in the Kemerovo Region [1].

For many decades the coal industry had been the main city-forming factor and contributed to high urbanization of the region. However, in the early 90s of the XX century the situation changed. From 1992 to 2016 the population of the Kemerovo region decreased by 12.5% [2]. And the natural decline of the population in that period was stable. The dynamics of demographic processes was affected by several major factors: the transformation of social-and-economic conditions; the impact of long-term effects of the Great Patriotic War; changes in reproductive goals; unfavorable ecological situation caused by the current structure of industrial production.

In these conditions long-term demographic and migration policies become particularly important. So, in 2001 the region adopted the Concept of demographic development in the Kemerovo region for the period up to 2005. Among the key provisions of the document should be highlighted: the problem of increasing life expectancy by reducing premature and preventable death, particularly in the working age; increasing the healthy (active) life by reducing disease and injury; improve the life quality of chronically ailing and disable people by providing them with conditions for realization of the existing (residual) health potential; creating conditions for increasing the birth rate by gradual transition from a predominantly small to medium type of family reproductive behavior; comprehensively strengthening the family institution as the most efficient form of life of the individual and his normal socialization; normalization of migration flows and attracting migrants, while reducing the migration flow and protection of the labor market from the illegal migration.

It is worth noting the fact that characteristics, affecting the practice of inclusion the migrants in the labor markets, are usually related with the migratory experience, motivation of trips, and orientation on the length of staying. Ultimately, it is connected with the peculiarities of the regional mentality, and existing and widespread "practice of double standards in the field of environmental policy establishes certain behavior models [3]. This is typical for the situations where rules and regulations are widely ignored, and life is moving under its own laws, if not like a "fight without rules" [4].

Considering mental grounds in law enforcement there is a problem of "environment" (space) – the locals who – whether they like it or not – need to establish optimal rules (conditions) for their livelihood. An important issue is the ability of "environment" to share the proposed rules of the game. Naturally there is a question about "the grounds for agreement [5], which usually identify two – mutual benefit and mutual humanitarian assistance of the parties, in which the benefit takes the form of moral satisfaction" [6]. In this connection we recognize the consent as the Russian civilized archetype. This is what the Russian multiculturalism is based on, and it is the fundamental basis and the main need of our society.

Regional characteristics (educational level, employment, the implementation of social safety nets by local authorities and management, crime rates, etc.) affect the minds and activities of daily living. For example, despite the fact that in recent years in the Kemerovo Region there has been intensive housing construction, improving housing affordability, and so on, Human Rights Commissioner in the Kemerovo region, states the following in his annual reports that the number of citizens' reports about issues related to the provision of housing, complaints about the quality of housing and public facilities, etc. remains unchanged. This fact indicates the "objective" problems in the field of law enforcement at both regional and municipal levels.

The reports have repeatedly emphasized that there are collective complaints of citizens in which violations of mass character are described as they do not relate to one person, but affect the rights of the large group of people. It concerns the problems of violation of citizens' environmental rights, respect human rights and freedoms in the field of migration, the rights of citizens in criminal and administrative law, etc.

According to R. Berger, "the intenseness of consciousness is the most in everyday life, that is, daily life installs itself in the mind in the most massive, persistent and intense way. In its imperative presence it is impossible either to ignore or to weaken it" [7]. Having been on the level of mentality, it (daily life) reproduces itself and functions as a kind of replacement of the mechanism of biological inheritance (succession).

The social aspect of law enforcement should be associated with an active response to the demand for any changes in usual style of communication, or as an attempt to reach an agreement with something new, unknown and, at first sight, sometimes repulsive [8]. Position "it is difficult, but possible to tolerate" should not be equated in any way with traditional Russian "unresponsive" patience: to a large extent it is linked with the expectation of "response", i.e. a certain result, personal or social. Russian patience was not always "unresponsive", rather we can speak about a "dotted line" of irresponsibility, broken by explosive discontent and despair acts.

The impact of information technologies promotes the development of the given situation, particularly in Russia [9-11]. An individual, as a rule, cannot tell if it is a rumor, a version, an opinion, a fact and so on. And here the dialogue would be particularly useful, even within the "imposed consensus (increased integration and, at the same time, the decline of differentiation), and analysis of enthymemes - this very old art of imposing food for mindless ingestion when implied is introduced as already known. In this case there is a risk of getting used by the consciousness to the low-somniculous mode in which, for example, meeting death (the facts of terrorist threats by destabilizing the social environment subjects, etc.) causes panic. All this leads to the disintegration of the whole society, its social and cultural environment and updates the law enforcement problem. Based on the equivalence of priorities, economic development and environmental conservation, it is necessary to promote environmental law enforcement within the country, to facilitate the interaction of responsible for the environmental policy departments and institutions, to extend financial support for environmental protection programs [12].

This is the equivalence of priority of authority subjects. In China, for example, funding the environmental protection directly depends on the local authorities. As a result, the orientation of the local authorities on economic growth neutralizes the autonomy of environmental institutions. In addition, for the last 20 years in China a number of independent environmental organizations had been established: "Friends of Nature", "Institute of Global Environmental Culture in Beijing", "Green Earth Volunteers", "Environmental Protection Beijing Fund". These organizations aim to environmental education of China's population, training volunteers, organizing social events, distribution of environmental information, and others. At the same time, environmental protection student groups are organized in many educational institutions. In Kemerovo Region (Novokuznetsk April 6, 2013) the following goal was set to consider the possibility of restoring a system of state control over the land remediation financing its activities from a variety of sources. As the sources of financing to consider the system (extra-budgetary) environmental funds at various levels replenished at the expense of environmental fees, fines and other payments (with the distribution of funds in the proportion of 60% – the municipal, 30% – the regional, 10% – the federal environmental funds), with the further financing of the environmental projects at the definite level on the competitive basis, etc.

Based on the activity principle, every action is possible to make due to the ability of the subject to the discursive and practical interpretation of nature and causes of his actions within the framework of not only subordination but fair law enforcement. In this regard for the practical consciousness it is peculiar "the sense of ontological security" (A. Giddens [13]), similar to "the sense of basic trust" by E. Erikson [14]. The sense of basic trust is defined as the definite relation to oneself and the world. It implies both the human own credulity and the sense of constant propensity to oneself of other people.

Everyone recognizes that the destruction of the conditions for the real equality is dangerous for freedom of all people. This judgment is true, because the maintenance of public order is understood as a necessary condition for achieving by everyone his goals (for example, the right to the healthy environment) whatever they may be (if they lie within the certain limits), as well as the realization of his own interpretation of moral obligations. As an example may be given the situation of restrictions on the freedom of conscience by some limits even indistinct imposed by the state, due to the limit, derived from the principle of public interest, the interest of the representative equal citizen. This limitation is due to the so-called paradox of freedom, which shows that freedom, in the sense of any kind of its limit, should lead to a significant restriction of it, so as not to allow the bully to enslave the meek. This manifests the dialectic of confrontation and cooperation, "weak" and "strong", and the person responsible only for himself, but everyone, without exception, is responsible for others to maintain the integrity in the society. Moreover, it is a distinctive feature of the formation of mental, social and cultural space.

The signs of stabilization of the demographic situation, which appeared in the first years of the XXI century appeared to be unstable and, to a large extent, caused by the impact of the structural demographic factors. Against the background of increasing the birth rate the main reason of negative natural population growth remained high mortality rate.

The implementation of the program "The improvement of environment and population of the Kemerovo region" according to which, in particular, during 1998-2000 it was possible to improve drinking water quality and bring into use the sewage treatment plants in the 14 areas, did not allow changing the situation radically. In terms of occupational illness the region took the second place in Russia, and air pollution in the large regional cities remained at the high level.

The analysis of results of population migration in the Kemerovo region shows wave-like nature of the process. Reducing migration turnover started in the 90-ies of the XX century continued in the first decade of the XXI century. The migration growth remained predominantly positive.

At the beginning of the second decade of the XXI century migration activity in the region increased, but the effectiveness of migration became negative. Taking into account the current situation, in 2016 the authorities of the Kemerovo Region approved a comprehensive regional program "Improving the demographic situation in the Kemerovo region" for 2016-2018 years.

Within "The state migration policy concept up to 2025" defining the strategy for the development of migration processes on the basis of estimates and the prospects of demographic and social-and-economic development of the region, it is necessary to consider that the possibilities for migrants' attraction in the nearest future may increase. By the end of 2018 it is supposed to ensure the reduction of migration loss in the Kemerovo region by attracting compatriots living abroad (qualified professionals and young people) for permanent residence. The interest of the compatriots in the Kemerovo region, according to the developers of the document, is constantly increasing. The measures of additional guarantees for compatriots - participants of "The state program of assisting the voluntary resettlement to the Russian Federation of compatriots living abroad" approved by the Presidential Decree dated June 22, 2006 № 637 should help to improve the efficiency of this process.

In our opinion, one of the factors having negative impact on the demographic situation in the Kemerovo region in the medium term will be the relocation of coal-mining center from Kuzbass to Eastern Siberia and the Far East, provided by "The program of Russian coal industry development for the period up to 2030".

According to the expert from the Institute of Energy Research of RAS the measures contributing to the stabilization of the current situation include the following:

- The consistent adjustment of regulatory requirements for the coal industry organizations in the sphere of environmental protection in accordance with the world standards;
- The planned liquidation of unpromising coal industry organizations within the measures of social protection of redundant workers;
- Encouraging environmental protection measures, including waste processing and reclamation of land [15].

4 Conclusion

Thus, in the mining region the demographic and migration policy depends on the interaction level of both federal and regional authorities, coordination of their interests and needs to increase coal production, create a favorable ecological environment for living, etc. For environment preservation and sustainable social-and-economic development the constructive cooperation, not only between authorities, but also with the population living in these areas is necessary in the framework of the territorial mentality and ecological consciousness. Ultimately, the formation of social and cultural environment and the solution of law enforcement problems depend on it.

References

1. M. Gasanov, S. Zhironkin, EpSBS, **VII**, 117-123 (2016)

2. S. Zhironkin, E. Gasanov, O. Zhironkina, E. Taran, SHS Web of Conferences, **28**, 01145 (2016)
3 V. M. Zolotukhin, V. A. Gogolin, M. Yu. Yazevich, M. I. Baumgarten, A. V. Dyagileva, IOP Conf. Ser.: Earth and Envir. Sci., **50**, 012027 (2017)
4. N. Ferguson Civilization. *The West and the Rest* (Penguin Books Publ., London 2011)
5. S. Zhironkin, V. Guzyr, EpSBS, **VII**, 160-166 (2016)
6. M. Gasanov, S. Zhironkin, M. Hellmer, EpSBS, **XIX**, 745-752 (2017)
7. R. L. Berger, Thomas Luckmann, *Die gesellschaftliche Konstruktion der Wirklichkeit* (Wissen., Freiburg, 1967)
8. A. Zhavoronok, M. Gasanov, S. Zhironkin, SHS Web of Conferences, **28**, 01144 (2016)
9. S. Zhironkin, M. Gasanov, O. Zhironkina, EpSBS, **VII**, 124-131 (2016)
10. S. Zhironkin, M. Gasanov, O. Zhironkina, EpSBS, **XIX**, 753-760 (2017)
11. S. Zhironkin, M. Gasanov, K. Kolotov, EpSBS, **XIX**, 761-770 (2017)
12. V. Trifonov, O. Loyko, D. Nesteruk, S. Zhironkin, E. Strekovtsova, AIP Conf. Proceed., **1800**, 050009 (2017)
13. A. Giddens, *The Politics of Climate Change* (Polity, Cambridge, 2009)
14. E. Erikson, *Insight and Responsibility* (Norton, London, 1994)
15. A.A. Khoreshok, S.A. Zhironkin, M.A. Tyulenev, G.A. Barysheva, V.Yu. Blumenstein, M.C. Hellmer, S.V. Potyagailov, IOP Conf. Ser.: Mater. Sci. Eng., **142**, 012122 (2016)

Cross-cultural analysis of the verbal conflict behavior of the graduate mining engineers

Inna Pevneva[1*], *Olga* Gavrishina[2], *Anastasia* Rolgayzer[1], *Marina* Agienko[1] and *Alexandr* Myaskov[3]

[1]Kemerovo institute (branch) of Plekhanov Russian University of Economics, Department of Humanities, 650992, 39 Kuznetsky av., Kemerovo, Russia
[2]Kemerovo State University, Institute of Sciences, UNESCO IT department, 650000, 6 Krasnaya st., Kemerovo, Russia
[3]National University of Science and Technology MISiS College of Mining, 119049, 4 Leninsky av., Moscow, Russia

Abstract. The article is devoted to the crucial issue of the interpersonal communication skills of engineering graduates and studies the verbal behavior of the graduates majoring in mining engineering in conflict professional communication considered in a cross-cultural aspect. The research is based on the needs that future mining engineers have for conducting successful communication, work in teams and run an effective discourse both verbally and in writing. Verbal communication involves a strategic process by which a speaker defines the language resources for its implementation. By choosing a strategy which should contribute to the goals and objectives of the interaction a speaker makes the process of communication either successful or leading to a communicative failure. The scientific importance of this work is in multidiscipline approach and cross-cultural study of ethnic and cultural influences, gender and other characteristics of the verbal behavior of Russian and American engineering graduates.

1 Introduction

There are currently several concepts and theories that consider the problem of cross-cultural interaction. According to some scholars globalization entails cultural homogenization where a stronger culture contributes to the formation of a homogeneous world or global culture. Consequently, the development of intercultural communication skills and the formation of a tolerant attitude to other people and cultures becomes an integral part of university education in the era of globalization [1-2].

In this regard we witness a special approach that criticizes a pure technical direction of engineer training and education [3-4]. Traditional engineering education considers a mining engineer only as a professional in his area but completely ignoring the future engineer as a developing personality that should be able to handle solve complex problems based on the synthesis of the various spheres [5-7]. Nowadays a mining engineer graduate needs to have

* Corresponding author: Inna_nag@mail.ru

ability to work in a team, conduct negotiations and preferable speak a foreign language. Thus a university education should help graduates to learn self-development and contribute to formation of their humanist values and tolerance.

Humanization of technical education is often discussed and this issue has taken an important place in the paradigm of modern higher education on the global level [8].

It might seem illogical, but being a purely technological in its essence, engineering presupposes a great part of communication involvement [9]. A number of research and practice reports show that many graduate engineers face difficulties in verbal and written communication. Companies lose substantial sums of money due to written communication problems of graduate engineers. The most common communication barriers turn out to be a lack of special training or communication skills, inability to formulate ideas correctly and poor listening skills. It is often taken for granted that engineers have skills to conduct an effective communication discourse and be able to transfer ideas to an interlocutor freely both verbally and in written form. However, a few findings in the sphere of engineering education in the USA, Canada and Australia show that the necessity to develop and perfect communicative skills is of prior concern. This is connected both with changes in modern engineering, sufficient increase of paper turnover, quality and safely control. Taking into account the internationalization and global economy trends, these changes require a good understanding of intercultural specifics as well [10-12].

The purpose of this research was to explore verbal conflict-handling differences Russian and American graduate students majoring in mining engineering and their understanding of conflict concept which directly defines the success or failure of communication. More specifically, we were interested in the choice of communicative strategies that graduates prefer to stick to in course of natural professional communication.

The research regarding the difference between Russian and American perception of conflict has been limited but the volume of tourism, trade, academic mobility and migration that now necessitates intercultural interaction and the potential for conflict gives rise to the need for greater understanding and tolerance of differences.

Definitions of conflict offered by Russian and Soviet researchers illustrate contradictions between interests, purposes or motives, but also present negative emotions and the desire to cause damage to opponents. The dichotomous concept is cooperation [13]. The Longman Dictionary of Contemporary English offers the following definition of conflict: "...when two ideas or opinions cannot exist together" [14]. American researchers such as Edmondson and Smith cite "personality differences and interpersonal tensions" as being at the heart of conflict [15]. Thus, it is possible to assume that in the Russian verbal consciousness, counteraction to the opponent is emphasized while for Americans the subject of the conflict (i.e. the belief or course of action) is in question.

Linguists who have conducted research on conflict have examined conflict-handling strategies from the point of view of the “self” in altercation with “another”

Miming engineers have to communicate to each other in international cooperation during new equipment installation, reporting on international conferences, attending training centers etc. [16-17]. It was found that conflict strategies can be classified as cooperation (constructive), distancing (neutral) or confrontation (destructive). The cooperative strategy of conflict-handling includes compromise (tactics of promise, agreement and compliment), cooperation (tactics of rational persuasion, agreement and support) and adjustment (tactics of concession, sympathy and changing the topic). Distancing strategy includes an attitude of active neutrality (concession) and passive neutrality (silencing and ignoring). The third strategy - confrontation - consists of manipulation (threat, pressure, appeal to authority), and verbal aggression (insult and criticism). Verbal aggression can also include auto-aggression, which is aggression directed not at another person, but toward the speaker him/herself. For the present study, four dependent variables were used: cooperation, distancing, confrontation (manipulation), verbal aggression (aggression toward self or other).

It is widely known that there are differences between men and women in terms of conflict-handling. Women, regardless of nationality, are less aggressive than men and in general they focus on establishing contact, building relationships and finding mutual understanding.

Men, on the other hand, often reveal competitive and dominating behaviors. We sought to clarify the differences in Americans' and Russians' conflict-handling strategies and to shed light on gender differences with regard to verbal communication choices and thus derived the following hypotheses:

H1 – Engineering graduates will use different verbal strategies in situations of conflict depending on nationality.

H2 – There are gender justified differences between Russian and American mining engineering graduates with regard to verbal behavior in conflict situations.

2 Methods

Two-hundred-and-fifty students (118 males and 132 females) from state universities in Kuzbass region and 248 students (107 males and 141 females) from a state university in the USA (PA state) have participated in this experiment. Three questionnaires were used to gather data: The Conflict Questionnaire, The Situation Response Questionnaire and a word association test.

Free-response (24 items), nominal (6 items) and multiple-choice questions (16 items) appeared on the questionnaires and, thus, the data was analyzed both qualitatively and quantitatively. The free-response items on The Conflict Questionnaire were content-analyzed by lexical units of conflict and emerging patterns were noted. Linguistic markers included, for example, one word phrases without emotional charge like "fine" or "OK, I agree" which show distancing. The lexical units were also examined for subject, object, motive, reason for and form of aggression, and attitude toward aggression. The Conflict Questionnaire also included items allowing students to choose responses that best described how they would most likely react in professional conflict interactions. In addition, participants completed the Situation Response Questionnaire and thus were given situational models of speech with five possible responses and asked to choose the verbal response that best represented how they would react in potential conflict situations. Each of the proposed variants of verbal reaction corresponds to a particular conflict-handling strategy including cooperation, confrontation, distancing, aggression (sometimes tested as autoaggression), or other.

3 Results

3.1 Verbal Choices in Situations of Conflict

Our first hypothesis was that Russian engineering graduates and American graduates would make different verbal choices in situations of conflict. We found support for this hypothesis as well for the strategies of confrontation and aggression. The analysis of the verbal reactions for the multiple choice questions showed that 39% of the Russian graduates preferred to use cooperation, while 37% of the Americans chose this strategy (see Table 1). Only 14% of the Russians and 11% of the Americans use the distancing strategy. However, 36% of the Russians chose the strategy of confrontation as opposed to 25% of the

American respondents and 11% of the Russians (vs. 27% of the Americans) chose aggression (Table 1).

One question, Item 11 on The Situation Response Questionnaire, was designed to reveal strategic differences specifically related to professional interactions. Table 2 reveals the results of this question.

Table 1. Significance Levels for Strategic Conflict-handling Choices of Russian and American engineering graduates.

Conflict Strategy	Russians (n_1=250)		Americans (n_2=248)		Fisher's criterion φ^* (value level)
Cooperation	39%	φ_1=1.349	37%	φ_2=1.308	0.457 (>0.1)
Distancing	14%	φ_1=0.767	11%	φ_2=0.676	1.015 (>0.1)
Confrontation	36%	φ_1=1.287	25%	φ_2=1.047	2.678 (0.003)
Autoaggression	11%	φ_1=0.676	27%	φ_2=1.093	4.653 (0.0001)

Table 2. Significance Levels for Strategic Conflict-handling Choices during professional communication.

Conflict Strategy	Russians (n_1=250)		Americans (n_2=248)		Fisher's criterion φ^* (value level)
Confrontation	17%	φ_1=0.850	8%	φ_2=0.574	3.08 (0.001)
Cooperation	52%	φ_1=1.611	63%	φ_2=1.834	2.48 (0.001)
Distancing	69%	φ_2=1.961	36%	φ_1=1.287	7.52 (0.0001)

Notes. Fisher's criterion (φ^*) is used to determine statistical significance such that the critical value of φ^* = 1.64 ($p \leq 0.05$ and 2.31 ($p \leq 0.01$).

3.2 Gender

We also hypothesized that there would be differences between Russian and American engineering graduate men and women with regard to choices in conflict situations and significant differences were found between sexes for each nationality for all four dimensions. In addition, significant differences were found between Russian and American men and between Russian and American women for distancing and aggression.

Russian women use the cooperation strategy far more often than do Russian males. Of the Russians who chose cooperation, this strategy was used by 66% of the females, but only 34% of the males. The gender difference for cooperation was almost identical for the Americans who chose it. Thirty-five percent of those were men and 65% were women.

Results showed that 100% of the Russians who use the distancing strategy are women, while 100% of the Americans who chose this strategy are men. Another significant effect was found for confrontation for Russians. Thirty-eight percent of those who chose it were men, while 62% are women. Finally, a very high percentage of Russian engineering graduate who chose the strategy of aggression were women (72%), while far fewer were men (28%). There was a significant difference between American men and women for this dimension as well, but it was not as pronounced (57% men vs. 42% women).

4 Discussion

The emerging lexical units for American engineers in describing conflict reveal a dynamic plane of handling conflict. That is, Americans focus more on the actions related to engaging in the conflict, rather than the cause and effect. The analysis showed that conflict is associated in the American language consciousness not only with the presence of opposition but also with the necessity of resolution. (*Conflict is... some uncomfortable situation that needs resolving, a situation that arises between two or more people which pushes them to seek a resolution, a situation in which two parties disagree usually has a resolution which may or may not be mutual*). Both Russian and American graduates stress negative emotional consequences resulting from conflict.

We also found that both American and Russian mining engineering graduates prefer the cooperation strategy not only most, but in almost equal proportions (39% of Russians and 37% of Russians prefer this strategy). This suggests that, in terms of conflict communication, there are more similarities than differences. Many participants expressed constructive ways of dealing with conflict. The results for distancing also revealed more similarities than differences between Russians and Americans (14% of Russians and 11% of Americans chose distancing). The greatest differences between Russians and Americans in terms of handling conflict occurred for confrontation and aggression. Russian graduates appear to be more comfortable with confrontation than are Americans (36% of Russians and 25% of Americans chose this strategy). A distinctive sign of conflict in Russian language consciousness is the characteristic of an aggressive line of conduct expressed in verbal or physical interaction, which is often followed by negative emotional or interpersonal consequences (anger, insult). Russian young people revealed confrontational verbal choices in the form of verbal aggression or physical threats. The finding of greatest statistical significance, though, was that Americans far more often than Russians actually choose the strategy of aggression (16% more Americans chose this strategy).

All in all, the findings of the research and the conditions of the sphere of engineering professional communication prove the necessity of the educational process and training that would compensate the lack of effective communication skills and that would contribute to further development of thinking, emotional control and ability to communicate for mining engineering graduates who is be able to adapt successfully at a new working place, work in a team and enhance personal growth.

References

1. M. Gasanov, S. Zhironkin, Procedia - Social and Behavioral Sciences, **166**, 97 (2015)
2. M. Gasanov, S. Zhironkin, EpSBS, **VII**, 117 (2016)
3. A. Csikosova, K. Teplicka, Procedia - Social and Behavioral Sciences, **46**, 2978 (2012).
4. G.V. Shevtsova, Int. Res. J., **3** (45), 52 (2016).
5. A. Zhavoronok, M. Gasanov, S. Zhironkin, SHS Web of Conferences, **28**, 01144 (2016)
6. M. Gasanov, S. Zhironkin, M. Hellmer, EpSBS, **XIX**, 745 (2017)
7. S. Zhironkin, M. Gasanov, O. Zhironkina, EpSBS, **XIX**, 753 (2017)
8. J. Rojter, *International Conference on Engineering Education and Research "Progress Through Partnership"* (2004)
9. C. Tenopir, D.W. Kin, *Communication Patterns of Engineers* (New York 2004).
10. G. Blake, IIE Solutions, **30**, 38-39 (1998).
11. S. Zhironkin, M. Gasanov, O. Zhironkina, EpSBS, **VII**, 124-131 (2016)
12. S. Zhironkin, V. Guzyr, EpSBS, **VII**, 160-166 (2016)
13. A.J. Antsupov, A.I. Shipilov, *Conflictology* (Moscow, 2001)

 DOI: 10.1051/e3sconf/20171504016

14. Longman Dictionary of Contemporary English (Pearson Education Limited, 2003).
15. A. Edmondson, D. Smith, California management review, **49**, 6 (2006).
16. S. Zhironkin, M. Gasanov, O. Zhironkina, EpSBS, **VII**, 124 (2016)
17. S. Zhironkin, E. Gasanov, O. Zhironkina, E. Taran, SHS Web of Conferences, **28**, 01145 (2016)

The impact of human factor on labor productivity at the mining enterprises

Galina Pinigina[1*], *Irina* Kondrina[1], *Svetlana* Smagina[2], *Viktor* Tatsienko[3], and *Anatoliy* Meshkov[4]

[1]T.F. Gorbachev Kuzbass State Technical University, Department of History, Philosophy and Social Sciences, 650000 Kemerovo, 28 Vesennya st., Russian Federation
[2]Kemerovo State University, Department of Psychology of Education, 650000 Kemerovo, 6 Krasnaya st., Russian Federation
[3]T.F. Gorbachev Kuzbass State Technical University, Institute of Industrial and Ecological Security, 650000 Kemerovo, 28 Vesennya st., Russian Federation
[4] JSC SUEK-Kuzbass, 652507 Leninsk-Kuznezkii, 1 Vasilieva st., Russian Federation

Abstract. The article describes the term "human factor" which implies a person involved in the organizational process in the diversity of his natural and socio-psychological characteristics. The necessity to identify the impact of human factor on labour productivity at the mining enterprises is proved. It is assumed that considering human factor can be one of the ways to increase labour productivity. A research technique of the complex – mechanized team in order to identify the impact of human factor on its productivity is described. Definite research results and analysis which strongly support the assumption are given. The stages at which the human factor should be considered are analyzed. Based on the fact that person's mood determines all his vital functions, the following interpretation of the human factor was propose: to consider the human factor means to take into account everything that might spoil the mood of a person starting from his coming to the place of work till the work is finished. If it is necessary to provide high productivity, take care of the human mind. This thesis does not require proof and justification, it is obvious.

1 Introduction

In the world psychological and management literature the term "human factor" had been described long ago and come along with such conceptual partners as "human relations", "management style" and "psychological climate". Like any other term it is necessary to describe the term "human factor". What is the context of its use? The answer is simple: in this case it is a person involved in the organizational process in the diversity of his natural and socio-psychological characteristics [1-6].

* Corresponding author: ikondrina@mail.ru

2 Materials and Method

It is difficult to overestimate the role of human factor in all spheres of activity. Even in 1936, the famous American expert in Management Dale Carnegie noted that even in such technical areas as engineering about 15% of financial success achieved by a specialist is due to purely technical knowledge and about 85% due to his art of human engineering, the personal characteristics of his character [7]. Western management theorists confirm that taking into account the human factor increases productivity [8-12]. It obtains a special importance at present post-industrial stage of technological development of the economy [13-15].

Unfortunately, we have been taking in stride the foreign experience in management issues, especially its social and psychological aspects. In recent years interesting papers on psychological management problems [16-18] have been published. As a result, a favorable situation for determining the influence of human factor on labour productivity at the mining enterprises [19-24] was created.

The survey covered 64 respondents – employees of the complex-mechanized mining team. Any professional activity to be implemented successfully requires certain personal qualities; we call them professionally-important (PIQ).

The work at the enterprises of coal-mining industry refers to the dangerous occupations that require human social-and-psychological readiness to work in extreme conditions, namely, the high level of such personal characteristics as responsibility, self-control, resoluteness and being a team player. The description of these professionally-important qualities is given below:

- *Responsibility* (people with the high level of this quality have a sense of duty, they are precise and accurate in affairs, like order in everything, conscious, conscientious and highly moral);
- *Self-control* (people with the high level of self-control are organized, able to control their emotions and behavior, to overcome the obstacles, bring to the end everything they start, aware of social demands and try to fulfill them);
- *Resoluteness* (people with the high level of resoluteness are brave, resolute, risk-taking, able to withstand heavy loads; do not get lost confronting with unexpected circumstances);
- *Being a team player* (people with the high level of this quality keep good relations in the team, they are people friendly, unenvious, compliant, careful to some extent).

To the list of professionally-important qualities we added two qualities which, in our opinion, could influence the result of work in conditions of risk for life. These are *self-confidence* and *restraint*:

- *Self-confidence* (people with the high level of self-confidence have adequate self-esteem, confident in their abilities, actively respond to all the events);
- *Restraint* (people with the high level of this quality control their emotions well; do not take hasty decisions; reasonable in assessing the events; balanced in dealing with people).

To measure these qualities in our respondents, we chose one of the most reliable and valid questionnaire in the world made by R.B. Cattell. Multivariate personality Cattell questionnaire is widely used in psycho-diagnostic practice. This test is reliable, versatile, practical, and provides comprehensive information about the individual. Questions reflect ordinary life situations.

In addition, this field of activity presupposes flexible nervous system of a person, allowing orient fast and right in extreme situations and mobilize for problems solution.

These features include the type of temperament and characteristics of human behavior. To identify the temperament of the respondents we chose the methodology of H. J. Eysenck.

The selection of this technique was due to the direct interconnection between conditions of dangerous profession and the dynamics of nervous processes in the human body. The more dangerous the work is the stronger the type of nervous system must be.

In modern society there is a task to improve the humanistic content of labor, i.e. to achieve maximum performance with optimum preservation of health and sufficient emotional and spiritual satisfaction. One of the factors that ensure the maximum performance of the employee is the psychological climate in the team, namely, satisfaction from working conditions and relationships, both between team members and between managers and subordinates. Accordingly, the respondents were offered a few questions, the answers to which allowed to analyze the psychological climate in the team.

The team consisted of six sections. Based on our research goals we ranked the sections by the results of output for 10 months in 2016. Then, to determine the impact of human factor on productivity, we ranked sections by the level of professionally-important qualities and qualities which could affect labour productivity of the respondents.

3 Results and discussion

Analysis of the results allowed stating the obvious influence of human factor on productivity of people working at the mining enterprises. Members of the section, which took the first place by the results of work for 10 months, have all professionally-important qualities and qualities which could affect the results of work and they were highly formed. Only the level of "being a team player" quality was not at the top, but it can be explained. All members of the section, including the section leader, were choleric, i.e. leaders. However, the psychological climate in the section was good: 63% of staff was satisfied with job and did not intend to change it; 68% were satisfied with relationships within the team and relationships with section leader.

Members of the section taken the second place by the results of work for 10 months had lower level of professionally-important qualities. However, the human factor was evident in the face of the section leader, whose professionally-important qualities were formed at a good level, and the analysis of the replies of section members to the questionnaire showed a great impact made by the section leader on the psychological climate: 100% of respondents were satisfied with job; 78% did not intend to change it; 89% were satisfied with relationships within the section and 100% were satisfied with the relationships with the section leader. Favorable psychological climate in the team and skillful leadership shown by the section leader made it possible to achieve good production results.

Three sections were practically at the same level by the working results. They were insignificantly different in professionally-important qualities. The level of professionally-important qualities of both members and section leaders were high enough.

The section taken the last place by the results of labor showed a disturbing pattern. All its members had low level of professionally-important qualities. The psychological climate in the section couldn't be called good: 35% were not happy with the relationships with the section leader, while 83% were satisfied with relationships with the section members; 58% of respondents said that they did not like the job and 75% would like to change it. With such atmosphere in the team there is a need in a strong section leader, but it cannot be said about the present one: his responsibility was not formed at the proper level; he had low self-control; and was not a good team player. The impact of human factor on the results of work in the section was obvious.

The analysis of study results confirmed the hypothesis: considering the human factor can be one of the ways to increase the labour productivity at the mining enterprises.

Two factors deserve attention. The first, all respondents were people with a strong type of nervous system. This supports that people with a weak nervous system do not choose the job connected with the risk for life. The second factor concerns the high level of self-control and resoluteness which had absolutely all respondents. This shows the influence of the job itself on the formation of personal qualities.

What stages should the human factor be considered on? First of all, at the recruitment stage. Every company should have job description for all positions which contains the description of the content and conditions of work and, more importantly, psychological requirements, that is, the list of professionally-important personal qualities. The psychologist determines the level of these qualities, and considers the possibility of taking the person to the workplace. It is interesting to work only when there is a maximum matching of professional skills to the professional requirements. If this condition is fulfilled the person will have pleasure from the labor process itself.

The next stage of considering the human factor is the formation of the team (group of workers, section, and so on.); appointment of the division head. The results of our study clearly confirmed this.

4 Conclusion

In conclusion we would like to underline the thesis: to consider the human factor means to take into account everything that might spoil the mood of a person starting from his coming to the place of work till the work is finished. Why is it mood? It is because the mood determines all human activity.

This thesis does not require proof and justification, it is obvious. Do you want a person to work hard, creative? Take care of his mood. Mood creates a desire to work. Convince the person in belonging to everything going on at the enterprise. It will be another drop in the formation of desire to do everything that depends on him personally in order to achieve overall success.

Is it necessary to convince that considering the human factor has a positive effect on the mood of all team? The quality of work and productivity always accompany a good socio-psychological climate in the team. Hence there should be an obvious psychological readiness of managers at mining enterprises.

References

1. H. Tichy, M. Devaney, *Leaders of reorganization* (Economics, 1990)
2. J. Sullivan, C. Snodgrass, Asia Pacific journal of management **8** (1991)
3. W.F. Ouchi, *Theory Z: How American Business Land Meet the Japanese Challenge (*1981)
4. M.R. Sethi, Q.B. Baloch, *How to develop self Confidence and influence people by public speaking, published by pocket books* (Simon and Shuster, 1956)
5. R.J. Deluga *«Can work groups be made more effective?» Yes!* (Academy of Management executive, August, 1994)
6. B.M. Bass, *Handbook of leadership. Theory, research and management applications* (New-York: The press division of Macmillan, 1990)
7. D. Carnegie, *How to win friends and influence people* (1998)
8. H. Cornelius, Sh. Faire, *Everyone can win* (Simon & Shuster Australia, 1989)
9. H. MacKay, *Swim with Sharks without being eaten: outself, outmanage, outmotivate and outnegotiate your competition* (2005)

10. K. Muangmukrafan, *Factors in job satisfaction contributing to retention of employees in organization*, (Thammasat University, Language Institute, 2006)
11. P. E. Hart, *"Theories of firms" growth and generations of jobs* (Review of Industrial Organization, 2000)
12. D. Leonard, W.C. Swap, *Deep Smarts: how to cultivate and transfer enduring business wisdom* (Boston, Harvard Business School, Press, 2005)
13. S. Zhironkin, E. Gasanov, O. Zhironkina, E. Taran, SHS Web of Conferences, **28**, 01145 (2016)
14. A. Zhavoronok, M. Gasanov, S. Zhironkin, SHS Web of Conferences, **28**, 01144 (2016)
15. S.A. Zhironkin, A.A. Khoreshok, M.A. Tyulenev, G.A. Barysheva, M. C. Hellmer, IOP Conf. Ser.: Mater. Sci. Eng., **142**, 012127 (2016)
16. Sh. Akhmetova, A. Sakhanova, Ahmet Yesevi University, **31** (2004)
17. F. Fallahi, S. Soujudi, Human Resources management journal, **3** (2012)
18. M. Alwany, P. Ahmadi, Modarres, **5** (2002)
19. M. Tyulenev, E. Garina, A. Khoreshok, O. Litvin, Y. Litvin, E. Maliukhina, IOP Conf. Ser.: Earth Environ. Sci., **50**, 012035 (2017)
20. M.Y. Blaschuk, A.A. Dronov, S.S. Ganovichev, IOP Conf. Ser. Mater. Sci. Eng., **142**, 012128 (2016)
21. A.V. Koperchuk, A.V. Murin, V.V. Filonov, IOP Conf. Ser. Mater. Sci. Eng., **127**, 012040 (2016)
22. M.Y. Blashchuk, A.A. Kazantsev, R.V. Chernukhin, App. Mech. Mat, **682**, 418 (2014)
23. A.B. Efremenkov, A.A. Khoreshok, S.A. Zhironkin, A.V. Myaskov, IOP Conf. Ser.: Earth Environ. Sci. **50**, 012009 (2017)
24. V.M. Zolotukhin, V.A. Gogolin, M.Yu. Yazevich, M.I. Baumgarten, A.V. Dyagileva, IOP Conf. Ser.: Earth and Envir. Sci., **50**, 012027 (2017)

Enhancing the Role of Educational Services of Higher Education System in the Competitive Specialists Training for Industry

Olga Kuznetsova[1], *Svetlana* Kuznetsova[1], *Egor* Yumaev[1], *Vladimir* Kuznetsov[1,*] and *Inna* Plotnikova[2]

[1]Omsk State Technical University, 644050, Omsk, Russia
[2]Tomsk Polytechnic University, 634050, Tomsk, Russia

Abstract. The paper shows determination of educational services effect of higher education system to the competitive specialists training for the industry. Universities should focus their efforts on the development of innovative scientific and educational environment of the global level if they want to supply modern industry with highly qualified specialists. Despite the good results achieved in training for industry, universities should always be a step ahead in enhancing the quality of educational services. According to the findings options for professional trainings are offered; also the regional specialized educational resources launch is suggested.

1 Introduction

In accordance with the human capital assets theory, investments into higher education are paying off in the form of work, resulting in the labor market [1]. However, in our opinion, not investments as they are will get the job done, but the so-called "right" investments into higher education: when the institution is able to provide educational services at a high level.

In today's world, universities must respond flexibly to the rapid development of technology, the emergence of new industrial enterprises, increasing complexity of production processes [2–4]. Great attention should be given to equipping of university laboratories with modern equipment [5, 6].

It is not an exaggeration to say that the quality of education determines the future of the industry and competitive perspective of the economy as a whole. To do this, the effectiveness of the higher education system should be improved, as well as the formation of innovative world-class universities network.

In the context of global competition tightening in the educational services market universities need to look for new opportunities to increase awareness of their brands, to develop a variety of directions, among them renovation of educational services for students has primary importance [7]. Reliable and positive reputation becomes a critical factor in the students' attraction to higher education. This trend has been intensified in the context of

[*]Corresponding author: mivladirvvk@rambler.ru

globalization, as universities compete with each other not only for the students of the country, but also for foreign ones [8].

Thus, the basis of highly-qualified sought-after in modern industrial enterprises specialists training improves the educational services quality. New opportunities should be searched for to improve the educational process with the purpose of staff training with the current set of competencies.

2 Analysis

Studies of foreign scientists show close connection of the University attraction and its brand recognition level [9]. At the same time, if the University takes care of its own brand, it needs a sustainable development program [10]. Its central element is the focus on attracting the best students to develop dynamic creative scientific atmosphere, fair competition, to raise the educational level in the institution.

Much attention in the economy of the developed countries is paid to revolutionizing educational space. The traditional approach to deliver lectures in large "in-line" classrooms, practical training in the group with the number of students from 20 to 30 is unacceptable for today's students, who prefer to transfer part of their studying home, where they have opportunity to be concentrated on training. Personal communication with peers, co-education is important to them, but now the opportunity to work in a small group in separate convenient space is at the forefront requirement. Large spaces for training large number of students lose popularity faster [11–13]. Thus, the education in advanced countries moves towards individualization of learning in higher education, active use of scientific and educational IP-based networks, distance learning technologies allow geographically remote from one to another students work together to solve the task. The emergence of online educational resources with free access (e.g., Massive open online courses and Small private online courses) changed forever the educational services market and dramatically accelerated the online education dynamics [14].

Another dominant trend in the world of higher education is to focus on the fact to interest the student, so the educational process should cultivate actual, latest knowledge, which are closely associated with the practice. The forms of carrying classes must also undergo change into the direction of discussions encouraging [15].

In modern conditions the development prospects are determined with the ability to deploy production industry 4.0 [16–18]. Employers require increased demands to the personnel: developing industrial enterprises need young, but upscaled and experienced specialists, i.e., having not one, but two or three specialties.

3 Results and Discussion

To find qualified specialists companies interact with universities and open the specialized Chairs. The specialized Chairs are departments of universities, organized at the initiative of the enterprise or organization on the territory of the latter. In order to improve the quality of training university education programs should be developed in cooperation with enterprises and organizations, provided that they will order a full training group. For a successful career each student is to find his way to his liking. Professional qualities of a man are determined with his inborn psychological type, strengths and weaknesses of the individual.

Students in higher education institutions can be divided into four groups depending on the degree of motivation to the development of the educational program. The task of the university as a whole, and in particular the graduate Chair is to provide an individual approach to each student.

First. The student is not satisfied with the choice of specialty and does not foresee his career, he receives low scores in training, does not attempt to find a job, he negatively predicts the possibility of successful employment, has no plans to continue his education, he does not monitor the situation on the labor market for the acquired profession. In this case, the Chair shall provide assistance in carrying out events for student career guidance.

Second. The student has an average rating of satisfaction with career choice, he does not attempt to find a job in an organization or independently employs in poorly satisfying his value preferences when choosing a job, desire for professional growth is limited, he gets an average score in training, does not monitor the situation on the labor market for the acquired profession. In this case, the Chair should conduct activities aimed at the awakening of interest in professional activities, facilitating a practical training period for an employer who is able to arise interest in professional self-realization.

Third. The student is satisfied with the choice of profession, has no intention of changing it, has plans for further professional growth, but very carefully simulates career move, shows high performance in the educational program, executive, he does not work during training and does not attempt to find a job. The Chair can help the student with skills of self-presentation, developing relationships through active methods, assisting in developing search work technology.

Fourth. The student is satisfied with the choice of profession, he is aimed at the acquisition of professional experience in the training period, he uses various employment possibilities, the dominant factor in the choice of employment is professional growth and opportunity for career advancement, he is characterized with medium or high performance during training, he is active in the acquisition of knowledge and participation in conferences, discussions, he has assured intentions for career advancement, he sets ambitious goals. In this case, the Chair is to motivate the students to continue their education in master and postgraduate studies, to maintain active links with the Chair, to prevent the young specialist professional "burnout".

To develop the skills of self-presentation and self-study to improve the competitiveness in the labor market, we recommend carrying out professional training to students starting from the first course. Students should be explained how to position themselves in the job market, how to set career goals and how to develop a plan to achieve those goals. Various trainings should be developed. For example:

Professional training "How to find a job in the crisis." Visiting it, students learn how to find their dream job, to diagnose a career crisis, where to look for job, how to move on to another industry due to changes in the labor market and how to overcome the risks of this switchover, how to prepare a CV and a cover letter, be interviewed, get increase in wages and positions. Questions for a career move into another city and how to do it as efficiently as possible are determined. General aspects of the economic crisis, including its stages, changes in the labor market will be covered.

Professional training "How to find a good job," will consist of two parts. The following issues will be discussed in the first part:

– Assessment of the situation in the labor market; opportunities in the crisis period.

– Evaluation of opportunities, resources of training participants as "players" in the labor market (students learn to formulate and express their personal strengths as competitive advantage in the job search).

– Setting goals when looking for work; the main stages of the search (students learn to develop personal job search strategy).

– Work with the sources of information about vacancies; non-standard ways of finding work.

– Communication with the employer by phone (development of the first contact - the ability to make an impression on the phone and to interest, students learn to negotiate with the employer and analyze vacancy announcements).

The second part will include:

– CV preparation, the types and styles of CVs; good and bad CVs; mistakes in CV writing; CV analysis; making a CV and a cover letter for a particular position.

– Interviews; preparation for the interview with the employer; why people are afraid of interviews; rules of conduct at the interview; dress code; types of interviews; the most frequently asked questions of the employer (the students learn to respond to the employer's questions, learn to ask questions to the employer).

– Workshop on the passing the interview by the training participants: the story of their experience (students learn to talk about their professional experience in a positive way).

Since the beginning of the global economic crisis 2008, all countries are faced with ever new problematic issues, one of which is employment, so to help the youth in a successful career start is necessary to search for new approaches.

To solve the problem of cooperation between universities and employers information system "Personnel reserve" on the basis of the University official web-portal can be developed. The new information system will allow students to present their achievements, skills, knowledge and professional experience. Employers will be able to select independently successful graduates and young professionals for internships, training and, of course, employment. The information system will be open, i.e., each student will be able to browse the portfolio of his classmates, fellow students, etc. There will be such sections as "Jobseeker", "Employer", "Student achievements", "Partner Company", "Monitoring of employment."

In the "Jobseeker" section students can put their CVs and see the current vacancies as in the region of residence, as in other regions.

In the "Employer" section the employer will be able to create an individual request according to certain criteria and have a list of potential candidates that meet the stated requirements.

In the "Student achievements" section we will trace the development trajectory of students in extra-curricular, scientific, social, sporting life. There will be a portfolio of both as students as each university department. Clicking on the link of the department the portfolio of all enrolled students and graduates can be viewed as well as the rating of the most successful students. Each student's portfolio will provide information about work experience, participation in contests and competitions, the level of PC skills, foreign languages, participation in social life, obtained diplomas and certificates, GPA, theme and evaluation of the final qualifying work.

In the "Partner Companies" section there is a list of companies which cooperate with the university.

In the "Employment Monitoring" section there is information about the number of students and in what areas and specialties University graduates; graduates of what specialties were employed at the end of training. There will be also an opportunity to see the migration situation movement of university graduates.

The "Personnel reserve" information system is not limited to the CV bank of students and graduates, but will provide students with opportunities in their positioning in the labor market. The database will collect information on the students competencies, obtained in the course of training and in the future. After evaluating these criteria, the employer may invite students/graduates for practical training session, training, interview. To ensure the monitoring of incoming information from the students portfolio materials of graduate Chair should be loaded, as well as the introduction of an internal exam to determine the level of foreign language knowledge with the results publication in the information system.

4 Conclusions

The "Personnel reserve" information system will increase the motivation of students to treat study, work, and self-actualization more seriously. Fair competition will be developed between the students who will be interested to collect as many achievements on their CVs, to attract the employer. Due to it, students will use all various resources of the University and graduate with the modern highly qualified level, possessing complementary skills. Each student will get a perfect opportunity to assert himself on the labor market with a simple and effective way - to locate his individual portfolio on the virtual site.

The possibility of students' feedback and especially university graduates is to be provided. To this end, the development of an information system should be conducted from the perspective of creating inter university local social network, its members will be not only the students and alumni, but staff and even to broaden participation by future students - it will strengthen the capabilities of the University in attracting talented and promising youth. Thus, as a result of self-presentation and self-employment training, the development of "Personnel reserve" information system, the students interest to the complete mastering of the educational program, active participation in extra-curricular, scientific and sports activities will increase and that allows students to prove them the best in the labor market. The implementation of the proposed events will improve the status and recognition of the university at the regional and international level. The final result will be the graduation of versatile educated professionals who will be in demand in the modern industrial productions, will be capable of generating new ideas, modernization of production processes, which will increase the sustainability of economic development of separate regions and entire countries, increase their competitiveness in the global economy.

References

1. P. Courtioux, V. Lignon, Economic Modelling **57**, 221 (2016)
2. Castillo, Journal of Cleaner Production **143**, 180 (2017)
3. Ch. Featherston, E. O'Sullivan, Technological Forecasting and Social Change **115**, 261 (2017)
4. F. Huang et al, Journal of the Energy Institute (2016)
5. M. Akçayır et al, Computers in Human Behavior **57**, 334 (2016)
6. H.B. Congdon, Journal of Interprofessional Education & Practice **5**, 53 (2016)
7. J. Hemsley-Brown et al, Journal of Business Research **69**, 3019 (2016)
8. C. Piewa et al, Journal of Business Research **69**, 3087 (2016)
9. Ch. Dennis et al, Journal of Business Research **69**, 3049 (2016)
10. W.L. Filho , C. Shiel , A. Paço, Journal of Cleaner Production **133**, 126 (2016)
11. R. Beckers, T. Voordt, G. Dewull, Building and Environment **104**, 243 (2016)
12. N.M. Natalinova et al, Proc. of EpSBS **XIX**, 476 (2017)
13. Trappey et al, Advanced Engineering Informatics, (2016)
14. Th.D. Oesterreich, F. Teuteberg, Computers in Industry **83**, 121 (2016)
15. E. Shutenko, A. Shutenko, Procedia – Social and Behavioral Sciences **214**, 332 (2015)
16. L. Huo, N. Song, Physica A: Statistical Mechanics and its Applications **461**, 73 (2016)
17. J. Stoutenborough, A. Vedlitz, Energy Policy **96**, 206 (2016)
18. Y.-S. Changetal, Computers in Human Behavior, **63**, 988 (2016)

Formation and Development of the Training System for Innovative Development of Regional Industry

Olga Kuznetsova[1], *Svetlana* Kuznetsova[1], *Egor* Yumaev[1], *Vladimir* Kuznetsov[1,*] and *Olga* Galtseva[2]

[1]Omsk State Technic University, 644050, Omsk, Russia
[2]Tomsk Polytechnic University, 634050, Tomsk, Russia

Abstract. The paper determines tendencies of modern economy development. The key role in the expanded reproduction of innovation processes in the regions in modern conditions belongs to the enhancement of human capital. Regions are actively increasing their efforts in creating innovative infrastructure, knowledge-intensive industries, while success of regional development is directly related to the effectiveness and cohesion of all innovation infrastructure elements. An indispensable condition for the successful development of innovative infrastructure and high-tech industries is the region economy saturation with highly qualified personnel, particularly mining, trained in view of the projected trends of innovative development.

1 Introduction

Innovation economy requires transformation of intelligence, the creative human potential into the most important factor in the economic growth of regions and national competitiveness. The production of new ideas, technologies, and social innovation becomes a source of high income.

To improve the modern economy efficiency and competitiveness requires "innovative person" –a citizen having a high degree of adaptability to the constant changes in his own life, economic development, the development of science and technology, he is an active initiator and producer of these changes, considering the constant development as an integral part of his life principles. The regional economy has no such businesses and organizations that would not benefit from innovation and innovation can be both as technical as human relationships involved into the provision of services in the service organizations [1]. The staff of the organization is in constant search for innovations, and it is characterized with positive attitude aimed at improving and achieving more and more ambitious goals [2]. To do this, a culture of innovation in the organization should be cultivated [3]. Current trends of modern foreign researches are the employee psychological types, as a reserve of organization innovative activity increase [4].

*Corresponding author: mivladirvvk@rambler.ru

Modern high-tech enterprises impose more and more new requirements to their employees [5], therefore, to ensure the regional economy innovative productions with necessary staff and graduates with interesting, well-paid work, learning environment must quickly adapt to new conditions.

An organization or enterprise that wishes to maintain a stable position in the market must innovate constantly, not episodically, and it requires continuous training of employees [6]. However, in our opinion, from the regional economy perspective it is a mistake to talk about lifelong learning only for adult employees. Continuous education should begin in kindergartens, teaching to think creatively and continuously seek effective solutions to non-standard tasks from an early age. Creativity of any enterprise is formed from the union of the employees' creative abilities [7]; therefore, the higher the share of innovation-minded people will be in the region, the higher the stability of the economy will be characterized as a whole.

Modern science affirms the idea that the main task of the higher education systems must be not just to train highly qualified specialists, but production of creative highly qualified professionals, in particular, for the mining industry, and they would feel more confident in the "creative industries" [8]. It affirms the increasing role of regional educational systems in the education of professionals with competencies that will be needed in future [9].

Thus, it is necessary to intensify the search for effective modernization of regional educational environment, and not only the higher education system, but the entire system of education and training, starting from kindergartens and finishing with higher education.

2 Analysis

The European Union countries made great efforts to modernize the educational environment, to increase labor mobility, to train sought-after specialists, to reduce tensions on the labor market in recent years.

Among the leading countries of the world economy, the experience of France should be noted where the number of employees engaged in science as from educational institutions as from industrial enterprises is steadily increasing in recent years. Cooperation between educational institutions and production facilities expands, more and more new forms of cooperation of academic science and practice are being introduced. In fact, France generated environment where there are no any boundaries between academia and industry.

The achievements made were possible largely due to significant state expenditure on scientific researches; in 2013 they reached almost 50 billion Euros. However, it is impossible to work only with investment funds, so the approach to the formation and development of the training system for innovative development of regional economies is necessary [10].

In Russia, the events aimed at accelerating the development of innovative industries are in all regions of the country. However, the limiting factor is the lack of qualified specialists for modern innovation-oriented industry. The most important problem is the issue of implementation of mass retraining programs that will allow employees of enterprises of any form to obtain new spheres of competence. [11]. The proposed measure is correct, but only at the expense of retraining programs, neither in Russia nor in other countries of the world the labor market saturation problem with sought-after specialists in the innovation productions, is resolved.

The development of modern regions takes place under conditions of competition growth for resources, including highly qualified specialists. Knowledge-based industry presents regional demand in the field of management, commercialization, protection and intellectual property protection, as well as innovators, inventors and other professionals. The idea of innovative development should be widely promoted, including through various international forums and conferences in the field of innovation. In the developed regions business incubators, techno parks, intercollegiate innovative business incubators become commonplace. Techno parks activities are aimed at the return of the engineering profession prestige, the formation of children and adolescents' professional competence and practical skills in the highly technical special fields: robotics, mechanics, electronics, automation, computer, printing and telecommunications sectors, economy. However, it is not enough to construct a building, the space is to be filled with real interaction, to attract innovators, and the economic results of their interaction will exceed investments in the construction of these buildings.

3 Results and Discussion

In the most diverse regions of the world in connection with the development of industry 4.0 it is required to step up the training of qualified personnel for the innovative development of the economy [12; 13]. In the course of study the weaknesses of qualified personnel training for the innovative development of regions were determined and proposals for solving problems were made.

1. Lack of coherent, formed region policy in the innovation sphere: it is necessary to conduct deliberate policy in the region in the field of innovation, promotion of innovative ideas development.

2. Undeveloped training system for innovation development of economy on the principle of continuous education: the creation of continuing professional education in the sphere of innovations "Kindergarten-School-University-Enterprise".

3. Inefficient use of foreign experience of training for the innovation development of economy: the decision is to promote international exchange and training of teachers in the field of innovation development in Europe, USA, Canada, Japan and other prosperous countries, with a view to practical application of positive foreign experience in the formation and development of the training system for the innovative development of the region's economy.

4. Low level of patent culture: it is required to adopt a program of development of patent culture in enterprises and society as a whole.

5. Inadequate integration of education, science and business: the problem can be solved through the development of effective integration ties of education, science and industry.

6. The presence of a language barrier: it is expedient to expand foreign language training system.

7 The problem of "innovative person "formation: carrying out purposeful work on formation in the period of study personal and professional skills needed to work in the sphere of innovations

While developing international exchange and training in innovation sphere, particular attention should be paid to the participation of students and teachers of higher educational institutions –the experience of foreign teaching should be not only shared , but also it should be acquainted in practice with the principles of conduct of innovation business in the advanced countries of the world that have achieved significant success in the formation of innovative industry and industry 4.0.

The links between education, science and business are essential for the commercialization of intellectual sphere objects. The practical orientation of specialists in the intellectual sphere training is equally important.

The economy and education depend on each other: education institutions train new qualified specialists for the innovative development of the economy, and the task of the regional authorities is to create conditions for the development of innovative industries, so that trained professionals would be in demand.

Higher education institutions have a special innovative mission - Universities are at the heart of technological transformation, activation of innovative activity in the economy without their participation is not productive [14]. A significant effect is obtained due to the introduction the specialty "Intellectual Property" in the largest research universities. The demand for specialists in the innovative development of regional economy increasingly manifests itself.

The area of professional activity of the graduate who has mastered the specialty "Intellectual Property" includes the creation of intellectual property; legal protection of intellectual property; control the results of creative activity; social and psychological support for innovation; study of social needs and the importance of intellectual property rights and socio-economic consequences of their use.

Objects of professional activity are the regional and local authorities whose activities are linked to the development of innovation; firms of patent attorneys; organizations of all forms of ownership associated with the creation and use of results of creative activity; societies for the collective management of rights; societies of appraisers; consulting companies.

Implementation of the basic educational program of the specialty must be provided with highly qualified scientific and pedagogical staff and teachers with practical experience of creation and development of innovative industries. Much attention should be paid to the use of distance education technologies. It is important for the project launching to have the support of regional and local authorities.

One obstacle in the training of qualified personnel for the innovative development of regional economy is the lack of continuing education coverage. The rate of emergence of new knowledge and technology has accelerated dramatically in the last decade [15;16], so apart from the systematic training and retraining the enterprises and organizations employees will lose the ability to generate innovations.

Continuing education, aimed at training the "creative person" must be effectively organized in all components of the system "Kindergarten-School-University-Enterprise". This is the key condition for progress and innovation-based economy, creating new high-tech industrial enterprises. The development of new, modern programs together with leading foreign educational institutions is of great interest.

At the stage of learning in kindergartens the purpose is to familiarize children with the basic terms of the innovation sphere in the process of immersion in fairy tales. At school students learn knowledge about inventions and innovations, master methods of obtaining new knowledge.

When organizing the work with schoolchildren authorities should promote the policy of popularization of the most promising areas of economic development in the long term. The requirement today is to ensure equal access to quality education for all students regardless of where they live and the level of material security. The main role is given to distance education technologies and the creation of a unified system of students online-learning in the region. In small towns and rural districts a resource center should be created on the basis of school for the implementation of one or more specialized areas (Physics and Mathematics, Philology, Chemical, etc). If the school education is implemented with the help of distance learning technologies, the training of students in remote areas to contests, competitions, and exams will be improved significantly. The development of distance learning technologies and change of thinking of authorities and high school teachers, whose participation is required for the creation of online education system, will allow all children from remote areas to gain knowledge at the level of the leading schools in large cities. It will eliminate the shortcomings of education in areas far from major cities, and it significantly enhances the potential of future skilled workers.

In future, it is necessary to secure and expand the activities of regional innovative infrastructure objects to work with schoolchildren and students. In large cities, specialized centers should be established where schoolchildren and students are helped to develop skills to create their own high-tech productions.

The accumulated positive experience of the past is to be accumulated with modern technologies. Foreign studies show that the online (e-learning) latest technology and mobile (m-learning) training can improve young people's interest in conducting research, stimulate the creative activity [17].

4 Conclusions

Thus, in regions a deliberate policy is to be carry out to develop children and young people skills of scientific research and new knowledge generation, as well as the ability to put into practice their own projects, to create small innovative enterprises to communicate effectively in a team of innovators. In recent years, many regions around the world spend a lot of preparatory work, there are positive results, and they set the stage for accelerating the development of innovative economy. In the course of study the recommendations were made and they will enhance the ability of the staff training system for the innovative development of industry in the region, also they will enhance the creation and development of new science-intensive industries, and in the end they will lead to the creation of highly-paid jobs and increase the share of high technology products in the gross regional product and increase of economic competitiveness of the region as a whole.

References

1. R.L. Dhar, Tourism Management **57**, 139 (2016)
2. H.-T. Chang et al, Personality and Individual Differences **96**, 260 (2016)
3. M. Ali, K. Park, Journal of Business Research **69**, 1669 (2016)
4. B.Wisse, D.Barelds, E.Rietzschel, Personality and Individual Differences **82**,158 (2015)
5. K. Charmondusit, S. Gheewala, T. Mungcharoen, Journal of Cleaner Production **134(B)**, 443 (2016)
6. D. Suarez, Research Policy **43**, 726 (2014)
7. Y.-S.Chang, K.-C. Yu, Computers in Human Behavior. **49**, 38 (2015)
8. N.Yachina, G.Fahrutdinova, Procedia - Social and Behavioral Sciences **177**, 213 (2015)
9. K. Hartley, J. Plucker, H. Long, Thinking Skills and Creativity **22**, 142 (2016)
10. V.Chassagnon, N. Haned, Technological Forecasting and Social Change **91**, 194 (2015)
11. J. Yeleneva et al, Procedia - Social and Behavioral Sciences **214**, 779 (2015)
12. A. Trappey et al, Advanced Engineering Informatics (2016)
13. Th.D. Oesterreich, F. Teuteberg, Computers in Industry **83**, 121 (2016)
14. E. Shutenko, A. Shutenko, Procedia–Social and Behavioral Sciences **214**, 332 (2015)
15. L. Huo, N. Song, Physica A: Statistical Mechanics and its Applications **461**, 73 (2016)
16. J. Stoutenborough, A. Vedlitz, Energy Policy **96**, 206 (2016)
17. Y.-S. Chang et al, Computers in Human Behavior **63**, 988 (2016)

Evaluating the Effectiveness of Internal Corporate Controls in Coal Mines Illustrated By the Example of JSC "SUEK-Kuzbass"

Elena Kucherova[1], *Tamara* Ponkratova[1], *Tatiana* Tyuleneva[1*], and *Natalia* Cherepanova[2]

[1]T.F. Gorbachev Kuzbass State Technical University, Department of Management Accounting and Analysis, Vesennaya street, 28, Kemerovo, Russian Federation
[2]JSC "SUEK-Kuzbass", Vasilieva street, 1, Leninsk-Kuznetsky, Kemerovo region, Russian Federation

Abstract. The article gives reasons for the need to develop the mechanism and tools of performance evaluation based on the implementation of specific management functions, as well as the integrated evaluation of the effectiveness of internal corporate controls. It presents the approbation of the proposed assessment methodology on the example of JSC "SUEK-Kuzbass". The monitoring role as one of the functions of management is constantly increasing in the market economy. The participation of Russian companies in the international economy requires the organisation of internal company control. Moreover, with the enacting of the Law "On Accounting» № 402-FZ, for the majority of economic agents the organising of internal control became mandatory. Thus, the internal control means the process carried out by the Board of Directors and other staff of the organisation aimed at providing reasonable assurance that the objectives of the organisation such as ensuring the effectiveness and efficiency of operations, reliability of financial reporting and compliance with the legislation in force will be achieved.

1 Introduction

The effective functioning of economic entities, including holding companies, largely depends on the well-organised internal corporate control, as it is designed to not only identify the shortcomings and violations, but also to prevent them, and to promote their timely elimination [1-3]. In order to finally evaluate the results of the functioning of the organised internal control it is necessary to develop an evaluation mechanism [4, 5].

Thus, the relevance of the research topic is determined by the need to improve the efficiency of the company through development and use of models and algorithms for the organisation of internal corporate monitoring and evaluation of its effectiveness due to the special role of control in the management of the corporate activities.

The internal corporate control presents a set of interrelated elements representing the target, objects and subjects of individual business processes, information technology, risks,

* Corresponding author: krukta@mail.ru

regulatory procedures and evaluation tools to quickly identify the causes of deviations from established standards and implement corrective actions to address them to ensure effective achievement of the objectives by the corporation management [6]. According to this definition, the mechanisms were developed both for organisation as well as for assessment of the effectiveness of internal corporate controls.

2 Results and discussion

The model of implementation of internal corporate control includes the mechanism of organisation of internal corporate controls and the mechanism of assessment of the effectiveness of internal corporate controls. The mechanism of the organisation of internal corporate control involves determination of significant risks and business processes that are inherent to coal mining operations [7]; choice of risk assessment criteria by the level of risk impact, its probability and occurrence rate; development of risk maps for business processes and types of evaluation criteria; formation of risks matrix of business processes for risk assessment and operational risk management; the development of the matrix of control procedures for the main business processes in order to minimize risks and losses due to their occurrence.

In turn, the mechanism of assessment of internal corporate control consists of two components. First, it is the evaluation of the effectiveness of internal corporate controls by specific control functions: a) the effectiveness and efficiency of operations, b) compliance of the activity with the current legislation, the performance of the legislative basis including compliance with IFRS and the reliability of reporting, c) optimization of the management costs. This mechanism is shown as the scheme (Fig. 1).

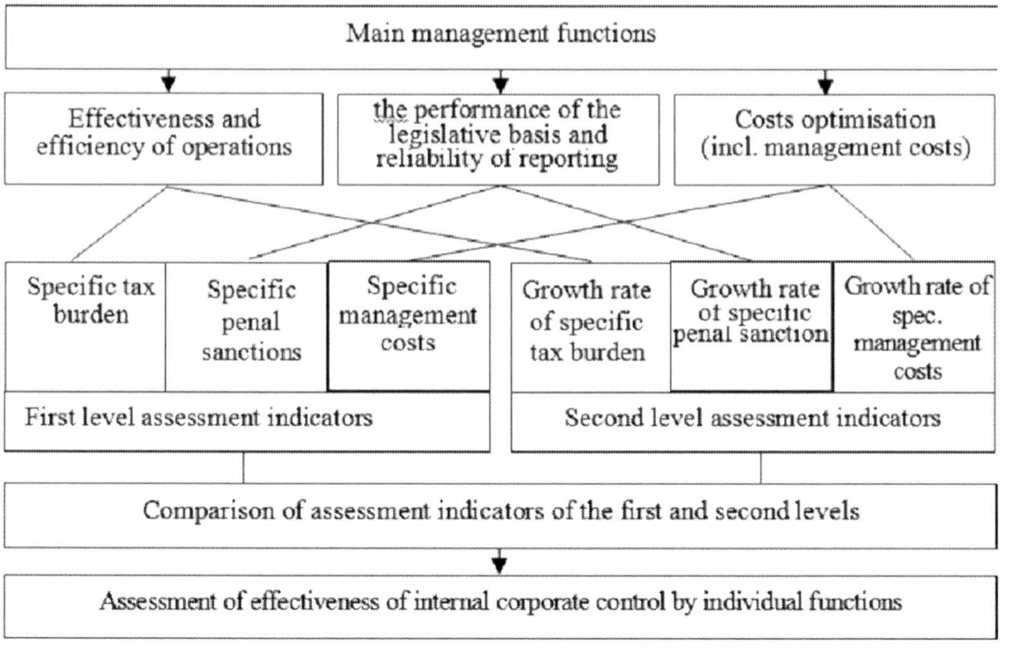

Figure 1. Mechanism of assessment of efficiency of internal corporate control by individual functions of control

Secondly, an integrated assessment of the effectiveness by the basic management functions through scoring method. This mechanism is shown in the scheme form (Fig. 2).

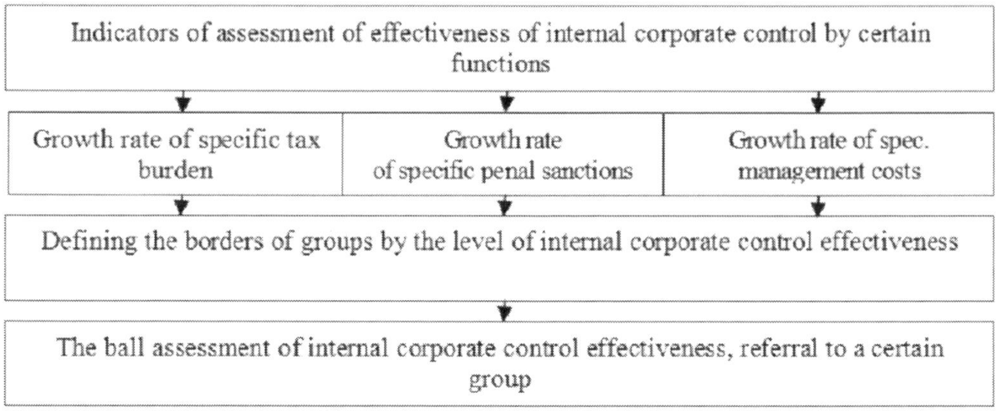

Figure 2. Mechanism of integral assessment of effectiveness of the internal corporate control

Assessment of the effectiveness of internal corporate controls by individual control functions at the beginning is carried out by the indicators of the first level (i.e. specific values that characterize the size of the tax revenue, penalties and administrative costs per ruble of sales), and then by the indicators of the second level (i.e. growth rates of specific values). The average values were used for the "threshold" values of the indicators. The "threshold" values of indicators are the basis for characterizing the deviations of actual indicators from the "threshold" ones and the calculation of the range of indicators' changes at the integrated (comprehensive) assessment [8].

In 2015, the new competencies of the organization and evaluation of the effectiveness of internal corporate controls proposed by the authors were implemented at JSC "SUEK-Kuzbass". Hence, the analysis of the indicators was carried out on the example of JSC "SUEK-Kuzbass".

Table 1. The analysis of first level indicators of JSC "SUEK-Kuzbass" (kop. / RUB.)

Indicators	**Threshold value**	**Including by years**				
		2012	**2013**	**2014**	**2015**	**2016**
1. Specific tax accruals	1.58	1.7	1.7	1.5	1.4	1.3
2. Specific penal sanctions	0.0685	0.081	0.061	0.062	0.060	0.057
3. Specific management costs	1.34	1.2	1.3	1.4	1.5	1.6

The analysis of the first-level indicator values suggests minor deviations by the years, this means that for a comprehensive assessment of the effectiveness of internal corporate controls it is required to use not only the specific indicators. Based on this, at the second level of assessment there were growth rates used, which gave a clearer picture of trends in the indicators' changes.

Table 2. The analysis of second level indicators of JSC "SUEK-Kuzbass"

Indicators	**Threshold value**	**Including by years**				
		2012/ 2011	**2013/ 2012**	**2014/ 2013**	**2015/ 2014**	**2016/ 2015**
1.Rate of changes in specific tax accruals	-3.05 %	6.25	0	-11.77	-6.67	-7.75
2. Rate of changes in specific penal sanctions	-5.27 %	5.19	-24.69	1.64	-3.23	-5.12
3. Rate of changes in specific management costs	+3.87 %	-7.69	8.33	7.69	7.14	5.67

The analysis of the rate of changes in the indicators suggests that with the minor deviations of specific indicators the growth rates provide a clearer description of the trends which allow us to judge how effective are the solutions of individual management tasks. The positive values of the indicator's growth rate suggest that adjustment of internal corporate controls activities is required in the performance of the given control function, while the negative values prove the effective implementation [9, 10]. In order to obtain a comprehensive assessment, the state of the indicators of the first and the second levels was compared. The methods of comparing the state of indicators of the first and second levels of performance assessment involves the following features: on the first level - corresponds to the "threshold" value (a), the value does not correspond to the range (b); on the second level - improvement of the indicators (1), the values are sustainable (2), deterioration of the indicators (3). Assessment of the indicators by this method led to the conclusion that in 2015 they all still do not meet the "threshold" value, but there is an improvement in the indicators characterizing the effectiveness of internal corporate control of the organization in connection with the introduction of new approaches to its organization and evaluation [11, 12].

The evaluation of the effectiveness based on the record of individual management and control functions allows us to characterize the success of performing the individual functions of management, this is necessary for adjustment and coordination of the functioning of the business-processes, though, unfortunately, this does not provide an integrated assessment of the effectiveness of the entire internal corporate control [13]. Hence, the integrated evaluation mechanism was developed. The proposed method uses the evaluation of the results based on the scoring, its distinguishing feature is the introduction of rating that reflects a comprehensive account of quantitative, qualitative and probabilistic indices. Also, it is possible to revise the rating of each constituent element of internal corporate controls, depending on the priority of the control task at the given stage of development of the enterprise (the mechanism of an integrated assessment of the effectiveness of internal corporate controls is shown in Fig. 2). The integral indicator of the effectiveness of internal corporate controls is computed by summing the points for each indicator. The total score is 100. The rating parameters, upper and lower criteria, as well as the conditions of decreasing the criterion were determined empirically on the basis of previously defined "threshold" values and evaluation in scores which in turn were established based of trial calculations and priority tasks at the given stage of coal mines development. The essence of this technique lies in the fact that any analyzed parameter can be assigned to a certain class depending on the "obtained" number of points, based on the actual values.

The first group (maximum 100 points) characterizes high effectiveness of the existing internal corporate control that does not require correction. The second group (85 points) suggests that the organization of internal corporate control is quite effective, and requires a slight adjustment by the individual functions, downside risks to the company are negligible. The third group (60 points) characterizes the boundary condition of the enterprise, the risk of inefficient management increases, a significant adjustment to the individual areas of the internal corporate controls is required. The fourth group (25 points) represents a high level of risk and inefficiency of management, it is required to adjust all the components of internal corporate controls, to implement their radical restructuring, since this organization is ineffective. The last group shows lack of internal corporate controls (or the need to revise the established evaluation criteria in the event of changes in the situation in the company or the economy).

Based on the data presented in Table 3, we see a tendency of increasing the effectiveness of internal corporate controls since 2015 after the beginning of the introduction of new concepts. Adjustments are required in certain areas of control related to the optimization of management costs (though there is some improvement). In general, from 2015 the internal control in the company can be characterized as quite efficient, because it refers to the second class of efficiency.

Table 3. The integral assessment of internal corporate control effectiveness at JSC "SUEK-Kuzbass"

Indicators	Threshold values	Including by years				
		2012/ 2011	**2013/ 2012**	**2014/ 2013**	**2015/ 2014**	**2016/ 2015**
1.Rate of changes in specific tax accruals	-3,05 %	6,25	0	-11,77	-6,67	-7.75
2. Rate of changes in specific penal sanctions	-5,27 %	5,19	-24,69	1,64	-3,23	-5,12
3. Rate of changes in specific management costs	+3,87 %	-7,69	8,33	7,69	7,14	5,67
Integral assessment, points	100	25	50	55	75	85

Thus, evaluation of the internal corporate controls effectiveness in using the proposed mechanism and certain instruments, will allow us to realize the main purpose of internal corporate that is improving the efficiency of the company performance [14].

The above tools are recommended for use in combination with the developed control procedures and the elements of the organization of internal corporate controls in the coal-mining companies, and they allow us to improve the organization of internal corporate controls and increase their efficiency.

We propose to characterize the effect of the implementation of internal corporate controls as obtaining an additional income from the effective functioning of internal corporate controls. Its value is made of the additional profit (loss) from changes in the value of the indicators used to assess the effectiveness: profit (loss) from changes in the size of the specific tax burden; profit (loss) from changes in the size of specific penalties; profit (loss) from changes in the size of the specific management costs [15].

Thus, by reducing the value of the specific tax burden and the specific penalties, but with the growth of specific management costs associated with the increase in the costs of implementing the new provisions of internal corporate controls in 2015, JSC "SUEK-Kuzbass" received additional profit of 1,443 thousand rubles, including due to reducing the size of the specific tax burden – 72,194 thousand rubles; due to reducing the size of the specific penalty – 1,467 thousand rubles.

3 Conclusion

The analysis of the problems of internal corporate controls development has shown that internal control is required for businesses in order to effectively manage business units performance, which determines the need for the development of theoretical, methodological and practical grounds not only for organization of control but also for evaluation of its effectiveness.

Based on the above, the mechanism and tools were developed for evaluation of the functioning of internal corporate control by individual tasks and its integrated assessment. They allow us to quickly adjust the functioning of internal corporate controls based on the assessment of the tasks execution by specific functions and to make management decisions on elimination of identified performance shortcomings. Also, the mechanism and tools were proposed for integrated evaluation of the effectiveness of internal corporate controls by taking into account implementation of individual management functions that allow us to give a general assessment of the effectiveness of internal corporate control, the effect of the introduction of internal corporate controls was defined.

Thus, evaluation of the effectiveness of internal corporate controls provides for presence of certain instruments, the use of which will allow us to realize the main purpose of internal corporate control that is improving the effectiveness of the enterprise performance.

References

1. T.A. Tyuleneva, Europaische Fachhochschule, **6,** 84 (2014)
2. W. Shewhart. *Economic Control of Quality of Manufactured Product* (Milwaukee, WI: ASQ Quality Press, 1931)
3. J. Christopher. *The adoption of internal audit as a governance control mechanism in Australian public universities – views from the CEOs* (Journal of Higher Education Policy and Management Vol. 34, No. 5, 529–541, October 2012)
4. H. Fayol. *General and Industrial Management* (London, Pitman, 1949)
5. M. Jensen, W. Meckling *Theory of the Firm: Managerial Behavior, Agency Costs and Ownership Structure* (Journal of Financial Economics, 1976)
6. E.V. Kucherova, T.A. Tyuleneva, N.A. Cherepanova, *Coal in the 21st Century: Mining, Processing and Safety*, 130 (2016)
7. R. Baxter, J.C. Bedard, R. Hoitash, A. Yezegel, *Enterprise Risk Management Program Quality: Determinants, Value Relevance, and the Financial Crisis* (Contemporary Accounting Research; Winter 2013)
8. M. Madhani Pankaj, *Corporate Governance and disclosure public sector vs private sector* (SCMS Journal of Indian Management, March, 2014)
9. Professional Accountants in Business Committee International Good Practice Guidance: Evaluating and Improving Internal Control in Organizations (2012)
10. COSO (2013): Internal Control –Integrated Framework: Framework, Committee on Sponsoring Organizations of the Treadway Commission, Jersey City (2013)
11. Handbook of international quality control, auditing, review, other assurance, and related services pronouncements (New York: International Federation of Accountants, 2010)
12. T. Aabo, J. Fraser, B. Simkins, Journal of Applied Corporate Finance, **17**, 62 (2005)
13. M. Beasley, D. Pagach, R. Warr, Journal of Accounting Auditing and Finance, **23**, 311 (2008)
14. C. Chapman, S. Ward, International Journal of Project Management, **22,** 619 (2004)
15. David Otley, *Performance Management: A Framework for Management Control Systems Research* (1999)

 DOI: 10.1051/e3sconf/20171504021

The communication aspect of specialists' professional competence

Lyudmila Znikina[1,*], *Natalia* Mamontova[1] and *Pavel* Strelnikov[1]

[1] Department of Foreign Languages, Kuzbass State Technical University, Kemerovo, Russia

Abstract. In the article the issues of specialists' professional competence in a multicultural context are considered from the perspective of information and communication concepts. It is proved that the main objective of communication process is the adequate interpretation of communicants' intentions. Communication competence of a specialist is seen as a resource to provide language and intercultural readiness for work, social flexibility and interaction on a professional level, including multicultural work environment, in the conditions of global academic mobility and expansion of business contacts.

1 Introduction

Modern society development, socio and economic realities of society integration into a universe economic and educational environment dictate the tough conditions for today's international labor market which requires professionals who are trained for functioning with multipurpose intercultural contacts. Developing communicative behavior strategy among partners in interpersonal and professional interaction, especially in the framework of intercultural communication, becomes the key aspect in the professional activity of multi-ambitious specialists. These are the specialists of a new level – the ones with high professional and socio-cultural competence, which provides them competitiveness in the international labor market, gives an opportunity to implement the society needs in the conditions of expanding economic environment and dynamic international cooperation. The politically correct ability of a communicator to convey the message to another person becomes really important.

In other words, one of the significant qualities is the ability to interact effectively with representatives of another culture including different stages of scientific and academic activities. They may include organizing domestic scientific conferences, participation in international scientific conferences, preparing documents for educational and scientific grants, learning how to use the terms and conditions of studying abroad, participating in business negotiations (as a member of a team of experts, a representative of a work department, a technical specialist of specialized aspects for contracts with enterprises from partner countries, etc.).

* Corresponding author: znikina@kuzstu.ru

Recently specialists have considered this issue not as a significant problem and would comment that 'professionals will find common ground independently of the cultural traditions they were brought-up in' [1]. But in recent time the importance and urgency of the problem of intercultural interactions is appreciated by a wider audience of politicians, entrepreneurs and all those who have to work with people from other cultures. On the contrary, unfamiliarity and sometimes unacceptable communication styles adopted in the host country due to cultural traditions significantly complicate the adaptation of foreigners to the culture and prevent the fruitful cooperation establishment.

2 The results and discussion

One of the requirements to the level of professional competence development of a specialist is simply the ability to communicate. 'Unfortunately, many of them have some patterns of thinking; they do not have enough knowledge of the world, patience and perseverance to negotiate with people of other cultures and other ways of thinking. They are just not international' [2].

One of the possibilities for internationalizing business relations is learning foreign languages. But in order to prepare professionals for effective intercultural contacts, knowledge of a foreign language must be supplemented by knowledge of the nature of intercultural interaction with all the features of verbal and non-verbal aspects. It's the time when it's possible to talk about the ability to integrate intercultural knowledge and intercultural communication experience in the course of professional activities. The abilities of modern specialist for intercultural communication may generally include:

- perceiving and interpreting different cultural values;
- considering the various cultural phenomena and other cultures representatives from the standpoint of empathy;
- being able to review and change their assessment of a 'foreign' culture in the process of improving skills and gaining experience in intercultural communication;
- changing self-evaluation of one's behaviour while comprehending a "foreign" culture;
- generalizing one's personal experience while participating in an intercultural dialogue.

Communication business differences are based on the conceptual difference in the approaches to both intercultural and linguistic interaction. The sequence of possible problem situations in intercultural communication can be summarized as follows: lack of basic intercultural knowledge → unawareness of communication peculiarities of other cultures representatives→ the communication gap or culture bump → the inability to achieve the communicative purpose→ stress → generalization and categorization of the problem → biased perception of communication (as well as language) situation→ refusal of potential direct contacts.

The concept of 'intercultural communication' (in other terms – cross-cultural communication, multicultural communication) does not currently have a precise definition. One thing is clear – it is a synthesis of 'culture' and 'communication' concepts. This may define intercultural communication as interpersonal interaction between members of different groups, who have different views on the level of knowledge inherent to these groups, and different language characters.

Speaking of differences, we cannot avoid mentioning the fact that the ambiguous attitude to other cultures existed throughout the human history. In a modified form, the dilemma persists in modern conditions and is leading to the debates and discussions of specialists on the content, object, subject and purpose of intercultural communication. This has resulted in several theories of intercultural communication, which consider various aspects of interactions of cultures. The most well-known theories of intercultural communication include the theory of the American anthropologist Edward T. Hall, the theory of cultural dimensions by the Dutch socio-psychologist and anthropologist Geert Hofstede, the theory of cultural literacy by the American educator E. Hirsch.

Anthropologist Edward T. Hall wrote extensively about context, he highlighted not only the cultural and communicative models of different groups [2], but developed a common typology in relation to their context [3]. According to his theory context is information surrounding and accompanying a particular cultural event. Cultures differ in their interpreting the context. Although he categorizes cultures as being either high-context or low-context, context really is a cultural dimension that ranges from high to low. In high-context cultures most of the information is either in the physical context or is internalized in the people who are a part of the interaction. Very little information is actually coded in the verbal message. In low-context cultures, however, most of the information is contained in the context or within the participants. In other words, low-context cultures are those in which social context surrounding a negotiation counts as nothing; all that counts are the actual written agreements. Those societies in which the social context that surrounds a formal, written document is far more important than the written, legal documentation are called high-context cultures. Hall also differentiated between monochromic or polychromic aspects of cultural differences.

In the theory of cultural dimensions, Geert Hofstede proves [4] that each person perceives the world in a special way under the influence of individual peculiarities of the psyche, social environment and specific features of an ethnic culture. The vast majority of social behaviour is formed in childhood, because it is in childhood when a person is most sensitive to learning processes. The result of a process of forming feelings, thoughts and behaviour are the so-called mental programs that can be explored using dimensions of culture according to the four indicators. Hofstede emphasises the predominance of certain traits in the national character:

- power distance – comparison of cultures according to the degree of concentration of power or its distribution on various levels of the organization;
- collectivism / individualism – encouragement of social communication as opposed to individual independence and reliance on one's own strength;
- masculinity / femininity – a clear delineation of gender roles;
- uncertainty avoidance - comparison of cultures according to the degree of acceptable deviations from established norms and values.

Although the Hofstede's theory has some kind of a restricted character, the given parameters of cultural differences allow to respond flexibly to unexpected actions of partners in intercultural communication and to avoid possible conflicts when dealing with other cultures.

The theory of cultural literacy by E.D. Hirsch, Jr. is based on the fact that a necessary condition for effective intercultural communication is an adequate level of cultural literacy, which involves understanding the background knowledge, values, psychological and social identity which are typical of a particular culture. For a successful language competence, deep knowledge of the various cultural symbols of the respective national culture is required. E. Hirsch in his theory believes that cultural literacy 'makes us masters of the standard instrument of knowledge and communication, thus allowing us to send and receive complex information orally and in writing, in time and space' [5]. Depending on the value and role of a particular type of communication in specific situations Hirsch distinguishes different levels of intercultural competence: the level required for 'survival'; the level sufficient for entry into a foreign culture; the level that ensures a real-life existence in a new culture – its 'appropriation'; the level, allowing to realize fully the identity of a language user.

An important feature of intercultural communication is that one of the partners involved in it has to use the second (foreign) language, which is not any version of their own. In general, (Fig. 1) the process of intercultural communication can be represented in the form of a communication model developed by Gudykunst and Kim [7].

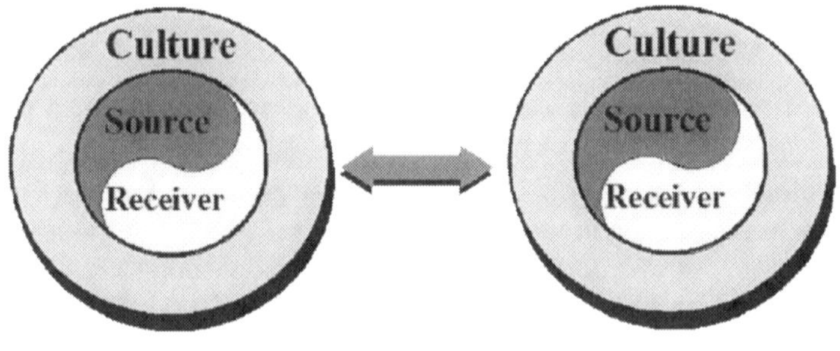

Fig. 1. Model of intercultural communication

Both communication partners are influenced by their perceiver filters on the basis of which they select and divide into categories the environment and information. The authors define this filter as a mechanism which helps us 'encode and decode a message' [7]. This global impact is due to the macro-system of the culture. It is also necessary to consider the impact of the middle level, which is the result of social relations and mental system of the individual. Each of these filters affects both the processes of encoding and decoding of the parties who fulfil simultaneously the roles either of the communicator or the recipient in the interactive process due to the circumstances. To convey a message to the recipient, the communicator must encode it specifically. There are many possible codes but verbal and non-verbal codes are of special interest here. The encoding processes, the act of the message interpretation obviously depend on the appropriate filters. Information transfer occurs when the conceptual filters of the parties are relatively similar (semantically tuned).

This model can serve as an example to identify characteristics and potential problems of intercultural communication in a professional context. Issues of intercultural communication and intercultural differences should be logically considered through the way of thinking and norms of behaviour of representatives of a particular culture that in fact may hinder communication. In communication, we behave subconsciously in accordance with social and communicative norms of our native culture: socio-historical background determines our values, behaviour and customs. That is why any message can be interpreted in different ways in a particular social or professional context in intercultural communication. As mentioned above verbal or nonverbal behaviour together with the lack of knowledge of the peculiarities of intercultural and communicative environment can lead to the wrong conclusions about the personal and professional qualities of a person in the communication process. According to the information theory, communication is the means of conveying information. Therefore, we consider it appropriate to refer to the relationship between the concepts 'information' and 'communication'.

If we analyse the frequency of use of certain concepts in scientific research or even in everyday life, then, undoubtedly, information tops the list and comprises everything that bears the traces of knowledge, messages, data, signals, etc. Information interactions are the basis of the existence of both the living and inanimate natures. Without information processes, social life is also unthinkable.

In scientific literature one can find dozens of definitions of information. However, it is impossible to give a complete definition to the concept of information since the term "information" has been widely used not only in scientific literature but also in other areas of our lives. So, information is commonly understood in the broadest sense as all knowledge, data, facts, messages, signals, which we deal with in everyday life and the manifestation of which we observe in nature and society.

Let's note, however, that the term 'information' comes from the Latin word 'informatio', which can be interpreted as the message, explanation, description. Moreover, we take the liberty to say that such interpretation of information is conceptually very close to the concepts of education, training. Thus, in the process of comprehension it is information that makes us 'correct' and 'upgrade' our knowledge. That is why the value of information is always subjective and is largely determined by the capabilities of information consumers, their purposes, also by the degree of their sensitivity and the level of their knowledge.

It's worth considering how the concepts of 'information', 'communication' and 'message' correlate and the latter concept is one of the most common forms of representation of information. In his fundamental work 'Informatics' F. L. Bauer introduces the concepts of 'communication' and 'information' as the basic concepts, meaning that all other informatics concepts are derived from them [8].

To define the message using the concept of information is not difficult: for example, a message is a set of signs or primary signal containing information. It is important to emphasize that the message in the communication process has completely different information for each category of people that suits the interests of the recipient.

Thus, the same message in the communication process (verbal or non-verbal) can be interpreted in different ways and carry different types of information: incoming and outcoming ones may be different. Meanwhile it can be argued that some kind of image which is either the result of an agreement or awareness, or is prescribed by the rule is decisive for the connection between the message and the information contained in it. This image is a rule of interpretation. The interpretation itself is determined by the interests of information recipients, their feelings, emotions and, speaking more formal language, just by their purpose. In this case, it is easy to imagine that one of the main conditions to extract the necessary information from the message is actually the purpose of this process.

In the context of our study the idea of communication as means of target sending and receiving information in the form of knowledge, data, facts, messages and signals coincides with the model of intercultural communication shown in Fig. 1. Thus we characterize communication as a phenomenon, supporting the processes of understanding, accumulation and systematization of knowledge, decision-making processes and management. Let's call this function of communication instrumental-functional.

Such communication rules in specific contexts are based on the system of expected behaviour patterns, which foster verbal forms of interaction among individuals. But it is decision making that precedes people's activities. This kind of decision making can be a carefully thought-out, conscious act which is based on a logical analysis. But it can also be based on some intuitive ideas and emotions, made on a whim. Moreover, the decision may be taken in a state of stresses and strains. In any case, the act of making decisions is a discharge of socio-psychological energy, which normally occurs as a result of communication in the form of information and signal effects. At the same time, communication is a tool of influence, causing the discharge of socio-psychological energy, which leads to making decisions and taking specific actions. Communication as a tool of signal impact can be schematically presented as follows: information→ acc umulation → signal impact→ energy discharge (actions, comprehension, decision making). Many factors are the reasons for such psychological reactions. Principal among them is the lack of awareness about the realities and negotiating style of a partner country; low language competence can be mentioned second.

Thus, communication performs on the one hand 'ensuring' function representing the method of transferring information, knowledge, content of messages that are interpreted on the basis of goals and objectives. On the other hand, communication can perform directly influencing function forming the basis for those signals which can stimulate the accumulation of energy, its transformation and, finally, discharge, transforming the system into a 'new' condition, and in some cases simply destroying it. This coincides perfectly with the previously presented chain of causes and sequences of possible problem situations in intercultural communication.

It is clear that for successful communication knowledge of linguistic code alone is not enough: in addition, one should know the fundamentals and characteristics of communication in native and foreign cultures. Moreover, the study of communication is interdisciplinary, since it requires joint efforts of representatives of different professions: sociologists, linguists, culturologists, psychologists, and teachers.

Dealing with the problems of communication has become particularly urgent in the business environment. In the business world, the significance of cultural differences and the role of intercultural communicator have been under discussion. 'These days there is not a discussion or a decision that does not have an international dimension. We would have to be blind not to see how critically important international experience is' [9]. As companies integrate their operations globally different national approaches can send conflicting messages to success-oriented communicators.

Meanwhile intercultural characteristics can become a barrier which may hinder communication and prevent communicators from realizing their communicative intentions. The challenges of business communication generally include developing negotiation and motivation strategies and tactics to avoid communication breakdowns [10]. It is necessary to take into consideration national leadership styles and avoid national stereotypes that can ruin communication. It is a well-known fact that technical knowledge and skills make up only 15% of business success while 85 % depends on the ability to communicate with other people. One can only imagine how much money businesses lose, and how many promising business ideas will never be implemented due to the fact that basic rules of communication are not observed.

English consultants in management psychology Mike Woodcock and Dave Francis noted that managers need, above all, the ability to manage themselves and influence others through the ability to communicate. The list of requirements can be added with the following: 'Ideally, it seems a global manager should have the stamina of an Olympic runner, the mental ability of an Einstein, the conversational skill of a professor of languages, the detachment of a judge, the tact of a diplomat, and the perseverance of an Egyptian pyramid builder. And that's not all. If they are going to measure up to the demands of living and working in a foreign country, they should also have a feeling for the culture; their moral judgment should not be too rigid; they should be able to merge with the local environment; and they should show no signs of prejudice' [11].

There is perhaps one main conclusion to be drawn from the above discussion concerning specialists' training: fundamental professional knowledge must be backed up by the knowledge of communication as a management tool. In our case we find it appropriate to present some findings [1, 12].

- Communication in a business context is initially focused on and directed to the target.
- Managers communicate not only with each other, but with representatives of external communities.
- In a business context there is no free choice of the communicative partners, there is a 'compulsive communication'.
- Communication is a prerequisite for functioning of the system 'Business'.

The following preliminary conclusion would be logical: the aspect of training a competitive specialist is reasonably important. The purpose of business communication training is to acquire competences of two-way communication necessary for identifying common interests and mutual understanding based on knowledge of communication rules and cultural awareness.

3 Conclusion

The demands of modern society and the priorities of socio-economic development of a society determine the level and content of competences required for specialists' professional training. The analysis of the socio-economic realities revealed a number of contradictions typical of the modern education system: first, between the need for education of the individual and public expectations; between the need for highly qualified specialists of a new level and training practices of these professionals in accordance with their further professional orientation. There is formation not only of knowledge and skills, but also the development of skills and social competences of a personality required to implement that knowledge. Intercultural communication is regarded as an important component of the overall professional competence of the specialist in the field of management. In the modern educational paradigm, values such as social capital, intercultural communication, tolerance and respect for cultural differences, knowledge of a foreign language have become important.

References

1. B. Lentz, Manager Magazin, **5**, 257 (1997)
2. E. Hall. Psychology Today, **6,** 67 (1976)
3. E. Hall, *The dance of life* (New York, 1983).
4. G. Hofstede, *Cultures and Organisations. Software of the Mind* (London, 1993)
5. E. Hirsh, *Cultural Literacy (What Every American Needs to Know)* (Boston, 1988)
6. K. Knapp, Zeitschrift für Fremdsprachenforschung, **1**, 66 (1990)
7. W. Gudykunst, Y. Kim, *Communicating with Strangers* (New York, 1992)
8. F. Baujer, G. Gooz, *Informatika* (Mir, 1990).
9. N. Adler, *International Dimentions of Organisational Behaviour* (South-Western College Publishing Ohio, 1997).
10. P. O'Connor, A. Pilbeam, F. Scott-Barrett, *Negotiating* (Longman Group UK Limited. 1996)
11. M. Woodcock, D. Fransis, *The unblocked manager. A practical guide to self development* (1991)
12. A. Pilbeam, *Market Leader. International Management* (Longman: Pearson Education Limited, 2003)

AUTHOR INDEX

AUTHOR INDEX

EDP Sciences – Web of Conferences
17, Avenue du Hoggar
Parc d'Activité de Courtabouf
BP 112
F-91944 Les Ulis Cedex A
France

ISBN 978-1-5108-3894-9

9 781510 838949

CURRAN ASSOCIATES INC.
proceedings.com

Regional Conference on Nuclear Physics (RCNP 2016)

EPJ Web of Conferences Volume 156 (2017)

Johor Bahru, Malaysia
15 - 16 December 2016

edp sciences